AF598331

# Methods in Molecular Biology™

*Series Editor*
John M. Walker
School of Life Sciences
University of Hertfordshire
Hatfield, Hertfordshire, AL10 9AB, UK

For other titles published in this series, go to
www.springer.com/series/7651

# Rat Genomics

## Methods and Protocols

Edited by

**Ignacio Anegon**

*INSERM, UMR 643, Nantes, France*

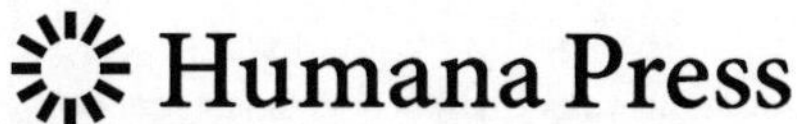

*Editor*
Ignacio Anegon
INSERM, UMR 643
Nantes, France
Ignacio.Anegon@univ-nantes.fr

ISSN 1064-3745 e-ISSN 1940-6029
ISBN 978-1-60327-388-6 e-ISBN 978-1-60327-389-3
DOI 10.1007/978-1-60327-389-3
Springer New York Dordrecht Heidelberg London

Library of Congress Control Number: 2009936793



Printed on acid-free paper

Humana Press is part of Springer Science+Business Media (www.springer.com)

*To:*
*My beloved parents, Ignacio and Hani*
*And to my wonderful and adorable family, Quica, Claudia, and Alex*

*"It's never too late to be who you might have been."*
*George Eliot*

# Preface

This book aims to provide practical information for researchers involved in genomic research in the rat along with a more contextual discussion about the usefulness of the rat in physiological or translational research in different organs and systems.

The rat has been a model of choice for physiological studies, and thus a great deal of information is already available. In addition, a large number of inbred, congenic, and transgenic rat lines have already been described. An exponential growth in rat genomics is taking place based on the recent availability of genomic tools. These include genome sequencing, quantitative trait loci mapping, and the identification of single nucleotide polymorphisms as well as the development of transgenic technologies such as nuclear cloning, lentiviral-mediated transgenesis, gene knock-down using RNA interference, gene knock-out by mutagenesis, and zinc finger nucleases plus exciting advances in the obtention of rat embryonic cell lines. All of these topics and others important for research in the rat are covered by world-wide experts in each field.

The convergence and integration of physiological and genomic data using the infrastructure of genome databases and strain depository centres is boosting translational research using rat models to improve human health.

Within this context, the edition of a book that thoroughly covers the techniques used and overviews the applications of the data obtained is very likely to be useful to the scientific community as a source of information, references and methods.

*Nantes, France* *Ignacio Anegon*

# Acknowledgments

The very efficient secretarial assistance of Valérie Chatellier. The financial support by the Région Pays de la Loire through Biogenouest (former OUEST-Genopole) and the program IMBIO, by the French government through the program IBiSA and by the Fondation Progreffe.

# Contents

## Contributors

TIMOTHY J. AITMAN • *Physiological Genomics and Medicine Group, Medical Research Council Clinical Sciences Centre, Section of Molecular Genetics and Rheumatology, Imperial College, Hammersmith Hospital, London, UK*

IGNACIO ANEGON • *INSERM, UMR643, Nantes, France; CHU Nantes, Institut de Transplantation et de Recherche en Transplantation, ITERT, Nantes, France; Faculté de Médecine, Université de Nantes, Nantes, France*

MICHAEL BADER • *Max-Delbrück-Center for Molecular Medicine (MDC), Berlin, Germany*

BETH A. BAUER • *Research Animal Diagnostic Laboratory, University of Missouri, Columbia, MO, USA*

JACQUES BEHMOARAS • *Physiological Genomics and Medicine Group, Medical Research Council Clinical Sciences Centre, Department of Histopathology, Imperial College, Hammersmith Hospital, London, UK*

ISABELLE BERNARD • *Institut National de la Santé et de la Recherche Médicale (INSERM) U563, Institut Fédératif de Recherche (IFR) 30, Hôpital Purpan and Université Paul Sabatier, Toulouse, France*

FELIX BODE • *Franz-Penzoldt-Center, Experimental Therapy, Friedrich-Alexander-University of Erlangen-Nürnberg, Erlangen, Germany*

ELIZABETH C. BRYDA • *Research Animal Diagnostic Laboratory, University of Missouri, Columbia, MO, USA*

ROLAND BUELOW • *Open Monoclonal Technology, Inc., Palo Alto, CA, USA*

YACINE CHERIFI • *genOway SA, Lyon, France*

GREGORY J. COST • *Sangamo BioSciences, Inc., Richmond, CA, USA*

JEAN COZZI • *genOway SA, Lyon, France*

XIAOXIA CUI • *Sigma-Aldrich Biotechnology, St. Louis, MO, USA*

EDWIN CUPPEN • *Hubrecht Institute, NIOB KNAW, Utrecht, The Netherlands*

SIMON-PIERRE DEMERS • *Centre for Research in Animal Reproduction, Faculty of Veterinary Medicine, University of Montreal, St-Hyacinthe, QC, Canada; Clonagen Inc., St-Hyacinthe, QC, Canada*

MARTINA DORSCH • *Institute for Laboratory Animal Science and Central Animal Facility, Hannover Medical School, Hannover, Germany*

GILBERT J. FOURNIÉ • *INSERM, U563, Toulouse, France; Université Toulouse III Paul Sabatier, Institut Claude de Préval (IFR30), Toulouse, France*

ALEXANDRE FRAICHARD • *genOway SA, Lyon, France*

JAMES C. FUSCOE • *Division of Systems Toxicology, National Center for Toxicological Research, Jefferson, AR, USA*

ARON M. GEURTS • *Human and Molecular Genetics Center, Department of Physiology, Medical College of Wisconsin, Milwaukee, WI, USA*

MICHAEL N. GOULD • *McArdle Lab for Cancer Research, University of Wisconsin-Madison, Madison, WI, USA*
LEI GUO • *Division of Systems Toxicology, National Center for Toxicological Research, Jefferson, AR, USA*
YOJI HAKAMATA • *Department of Basic Science, School of Veterinary Nursing and Technology, Faculty of Veterinary Science, Nippon Veterinary and Life Science University, Tokyo, Japan*
MASUMI HIRABAYASHI • *National Institute for Physiological Sciences, Okazaki, Aichi, Japan; The Graduate University for Advanced Studies, Okazaki, Aichi, Japan*
SHINICHI HOCHI • *Faculty of Textile Science and Technology, Shinshu University, Ueda, Nagano, Japan*
LOUIS-MARIE HOUDEBINE • *Département de Physiologie Animale, Institut National de la Recherche Agronomique, Nouzilly, France*
JUNYA ITO • *Laboratory of Animal Reproduction, School of Veterinary Medicine, Azabu University, Kanagawa, Japan*
HOWARD J. JACOB • *Human & Molecular Genetics Center, Departments of Physiology and Pediatrics, Medical College of Wisconsin, Milwaukee, WI, USA*
CHRISTELLE JACQUET • *genOway SA, Lyon, France*
NAOMI KASHIWAZAKI • *Laboratory of Animal Reproduction, School of Veterinary Medicine, Azabu University, Kanagawa, Japan*
MASAKI KAWAMATA • *Section for Studies on Metastasis, National Cancer Center Research Institute, Tokyo, Japan*
EIJI KOBAYASHI • *Division of Organ Replacement Research, Center for Molecular Medicine, Jichi Medical University, Tochigi, Japan*
DOMINIQUE LAGRANGE • *INSERM, U563, Toulouse, France; Université Toulouse III Paul Sabatier, Institut Claude de Préval (IFR30), Toulouse, France*
VLADIMÍR LANDA • *Institute of Physiology, Czech Academy of Sciences, Prague, Czech Republic*
EDWARD K. LOBENHOFER • *Scientific Affairs, Cogenics, a Division of Clinical Data, Inc., Morrisville, NC, USA*
NAOKI MAEDOMARI • *Institute of Laboratory Animals, Graduate School of Medicine, Kyoto University, Kyoto, Japan*
NAN MEI • *Division of Genetic and Reproductive Toxicology, National Center for Toxicological Research, Jefferson, AR, USA*
SÉVERINE MÉNORET • *INSERM, UMR643, Nantes, France; CHU Nantes, Institut de Transplantation et de Recherche en Transplantation, ITERT, Nantes, France; Université de Nantes, Faculté de Médecine, Nantes, France*
MARIE-PIERRE MOISAN • *INRA, UMR 1286 PsyNuGen, CNRS, UMR 5226, Université de Bordeaux 2, Bordeaux, France*
HUU PHUC NGUYEN • *Department of Medical Genetics, University of Tübingen, Tübingen, Germany*
TUAN HUY NGUYEN • *INSERM, U948, Biothérapies Hépatiques, CHU Hotel-Dieu, Nantes cedex, France*
TAKAHIRO OCHIYA • *Section for Studies on Metastasis, National Cancer Center Research Institute, Tokyo, Japan*

SILVÈRE PETIT • *genOway SA, Lyon, France*
ENRICO PETRETTO • *Physiological Genomics and Medicine Group, Medical Research Council Clinical Sciences Centre, Imperial College, Hammersmith Hospital, London, UK*
MICHAL PRAVENEC • *Institute of Physiology, Czech Academy of Sciences, Prague, Czech Republic*
ANDRÉ RAMOS • *Lab Genética do Comportamento, Departamento de Biologia Celular, Embriologia e Genética, CCB, Universidade Federal de Santa Catarina, Florianópolis, SC, Brazil*
SÉVERINE REMY • *INSERM, UMR643, Nantes, France; CHU Nantes, Institut de Transplantation et de Recherche en Transplantation, ITERT, Nantes, France; Université de Nantes, Faculté de Médecine, Nantes, France*
OLAF RIESS • *Department of Medical Genetics, University of Tübingen, Tübingen, Germany*
ABDELHADI SAOUDI • *Institut National de la Santé et de la Recherche Médicale (INSERM) U563, Institut Fédératif de Recherche (IFR) 30, Hôpital Purpan and Université Paul Sabatier, Toulouse, France*
YASUNARI SEITA • *Laboratory of Animal Reproduction, School of Veterinary Medicine, Azabu University, Kanagawa, Japan*
TADAO SERIKAWA • *Institute of Laboratory Animals, Graduate School of Medicine, Kyoto University, Kyoto, Japan*
LAWRENCE C. SMITH • *Centre for Research in Animal Reproduction, Faculty of Veterinary Medicine, University of Montreal, St-Hyacinthe, QC, Canada; Clonagen Inc., St-Hyacinthe, QC, Canada*
BART M. G. SMITS • *McArdle Lab for Cancer Research, University of Wisconsin-Madison, Madison, WI, USA*
ALEXANDER J. STODDARD • *Human & Molecular Genetics Center, Medical College of Wisconsin, Milwaukee, WI, USA*
CLAUDE SZPIRER • *Institut de Biologie et de Médecine Moléculaires, Université Libre de Bruxelles, Gosselies, Charleroi, Belgium*
RI-ICHI TAKAHASHI • *Phoenixbio Co. Ltd., Tochigi, Japan*
AKIKO TAKIZAWA • *Institute of Laboratory Animals, Graduate School of Medicine, Kyoto University, Kyoto, Japan*
LAURENT TESSON • *INSERM, UMR643, Nantes, France; CHU Nantes, Institut de Transplantation et de Recherche en Transplantation, ITERT, Nantes, France; Université de Nantes, Faculté de Médecine, Nantes, France*
KADER THIAM • *genOway SA, Lyon, France*
MASATSUGU UEDA • *Phoenixbio Co. Ltd., Tochigi, Japan*
YVONNE K. URBACH • *Franz-Penzoldt-Center, Experimental Therapy, Friedrich-Alexander-University of Erlangen-Nürnberg, Erlangen, Germany*
CLAIRE USAL • *INSERM, UMR643, Nantes, France; CHU Nantes, Institut de Transplantation et de Recherche en Transplantation, ITERT, Nantes, France; Université de Nantes, Faculté de Médecine, Nantes, France*
RUBEN VAN BOXTEL • *Hubrect Institute, Niob Knaw, Utrecht, The Netherlands*
BIRGER VOIGT • *Institute of Laboratory Animals, Kyoto University, Kyoto, Japan*
STEPHEN VON HÖRSTEN • *Franz-Penzoldt-Center, Experimental Therapy, Friedrich Alexander University of Erlangen-Nürnberg, Erlangen, Germany*

ERYAO WANG • *genOway SA, Lyon, France*
DIRK WEDEKIND • *Institute for Laboratory Animal Science and Central Animal Facility, Hannover Medical School, Hannover, Germany*
ELIZABETH A. WORTHEY • *Human & Molecular Genetics Center, Medical College of Wisconsin, Milwaukee, WI, USA*
CHANA YAGIL • *Laboratory for Molecular Medicine, Israeli Rat Genome Center and Department of Nephrology and Hypertension, Ben-Gurion University, Ashkelon, Israel*
YORAM YAGIL • *Laboratory for Molecular Medicine, Israeli Rat Genome Center and Department of Nephrology and Hypertension, Ben-Gurion University, Ashkelon, Israel*
QI ZHOU • *Institute of Zoology, CAS, Beijing, P R China*
VÁCLAV ZÍDEK • *Institute of Physiology, Czech Academy of Sciences, Prague, Czech Republic*

Chapter 1

# The Rat: A Model Used in Biomedical Research


## Howard J. Jacob



## Abstract

The rat evokes fear and disgust in a large percentage of people around the world. Yet, other people are fascinated by this amazing creature that is raised as a pet, has an important place in several religions, and is a prominent model for biomedical research. This book focuses on a variety of methodologies that can be used in this remarkable model. This chapter sets the stage by providing a perspective on why the rat remains an important model in biomedical research.



**Key words:** Rats, Model, History, Outlook, Biomedical


## 1. Introduction

The rat has had a long association with humans; in some cases, it has been the "Lapdog of the Devil" (1), and in others it is the model system through which drugs used to treat human disease are tested. The rat (*Rattus rattus* and *Rattus novegicus*) conquered the world sequentially with *Rattus rattus* coming through first. Genghis Khan's invading armies may have brought the rat (*Rattus rattus*) to Europe. The migration of this rat in Europe can be followed by following the bubonic plague. *Rattus novegicus* (the laboratory rat) arrived in Europe in the early 1700s and quickly traveled around the world, replacing, to a large extent, the smaller *Rattus rattus* (2). The plague and the rat's reputation for eating upto one-fifth of the food destined for human consumption (1) has tarnished the reputation of the amazing laboratory rat (*Rattus novegicus*). Indeed, even with scientists, the rat evokes a sense of foreboding and is viewed differently than other model organisms; consider the cover art of several other sequenced genomes (Fig. 1.1). The rat is viewed as being a sewer rat, not a


I. Anegon (ed.), *Rat Genomics: Methods and Protocols*, Methods in Molecular Biology, vol. 597
DOI 10.1007/978-1-60327-389-3_1, © Humana Press, a part of Springer Science+Business Media, LLC 2010

Fig. 1.1. Cover art from *Nature* for some of the animal models sequenced. The *center* cover shows a rat in a sewer pipe, leaving a very different impact on the reader than the other covers. Yet, the rat is the most commonly used animal model.

saver of lives, whereas other species are viewed much more favorably.

The role of rats in research is probably the direct result of rat fanciers (http://ratfanciers.com/ratfanciers/; http://www.nfrs.org; http://www.aprac.fr), who love the rat and collect it for a variety of morphological, and in some cases, behavioral attributes. These rat fanciers can be traced back to the 1700s in Japan, where the rodent was bred for its unique coat colors, and there was even

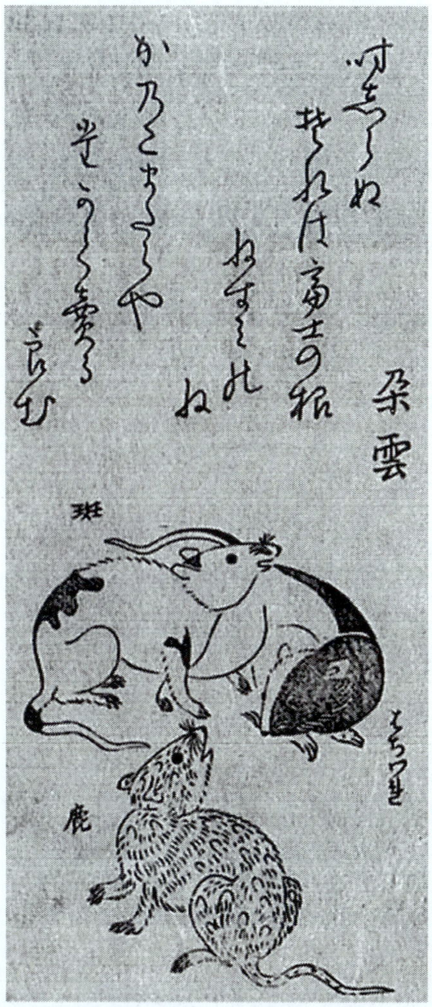


Fig. 1.2. A figure from the 1781 Guide Book on Fancy Rodents. The text in the figure (as translated by Dr. Tadao Serikawa): In any age, Mt. Fuji and fancy rodents are both around. In particular, precious ones with deer or spotted coat colors would bring wealth. Indeed, coat color were some of the first genetic markers used by biomedical researchers.

a guide book (3) published about rats (Fig. 1.2). It is most likely that the laboratory rat developed from these stocks . Indeed, it is likely that the genetic code of the original strains in Japan still course through the laboratory rat genome. While important to the rat's history, it is not the reason why the rat has become a dominant model. It was the first mammalian species domesticated for scientific research, with work dating back to before 1850 (4). The first genetic studies were carried out by Crampe from 1877 to 1885 and focused on the inheritance of coat color (4). Hugo De Vries, Karl Correns and Erich Tschermak rediscovered

Mendel's laws at the turn of the century, and Bateson used these concepts in 1903 to demonstrate that rat coat color is a Mendelian trait (4). The first inbred (20 generations of brother-sister mating and considered homozygous at all loci) rat strain, PA, was established by King in 1909 – the same year that inbreeding began for the first inbred strain of mouse, DBA (5). The rise in rat use is largely attributed to the success of the Wistar Institute because of its development of the Wistar Albino rat. This outbred strain is still commonly used today, and many laboratory strains used today can be traced back to being derived from this outbred rat strain (4).

The prevalence of the rat in biomedical research is second only to humans; there are more scientific publications using the rat than any other model system (Table 1.1). Yet, if you were to ask most scientists what the most widely used animal model is, the answer would very likely be the mouse, dog, or monkey, and only then the rat. As discussed in detail in this book, the rat has an incredible tool box available for investigators interested in using this remarkable animal for biomedical research. Perhaps, the benefit of this tool box and the usefulness of the rat in helping humans develop new drugs and increase our understanding of common disease has prevailed, or will prevail over the death and destruction cause by this Dr. Jekyll and Mr. Hyde model system of human disease.

**Table 1.1**
**Number of publications by species. PubMed was queried using *genus species* names**

| Organism | # of papers |
|---|---|
| *Homo sapiens* (human) | 10,365,775 |
| *Rattus norvegicus* (rat) | 1,255,263 |
| *Mus musculus* (mouse) | 919,934 |
| *Oryctolagus cuniculus* (rabbit) | 301,740 |
| *Canis lupus familiaris* (dog) | 271,243 |
| *Saccharomyces cerevisiae* (yeast) | 82,368 |
| *Drosophila melanogaster* (fruit fly) | 31,187 |
| *Danio rerio* (zebrafish) | 10,637 |

## 2. The Rat Toolbox

### 2.1. Biochemistry, Nutrition, Pharmacology, and Physiology

In the ensuing years, rat strains tended to be developed by biomedical researchers in their laboratories, selecting and inbreeding the strains for traits of biomedical interest. There are inbred strains of rats used for research in various areas: addiction, aging, anatomy, autoimmune diseases, behavior and neurodegenerative diseases (Chap. 25), blood diseases, cardiovascular diseases (Chap. 28), dismorphogenesis, cancer (Chap. 32), dental diseases, diseases of the skin and hair, endocrinology, eye disorders, growth, hematologic disorders, histology, inflammation and immunity (Chap. 27), renal diseases (Chap. 30), metabolic disorders (Chap. 29), neuroscience (Chap. 26), nutrition, pathophysiology, pharmacology, pulmonary diseases, physiology, reproductive disorders, skeletal disorders, sleep apnea, transplantation, toxicology, and urological disorders (6, 7). The advent of molecular genetics and the ability to modify the genome of the rat through a variety of techniques (all discussed in later chapters) has resulted in an explosion of strains, substrains and genetically modified strains. The Rat Genome Database (http://www.rgd.mcw.edu) currently lists 594 strains and substrains (personal communication, Mary Shimoyama from RGD).

Many of the rat strains have been extensively studied since the 1960s when the research community began to move away from the dog toward the rat. Most of these publications are related to mechanism, function or drug responses. For example, much of the early work on essential amino acids, immunology, neurobehavior, and toxicology was carried out using rats. Accordingly, the physiology of the rat is extremely well characterized across all major organ systems, the whole animal being ideal for work related to systems biology. An extensive literature on the comparative physiology and the rich biological characterization of the rat, compiled over the last four decades adds extra value to this model system; it has led to important decisions in building the molecular genetics tool box. With these in place, it is likely that the rat will help to further advance biomedical research.

### 2.2. Molecular Genetic Tools

Much of the needed infrastructure for the rat molecular genetic tool box was outlined in the following report from the National Institutes of Health in USA. Support was organized for a national strain repository in the US (complementing the ones in Japan and Germany), to facilitate efforts toward genetic modification, strengthening genomic tools, and hopefully, piloting genomic sequencing (http://www.nhlbi.nih.gov/resources/docs/ratmt-gpg.htm). All goals have now been completed. Since the draft sequence of the rat was published in 2004, the genomic tools have continued to expand. There have been many groups working on the rat genome project, with the bulk of the work coming from the EU, Japan and the USA. Many of the chapters in this book are a direct result of the molecular genetic tool box and investment into the infrastructure for the rat.

#### *2.2.1. Genetic Markers*

The rat has a large set of genetic markers in a variety of formats. Coat color and protein isoforms, have given way to molecular markers. The molecular markers have evolved from anonymous markers to microsatellites (simple sequence length polymorphisms), and currently, to Single Nucleotide Polymorphisms (SNPs). The SNPs are likely to be the last version of genetic markers, but the way they are assayed will continue to evolve and move towards re-sequencing. The different classes of genetic markers have played a key role in mapping quantitative trait loci (QTL) and positional cloning (Chap. 3).

A European STAR consortium, together with its Japanese colleagues, used DNA from six strains (SD, SS/Jr, GK/Ox, WKY/Mdc, F344 and SHRSP/Mdc) to identify 2.9 M SNPs, and 20,000 of these SNPs have been genotyped in over 300 inbred and hybrid strains (8). Eight strains (PVG, F344, SS, LEW, BB, FHH, DA, and SHR) have been sequenced by the U.S. SNP discovery effort at Baylor College of Medicine. The EURATools project has built a haplotype map using 167 inbred strains, two recombinant inbred strains, and an F2 intercross. The databases and tools available for working with these SNP resources are described below.

#### *2.2.2. Genetic Sequence*

The rat genomic sequence was published in 2004 (9). The sequencing strategy was a combination of Whole Genome Shotgun Sequencing and the bacterial artificial chromosome hierarchical approach. The rat strain selected for the sequencing was the BN/SsNHsd/Mcw (Brown Norway), and sequencing covered 90% of the rat genome. The rat genome is estimated to be 2.75 Gb long, distributed in 21 chromosomes (the Y chromosome has not been sequenced), and is predicted to encode for approximately 20,973 genes, with 28,516 transcripts and 205,623 exons (9). The exact number of genes and transcripts will take several more years to resolve; however, the current data provides the researcher with a precise knowledge of the rat gene content, with improved physical and genetic mapping. However, it must be noted that various lines of evidence will be required, including genetic linkage analysis or other forms of mapping to ensure that the region of the genome under investigation has been constructed correctly.

It is expected that in the coming years, the field will see an acceleration in the rate of gene discovery for a host of traits, reflecting the availability of the rat sequence and the accessibility to high-throughput sequencing for the search of sequence variants;

these resources will greatly facilitate the identification of genes underlying the hundreds of QTLs mapped for complex diseases and phenotypes. The advent of next generation sequencing, and the complete sequencing of several other strains will accelerate discovery even more (Chap. 4).

#### 2.2.3. Other Genomic Tools

The rat genome project has yielded a tremendous wealth of genomic resources, including genetic maps; radiation hybrid (RH) cell lines and the associated RH maps (24,437 genetic markers: 5,035 microsatellites, and 19,241 ESTs mapped); cDNA libraries generating more than 1,009,817 ESTs (with more being generated) clustered into over 63,440 UniGenes. Most of these data can be found in the public domain through the Rat Genome Database (RGD), RatMap, and the National Bio Resource Project for the Rat (NBRP). Mammalian Gene Collection (10) (full length cDNA project) has recently completed 6,846 full-length genes from the same BN strain that was sequenced. RGD has annotations on 35,427 genes, splice variants, and pseudogenes. Clearly, the infrastructure for the rat genome has come a long way since 1987, when Robinson reported that the rat genome consisted of 10-linkage groups and 4 named chromosomes constructed with 39 phenotypes (coat color, eye color, growth, tumors, teeth, etc.), and 33 electophoretic markers and coat color markers (2).

#### 2.2.4. Genetic Modifications

The increased interest in using genetic strategies and the development of dense genetic markers has facilitated their use building a variety of other types of genetic rat strains. The notable examples of these "new" genetic models include congenic animals, in which a single region of a genome is transferred to the genomic background of another strain using a series of backcrosses; consomics, in which an entire chromosome is transferred by backcrossing; recombinant inbred strain where a collection of F2 animals are inbred by brother-sister mating for 20 generations (Chap. 18)

Using genetic crosses is only one way of generating new animal models. Many of the chapters in this book discuss a variety of strategies: nuclear transfer (Chap. 10); transgenic animals created by pronuclear injection of DNA (Chaps. 6, 7, and 8), via the chemical mutagen N-ethyl-N-nitrosourea ENU mutatgenesis (Chap. 11); transposon mediated mutagenesis (Chap. 14); gene targeting via zinc finger nucleases (Chap. 16); and two new strategies using embryonic stem cells (Chap. 12) and inducible pluripotent stem cells (Chap. 13). The deployment of many of these different types of genetically modified strains has already permeated all the fields that use rats, and it can be anticipated that the use of rats will continue to increase/multiply with the availability of targeted genes. How fast the development will be is difficult to predict, but publications on the mouse doubled in 5 years after gene targeting became possible (Mary Shimoyama from RGD, personal communication).

## 3. Bioinformatic Tools

There has been an explosion not only in the amount of genetic and genomic data for the rat, but also in the number of bioinformatic tools for the rat. While it is not possible to even outline all those available in this introductory chapter, or even the entire book, it is possible to categorize the different types of bioinformatic tools.

At the most basic level is a database, which is simply a structured collection of data: in this case, information about the rat, stored in a computer system. Typically, the database also provides various levels of functionality to the data collected. Databases can exist within an investigator's laboratory, at a project, institution or international level. Many investigators obtain genomic data from large genome browser groups, for example, Ensembl (http://www.ensembl.org), National Center for Biotechnology Information (http://www.ncbi.nih.gov), and University of California at Santa Cruz (http://genome.ucsc.edu/). These large database groups provide pipelines, gene assembly and a host of other resources for the investigator, but they are not species dependent and focus on a broad range of activities. What is not appreciated by many is that much of the rat data in these major databases comes from other sources, such as National BioResource Project for the rat (http://www.anim.med.kyoto-u.ac.jp/NBR/); RATMAP (http://www.ratmap.org) and the Rat Genome Database (http://www.rgd.mcw.edu). Chapter 3 and a recent article by Twigger et al. (11) highlight many of these resources and provide links to the various sites in the supplemental materials.

Once data is stored, there is a need to retrieve, analyze and visualize it. These needs are accomplished by a host of tools designed to accomplish such tasks. While the data in a database tends to expand over time, tools come and go depending upon why they are needed. It is not possible to provide a detailed overview of all the tools available. Much of the rat centric information can be found at RGD, which coordinates all nomenclature for genes, strains, and QTLs and provides access to this data to support research using the rat as a genetic model for the study of human disease; it was created to serve as a repository of rat genetic and genomic data, as well as mapping, strain, and physiological information. It also facilitates investigators' research efforts by providing tools to search, mine, and analyze these data sets.

## 4. Outlook

There will be an explosive growth in the use of the rat over the next 5 years for the following reasons: First, the ability to clone genes by position is poised for a quantum jump based on the availability of the genomic sequence for several strains used for many QTL and the large number of genetic mapping studies, and congenic strains already completed. Second, the many Genome-Wide Association Studies (GWAS) in humans are identifying several new genes and regions that will require follow-up studies, including those that are mechanism based. Comparative genomics and the numerous ways to genetically modify the rat will continue to attract other investigators. Third, the long history of using the rat to develop new therapeutic agents and to understand toxicity will be enhanced by the genomic tools and new animal models. Finally, comparative medicine will become a more common field of research by the end of the decade. The availability of dense genetic maps in multiple species, the genomic sequence for 38 mammals, and the strong evidence of QTLs being conserved across multiple species suggest that comparative medicine/genomics will begin to unravel the biological basis of common complex disease. We can anticipate that rats that carry human genes, will be constructed and likely even human organs to facilitate the field's understanding of a mechanism at a whole organism level. The possibility of having sequenced genomes from individual strains will truly potentiate comparative medicine within and between species. With this, the Rosetta stone in genomics will finally be complete. However, the interpretation of the stone with respect to a gene's function and role in the pathogenesis of disease will still be required. Now is the time to reconsider how we use animal models to understand human disease

In 2009, animal models are selected primarily on the basis of one of two criteria: (1) a gene or its expression is modified, or can be modified and its implication in a human disease can be studied in a particular model system; (2) the physiological characteristics of the model systems reflect some aspect of the clinical picture. There has been some movement towards other considerations, such as genome background and effects of the environment (e.g., diet). It appears that the added benefit of comparative genomics is seldom considered when a model is selected. With a large number of QTL genetically mapped in many different species, and several of them mapping to the same evolutionarily conserved regions, it seems reasonable that many of the same genes will play a role in the same disease process in more than one species. While comparative genomics research has built a powerful tool set for identifying the genes and the regulatory elements contained within the genomes, it is the nascent development of "comparative medicine" that is the likely future of comparative genomics, which will lead to improved models aiding in the development of new therapeutics. If this occurs, perhaps humans will no longer think of the rat as the "lapdog of the devil."

## Acknowledgments

I am deeply indebted to Dr. Tadao Serikawa, for providing me information about the start of fancy rat collections in Japan, and the information and picture used for Fig. 1.2 from the Guide of Fancy Rats. My thanks are also due to the many Students, Staff, Postdoctoral Fellows and Faculty who have worked with me for the last 20 years on projects using the rat as a biomedical model.

### References

1. Canby TY, Stanfield JL (1977) The rat: lapdog of the devil. Natl Geogr 152(1):60–87
2. Robinson R (1965) Genetics of the Norway rat. Pergamon Press, Oxford, UK
3. Unknown (1775) Yosotamanokakehashi. Toto Publisher and Seppu Publisher, Japan
4. Lindsey JR, Baker HJ (2006) Historical foundations. In: Suckow MA, Weisbroth SH, Franklin CL (eds) The laboratory rat. Elsevier Academic Press, pp. 1–52
5. Nishioka Y (1995) The origin of common laboratory mice. Genome 38(1):1–7
6. Greenhouse DD, Festing MFW, Hasan S, Cohen AL (1990) Catolgue of inbred strains. In: Hedrich HJ (ed) Genetic monitoring of inbred strains of rats. Gustav Fischer Verlag, Stuttgart, New York, pp 410–521
7. Gill TJ 3rd, Smith GJ, Wissler RW, Kunz HW (1989) The rat as an experimental animal. Science 245(4915):269–276
8. Saar K, Beck A, Bihoreau MT, Birney E, Brocklebank D, Chen Y, Cuppen E, Demonchy S, Dopazo J, Flicek P, Foglio M, Fujiyama A, Gut IG, Gauguier D, Guigo R, Guryev V, Heinig M, Hummel O, Jahn N, Klages S, Kren V, Kube M, Kuhl H, Kuramoto T, Kuroki Y, Lechner D, Lee YA, Lopez-Bigas N, Lathrop GM, Mashimo T, Medina I, Mott R, Patone G, Perrier-Cornet JA, Platzer M, Pravenec M, Reinhardt R, Sakaki Y, Schilhabel M, Schulz H, Serikawa T, Shikhagaie M, Tatsumoto S, Taudien S, Toyoda A, Voigt B, Zelenika D, Zimdahl H, Hubner N (2008) SNP and haplotype mapping for genetic analysis in the rat. Nat Genet 40(5):560–566
9. Gibbs RA, Weinstock GM, Metzker ML, Muzny DM, Sodergren EJ, Scherer S, Scott G, Steffen D, Worley KC, Burch PE, Okwuonu G, Hines S, Lewis L, DeRamo C, Delgado O, Dugan-Rocha S, Miner G, Morgan M, Hawes A, Gill R, Celera, Holt RA, Adams MD, Amanatides PG, Baden-Tillson H, Barnstead M, Chin S, Evans CA, Ferriera S, Fosler C, Glodek A, Gu Z, Jennings D, Kraft CL, Nguyen T, Pfannkoch CM, Sitter C, Sutton GG, Venter JC, Woodage T, Smith D, Lee HM, Gustafson E, Cahill P, Kana A, Doucette-Stamm L, Weinstock K, Fechtel K, Weiss RB, Dunn DM, Green ED, Blakesley RW, Bouffard GG, De Jong PJ, Osoegawa K, Zhu B, Marra M, Schein J, Bosdet I, Fjell C, Jones S, Krzywinski M, Mathewson C, Siddiqui A, Wye N, McPherson J, Zhao S, Fraser CM, Shetty J, Shatsman S, Geer K, Chen Y, Abramzon S, Nierman WC, Havlak PH, Chen R, Durbin KJ, Egan A, Ren Y, Song XZ, Li B, Liu Y, Qin X, Cawley S, Cooney AJ, D'Souza LM, Martin K, Wu JQ, Gonzalez-Garay ML, Jackson AR, Kalafus KJ, McLeod MP, Milosavljevic A, Virk D, Volkov A, Wheeler DA, Zhang Z, Bailey JA, Eichler EE, Tuzun E (2004) Genome sequence of the Brown Norway rat yields insights into mammalian evolution. Nature 428(6982):493–521
10. Gerhard DS, Wagner L, Feingold EA, Shenmen CM, Grouse LH, Schuler G, Klein SL, Old S, Rasooly R, Good P, Guyer M, Peck AM,

Derge JG, Lipman D, Collins FS, Jang W, Sherry S, Feolo M, Misquitta L, Lee E, Rotmistrovsky K, Greenhut SF, Schaefer CF, Buetow K, Bonner TI, Haussler D, Kent J, Kiekhaus M, Furey T, Brent M, Prange C, Schreiber K, Shapiro N, Bhat NK, Hopkins RF, Hsie F, Driscoll T, Soares MB, Casavant TL, Scheetz TE, Brown-stein MJ, Usdin TB, Toshiyuki S, Carninci P, Piao Y, Dudekula DB, Ko MS, Kawakami K, Suzuki Y, Sugano S, Gruber CE, Smith MR, Simmons B, Moore T, Waterman R, Johnson SL, Ruan Y, Wei CL, Mathavan S, Gunaratne PH, Wu J, Garcia AM, Hulyk SW, Fuh E, Yuan Y, Sneed A, Kowis C, Hodgson A, Muzny DM, McPherson J, Gibbs RA, Fahey J, Helton E, Ketteman M, Madan A, Rodrigues S, Sanchez A, Whiting M, Madari A, Young AC, Wetherby KD, Granite SJ, Kwong PN, Brinkley CP, Pearson RL, Bouffard GG, Blakesly RW, Green ED, Dickson MC, Rodriguez AC, Grimwood J, Schmutz J, Myers RM, Butterfield YS, Griffith M, Griffith OL, Krzywinski MI, Liao N, Morin R, Palmquist D, Petrescu AS, Skalska U, Smailus DE, Stott JM, Schnerch A, Schein JE, Jones SJ, Holt RA, Baross A, Marra MA, Clifton S, Makowski KA, Bosak S, Malek J, MGC Project Team (2004) The status, quality, and expansion of the NIH full-length cDNA project: the Mammalian Gene Collection (MGC). Genome Res 14(10B): 2121–2127

11. Twigger SN, Pruitt KD, Fernandez-Suarez XM, Karolchik D, Worley KC, Maglott DR, Brown G, Weinstock G, Gibbs RA, Kent J, Birney E, Jacob HJ (2008) What everybody should know about the rat genome and its online resources. Nat Genet 40(5):523–527

# Chapter 2

# Genetic Mapping and Positional Cloning


## Timothy J. Aitman, Enrico Petretto, and Jacques Behmoaras



## Abstract

Genetic mapping and positional cloning of genetically complex traits in the laboratory rat (*Rattus norvegicus*) has recently led to the identification of various susceptibility genes in different rat models. Rat genetics has benefited from revolutionary advances in molecular biology, genetics, genomics and informatics and provide an unparalleled resource for molecular genetic investigation of mammalian physiopathology and its underlying complex genetic architecture. In this review, we will consider different strategies that are being used in the successful positional cloning of rat complex trait genes in the context of recent progress in rodent and human genetics.



**Key words:** Rat, Genetics, Positional cloning, QTL


## 1. Why Map Genes in the Rat?

The genetic characterisation of model organisms has progressed from the study of genetic modes of inheritance to establish the causal relationship between genes and both physiological and pathophysiological traits. Among these, monogenic traits are inherited in a Mendelian fashion in which single genotypic changes result in robust and discrete phenotypic differences. However, the expression of many Mendelian traits is also influenced by the actions of modifier genes and the segregation of phenotypic traits in a Mendelian fashion is the exception, rather than the rule. Most phenotypic differences between individuals in morphology, physiology, growth, behaviour and disease susceptibility are quantitative in nature, exhibiting a continuous and nearly normal phenotypic distribution (1). These genetically complex traits arise from the interaction of multiple segregating genetic variants together with the environmental factors. In this chapter, we shall focus mainly on genetic mapping and positional


I. Anegon (ed.), *Rat Genomics: Methods and Protocols*, Methods in Molecular Biology, vol. 597
DOI 10.1007/978-1-60327-389-3_2, © Humana Press, a part of Springer Science+Business Media, LLC 2010

cloning of genetically complex traits in the laboratory rat (*Rattus norvegicus*).

Rat models have been used in the elucidation of human physiology, pharmacology, toxicology, nutrition, behaviour, immunology and neoplasia for over 150 years (2). Because of its size, ease of manipulation and breeding characteristics, physiological research on the rat has generated a wealth of experience and methodological sophistication for the accurate determination of quantitative phenotype measurements. Although mapping genes in the rat has also been successfully applied in the discovery of Mendelian traits, most of the positionally cloned genes in rats underlie genetically complex phenotypes.

Genetic mapping of complex traits offers a powerful and complete approach when compared to candidate gene-based cell biology in the elucidation of disease susceptibility mechanisms. Complex traits are controlled by loci that have quantitative effects on the phenotype and the observed ubiquitous quantitative variation in biology is often characterised by a strong genetic component. Lander and Botstein first described the use of interval mapping based on DNA markers in order to genetically localise quantitative trait loci (QTL) in natural or experimental populations (3). Guidelines were then established for interpreting the genome-wide linkage results that were reported in numerous studies, including those undertaken in segregating rodent populations (4). Since then, the road from QTL discovery to causative gene identification has been described as "long and bumpy" (5). The challenging attempts to characterise the molecular basis of QTLs in rodents led to the consideration of alternative approaches such as forward genetic approaches using mutagenesis in mice (6, 7). Today, it is generally accepted that the mutagenesis approach is complementary to the QTL approach and will significantly increase the identification of complex trait genes in rat models (8, 9). More generally, rat genetics has benefited from revolutionary advances in molecular biology, genetics and genomics; some of the most significant ones being the sequencing of the complete rat genome with >90% coverage (10), dense single nucleotide polymorphism (SNP) and haplotype mapping allowing surveys of genetic variation (11) and characterisation of distribution and functional impact of DNA copy number variation in rats (12). Since 2001, 16 additional genes have been positionally cloned in various rat models (8).

Human disease gene discovery has entered a new era where two new technological approaches are currently used in the genetic dissection of common diseases: genome-wide association studies (GWAS) and deep resequencing. Recent studies on common diseases using these state-of-the-art techniques confirmed the complexity of their underlying genetic architecture (13, 14). Indeed, low-effect genomic variants identified by GWAS studies and designated as "common variants" constitute the genetic hallmark of many human common diseases. In September 2008, studies presented on "The Genomics of Common Diseases" meeting held at the Broad Institute of MIT and Harvard in Boston revealed that the rare variants with high penetrance identified using deep re-sequencing add to the complex architecture of human disease. Given the increased heritability, flexibility and statistical power of experimental rodent crosses over the corresponding studies in humans, rodent models provide an unparalleled resource for molecular genetic investigation of mammalian physiopathology and its underlying complex genetic architecture. In addition, the genome resources currently available for rat models form a solid basis for comparative genomics in order to study human physiology and disease (15).

In this review, we will consider different strategies used in the successful positional cloning of rat complex trait genes in the context of recent progress in genetics and genomics, and new statistical approaches complementing a range of high throughput techniques. We will also highlight the importance of rat positional cloning in terms of translational studies in humans.

## 2. Successful Positional Cloning in the Rat

A total of 22 positionally cloned genes over a 10 year period (Table 2.1) demonstrate the difficulty of positional cloning of genes underlying rat QTLs when one considers the nearly 1,000 QTLs identified in rat models. Table 2.1 shows that until 2004, positional cloning was mostly achieved by genetic mapping techniques (linkage and/or physical mapping) with functional in vivo complementation for some of these studies. After the publication of the genome sequence of the Brown Norway rat in 2004 (10), rat positional cloning was complemented with additional approaches such as computational analysis of the rat genome (16), association and functional studies in humans (17), expression QTL (eQTL), quantitative trait transcript (QTT) analyses (18–20) and RNA interference (RNAi) as an alternative approach for in vivo complementation (21). Recent studies using these relatively novel strategies have showed that the human orthologues of the rat positionally cloned genes are also susceptibility factors for human left ventricular mass (LVM) (19) and heart failure (18).

Standards of proof of gene discovery in complex traits have been previously described (22, 23) and all of the successfully positionally cloned genes in rat models (described in Table 2.1) fulfil the working criteria described in Glazier et al. (23). The burden of proof for complex trait mapping includes linkage, association

**Table 2.1**
**List of positionally cloned rat genes between 1999 and 2008 and their positional cloning strategy. All underlined genes were studied in humans in parallel with rat models. Until 2002 positional cloning was achieved mainly by genetic mapping, and positional cloning of Pkhdl is the only example of translation of findings in rat models into human disease**

| Gene name | Positional cloning strategy | Year | Reference |
|---|---|---|---|
| *Cd36* | Linkage analysis, expression profiling, transgenic rescue | 1999 | (24, 74) |
| *Aspa* | Genome shift approach, biochemical analysis | 2000 | (75) |
| *Mertk* | Genetic mapping, adenovirus mediated gene transfer | 2000, 2001 | (76, 77) |
| *Atrn* | Genetic mapping, transgenic rescue | 2001 | (78) |
| *Cyp11b1* | High-resolution substitution mapping | 2001, 2003 | (25, 26) |
| *Cblb* | Genetic and physical mapping, transgenic rescue | 2002 | (38) |
| *Gimap5* | High resolution physical mapping | 2002 | (79, 80) |
| *Pkhd1* | Linkage analysis, comparative genomics between rat and human | 2002 | (56) |
| *Ncf1* | Physical mapping, in vitro and in vivo functional complementation | 2003, 2004 | (28, 39) |
| *Rab38* | Linkage analysis, high resolution substitution mapping | 2004, 2005 | (30, 81) |
| *Ciita* | Linkage analysis, expression profiling, haplotype analysis, association and functional studies in humans | 2005 | (17) |
| *Gstm1* | Linkage analysis, expression profiling | 2005 | (27) |
| *Anks6* | Physical and genetic mapping, computational analyses of the rat genome | 2005 | (16) |
| *Fcgr3* | Linkage analysis, haplotype mapping, association studies in humans using genomic structural variation | 2006 | (33) |
| *Tmem67* | Linkage analysis in rats and humans, comparative genomics with humans | 2006 | (57) |
| *Fbx10, Frmpd1* | Linkage analysis, association studies in humans | 2007 | (31) |
| *Ephx2* | Linkage analysis, eQTL analysis, in vivo complementation using KO mice | 2008 | (18) |
| *Ogn* | Linkage and eQTL analysis, QTT analysis in rats and humans, in vivo complementation using KO mice | 2008 | (19) |
| *Jund* | Linkage analysis, expression profiling, in vitro RNA interference | 2008 | (21) |
| *Cd36*[a] | Linkage analysis, eQTL, QTT analyses, transgenic rescue | 2008 | (20) |
| *Igl* | High resolution physical mapping | 2008 | (36) |
| *Srebf1* | Linkage analysis, transgenic rescue | 2008 | (29) |

[a]*Cd36* was initially positionally cloned as an insulin resistance gene in 1999 and identified as a blood pressure susceptibility gene using a different study design in 2008

and fine mapping followed by the functional tests of candidate genes in relevant animal models. We will first describe the gene identification procedures used in rodents that are applied after initial QTL mapping, emphasising the use of congenic, subcongenic, advanced intercross lines (AIL) and haplotype mapping in successful positional cloning. We will then focus on recent advances in fine mapping and functional tests of candidate genes underlying QTLs by highlighting the inflection points in rat genetics such as the sequencing of the rat genome, underlining the importance of recent approaches in rat positional cloning.

### 2.1. Fine Mapping: The Success of Congenic, Subcongenic, Advanced Intercross Lines and Haplotype Mapping

Once an initial linkage of a given trait to a discrete chromosomal location has been established, fine mapping is necessary to reduce the linkage interval and to test the plausible biological candidates. Congenic strains have been widely used in the fine mapping of rodent QTLs in various studies using rat models and they still constitute a powerful tool in QTL positional cloning (20, 21, 24–31). By repeatedly backcrossing one strain onto another, it is possible to produce rats that have a particular genomic region from one strain and the remainder of their genome from the other (*see* also chapter 17). The effect of the introgressed genetic region derived from one strain (generally corresponding to a QTL) can then be specifically tested on the genetic background of the other strain. Congenic strains are used primarily to:

- Confirm the established QTL.
- Fine map candidate gene(s) responsible for the phenotypic variance within the QTL.
- Characterise the biology of the QTL gene(s).

Human essential hypertension is one of the most extensively studied complex traits in rat models (32). According to the rat genome database, (http://rgd.mcw.edu/), 325 distinct blood pressure (BP) QTLs have been identified using various rat crosses. Congenic strains for blood pressure can be constructed in two ways with a given pair of parental strains. The low-BP strain can be the donor and the high-BP strain the recipient, or vice versa. The main difference between the two is the genetic background. In Dahl rats, for example, the genetic background of the Dahl salt-resistant rats is not very permissive for expressing BP differences and so most congenic strains using Dahl rats have been made on a salt-sensitive genetic background. On the other hand, successful congenic lines have been made in both directions with spontaneously hypertensive rat (SHR) and Wistar Kyoto (WKY) (32).

When complex traits map to a single locus, the use of congenic strains becomes crucial in the molecular identification of QTLs. The identification of *Cd36* as an insulin-resistance gene causing defective fatty acid and glucose metabolism is a striking example showing the complete phenotypic effect of a QTL using congenic strains. For the defect in isoproterenol-mediated lipolysis in the SHR rat, Aitman and colleagues showed a complete rescue in the congenic strain carrying the region of the chromosome 4 QTL (linked to the defective catecholamine action) derived from the Brown Norway rat on an SHR genetic background (24). Similarly, Olofsson et al. showed a robust reduction in arthritis susceptibility using a congenic strain carrying the pristine-induced arthritis QTL 4 (*Pia4*) from the resistant E3 strain on a DA genetic background. The *Pia4* region has been found to be the only QTL associated with arthritis severity and joint erosions during the entire disease course (28). In addition, *Pia4* has been identified in models of multiple sclerosis and uveitis suggesting that it has a role in regulation of several inflammatory diseases. In contrast, when several loci are identified for a given phenotype, congenic strains do not show complete phenotypic correction. In a genome-wide linkage analysis for crescentic glomerulonephritis (*Crgn*), Aitman et al. identified seven QTLs (*Crgn1–7*) (33), and a follow-up study reported only modest phenotypic correction for *Crgn2* (Lod > 8) derived from the *Crgn* resistant Lewis rat introgressed into the genetic background of the susceptible WKY rat (21). Altogether, these studies suggest that although phenotypic effects of QTLs vary according to rat disease model, congenic strains are powerful tools to confirm the QTL effects.

Congenic strains can also be used for the identification of interacting loci within a QTL and the elegant work by Samuelson et al. shows that *Mcs5a* locus is a compound QTL with at least two non-coding interacting elements (*Mcs5a1* and *Mcs5a2*) in a rat mammary carcinogenesis model (31).

Another application of congenic strains is the use of comparative microarray based expression analyses. Congenic strains differ from their parental progenitors by the introgressed genetic interval containing the QTL. Thus, comparative microarray analysis between congenic and parental strains may be informative on potentially and differentially expressed genes encoded by the introgressed interval. This approach was adopted by McBride and colleagues and led to the identification of *Gstm1*, which plays a pathophysiological role in hypertension and oxidative stress in the stroke-prone of spontaneously hypertensive rat (SHRSP), a well-characterized experimental model for essential hypertension and endothelial dysfunction (27, 34).

A subcongenic (or interval-specific congenic) strain is obtained by backcrossing the congenic strain to the parental strain until recombination events reduce the initial congenic interval. The limits of a QTL can be narrowed by constructing subcongenic lines retaining the QTL within progressively smaller amounts of donor chromosome. Subcongenic strains are generally used to fine map large

QTLs and have been successfully used to reduce a QTL to a single gene resolution in high-resolution mapping studies (25, 26).

AIL are obtained when two parental strains are crossed to produce an F1 which is then intercrossed to produce an F2. Subsequent generations are produced by intercrossing the F2 individuals according to a pseudo-random breeding protocol. This results in high recombination rates and genetic variation in the population (35). Undoubtedly, AIL offer high resolution mapping when compared to F2 crosses and Swanberg and colleagues demonstrated the advantageous use of AIL derived from DA and PVG rats in the positional cloning of *Mhc2ta,* in the genetic mapping of strain differences in expression of MHC class II molecules after nerve injury (*VRA* ventral root avulsion) (17). In this study, the high recombination frequency in a relatively small genetic interval in the AIL was used to narrow down the *Vra4* QTL by correlating the expression of *Cd74*, a marker of MHC class II molecules, and genotypes in the recombination interval. This approach reduced the number of candidate genes to 13 in *Vra4* (17). Similarly, the positional cloning of the *Igl* genes controlling rheumatoid factor production and allergic bronchitis in rats was achieved by highly recombinant (19–21 generations of intercrossing) AIL derived from F344 and GK rats (36).

Haplotype-based mapping (or haplotype mapping) uses sets of closely linked genetic variants (SNPs or microsatellite markers) between inbred rat strains to identify regions that co-segregate with a given phenotype. This approach requires having phenotypic and genotypic information for a panel of inbred rat strains and assumes that the genetic variants shared by a set of inbred rat strains are identical by descent. Haplotype mapping was used in the positional cloning of *Fcgr3* in an F2 population derived from glomerulonephritis-susceptible WKY and –resistant Lewis rats (33). This study showed an enhanced macrophage activation in the WKY rats when compared to Lewis rats. Macrophage activity was then delineated to a minimal genetic interval that contained only *Fcgr3* within the QTL where genetic variants at the *Fcgr3* locus co-segregated with macrophage activity in a panel of inbred rat strains showing either high or low macrophage activation (33). While *Fcgr3* was positionally cloned on the rat chromosome 13 QTL (*Crgn1*), a promoter polymorphism found in the AP-1 transcription factor *Jund* co-segregated with macrophage infiltration on the chromosome 16 QTL (*Crgn2*) in this rat model of human *Crgn* (21). The promoter polymorphism and other microsatellite markers were then used in a haplotype map and defined a region of 130 kb containing *Jund* where genetic variants co-segregated with macrophage infiltration.

The fine-mapping of QTLs can also be achieved by using an outbred population of rats. By generating a genetically heterogeneous stock (HS) of rats, Johannesson and colleagues mapped a locus contributing to variation in a fear-related measure (two-way active avoidance in the shuttle box) to a region on chromosome 5 containing nine genes. By establishing a protocol measuring multiple phenotypes including immunology, neuroinflammation and haematology as well as cardiovascular, metabolic and behavioural traits, they established the rat HS as a new, powerful resource for the fine-mapping of QTLs contributing to variation in complex traits of biomedical relevance (37).

In summary, congenic, subcongenic, AIL and haplotype mapping have been successfully used in fine mapping of complex traits in various rat models of human disease. Given the importance of congenic strains in the verification of QTL-phenotype effects, we believe that in the future, positional cloning in the rat will still require a congenic strain strategy in combination with approaches allowing high recombination frequencies (subcongenic and AIL) and haplotype mapping in the genetic dissection of QTLs.

### 2.2. Recent Advances in Fine Mapping and Functional Tests of Candidate Genes

Generation of a high-quality draft of the sequence of the Brown Norway (BN) rat in 2004 was an important genome resource for rat geneticists (10). Rat positional cloning strategies have benefited from the generation of this comprehensive 7.5× sequence of the BN rat. Indeed, the positional cloning of *Pkdr1* in a rat model of polycystic kidney disease (PKD) combined traditional physical and genetic mapping methods with computational analyses of the emerging sequence of the BN genome (16). After fine-mapping and the determination of the critical interval responsible for spontaneous PKD in the *cy/+* rat, Brown and colleagues identified two genes by *in silico* mapping annotation. Following the identification of a C to T transition that replaces an arginine with a tryptophan at amino acid 823 in the protein sequence of SamCystin encoded by *Pkdr1*, the authors established the correct cDNA and protein sequences of the product. They then highlighted significant differences with all predicted *in silico* annotations of the *Pkdr1* gene and determined the expected sequences of the mouse and human orthologs (16). The publicly available rat sequence was also used in the fine mapping poorly annotated regions that are the subject to structural variations such as the *Fcgr3* locus on rat chromosome 13 (33). New informative microsatellite markers flanking the *Fcgr3* gene were identified from the rat genome sequence. Given the new sequencing technologies available, the sequence of other commonly used inbred rat strains will provide a basis for comparative genomics. The SHR rat is the most widely used strain for QTL mapping as a model of human metabolic syndrome, and a 10× sequencing of the SHR genome using massively parallel paired-end sequencing is currently being carried out as part of the EU-funded EURATools consortium. A complete sequence of the SHR genome will constitute an important genomic resource for positional cloning studies in crosses where SHR rats have been bred with the reference BN strain in order to identify various QTLs related to metabolic syndrome.

As mentioned previously, genetic variation between inbred rat strains, usually assayed by using a limited set of microsatellite markers, can be used for testing co-segregation with a wide range of disease phenotypes in different rat strains. However, a dense set of polymorphic markers such as SNPs provide a more powerful tool for high-resolution mapping when compared to microsatellite mapping. The international STAR consortium reported a survey of genetic variation based on almost three million newly identified SNPs and obtained accurate and complete genotypes for a subset of 20,238 SNPs across 167 distinct inbred rat strains, two rat recombinant inbred panels and an F2 intercross (11). This detailed SNP map was used for fine mapping of QTLs in two recent studies: the positional cloning of *Ogn* as a regulator of LVM used SNP-based fine mapping to exclude *Hbld2* as a positional candidate in the refined LVM QTL on rat chromosome 17 (19). Another example was the identification of soluble epoxide hydrolase (*Ephx2*) that was achieved by SNP genotyping of F2 crosses between spontaneously hypertensive heart failure (SHHF) rats and reference strains (18).

### 2.3. Functional Tests of Candidate Genes

The availability of the BN genome sequence together with recent advances in SNP-genotyping in inbred rat strains and eQTL mapping in recombinant inbred (RI) strains has allowed the positional cloning of many genes in various disease models. Although these approaches enable the identification of sequence variants in candidate genes, the biological testing of causative variants remains as the biggest challenge in QTL mapping. Functional testing of candidate genes until 2002 was achieved by transgenic rescue in rats (24, 38). Nonsense mutations in *Cd36* and *Cblb* associated with insulin resistance and type 1 diabetes respectively resulted in strong reduction or complete loss of function of the gene products, and these were therefore functionally complemented with transgenesis of the wild type copy of the genes (24, 38). However, the biggest obstacle for functional testing of candidate genes is the lack of rat pluripotent embryonic stem cells, making the generation of stable knockouts by homologous recombination not possible. Although several positional cloning studies in rat models tested the candidacy of positionally cloned genes in mouse knockouts (18, 19, 39), recent advances in molecular biology offer alternative approaches for generating knockout rats and enable testing of the candidate genes in rat models. The use of somatic cell nuclear transfer to develop cloned rats as an alternative to using embryonic stem cells has been reported, but remains technically challenging (40). N-ethyl-N-nitrosourea (ENU) mutagenesis followed by a screening method to detect single-nucleotide substitutions within the targeted gene was previously reported by several groups (41–43). However, the low induced mutation frequency, as well as the efficiency and throughput of the screening methods are the important drawbacks in generation of knockout rats by ENU mutagenesis. Indeed, most of the ENU-mutagenised rats are discarded within weeks after target genes have been screened, and the screening methods are based either on yeast (41, 43) or on high throughput resequencing that are currently expensive. A recent study by Mashimo and colleagues overcame these difficulties by combining a high throughput and low cost screening assay that uses phage Mu transposition reaction and intracytoplasmic sperm injection (ICSI) for the recovery of the rare heterozygous genotypes from a newly generated frozen sperm repository (44). This new approach is cost and resource effective and will allow functional testing of many candidate genes underlying QTLs in rat models.

RNAi is increasingly and widely used after the first description of this strategy in mammalian cells by introducing small double-stranded RNAs comprising 19–21 nucleotide complementary sequences (called small interfering RNAs, siRNAs) to silence gene expression with high specificity and without activating an interferon response (45). The positional cloning of vitamin K epoxide reductase (VKOR) is an outstanding example of the use of siRNA in gene identification (46). Although VKOR activity was first reported in 1974, the gene encoding VKOR was not identified until 2004. The previously mapped warfarin resistance (*Rw*) trait on rat chromosome 1 (47) and the mapping of a locus for combined deficiencies of vitamin-K-dependant (VKD) proteins to human chromosome 16, syntenic to the rat region (48), led the authors to focus on human chromosome 16p12-q21 for positional cloning. The 20.3 Mb region contained 190 predicted coding sequences and the authors prioritised 13 genes as they code for integral membrane proteins. An RNAi approach by using siRNA in vitro led to the identification of VKOR gene as only one gene among the 13 tested showed reduced VKOR activity following siRNA knockdown in A549 cells (46).

## 3. Translation to Humans

The ultimate goal of positional cloning using rat models is translation to related phenotypes in humans. The various existing examples in the literature show how findings deriving from rat models can be used to advance understanding of the genetic basis of complex human disease. Over the past 3 years, studies combining microarray and sequencing technologies have revealed that the human genome is characterised by extensive and complex structural variation contributing to structural diversity among individuals (49–53). Naturally occurring variation in gene copy number is increasingly recognised as a heritable source of susceptibility to genetically complex diseases. The positional cloning of *Fcgr3* in a rat model of human *Crgn* gave new insights into the importance of structural variants in common human disease (33). In humans, low copy number of *FCGR3B*, an orthologue of rat *Fcgr3*, was associated with systemic, but not organ-specific, autoimmunity (54, 55).

Translational studies in humans can also be achieved by comparative genomics between rat and humans. Genetic analysis of a rat with recessive PKD revealed an orthologous relationship between the rat locus and the autosomal recessive polycystic kidney disease (ARPKD) region in humans. The identification of a mutation in the rat gene led to the screening of 66 coding exons of the human ortholog (PKHD1) in 14 probands with ARPKD and revealed 6 truncating and 12 missense mutations; 8 of the affected individuals were compound heterozygotes (56). A similar comparative genomic approach was undertaken in the gene identification for Meckel-Gruber syndrome (MKS), a severe autosomal recessive disorder characterized by the bilateral renal cystic dysplasia, central nervous system malformations, hepatic abnormalities and polydactyly (57). One of the MKS loci maps in humans to chromosome 8 (MKS3) and is syntenic to the Wpk locus in rat, which is a model with PKD, agenesis of the corpus callosum and hydrocephalus. Positional cloning of the Wpk gene suggested a MKS3 candidate gene, TMEM67, in which the authors identified pathogenic mutations in five MKS3-linked consanguineous families (57).

Genes underlying polygenic complex traits have also been successfully translated to humans and this includes the rat breast cancer susceptibility locus *Mcs5a,* as well as *Ogn* and *Ephx2* as risk factors for increased LVM and heart failure respectively (18, 19, 31). However, following the positional cloning in rat models, the translation studies in humans were achieved by association studies (*Mcsa5*) or by correlation with phenotype (*Ogn*, *Ephx2*) underlying the role of these genes as susceptibility factors rather than highly penetrant causative variants.

In high resolution genetic mapping studies carried out by John Rapp and colleagues, the *Cyp11b1* locus was identified as being responsible for the blood pressure difference between Dahl SS/Jr and SR/Jr rats in a minimal congenic strain. Although mutations in the *CYP11B1* gene in humans were initially well known to cause rare monogenic forms of inherited hypertension resulting from 11b-hydroxylase deficiency (58), common or rare variants within the *CYP11B1* locus contributing to essential hypertension have not been reported so far. None of the variants previously associated with hypertension showed evidence for association in the recent Wellcome Trust Case Control Consortium (WTCCC) study (59) but deep resequencing have started to identify highly penetrant rare variants affecting blood pressure in humans (60).

## 4. New Approaches in Rat Positional Cloning

Segregating populations allow efficient identification of the genetic determinants of gene expression which may throw light on the molecular basis of complex traits. Quantitative variation in gene expression levels acts as an endophenotype (or intermediate phenotype) situated between the genomic DNA sequence variation and more complex physiological phenotypes. A number of studies indicate that the individual variation in gene expression (i.e. transcript abundance) is heritable in segregating populations (61–65) and gene expression levels can therefore be mapped to the genome using linkage methods, allowing identification of eQTLs. This approach has been termed genetical genomics (66, 67) as eQTLs represent genomic regions for the genetic control of gene expression. Although gene expression profiling is not a novel methodology for identifying causative genes in QTLs by itself, when combined with genetic linkage mapping, it can lead to new insights into the genes and regulatory pathways underlying a widespread range of complex whole-body phenotypes by elucidating metabolic, regulatory, and developmental pathways (67). eQTL analysis is particularly advantageous as it can discriminate between *cis*- and *trans*-acting influences on gene expression, which can help in identifying candidate genes whose expression is under local regulatory control (*cis*-eQTLs) and in dissecting complex regulatory networks comprised of multiple *trans*-eQTLs. A *cis*-acting eQTL maps to the physical location of the gene itself, whereas a *trans*-acting eQTL maps to a genomic region that is distant from the physical location of the gene being transcribed. The eQTL approach can be used as a complementary approach to the traditional physiological QTL (pQTL) mapping. Recent studies showed that the eQTL approach was indeed successful for the positional cloning of different genes in rat models. Thus, genome-wide expression profiling can be used as a trait in a segregating population allowing identification of eQTLs; and if the same study includes pQTL data, the combined approach significantly facilitates the molecular dissection of pQTLs. The key work by Hubner and colleagues integrating genome-wide expression profiling with linkage analysis in RI rat strains (68) led 3 years later to the positional cloning of *Ogn*, *Ephx2* and *Cd36* as susceptibility factors for increased LVM, heart failure and blood pressure, respectively (18–20).

RI strains are derived by crossing two inbred rat strains in order to obtain an F2 population and following 20 or more brother–sister matings between F2 individuals, a panel of inbred animals each with a different combination of progenitor genomes is produced. In the early 1980s, the SHR strain was crossed with the normotensive Brown Norway (BN) strain to generate the BXH/HXB panel of RI strains (69–71). Although rat RI panels are powerful and renewable resources for genetic mapping that offer the opportunity to accumulate genetic and physiological data over time, to date, there is no reported study on the positional cloning of pQTL genes using exclusively linkage analysis for different metabolic phenotypes in RI strains. However, the combined eQTL–pQTL approach has been successfully used in positional cloning using RI strains as eQTLs identified in the heart highlighted *Ogn* as a candidate as it was the only eQTL within the chromosome 17 pQTL for LVM (19).

On the horizon, however, is an important change of perspective: from single gene(s) to gene networks and pathways. The relevance of a "systems biology" approach to identify pathways operational in common disorders has been suggested by Petretto et al. (19) where they inferred that *Ogn* may influence LVM through modulation of the TGF-β pathway. The role for the TGF-β signalling pathway in the regulation of LVM was supported by human data that show an association of LVM in humans with several genes that are important in this pathway. To date, integrated gene network approaches to positionally clone disease genes have been successfully carried out mainly in the mouse (72, 73) where perturbation signatures in gene expression data and networks were used to map genes for complex traits. These recent examples in the mouse raise the importance of gene networks in the pathogenesis of disease, suggesting that similar approaches can be employed to dissect the networks of *cis* and *trans* interactions involved in quantitative traits. The large QTL, eQTL and QTT studies published in the rat reveal that the datasets and study approaches are now also at hand in the rat model to accelerate identification of QTL genes through the integration of gene network and pathway analyses.

## 5. The Future of Genetic Mapping and Positional Cloning

Figure 2.1 summarises the present and future strategies in rat positional cloning of complex trait genes. It is noteworthy that a combination of different approaches such as eQTL–pQTL–QTT has been successfully applied in the identification of risk variants

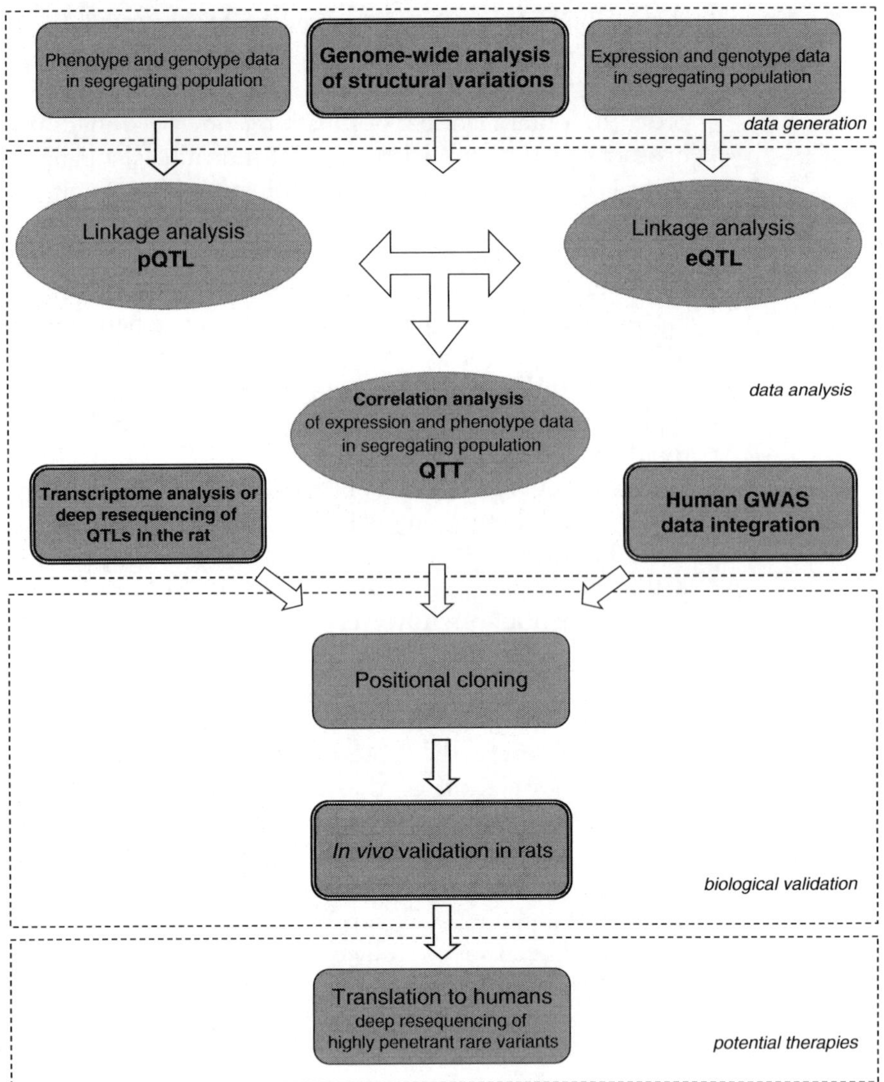


Fig. 2.1. The future of positional cloning in the rat. The combined pQTL–eQTL–QTT approaches were recently and successfully used in positional cloning of several complex trait genes (*see* text). The emerging deep resequencing, transcriptome analysis, genome-wide association (GWAS) and structural variation studies will allow integration of additional tools for positional cloning of complex traits in rat models. Given the recent advances in generating knock out rats, the in vivo validation is expected to be mostly carried out in the rat. The rat will constitute a valuable "reagent" for in vivo validation of genetic variants identified by human GWAS. Furthermore, rat positional cloning could provide rare genetic variants to be tested in human populations by deep resequencing (*pQTL*, physiological QTL; *eQTL,* expression QTL; *QTT,* quantitative trait transcript).

for different disease models. Figure 2.1 does not show the fine mapping strategies that could also lead to positional cloning (described in detail in "Successful positional cloning in the rat"). To date, translation to humans have been achieved by association or correlation studies; however we believe that in the near future, rat models will be a valuable "reagent" for the biological testing of genetic variants identified in GWAS in human populations (Fig. 2.1). In addition to the common variants identified by these high throughput techniques, the emerging "rare variant hypothesis" of complex human diseases highlights the importance of rat genetics to facilitate elucidation of the underlying molecular basis of common human disease. Data from genetic studies in rodent models will suggest that genes whose orthologs should be prioritised for deep re-sequencing in order to identify rare variants that can be tested for disease association in human populations.

## References


1. Falconer DS, Mackay TFC (1996) Introduction to quantitative genetics, 4th edn. Longman, Essex, England
2. Jacob HJ (1999) Functional genomics and rat models. Genome Res 9:1013–1016
3. Lander ES, Botstein D (1989) Mapping mendelian factors underlying quantitative traits using RFLP linkage maps. Genetics 121: 185–199
4. Lander E, Kruglyak L (1995) Genetic dissection of complex traits: guidelines for interpreting and reporting linkage results. Nat Genet 11:241–247
5. Nadeau JH, Frankel WN (2000) The roads from phenotypic variation to gene discovery: mutagenesis versus QTLs. Nat Genet 25:381–384
6. Battey J, Jordan E, Cox D, Dove W (1999) An action plan for mouse genomics. Nat Genet 21:73–75
7. Nolan PM, Peters J, Strivens M, Rogers D, Hagan J, Spurr N, Gray IC, Vizor L, Brooker D, Whitehill E, Washbourne R, Hough T, Greenaway S, Hewitt M, Liu X, McCormack S, Pickford K, Selley R, Wells C, Tymowska-Lalanne Z, Roby P, Glenister P, Thornton C, Thaung C, Stevenson JA, Arkell R, Mburu P, Hardisty R, Kiernan A, Erven A, Steel KP, Voegeling S, Guenet JL, Nickols C, Sadri R, Nasse M, Isaacs A, Davies K, Browne M, Fisher EM, Martin J, Rastan S, Brown SD, Hunter J (2000) A systematic, genome-wide, phenotype-driven mutagenesis programme for gene function studies in the mouse. Nat Genet 25:440–443
8. Aitman TJ, Critser JK, Cuppen E, Dominiczak A, Fernandez-Suarez XM, Flint J, Gauguier D, Geurts AM, Gould M, Harris PC, Holmdahl R, Hubner N, Izsvak Z, Jacob HJ, Kuramoto T, Kwitek AE, Marrone A, Mashimo T, Moreno C, Mullins J, Mullins L, Olsson T, Pravenec M, Riley L, Saar K, Serikawa T, Shull JD, Szpirer C, Twigger SN, Voigt B, Worley K (2008) Progress and prospects in rat genetics: a community view. Nat Genet 40:516–522
9. Smits BM, Cuppen E (2006) Rat genetics: the next episode. Trends Genet 22:232–240
10. Gibbs RA, Weinstock GM, Metzker ML, Muzny DM, Sodergren EJ, Scherer S, Scott G, Steffen D, Worley KC, Burch PE, Okwuonu G, Hines S, Lewis L, DeRamo C, DelgadoO, Dugan-Rocha S, Miner G, Morgan M, Hawes A, Gill R, Celera Holt RA, Adams MD, Amanatides PG, Baden-Tillson H, Barnstead M, Chin S, Evans CA, Ferriera S, Fosler C, Glodek A, Gu Z, Jennings D, Kraft CL, Nguyen T, Pfannkoch CM, Sitter C, Sutton GG, Venter JC, Woodage T, Smith D, Lee HM, Gustafson E, Cahill P, Kana A, Doucette-Stamm L, Weinstock K, Fechtel K, Weiss RB, Dunn DM, Green ED, Blakesley RW, Bouffard GG, De Jong PJ, Osoegawa K, Zhu B, Marra M, Schein J, Bosdet I, Fjell C, Jones S, Krzywinski M, Mathewson C, Siddiqui A, Wye N, McPherson J, Zhao S, Fraser CM, Shetty J, Shatsman S, Geer K, Chen Y, Abramzon S, Nierman WC, Havlak PH, Chen R, Durbin KJ, Egan A, Ren Y, Song XZ, Li B, Liu Y, Qin X, Cawley S, Worley KC, Cooney AJ, D'Souza LM, Martin K, Wu JQ, Gonzalez-Garay ML, Jackson AR, Kalafus KJ, McLeod MP, Milosavljevic A, Virk D, Volkov A, Wheeler DA, Zhang Z, Bailey JA, Eichler EE, et a. (2004) Genome sequence of the Brown Norway rat yields insights into mammalian evolution. Nature 428:493–521
11. Saar K, Beck A, Bihoreau MT, Birney E, Brocklebank D, Chen Y, Cuppen E, Demonchy S, Dopazo J, Flicek P, Foglio M, Fujiyama A, Gut IG, Gauguier D, Guigo R, Guryev V, Heinig M, Hummel O, Jahn N, Klages S, Kren V, Kube M, Kuhl H, Kuramoto

T, Kuroki Y, Lechner D, Lee YA, Lopez-Bigas N, Lathrop GM, Mashimo T, Medina I, Mott R, Patone G, Perrier-Cornet JA, Platzer M, Pravenec M, Reinhardt R, Sakaki Y, Schilhabel M, Schulz H, Serikawa T, Shikhagaie M, Tatsumoto S, Taudien S, Toyoda A, Voigt B, Zelenika D, Zimdahl H, Hubner N (2008) SNP and haplotype mapping for genetic analysis in the rat. Nat Genet 40:560–566

12. Guryev V, Saar K, Adamovic T, Verheul M, van Heesch SA, Cook S, Pravenec M, Aitman T, Jacob H, Shull JD, Hubner N, Cuppen E (2008) Distribution and functional impact of DNA copy number variation in the rat. Nat Genet 40:538–545
13. Manolio TA, Brooks LD, Collins FS (2008) A HapMap harvest of insights into the genetics of common disease. J Clin Invest 118: 1590–1605
14. Petretto E, Liu ET, Aitman TJ (2007) A gene harvest revealing the archeology and complexity of human disease. Nat Genet 39:1299–1301
15. Moreno C, Lazar J, Jacob HJ, Kwitek AE (2008) Comparative genomics for detecting human disease genes. Adv Genet 60: 655–697
16. Brown JH, Bihoreau MT, Hoffmann S, Kranzlin B, Tychinskaya I, Obermuller N, Podlich D, Boehn SN, Kaisaki PJ, Megel N, Danoy P, Copley RR, Broxholme J, Witzgall R, Lathrop M, Gretz N, Gauguier D (2005) Missense mutation in sterile alpha motif of novel protein SamCystin is associated with polycystic kidney disease in (cy/+) rat. J Am Soc Nephrol 16:3517–3526
17. Swanberg M, Lidman O, Padyukov L, Eriksson P, Akesson E, Jagodic M, Lobell A, Khademi M, Borjesson O, Lindgren CM, Lundman P, Brookes AJ, Kere J, Luthman H, Alfredsson L, Hillert J, Klareskog L, Hamsten A, Piehl F, Olsson T (2005) MHC2TA is associated with differential MHC molecule expression and susceptibility to rheumatoid arthritis, multiple sclerosis and myocardial infarction. Nat Genet 37:486–494
18. Monti J, Fischer J, Paskas S, Heinig M, Schulz H, Gosele C, Heuser A, Fischer R, Schmidt C, Schirdewan A, Gross V, Hummel O, Maatz H, Patone G, Saar K, Vingron M, Weldon SM, Lindpaintner K, Hammock BD, Rohde K, Dietz R, Cook SA, Schunck WH, Luft FC, Hubner N (2008) Soluble epoxide hydrolase is a susceptibility factor for heart failure in a rat model of human disease. Nat Genet 40:529–537
19. Petretto E, Sarwar R, Grieve I, Lu H, Kumaran MK, Muckett PJ, Mangion J, Schroen B, Benson M, Punjabi PP, Prasad SK, Pennell DJ, Kiesewetter C, Tasheva ES, Corpuz LM, Webb MD, Conrad GW, Kurtz TW, Kren V, Fischer J, Hubner N, Pinto YM, Pravenec M, Aitman TJ, Cook SA (2008) Integrated genomic approaches implicate osteoglycin (Ogn) in the regulation of left ventricular mass. Nat Genet 40:546–552
20. Pravenec M, Churchill PC, Churchill MC, Viklicky O, Kazdova L, Aitman TJ, Petretto E, Hubner N, Wallace CA, Zimdahl H, Zidek V, Landa V, Dunbar J, Bidani A, Griffin K, Qi N, Maxova M, Kren V, Mlejnek P, Wang J, Kurtz TW (2008) Identification of renal Cd36 as a determinant of blood pressure and risk for hypertension. Nat Genet 40: 952–954
21. Behmoaras J, Bhangal G, Smith J, McDonald K, Mutch B, Lai PC, Domin J, Game L, Salama A, Foxwell BM, Pusey CD, Cook HT, Aitman TJ (2008) Jund is a determinant of macrophage activation and is associated with glomerulonephritis susceptibility. Nat Genet 40:553–559
22. Abiola O, Angel JM, Avner P, Bachmanov AA, Belknap JK, Bennett B, Blankenhorn EP, Blizard DA, Bolivar V, Brockmann GA, Buck KJ, Bureau JF, Casley WL, Chesler EJ, Cheverud JM, Churchill GA, Cook M, Crabbe JC, Crusio WE, Darvasi A, de Haan G, Dermant P, Doerge RW, Elliot RW, Farber CR, Flaherty L, Flint J, Gershenfeld H, Gibson JP, Gu J, Gu W, Himmelbauer H, Hitzemann R, Hsu HC, Hunter K, Iraqi FF, Jansen RC, Johnson TE, Jones BC, Kempermann G, Lammert F, Lu L, Manly KF, Matthews DB, Medrano JF, Mehrabian M, Mittlemann G, Mock BA, Mogil JS, Montagutelli X, Morahan G, Mountz JD, Nagase H, Nowakowski RS, O'Hara BF, Osadchuk AV, Paigen B, Palmer AA, Peirce JL, Pomp D, Rosemann M, Rosen GD, Schalkwyk LC, Seltzer Z, Settle S, Shimomura K, Shou S, Sikela JM, Siracusa LD, Spearow JL, Teuscher C, Threadgill DW, Toth LA, Toye AA, Vadasz C, Van Zant G, Wakeland E, Williams RW, Zhang HG, Zou F (2003) The nature and identification of quantitative trait loci: a community's view. Nat Rev Genet 4:911–916
23. Glazier AM, Nadeau JH, Aitman TJ (2002) Finding genes that underlie complex traits. Science 298:2345–2349
24. Aitman TJ, Glazier AM, Wallace CA, Cooper LD, Norsworthy PJ, Wahid FN, Al-Majali KM, Trembling PM, Mann CJ, Shoulders CC, Graf D, St Lezin E, Kurtz TW, Kren V, Pravenec M, Ibrahimi A, Abumrad NA,

Stanton LW, Scott J (1999) Identification of Cd36 (Fat) as an insulin-resistance gene causing defective fatty acid and glucose metabolism in hypertensive rats. Nat Genet 21: 76–83

25. Cicila GT, Garrett MR, Lee SJ, Liu J, Dene H, Rapp JP (2001) High-resolution mapping of the blood pressure QTL on chromosome 7 using Dahl rat congenic strains. Genomics 72: 51–60
26. Garrett MR, Rapp JP (2003) Defining the blood pressure QTL on chromosome 7 in Dahl rats by a 177-kb congenic segment containing Cyp11b1. Mamm Genome 14: 268–273
27. McBride MW, Brosnan MJ, Mathers J, McLellan LI, Miller WH, Graham D, Hanlon N, Hamilton CA, Polke JM, Lee WK, Dominiczak AF (2005) Reduction of Gstm1 expression in the stroke-prone spontaneously hypertension rat contributes to increased oxidative stress. Hypertension 45:786–792
28. Olofsson P, Holmberg J, Tordsson J, Lu S, Akerstrom B, Holmdahl R (2003) Positional identification of Ncf1 as a gene that regulates arthritis severity in rats. Nat Genet 33:25–32
29. Pravenec M, Kazdova L, Landa V, Zidek V, Mlejnek P, Simakova M, Jansa P, Forejt J, Kren V, Krenova D, Qi N, Wang JM, Chan D, Aitman TJ, Kurtz TW (2008) Identification of mutated Srebf1 as a QTL influencing risk for hepatic steatosis in the spontaneously hypertensive rat. Hypertension 51:148–153
30. Rangel-Filho A, Sharma M, Datta YH, Moreno C, Roman RJ, Iwamoto Y, Provoost AP, Lazar J, Jacob HJ (2005) RF-2 gene modulates proteinuria and albuminuria independently of changes in glomerular permeability in the fawn-hooded hypertensive rat. J Am Soc Nephrol 16:852–856
31. Samuelson DJ, Hesselson SE, Aperavich BA, Zan Y, Haag JD, Trentham-Dietz A, Hampton JM, Mau B, Chen KS, Baynes C, Khaw KT, Luben R, Perkins B, Shah M, Pharoah PD, Dunning AM, Easton DF, Ponder BA, Gould MN (2007) Rat Mcs5a is a compound quantitative trait locus with orthologous human loci that associate with breast cancer risk. Proc Natl Acad Sci U S A 104:6299–6304
32. Rapp JP (2000) Genetic analysis of inherited hypertension in the rat. Physiol Rev 80:135–172
33. Aitman TJ, Dong R, Vyse TJ, Norsworthy PJ, Johnson MD, Smith J, Mangion J, Roberton-Lowe C, Marshall AJ, Petretto E, Hodges MD, Bhangal G, Patel SG, Sheehan-Rooney K, Duda M, Cook PR, Evans DJ, Domin J, Flint J, Boyle JJ, Pusey CD, Cook HT (2006) Copy number polymorphism in Fcgr3 predisposes to glomerulonephritis in rats and humans. Nature 439:851–855
34. McBride MW, Carr FJ, Graham D, Anderson NH, Clark JS, Lee WK, Charchar FJ, Brosnan MJ, Dominiczak AF (2003) Microarray analysis of rat chromosome 2 congenic strains. Hypertension 41:847–853
35. Darvasi A, Soller M (1995) Advanced intercross lines, an experimental population for fine genetic mapping. Genetics 141:1199–1207
36. Rintisch C, Ameri J, Olofsson P, LuthmanH, Holmdahl R (2008) Positional cloning of the Igl genes controlling rheumatoid factor production and allergic bronchitis in rats. Proc Natl Acad Sci U S A 105(37):14005–14010
37. Johannesson M, Lopez-Aumatell R, Diez M, Tuncel J, Blazquez G, Martinez E, Canete A, Vicens-Costa E, Graham D, Copley R, Hernandez-Pliego P, Beyeen A, Stridh P, Ockinger J, Fernandez C, Tobena A, Guitart-Masip M, Gimenez-Llort L, Dominiczak A, Holmdahl R, Gauguier D, Olsson T, Mott R, Valdar W, Redei E, Fernandez-Teruel A (2008) and Flint, J. The NIH heterogeneous stock. Genome Res, A resource for the simultaneous high-resolution mapping of multiple quantitative trait loci in rats
38. Yokoi N, Komeda K, Wang HY, Yano H, Kitada K, Saitoh Y, Seino Y, Yasuda K, Serikawa T, Seino S (2002) Cblb is a major susceptibility gene for rat type 1 diabetes mellitus. Nat Genet 31:391–394
39. Hultqvist M, Olofsson P, Holmberg J, Backstrom BT, Tordsson J, Holmdahl R (2004) Enhanced autoimmunity, arthritis, and encephalomyelitis in mice with a reduced oxidative burst due to a mutation in the Ncf1 gene. Proc Natl Acad Sci U S A 101:12646–12651
40. Zhou Q, Renard JP, Le Friec G, Brochard V, Beaujean N, Cherifi Y, Fraichard A, Cozzi J (2003) Generation of fertile cloned rats by regulating oocyte activation. Science 302:1179
41. Amos-Landgraf JM, Kwong LN, Kendziorski CM, Reichelderfer M, Torrealba J, Weichert J, Haag JD, Chen KS, Waller JL, Gould MN, Dove WF (2007) A target-selected Apc-mutant rat kindred enhances the modeling of familial human colon cancer. Proc Natl Acad Sci U S A 104:4036–4041
42. Homberg JR, Olivier JD, Smits BM, Mul JD, Mudde J, Verheul M, Nieuwenhuizen OF, Cools AR, Ronken E, Cremers T, Schoffelmeer AN, Ellenbroek BA, Cuppen E (2007) Characterization of the serotonin transporter knockout rat: a selective change in the functioning of the serotonergic system. Neuroscience 146:1662–1676

43. Zan Y, Haag JD, Chen KS, Shepel LA, Wigington D, Wang YR, Hu R, Lopez-Guajardo CC, Brose HL, Porter KI, Leonard RA, Hitt AA, Schommer SL, Elegbede AF, Gould MN (2003) Production of knockout rats using ENU mutagenesis and a yeast-based screening assay. Nat Biotechnol 21:645–651

44. Mashimo T, Yanagihara K, Tokuda S, Voigt B, Takizawa A, Nakajima R, Kato M, Hirabayashi M, Kuramoto T, Serikawa T (2008) An ENU-induced mutant archive for gene targeting in rats. Nat Genet 40:514–515

45. Elbashir SM, Harborth J, Lendeckel W, Yalcin A, Weber K, Tuschl T (2001) Duplexes of 21-nucleotide RNAs mediate RNA interference in cultured mammalian cells. Nature 411:494–498

46. Li T, Chang CY, Jin DY, Lin PJ, Khvorova A, Stafford DW (2004) Identification of the gene for vitamin K epoxide reductase. Nature 427:541–544

47. Kohn MH, Pelz HJ (2000) A gene-anchored map position of the rat warfarin-resistance locus, Rw, and its orthologs in mice and humans. Blood 96:1996–1998

48. Fregin A, Rost S, Wolz W, Krebsova A, Muller CR, Oldenburg J (2002) Homozygosity mapping of a second gene locus for hereditary combined deficiency of vitamin K-dependent clotting factors to the centromeric region of chromosome 16. Blood 100:3229–3232

49. Iafrate AJ, Feuk L, Rivera MN, Listewnik ML, Donahoe PK, Qi Y, Scherer SW, Lee C (2004) Detection of large-scale variation in the human genome. Nat Genet 36:949–951

50. Korbel JO, Urban AE, Affourtit JP, Godwin B, Grubert F, Simons JF, Kim PM, Palejev D, Carriero NJ, Du L, Taillon BE, Chen Z, Tanzer A, Saunders AC, Chi J, Yang F, Carter NP, Hurles ME, Weissman SM, Harkins TT, Gerstein MB, Egholm M, Snyder M (2007) Paired-end mapping reveals extensive structural variation in the human genome. Science 318:420–426

51. Redon R, Ishikawa S, Fitch KR, Feuk L, Perry GH, Andrews TD, Fiegler H, Shapero MH, Carson AR, Chen W, Cho EK, Dallaire S, Freeman JL, Gonzalez JR, Gratacos M, Huang J, Kalaitzopoulos D, Komura D, MacDonald JR, Marshall CR, Mei R, Montgomery L, Nishimura K, Okamura K, Shen F, Somerville MJ, Tchinda J, Valsesia A, Woodwark C, Yang F, Zhang J, Zerjal T, Zhang J, Armengol L, Conrad DF, Estivill X, Tyler-Smith C, Carter NP, Aburatani H, Lee C, Jones KW, Scherer SW, Hurles ME (2006) Global variation in copy number in the human genome. Nature 444:444–454

52. Sebat J, Lakshmi B, Troge J, Alexander J, Young J, Lundin P, Maner S, Massa H, Walker M, Chi M, Navin N, Lucito R, Healy J, Hicks J, Ye K, Reiner A, Gilliam TC, Trask B, Patterson N, Zetterberg A, Wigler M (2004) Large-scale copy number polymorphism in the human genome. Science 305:525–528

53. Tuzun E, Sharp AJ, Bailey JA, Kaul R, Morrison VA, Pertz LM, Haugen E, Hayden H, Albertson D, Pinkel D, Olson MV, Eichler EE (2005) Fine-scale structural variation of the human genome. Nat Genet 37:727–732

54. Fanciulli M, Norsworthy PJ, Petretto E, Dong R, Harper L, Kamesh L, Heward JM, Gough SC, de Smith A, Blakemore AI, Froguel P, Owen CJ, Pearce SH, Teixeira L, Guillevin L, Graham DS, Pusey CD, Cook HT, Vyse TJ, Aitman TJ (2007) FCGR3B copy number variation is associated with susceptibility to systemic, but not organ-specific, autoimmunity. Nat Genet 39:721–723

55. Willcocks LC, Lyons PA, Clatworthy MR, Robinson JI, Yang W, Newland SA, Plagnol V, McGovern NN, Condliffe AM, Chilvers ER, Adu D, Jolly EC, Watts R, Lau YL, Morgan AW, Nash G, Smith KG (2008) Copy number of FCGR3B, which is associated with systemic lupus erythematosus, correlates with protein expression and immune complex uptake. J Exp Med 205:1573–1582

56. Ward CJ, Hogan MC, Rossetti S, Walker D, Sneddon T, Wang X, Kubly V, Cunningham JM, Bacallao R, Ishibashi M, Milliner DS, Torres VE, Harris PC (2002) The gene mutated in autosomal recessive polycystic kidney disease encodes a large, receptor-like protein. Nat Genet 30:259–269

57. Smith UM, Consugar M, Tee LJ, McKee BM, Maina EN, Whelan S, Morgan NV, Goranson E, Gissen P, Lilliquist S, Aligianis IA, Ward CJ, Pasha S, Punyashthiti R, Malik Sharif S, Batman PA, Bennett CP, Woods CG, McKeown C, Bucourt M, Miller CA, Cox P, Algazali L, Trembath RC, Torres VE, Attie-Bitach T, Kelly DA, Maher ER, Gattone VH 2nd, Harris PC, Johnson CA (2006) The transmembrane protein meckelin (MKS3) is mutated in Meckel-Gruber syndrome and the wpk rat. Nat Genet 38:191–196

58. Lifton RP, Dluhy RG, Powers M, Rich GM, Cook S, Ulick S, Lalouel JM (1992) A chimaeric 11 beta-hydroxylase/aldosterone synthase gene causes glucocorticoid-remediable aldosteronism and human hypertension. Nature 355:262–265

59. Wellcome Trust Case Control Consortium (2007) Genome-wide association study of

14,000 cases of seven common diseases and 3,000 shared controls. Nature 447:661–678

60. Ji W, Foo JN, O'Roak BJ, Zhao H, Larson MG, Simon DB, Newton-Cheh C, State MW, Levy D, Lifton RP (2008) Rare independent mutations in renal salt handling genes contribute to blood pressure variation. Nat Genet 40:592–599

61. Brem RB, Yvert G, Clinton R, Kruglyak L (2002) Genetic dissection of transcriptional regulation in budding yeast. Science 296: 752–755

62. Cheung VG, Spielman RS, Ewens KG, Weber TM, Morley M, Burdick JT (2005) Mapping determinants of human gene expression by regional and genome-wide association. Nature 437:1365–1369

63. Morley M, Molony CM, Weber TM, Devlin JL, Ewens KG, Spielman RS, Cheung VG (2004) Genetic analysis of genome-wide variation in human gene expression. Nature 430:743–747

64. Schadt EE, Monks SA, Drake TA, Lusis AJ, Che N, Colinayo V, Ruff TG, Milligan SB, Lamb JR, Cavet G, Linsley PS, Mao M, Stoughton RB, Friend SH (2003) Genetics of gene expression surveyed in maize, mouse and man. Nature 422:297–302

65. Yvert G, Brem RB, Whittle J, Akey JM, Foss E, Smith EN, Mackelprang R, Kruglyak L (2003) Trans-acting regulatory variation in Saccharomyces cerevisiae and the role of transcription factors. Nat Genet 35:57–64

66. Jansen RC, Nap JP (2001) Genetical genomics: the added value from segregation. Trends Genet 17:388–391

67. Li J, Burmeister M (2005) Genetical genomics: combining genetics with gene expression analysis. Hum Mol Genet 14(2):R163–R169

68. Hubner N, Wallace CA, Zimdahl H, Petretto E, Schulz H, Maciver F, Mueller M, Hummel O, Monti J, Zidek V, Musilova A, Kren V, Causton H, Game L, Born G, Schmidt S, Muller A, Cook SA, Kurtz TW, Whittaker J, Pravenec M, Aitman TJ (2005) Integrated transcriptional profiling and linkage analysis for identification of genes underlying disease. Nat Genet 37:243–253

69. Pravenec M, Klir P, Kren V, Zicha J, Kunes J (1989) An analysis of spontaneous hypertension in spontaneously hypertensive rats by means of new recombinant inbred strains. J Hypertens 7:217–221

70. Pravenec M, Kren V (2005) Genetic analysis of complex cardiovascular traits in the spontaneously hypertensive rat. Exp Physiol 90:273–276

71. Pravenec M, Zidek V, Landa V, Simakova M, Mlejnek P, Kazdova L, Bila V, Krenova D, Kren V (2004) Genetic analysis of "metabolic syndrome" in the spontaneously hypertensive rat. Physiol Res 53(Suppl 1):S15–S22

72. Chen Y, Zhu J, Lum PY, Yang X, Pinto S, MacNeil DJ, Zhang C, Lamb J, Edwards S, Sieberts SK, Leonardson A, Castellini LW, Wang S, Champy MF, Zhang B, Emilsson V, Doss S, Ghazalpour A, Horvath S, Drake TA, Lusis AJ, Schadt EE (2008) Variations in DNA elucidate molecular networks that cause disease. Nature 452:429–435

73. Mehrabian M, Allayee H, Stockton J, Lum PY, Drake TA, Castellani LW, Suh M, Armour C, Edwards S, Lamb J, Lusis AJ, Schadt EE (2005) Integrating genotypic and expression data in a segregating mouse population to identify 5-lipoxygenase as a susceptibility gene for obesity and bone traits. Nat Genet 37:1224–1233

74. Pravenec M, Landa V, Zidek V, Musilova A, Kren V, Kazdova L, Aitman TJ, Glazier AM, Ibrahimi A, Abumrad NA, Qi N, Wang JM, St Lezin EM, Kurtz TW (2001) Transgenic rescue of defective Cd36 ameliorates insulin resistance in spontaneously hypertensive rats. Nat Genet 27:156–158

75. Kitada K, Akimitsu T, Shigematsu Y, Kondo A, Maihara T, Yokoi N, Kuramoto T, Sasa M, Serikawa T (2000) Accumulation of N-acetyl-L-aspartate in the brain of the tremor rat, a mutant exhibiting absence-like seizure and spongiform degeneration in the central nervous system. J Neurochem 74:2512–2519

76. D'Cruz PM, Yasumura D, Weir J, Matthes MT, Abderrahim H, LaVail MM, Vollrath D (2000) Mutation of the receptor tyrosine kinase gene Mertk in the retinal dystrophic RCS rat. Hum Mol Genet 9:645–651

77. Vollrath D, Feng W, Duncan JL, Yasumura D, D'Cruz PM, Chappelow A, Matthes MT, Kay MA, LaVail MM (2001) Correction of the retinal dystrophy phenotype of the RCS rat by viral gene transfer of Mertk. Proc Natl Acad Sci U S A 98:12584–12589

78. Kuramoto T, Kitada K, Inui T, Sasaki Y, Ito K, Hase T, Kawagachi S, Ogawa Y, Nakao K, Barsh GS, Nagao M, Ushijima T, Serikawa T (2001) Attractin/mahogany/zitter plays a critical role in myelination of the central nervous system. Proc Natl Acad Sci U S A 98: 559–564

79. Hornum L, Romer J, Markholst H (2002) The diabetes-prone BB rat carries a frameshift mutation in Ian4, a positional candidate of Iddm1. Diabetes 51:1972–1979

80. MacMurray AJ, Moralejo DH, Kwitek AE, Rutledge EA, Van Yserloo B, Gohlke P, Speros SJ, Snyder B, Schaefer J, Bieg S, Jiang J, Ettinger RA, Fuller J, Daniels TL, Pettersson

A, Orlebeke K, Birren B, Jacob HJ, Lander ES, Lernmark A (2002) Lymphopenia in the BB rat model of type 1 diabetes is due to a mutation in a novel immune-associated nucleotide (Ian)-related gene. Genome Res 12:1029–1039

81. Oiso N, Riddle SR, Serikawa T, Kuramoto T, Spritz RA (2004) The rat Ruby (R) locus is Rab38: identical mutations in Fawn-hooded and Tester-Moriyama rats derived from an ancestral Long Evans rat sub-strain. Mamm Genome 15:307–314

# Chapter 3

# Sequencing of the Rat Genome and Databases

**Elizabeth A. Worthey, Alexander J. Stoddard, and Howard J. Jacob**


## Abstract

The rat is an important system for modeling human disease. Four years ago, the rich 150-year history of rat research was transformed by the sequencing and annotation of the rat genome, ushering in an era of exceptional opportunity for identifying genes and pathways underlying disease phenotypes. With the genome sequence in place, there is the prospect of not only linking the extensive literature of mechanistic and pharmacological studies in the rat to its genome, but by using comparative genomics to other organisms as well. Genome-wide association studies (GWAS) in human populations have recently provided a direct approach for finding robust genetic associations in common diseases, but identifying the precise genes and their mechanisms of action remains challenging.

The explosion of genomic tools and sequence over the last decade has created a wealth of data. Along with the data has arisen a need to manage it and to make it usable to scientists with a wide-range of research interests. This chapter is designed to overview the existing sequence and its utility, as well as provide a glimpse of some of the databases and bioinformatic tools available to the investigator.


**Key words:** Rats, Model, Database, Sequence, Gene, Genomics

## 1. Introduction

The laboratory rat (*Rattus norvegicus*) has been used as an animal model for physiology, pharmacology, toxicology, nutrition, behavior, immunology and neoplasia for over 150 years (1). Because of its size, ease of manipulation and breeding characteristics, it remained the preferred choice for most of these fields throughout the twentieth century, while the mouse became the leading mammal for experimental genetics. As a result, when the Human Genome Project was developing, the mouse became the pilot project for additional mammalian sequencing rather than the rat.

Sequencing the rat or any other organisms not on the "genetics security council" (coined by Dr. Gerry Fink from the Whitehead Institute) was not imaginable in the mid- to late 1990s. However,

I. Anegon (ed.), *Rat Genomics: Methods and Protocols*, Methods in Molecular Biology, vol. 597
DOI 10.1007/978-1-60327-389-3_3, © Humana Press, a part of Springer Science+Business Media, LLC 2010

it was recognized that building genomic tools for the rat was a priority. Led by the National Heart, Lung and Blood Institute (NHLBI) and 12 other Institutes and Centers at the National Institutes of Health (NIH) in the USA, the first step in what would evolve into the eventual sequencing effort; the Rat Expressed Sequence Tag (EST) Project, was initiated in 1997. At this time, production of large numbers of ESTs and subsequent genomic mapping was viewed as the means to integrate rat data with the mouse and human genomes using comparative genomics. Planning for the production of a pilot rat genome sequence began in earnest in 1999 at the Rat Priority Setting Meeting (http://www.nhlbi.nih.gov/resources/docs/ratmtgpg.htm).

At the 2000 Cold Spring Harbor Laboratory meeting on Genomic Sequence and Biology, a series of discussions took place regarding sequencing of the rat genome. The National Heart, Lung, and Blood Institute (NHLBI) and the National Human Genome Research Institute (NHGRI) decided to split the projected $100 M price tag; it was clear that to reap the largest gains, the project needed to start within a short period of time. The timing of the initiation of this project was fortuitous in that there was an increased and available capacity at some of the countries large genomic sequencing facilities, and a need to rapidly fill this capacity. The two large sequencing centers selected to sequence the rat were the Baylor College of Medicine (BCM-HGSC) and Celera. The irony of this pairing was that the two groups had just been competing against each other for the human genomic sequence. Collectively, the two centers proposed a novel strategy that combined Whole Genome Shotgun (WGS) sequencing with a bacterial artificial chromosome (BAC) based strategy.

With the basic strategy in place, the project moved on to selection of a single reference rat strain to sequence. The challenge was to select a strain, out of over 300 available inbred strains, that already had substantial resources available in order to ensure the broad utility of the genome sequence. In 2000, only 48 strains had been genotyped to any significant depth (4,328 genetic markers) (2). A significant number of these strains were model systems for various disease areas with dedicated, but relatively small followings; others were excellent candidates for genome sequencing, but had too much heterozygosity. It was also recognized that in order to return the largest benefit from the $100 M investment, it would be critical to capture genetic variation for the development of additional genetic markers. The BN (Brown Norway) strain was genetically diverse, inbred, found in a recombinant inbred panel as well as two panels of consomic rats, in addition to being used as the control strain for many different fields and studies; as such, it was selected as the strain to be sequenced. The rat genome was the third mammalian genome sequenced.

In this chapter, we discuss the sequencing of the rat genome, available genomic data for the rat, and provide a few examples of the tools and approaches that can be used to analyze rat genomic data. We provide details on some of the databases and other data repositories where this data can be accessed and can be concluded with a brief discussion of the up and coming release of genome sequence from additional rat strains and the implications of the availability of this additional genomic sequence.

## 2. Reference Genome

### *2.1. Generation of the Reference Assembly*

The strain of BN (BN/SsNHsd/Mcwi) sequenced was obtained by the Medical College of Wisconsin (MCW). Based on microsatellite genotyping, when originally derived from Harlan Sprague-Dawley, this strain was not completely inbred, but was subsequently maintained at MCW for 13 additional generations before being nominated to be sequenced (3). A network of centers, led by the BCM-HGSC and Celera, and including the Genome Therapeutics Corporation, the British Columbia Cancer Agency Genome Sciences Centre, The Institute for Genomic Research, the University of Utah, the Medical College of Wisconsin, the Children's Hospital of Oakland Research Institute, and the Max Delbrück Center for Molecular Medicine, Berlin generated data and resources. Collectively, this group generated over 44 million DNA sequence reads. The next step was the production of a sequence assembly.

The assembly took advantage of the combined BAC and WGS sequencing approach (3). Following the removal of low-quality reads and vector contaminants, 36 million reads were used for Atlas assembly (4), which retained 34 million reads. In brief, WGS data were progressively melded with light sequence coverage of individual BACs (BAC skims) to yield intermediate products called "enriched BACs" (eBACs). eBACs covering the whole genome were then joined into longer structures (bactigs). Bactigs were joined to form larger structures: super-bactigs, and then ultra-bactigs. Additional data including BAC end sequences, DNA fingerprints and other long-range information such as genetic markers, and synteny information, were used to refine the assembly. Slightly different estimates of the depth of coverage achieved came from considering the entire "trimmed" length of the sequence data (7.3x), or only high quality reads (6.9x). 60% of the coverage was provided by WGS and 40% from the BAC sequencing. The Y chromosome was not included in the Rat Genome Sequencing Consortium (RGSC) assembly. The initial assembly had 128,000 contigs and therefore a significant number of gaps. In 2004, this was the most complete draft sequence assembled, there had been no plans bringing the rat genome to a "finished" level.

The rat genome was found to be smaller than that of the human genome, but larger than the mouse (3). It encodes similar numbers of genes to these two other mammalian species, although species-specific expansions and deletions were identified. Intronic structures were well conserved across these species. This initial analysis indicated that somewhere in the range of 90% of rat genes possess strict orthologs in both mouse and human genomes. Orthologs for almost all human genes known to be associated with disease were found in the rat genome; rates of synonymous substitution were found to be significantly different between these and the remaining rat genes (3).The rat genome was found to be highly repetitive; a considerable portion of the genome was found to be made up of simple sequence repeats, and many rodent-specific and rat-specific interspersed repeats are also present. About 3% of the rat genome was found to consist of large segmental duplications; a percentage intermediate between the mouse (1–2%) and human (5–6%) (3).

After the initial publication of version RGSC v3.1, there was another round of assembly work leading to the current RGSC v3.4 (5). This upgrade includes reassembly of the genome with the latest version of the Atlas assembler, targeted finishing of problematic regions, and a draft sequence of the Y chromosome. Although good, some difficult-to-assemble regions remained to be addressed in RGSC v3.4, and most significantly, the number of gaps was not significantly reduced in this new assembly.

### 2.2. Annotation of the Rat Genome

Following production of a genome assembly, the next step is annotation of the sequence with genomic features, such as genes and regulatory elements. What is not often clear to those outside the genomics field is that the annotation of genes is carried out initially using a variety of comparative (sequence identity to previously identified protein-coding genes) and ab initio (identification of signs of protein-coding status using the DNA sequence alone) gene-prediction programs, which predict the positions and structure of genes in the genome. The functions of these predicted genes (often referred to as gene models) are then assessed using additional automated analysis; functional annotations from other genomes and genes submitted to the large centralized databases such as GenBank and Ensembl are transferred to the genome being annotated based on the sequence identity. Finally, the ongoing process of manual curation is carried out to correct any errors or inaccuracies in the automated genome annotations. The significance of these steps is not always readily apparent. As with all science, there are differences in how the analysis is carried out at different sites. For the rat genome project, annotation of rat genes is carried out by three groups: Ensembl is a joint project between the European Bioinformatics Institute (EBI), an outstation of the European Molecular Biology Laboratory (EMBL), and the Wellcome Trust Sanger Institute (WTSI) (6); the National Center for Biotechnology Information (NCBI) (7) and the Rat Genome Database (RGD) (8). Each of these groups annotates on the latest genome assembly (currently the RGSC v3.4 2007).

The Ensembl Rat release (http://www.ensembl.org/Rattus_norvegicus/Info/Index) identifies genes using the Ensembl automatic analysis pipeline (9). In brief, this pipeline makes use of mRNA and protein data available in public scientific databases, Open Reading Frame (ORF) predictions from a variety of computational tools including ab initio gene predictions, homology, and gene prediction Hidden Markov models (which are used to identify the appropriate intron/exon structure based on modeling of known sequences), supported by protein, cDNA or EST evidence, as well as manually curated datasets from the Rat Genome Database (RGD), to predict gene models. At the time of the release of the draft genome sequence, this pipeline predicted 20,973 genes. Since this time, enhancements to the pipeline as well as the incorporation of additional sequence data have resulted in an increase in gene numbers. 22,938 protein coding genes are currently identified at Ensembl, together with 1,750 pseudogenes, and 3,424 non-protein coding RNA genes. Genes are functionally annotated in another automated pipeline that uses gene descriptions from the large sequence databases and RGD. A variety of annotation information on each rat gene is available, including ortholog and paralog assignments, transcript information, conserved domains, and a variety of links to additional data sources.

At the NCBI, an alternative set of computational tools are applied to construct rat gene models. In brief, NCBI aligns all known genes from the Reference Sequence database (RefSeq; see below) and from GenBank mRNA sequences to the genomic sequence (10). Additional genes are reported based on GenomeScan (11) predictions, which combine exon–intron composition modeling and splice signal models with sequence similarity to known genes in a single integrated model. The predicted genes are then functionally annotated using existing gene descriptions from the large sequence databases and RGD. Although the sequence being annotated is the same as at Ensembl, the differing gene prediction algorithms used gives rise to a different estimate of the number of genes; this pipeline currently predicts 17,141 protein coding genes, and 19,245 genes total. Akin to the Ensembl genes, the degree of support for these gene models varies. Access to the functional annotations is provided through NCBIs EntrezGene; 36,436 (http://www.ncbi.nlm.nih.gov/projects/Gene/gentrez_stats.cgi?SNGLTAX=10116). EntrezGene also reports assigned Gene Ontology (GO) terms (in this context, an ontology means a controlled vocabulary of well-defined terms with specified relationships between them, capable of interpretation by both humans and computer programs), cytogenetic locations, names and symbols, pathways, protein interactions, publications (including the GeneRIF annotated bibliography), sequences (GenBank and RefSeq) and links to numerous NCBI tools and other resources.

The NCBI RefSeq database (http://www.ncbi.nlm.nih.gov/RefSeq/) is an integrated, non-redundant, and well-annotated set of references sequences (10). RefSeq genes are built from the sequence data available in the GenBank database, but gene models are independent of any particular genome build, and are independently manually curated in an ongoing basis by NCBI staff and collaborating groups. Gene-specific RefSeqs for RNAs and proteins include curated records based on the submissions to GenBank and predicted records that are generated as a product of computing annotation for the genome assemblies. In addition to protein-coding genes, RefSeq includes reference genomic DNA, and transcripts. 14,571 rat protein-coding RefSeqs currently exist, together with 928 curated pseudogenes and 16 non-protein coding genes.

Curation of RefSeq transcripts and proteins for the rat is a continuous process and serves to ensure accurate, full-length sequence for the complete set of transcripts and proteins, including *loci* that use selenocysteine or non-AUG codons. In addition, the transcript-based curated RefSeq collection represents a high-quality complement to genome annotation because it can be used to identify genes that are not well represented in reference genomic assemblies.

As mentioned previously, although Ensembl and NCBI annotate genes from the same genome assembly version, there are differences in the gene prediction tools used at the two sites; hence, there are differences in the gene models predicted at each site. An important role of the RGD is to identify conflicts between these two sets of gene models; all conflicts and assignment of official nomenclature to rat genes are resolved manually. Conflict resolution is accomplished via gene withdrawals and merges, which are communicated to the originating sources. This process gives rise to a single set of protein-coding genes, which are maintained and further curated at RGD. RGD currently maintains data on 21,067 protein coding genes and 5,424 pseudogenes..

The RGD Gene reports include a comprehensive description of functional and biological process, as well as disease, expression, regulation and phenotype information. RGD Gene records include symbol, name, retired symbols and names, and genomic and genetic mapping data. Genes are further categorized using a variety of ontologies including the three aspects of the Gene Ontology (GO); Molecular Function, Biological Process, and Cellular Component, as well as Disease, Pathway and Mammalian Phenotype Ontologies.

It is important to note that the assembly and annotations continue to change with time as errors are identified, corrected and the genome assembly modified either through the addition of further sequence data or through enhancements to the assembly algorithms. A new assembly of the rat reference sequence is currently in progress; accordingly, the entire gene prediction and annotation process at each site will be repeated. The reference sequence and annotations will continue to evolve and therefore should be viewed as neither static nor error free.

### 2.3. Annotation of Other Elements on the Rat Genome

Gene models are not the only genomic features that require identification and annotation (automated and manual). Another important class of features is genetic markers. The field is currently transitioning from use of genetic maps and genotyping platforms based on microsatellite (simple sequence length polymorphism) data, to techniques based on single nucleotide variations SNVs; commonly referred to as single nucleotide polymorphisms (SNPs) (12). This transition requires existing databases and data analysis and visualization tools to be modified to support the use of these new markers. This progress also drives the development of new tools. The discovery of additional functional genomic features such as copy number variations (CNVs) necessitates significant modifications to the existing databases and tools to support not only their annotation, but also their use for discovery searches.

#### 2.3.1. SNPs

Within the rat genome, a SNP is defined as a DNA sequence variation occurring when a single nucleotide in the genome differs between inbred strains or individuals in outbred strains. The European STAR consortium, together with Japanese colleagues, identified 2.9 million SNPs from a number of rat strains (13). Twenty thousand of these SNPs have been genotyped in more than 300 inbred and hybrid strains. The complete set of SNPs and the entire set of genotypes across all rat strains were submitted to, and are publicly accessible through Ensembl. SNPs were also deposited in dbSNP and sequence traces in the NCBI Trace Archive. Processing of SNP data into a usable form requires a series of steps.

SNPs are submitted to the Single Nucleotide Polymorphism database (dbSNP) at NCBI (7). dbSNP assigns SNPs unique stable identifiers (ss numbers), and is responsible for the clustering SNPs of the same variant to reference SNPs (rs numbers), as well as mapping of reference SNPs to the reference genome based on flanking sequence. Submissions are clustered periodically. dbSNP is currently on build 126 for rat SNPs, mapped to the V3.4 assembly, and has 43,628 RefSNP Clusters, and a large number of unclustered SNPs, which should be clustered shortly. dbSNP allows researchers to search for SNPs, and provides access to a SNP Cluster Report, which details a variety of submitter information, annotation (including alleles, validation status and genomic location), map data and population data. This data is disseminated to the other databases.

Ensembl provides the Ensembl SNP repository, which provides access to dbSNP imported data as well as additional Ensembl annotation (14). A Gene SNP tool is available to aid researchers in visualizing the localization of a SNP within a gene, and in relation to other genomic features. Access to Sanger and Ensembl identified rat SNPs is provided in the Transcript View tool, which allows researchers to compare SNP variation data among the different rat strains with available genotype data. Sequences from these strains can be visualized using the SequenceAlignView, highlighting SNPs that exist between a particular strain and the reference assembly. At present, the data is available from the Sprague-Dawley, SS/Jr, GK/Ox, WKY/Mdc, F344 and SHRSP/Mdc strains, along with the reference strain BN/NHsdMcwi. Ensembl currently has data on 2,862,253 Rat SNPs.

The RGD GBrowse provides visualization of the SNPs location and proximity to other elements such as protein-coding genes, non-coding RNAs (ncRNAs) and Quantitative trait Loci (QTLs). A large amount of annotation is available for these SNPs including their type (e.g., intronic/coding/intergenic), identified alleles, and strain associations. Links are provided to the appropriate records in Ensembl or dbSNP.

RGD SNPlotyper is a visualization and analysis tool for Rat SNP and microsatellite data, which contains data for over 300 strains, 20,000 SNPs and 4,200 Microsatellites to help the investigators find informative markers. This tool also enables the researchers to view haplotype blocks shared between strains and identify informative (polymorphic) markers between multiple strains.

#### *2.3.2. CNVs*

A Copy Number Variant (CNV) is generally defined as a DNA segment, larger than 1 kb, with variable frequency in an otherwise equivalent genome region in related genomes. Recent estimates indicate that between 2 and 20% of mammalian genomes is comprised of such variants (15, 16). Since it has been estimated that these elements may be responsible for up to 30% of detected genetic variation in gene expression (17–19), it is critical to annotate these new features.

A 2008 study at the Hubrecht Institute used computational and experimental methods to catalog CNVs in the rat (20), and these data can be viewed using the RGD Genome Browser. Links are provided from the RGD GBrowse to the Hubrecht Institute's CNV pages to provide additional information on CNVs of interest. This same dataset is also available as a track within the Ensembl genome browser.

### *2.4. Expression Data*

The availability of the rat genome sequence has facilitated the design of microarray platforms for high-through-put experiments measuring gene expression on up to a genome-w ide scale (as well as comparative genome hybridization (CGH) studies). Researchers can choose amongst multiple commercial platforms or in house manufactured arrays. Manufacturer websites (*see* Table 3.1) generally provide microarray feature mapping and summarizing functional data (e.g., the NetAffx database at the Affymetrix website).

**Table 3.1**
**Tools and resources for rat data**

| Organization | Tool/Resource | URL |
|---|---|---|
| NCBI | Home Page | http://www.ncbi.nlm.nih.gov |
| | Map Viewer | http://www.ncbi.nlm.nih.gov/mapview |
| | Entrez Gene | http://www.ncbi.nlm.nih.gov/gene |
| | UniSTS | http://www.ncbi.nlm.nih.gov/unists |
| | dbSNP | http://www.ncbi.nlm.nih.gov/projects/SNP |
| | Gene Expression Omnibus | http://www.ncbi.nlm.nih.gov/geo |
| RGD | Home Page | http://rgd.mcw.edu |
| | Synteny browser | http://rgd.mcw.edu/gb/gbrowse/rgd_904/ |
| | Interactive Pathway Diagrams | http://rgd.mcw.edu/wg/pathway |
| | RGD Biomart | http://biomart.mcw.edu:9999/tutorials/rgdUniProt6.html |
| | Phenotype query tool | http://rgd.mcw.edu/phenotypes |
| EMBL-EBI | Home Page | http://www.ebi.ac.uk |
| | Ensembl Genome Browser | http://www.ensembl.org |
| | ArrayExpress | http://www.ebi.ac.uk/microarray-as/ae/ |
| | Biomart | http://www.biomart.org |
| UCSC Genome Browser | Home Page | http://genome.ucsc.edu |
| Galaxy analysis tool | Home Page | http://galaxy.psu.edu |
| | Galaxy tutorials | http://galaxy.psu.edu/screencasts.html |
| STAR SNP Consortium | Home Page | http:///www.snp-star.eu |
| Illumina (microarray) | Home Page | http://www.illumina.com |
| | Annotation Files | http://www.switchtoi.com/annotationfiles.ilmn |
| Affymetrix (microarray) | Home Page | http://affymetrix.com/index.affx |
| | NetAffx (annotation) | http://www.affymetrix.com/analysis/index.affx |
| Agilent (microarrays) | Gene Lists and Annotations | http://www.chem.agilent.com/en-us/products/instruments/dnamicroarrays/pages/gp58802.aspx |
| CodeLink (microarrays) | Home Page | http://www.appliedmicroarrays.com/ |

This table provides a reference for researchers seeking to utilize the genome tools discussed in this chapter. An updated and expanded set of resource links are maintained at RGD (http://rgd.mcw.edu/wg/resource-links)

Expression profiling data is submitted to the online expression databases, notably to GEO at the NCBI (http://www.ncbi.nlm.nih.gov/geo) and ArrayExpress from the EBI (http://www.ebi.ac.uk/microarray-as/ae/). The experiments are described by the source material, sample preparation and data normalization methods using MIAME (Minimal Information about a Microarray Experiment) format.

### *2.5. Comparative Genomics*

Having the genomic sequence enables the investigator to compare genes, and other genomic elements (down to the nucleotide level) between multiple species. As a result, structure and function studies of orthologous genes can be translated to human and other organisms. Comparative analysis also allows study of evolutionary changes among organisms, and speciation. As with other data sets, there are a variety of databases and viewers available (*see* Table 3.1 for some examples).

## 3. Major Databases and Bioinformatic Tools

From the basic genomic sequence to annotated features and comparative data, a wealth of data is assigned to the rat genome. The number of datasets and tools available are too numerous and rapidly changing to adequately cover them in this chapter. Therefore, this section provides a few highlights, a use-case showing interoperability between some tools (Use Case 1), and a table with URLs for some commonly used tools (Table 3.1). A frequently updated page of resource links is maintained at RGD (http://rgd.mcw.edu/wg/resource-links). It will be obvious to the reader that the data sets are shared across the various databases and that it is the strength of this collective that provides the wealth of data and tools to the investigator who is interested in rat.

### *3.1. The National Center for Biotechnology Information*

The Rat Genome Resources web page provides an up-to-date portal to access rat-specific data. Data is integrated from, and links provided to, databases such as RGD, RATMAP, Ensembl and UniProt. Links are also provided to a number of genome viewer tools (e.g., the NCBI's MapViewer and SequenceViewer; and the UCSC Genome Browser), allowing investigation of particular genomic locations and neighboring features. A variety of NCBI databases, including Entrez Gene, and Entrez Genome, as well as other databases such as Online Mendelian Inheritance in Animals records and Taxonomy, can be queried for rat data using the Entrez interactive query system. Many of these datasets are also accessible through Entrez Programming Utilities (E-utilities); tools that provide access to Entrez data outside of the regular web query interface.

Sequence-based data presented in the Map Viewer includes the annotated genome at the gene and transcript level, and an array of additional features including repeats, STS markers, CpG islands, and human, mouse and rat transcript sequences. Alternative displays provide tabular reports and downloadable data views, as well as presenting the alignments supporting the annotation (Evidence Viewer) or support using transcript alignments to generate alternative transcript models for further evaluation (Model Maker). The NCBI Map Viewer also supports comparative displays of human, mouse and rat data, as well as presenting the order and orientation of assemblies based on the placement of markers common to the sequence, genetic and radiation-hybrid maps.

A rat-specific BLAST facilitates access to several datasets including the reference genome assembly, alternate assemblies (including the Celera assembly), and a variety of gene, RNA, and protein subsets; the trace reads from the GWS and EST Trace Archives can also be queried (http://www.ncbi.nlm.nih.gov/genome/seq/BlastGen/BlastGen.cgi?taxid=10116). Results for transcript and protein queries return links to the Gene, UniGene and/or GEO databases when an accession for that database is available.

### 3.2. Ensembl

Ensembl's primary focus for the rat is to provide annotation of genomic features as well as support for comparative genome analyses with additional sequenced genomes. Summary pages are provided for rat genes, providing annotation on genes, transcripts, and variations, as well as links out to other resources and databases of interest to rat researchers. The Ensembl Rat Genome Browser provides researchers with the ability to explore the genomic region containing their gene or other feature of interest. Transcript views provide information specific to individual transcripts such as the cDNA and CDs sequences, and protein domain annotation. Gene views provide displays for data associated at the gene level such as orthologs and paralogs, regulatory regions and splice variants. Ensembl also provides access to synteny data (derived from alignments of confirmed or putative functional elements) via the Rat ContigViewer including rat to human synteny data, four species alignment data, and seven species alignment data. The AlignSpliceView tool allows the researcher to display rat alongside the syntenic mouse or human regions, showing homologous genes in their genomic context, as well as regions conserved between the species. Homology relationships at the gene level can be visualized in the GeneTreeView.

Ensembl also enables a wealth of data available outside the Ensembl system to be brought together using two strategies: (1) The Distributed Annotation System (DAS) integrates external resources into the Ensembl Genome Browser allowing any

DAS-compatible sources to be displayed alongside the primary Ensembl annotation. (2) The Ensembl BioMart provides a data-mining tool to interact with the Ensembl database. For rat, this contains the current Ensembl genome data sets, an SNP data set and EURAMart; a compendium of gene expression data populated with the Gene Expression Atlas from the Genomic Institute of the Novartis Research Foundation. EURAMart can act as a bridge between Ensembl and RGD data, allowing the integration of expression data (from EURAMart) with phenotype and disease information (from RGD).

### 3.3. RGD

Unlike the other databases that focus on numerous organisms including rat, RGD focuses solely on the rat, and is the source of the vast majority of manual curation for the rat. In addition to the Gene report pages, which provide a wealth of information pertaining to a particular gene, RGD gene models can be viewed in the RGD Genome Browser (GBrowse) (21). This tool allows researchers to view genes and other genomic features in the context of the linear genome sequence, allowing identification of neighboring genomic features. The RGD Gbrowse also provides a human, mouse, and rat synteny viewer, enabling researchers to view organizational differences and similarities in genomic features amongst these three genomes. RGD provides links to GEO for expression data for RGD genes, as well as annotating RGD genes with microarray probe-set identifiers for Affymetrix, Codelink, and Agilent. User-generated custom tracks can be uploaded and viewed in the Genome Browser to provide integration with users own data; these custom tracks also provide an ideal medium for visualizing the data submitted for publication.

The RGD Gviewer provides a whole genome view of the location of rat genes and the other genomic features, and enables investigators to search for genes using gene annotation as well as matching terms from the Gene Ontology, Mammalian Phenotype Ontology, Disease Ontology or Pathway Ontology.

RGD provides another layer of annotation data for the rat genome by assigning, validating and curating QTLs (and therefore associating phenotypic data) to the genome. To facilitate and guide efficient access for researchers focusing on complex diseases, portals for a variety of disease classes have been created at RGD including Neurophysiologic, Cardiovascular, Obesity and Metabolic Syndromes, and Cancer portals. Researchers are able to view gene and QTL data for a particular disease; data on phenotypes, biological processes and pathways; and link to Strain Report pages detailing rat strains associated with particular diseases.

To complement the RGD disease portals, a Phenotypes and Models Portal exists, which houses strain and phenotype data along with tools to integrate the genomic and phenotypic datasets.

The portal includes phenotype data for cardiovascular, pulmonary, blood biochemistry, renal, vascular and general morphological results, as well as strain availability, disease models and other strain details. Integration of large phenotype datasets from many rat strains is based on the implementation of a variety of biological ontologies; more complex data structures are used to support the integration of multiple ontology annotations, as well as qualifiers and quantitative results.

Interactive pathway diagrams are available at RGD, providing detailed, expert curated information on the proteins involved in the pathways, the relationships between them, small molecules interactions, and links to the related pathways. In addition to links to RGD gene reports, a brief overview of each pathway is provided, with a link to the pathway ontology report including a complete downloadable list of annotated genes involved in the pathway. As applicable, the diagrams also provide links to downstream activated pathways.

### 3.4. UCSC

The University of California, Santa Cruz (UCSC), Genome Browser Database (GBD) provides integrated sequence and annotation data for many vertebrate and model organism genomes (22). A broad set of annotations generated or acquired by UCSC and its collaborators is available for each rat assembly, including gene models, mapping and sequencing data, phenotype and disease association data, mRNA and EST data, regulation data, and variations and repeats (http://genome.ucsc.edu/). A variety of web-based resources for displaying, querying, analyzing, and downloading the assembly sequence and annotations are available.

The UCSC November 2004 Rat genome assembly is based on the RGSC v3.4 assembly. UCSC predicts their own set of protein-coding genes; UCSC Known Genes based on protein data from UniProt and mRNA data from RefSeq and GenBank. The UCSC Genome Browser allows researchers to zoom and scroll over chromosomes, reviewing not only gene annotations, but also other classes of genome data including assembly gaps, and repetitive elements. The UCSC Genome Browser provides alignment net and chain alignment tracks, allowing identification of orthologous genes and analysis of genome rearrangement. Probe positions for a number of the Rat Genome CGH microarrays from Agilent Technologies, and for a number of Affymetrix arrays for expression data can also be displayed on the browser.

The UCSC Blat tool offers a fast method for quickly mapping rat sequence to the genome. The Genome Graphs tool can be used to upload and view genome-wide datasets such as QTL mapping and linkage studies. The Proteome Browser allows the researchers to examine many characteristics (such as exon structure, domains and structural features) of proteins associated with the gene.

The UCSC Table Browser provides a graphical interface for downloading and manipulating the data underlying the rat Genome Browser annotations. Custom tracks of researchers' own data can be viewed in the Genome Browser and can be manipulated using the functionality of the Table Browser. Researchers can view data tables, filter the output based on one or more criteria, intersect or correlate data from more than one table, and restrict the output to specific coordinate ranges or lists of data elements. For more complex queries the Table Browser output can be exported to Penn State's Galaxy tool (http://main.g2.bx.psu.edu/) for further processing.

### 3.5. Use Case 1

*See* Fig. 3.1.

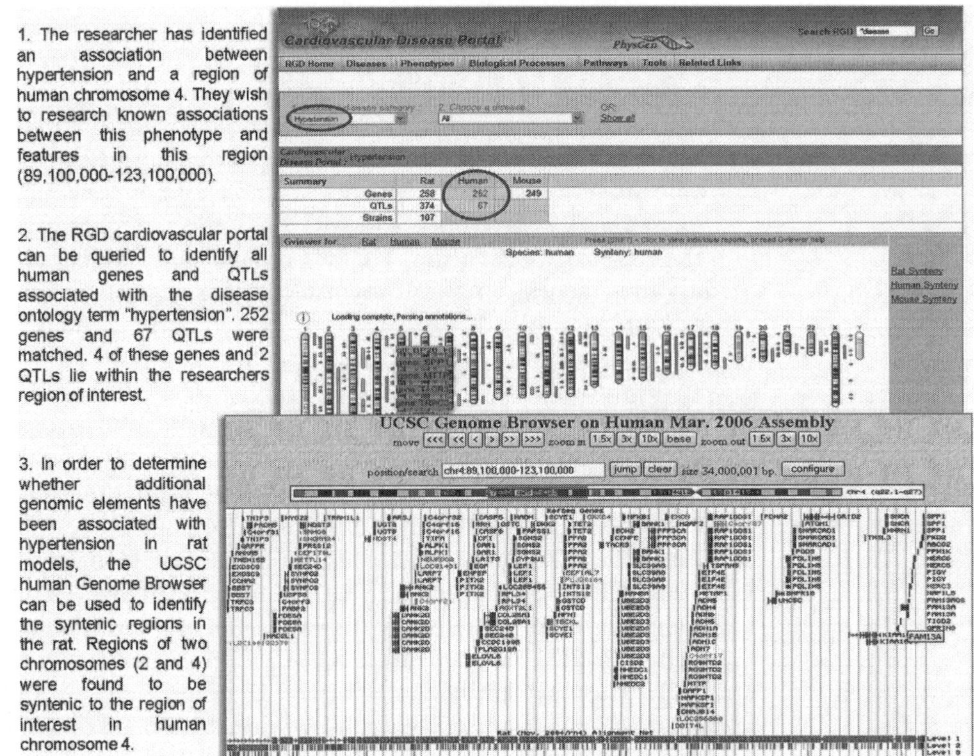


Fig. 3.1. A graphical workflow for exploration of a candidate genomic region for hypertension.

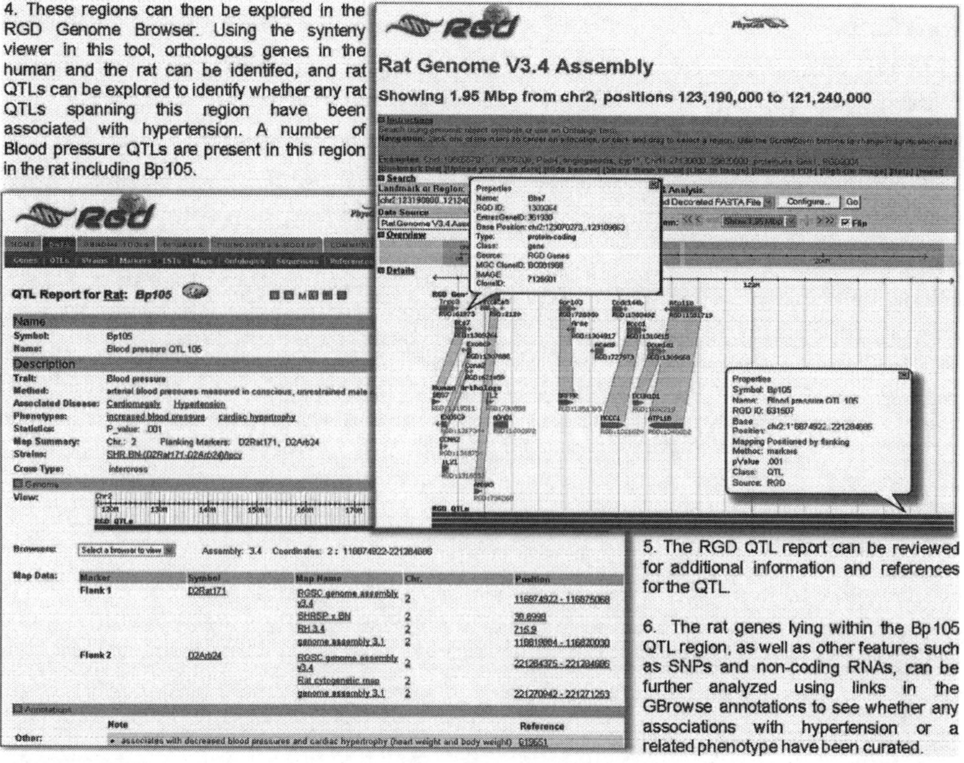


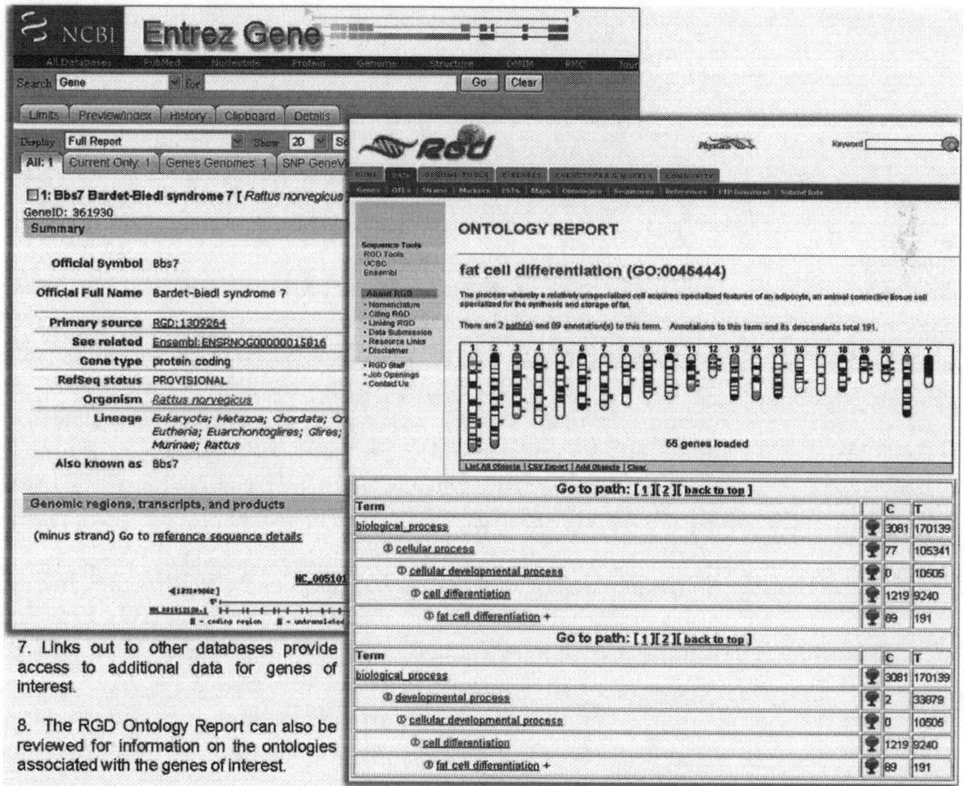


Fig. 3.1. (continued)

## 4. Future Role of Genomic Sequence and Bioinformatics for the Rat

With genomic sequencing rapidly becoming faster and cheaper due to advances in sequencing methodologies (23), we expect to see a dramatic increase in the number of rat strains sequenced. Indeed genome sequence data from a handful of additional rat strains are scheduled to be released by the start of 2010.

Researchers at the MCW and the Max Delbrück Center for Molecular Medicine, Berlin collaborate to generate ~10X sequencing of the Fawn Hooded hypertensive (FHH) rat and ~10X sequencing of the SS/JrHsdMcwi rat. Both of these strains are parental lines, crossed with the sequenced BN, to generate the consomic (whole chromosome substitution) strains built at the MCW (http://pga.mcw.edu).

The European Union funded EURATools consortium along with the University of British Columbia are in the process of completing the ~10X sequencing of the spontaneously hypertensive rat (SHR) genome using Illumina sequencing technology (personal communication from Dr. Tim Aitman, MRC CSC, UK). The SHR strain is one of the most widely studied inbred strains of rats. The SHR and BN are used to generate a series of recombinant congenics (24).

The deep genomic resources now become available: multiple rat strains, strains of other organisms used for biomedical models, and multiple individual human genome sequences, require integration and comparison to reach their true potential in clinical research and medicine. These goals cause a rapid expansion in the development of strategies and comparative tools to integrate the data across the currently disparate data sets.

### *4.1. Integrating Genomic Data*

Any one of the genome resources already introduced contains a wealth of data. However, generating and testing new biological hypotheses requires integration of this data. For example; data from multiple sources are needed to address questions such as "Which SNPs representing non-synonymous mutations might account for heart disease associated phenotypes associated with a region of chromosome 4?" or "Is such-and-such a strain a good candidate for testing a novel drug to treat Alzheimer's disease?" Work is progressing to enable more facile manipulation and querying of the data, and it is now possible to envisage a "cyber-infrastructure for the biological sciences" (25). In the short term, however, large-scale studies will likely continue to be the domain of multidisciplinary teams.

On the other hand, there already exist tools that can greatly assist the individual investigator in exploring particular genomic regions by combining multiple sources of information. Ongoing work to systematize gene function, disease, phenotype, and pathway information can also greatly accelerate the identification of candidate genes or biological pathways in genomic regions of interest.

#### *4.1.1. Curation-Based Integration*

The most basic level of data integration is the cross-mapping of unique identifiers for genes, proteins, SNPs, microarray probes etc.; the main genome resources cross-reference one another, as evidenced by various databases housing common data. RGD, as the designated community database and rat strain and gene naming authority, endeavors to provide a very rich set of cross-referenced identifiers and aliases for rat genome features. Individual queries can easily be made through the search interfaces at the respective database websites. Batch querying and the interconversion of lists of identifiers are enabled by tools such as Biomart.

Biology and bioinformatics make ever-increasing use of ontologies to capture and integrate biological knowledge (26). Data that are described by ontology terms allow accurate searching and reasoning by software tools, and enable the researchers to be confident of the context in which terms are used. Several ontologies make use of evidence codes to indicate annotation logic for example inference of a gene's function from observation of a mutant phenotype. Genomic features are annotated with ontology terms from the primary literature by the curation team (27, 28) and through computationally based methods (29).

RGD combines ontologies to describe genes, strains and QTLs with the goal of connecting phenotype to genotype data. Querying terms with different granularity enables the researchers to rapidly focus on areas of interest or expand their view to encompass related experiments.

#### *4.1.2. Programmatic Integration*

The information needed to address a specific biological hypothesis may not all be contained within one database or made available at one website. Tools that allow the integration and querying of such decentralized data are therefore required.

The Distributed Annotation System (DAS) (30) is implemented in the RGD, UCSC and Ensembl genome browsers to share data tracks. Smaller or more specialized datasets can also be exposed as DAS sources. Use Case 2 provides an example of programmatic integration.

Biomart (31) has already been mentioned as a tool that integrates data on genomic features. Information across multiple Biomart sources can be queried simultaneously with a unified web interface. Biomart has now been expanded to provide data to other tools (32), for example, biomart data sources can now also act as DAS servers.

The Galaxy system (http://galaxy.psu.edu) provides a web browser based set of tools through which results of queries from multiple genomic databases can be analyzed together. Available data sources for Galaxy include results from the UCSC table browser and the RGD and Ensembl biomarts. Galaxy also provides tools to manipulate retrieved data, for example to expand a set of gene coordinates obtained from one data source to include upstream regions to allow querying of potential regulatory sites. Results from a Galaxy analysis can then be exported and visualized as a custom track at the UCSC genome browser.

### 4.2. Use Case 2

*See* Fig. 3.2.

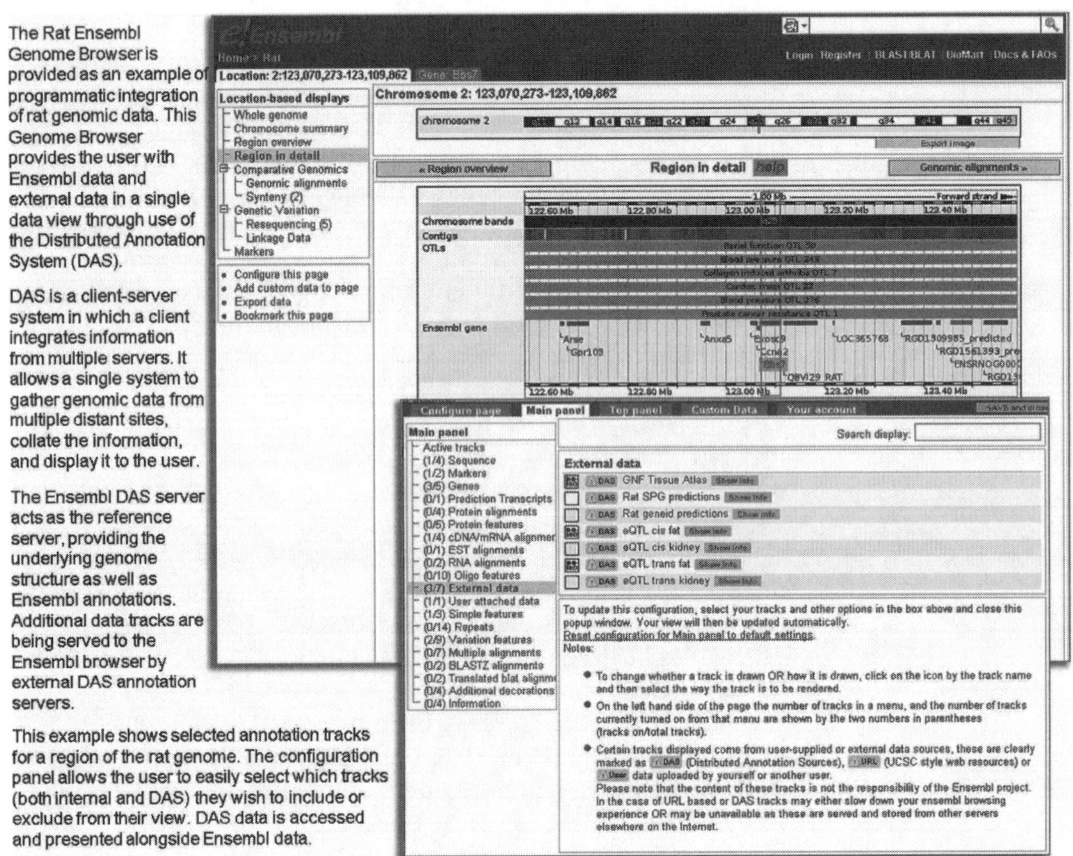


Fig. 3.2. Programmatic integration in the Ensembl Genome Browser.

## Acknowledgments

This chapter is predicated on the existence of a rich set of interconnected rat data resources made compatible through collaboration, both formal and informal, between multiple data producing, development, and curation teams. We gratefully acknowledge the ongoing contributions of groups at UCSC, RGD, NCBI, and Ensembl and the authors of tools interfacing with these resources, to which this chapter provides a very brief introduction. In particular, we acknowledge the contributions from Kim Pruitt, Donna Maglott, and Garth Brown and teams at NCBI; Ewan Birney, Steve Searle, and Xosé Fernández-Suárez and their teams at Ensembl; and Jim Kent and team at UCSC. In addition, we acknowledge the ongoing support of rat genomics by Richard Gibbs, George Weinstock, and Kim Worley and their teams at BCM-HGSC.

Funding for the Rat Genome Database is provided by the National Heart Lung and Blood Institute grant (HL64541) to Howard J. Jacob.

## References

1. Jacob HJ (1999) Functional genomics and rat models. Genome Res 9:1013–1016
2. Steen RG, Kwitek-Black AE, Glenn C, Gullings-Handley J, Van Etten W, Atkinson OS, Appel D, Twigger S, Muir M, Mull T, Granados M, Kissebah M, Russo K, Crane R, Popp M, Peden M, Matise T, Brown DM, Lu J, Kingsmore S, Tonellato PJ, Rozen S, Slonim D, Young P, Jacob HJ (1999) A high-density integrated genetic linkage and radiation hybrid map of the laboratory rat. Genome Res 9:AP1-8, insert
3. Gibbs RA, Weinstock GM, Metzker ML, Muzny DM, Sodergren EJ, Scherer S, Scott G, Steffen D, Worley KC, Burch PE, Okwuonu G, Hines S, Lewis L, DeRamo C, Delgado O, Dugan-Rocha S, Miner G, Morgan M, Hawes A, Gill R, Celera, Holt RA, Adams MD, Amanatides PG, Baden-Tillson H, Barnstead M, Chin S, Evans CA, Ferriera S, Fosler C, Glodek A, Gu Z, Jennings D, Kraft CL, Nguyen T, Pfannkoch CM, Sitter C, Sutton GG, Venter JC, Woodage T, Smith D, Lee HM, Gustafson E, Cahill P, Kana A, Doucette-Stamm L, Weinstock K, Fechtel K, Weiss RB, Dunn DM, Green ED, Blakesley RW, Bouffard GG, De Jong PJ, Osoegawa K, Zhu B, Marra M, Schein J, Bosdet I, Fjell C, Jones S, Krzywinski M, Mathewson C, Siddiqui A, Wye N, McPherson J, Zhao S, Fraser CM, Shetty J, Shatsman S, Geer K, Chen Y, Abramzon S, Nierman WC, Havlak PH, Chen R, Durbin KJ, Egan A, Ren Y, Song XZ, Li B, Liu Y, Qin X, Cawley S, Cooney AJ, D'Souza LM, Martin K, Wu JQ, Gonzalez-Garay ML, Jackson AR, Kalafus KJ, McLeod MP, Milosavljevic A, Virk D, Volkov A, Wheeler DA, Zhang Z, Bailey JA, Eichler EE, Tuzun E, Birney E, Mongin E, Ureta-Vidal A, Woodwark C, Zdobnov E, Bork P, Suyama M, Torrents D, Alexandersson M, Trask BJ, Young JM, Huang H, Wang H, Xing H, Daniels S, Gietzen D, Schmidt J, Stevens K, Vitt U, Wingrove J, Camara F, Mar Alba M, Abril JF, Guigo R, Smit A, Dubchak I, Rubin EM, Couronne O, Poliakov A, Hubner N, Ganten D, Goesele C, Hummel O, Kreitler T, Lee YA, Monti J, Schulz H, Zimdahl H, Himmelbauer H, Lehrach H, Jacob HJ, Bromberg S, Gullings-Handley J, Jensen-Seaman MI, Kwitek AE, Lazar J, Pasko D, Tonellato PJ, Twigger S, Ponting CP, Duarte JM, Rice S, Goodstadt L, Beatson SA, Emes RD, Winter EE, Webber C, Brandt P, Nyakatura G, Adetobi M, Chiaromonte F, Elnitski L, Eswara P, Hardison RC, Hou M, Kolbe D, Makova K, Miller W, Nekrutenko A,

Riemer C, Schwartz S, Taylor J, Yang S, Zhang Y, Lindpaintner K, Andrews TD, Caccamo M, Clamp M, Clarke L, Curwen V, Durbin R, Eyras E, Searle SM, Cooper GM, Batzoglou S, Brudno M, Sidow A, Stone EA, Payseur BA, Bourque G, Lopez-Otin C, Puente XS, Chakrabarti K, Chatterji S, Dewey C, Pachter L, Bray N, Yap VB, Caspi A, Tesler G, Pevzner PA, Haussler D, Roskin KM, Baertsch R, Clawson H, Furey TS, Hinrichs AS, Karolchik D, Kent WJ, Rosenbloom KR, Trumbower H, Weirauch M, Cooper DN, Stenson PD, Ma B, Brent M, Arumugam M, Shteynberg D, Copley RR, Taylor MS, Riethman H, Mudunuri U, Peterson J, Guyer M, Felsenfeld A, Old S, Mockrin S, Collins F (2004) Genome sequence of the Brown Norway rat yields insights into mammalian evolution. Nature 428:493–521

4. Havlak P, Chen R, Durbin KJ, Egan A, Ren Y, Song XZ, Weinstock GM, Gibbs RA (2004) The Atlas genome assembly system. Genome Res 14:721–732

5. Worley KC, Weinstock GM, Gibbs RA (2008) Rats in the genomic era. Physiol Genomics 32:273–282

6. Hubbard TJ, Aken BL, Ayling S, Ballester B, Beal K, Bragin E, Brent S, Chen Y, Clapham P, Clarke L, Coates G, Fairley S, Fitzgerald S, Fernandez-Banet J, Gordon L, Graf S, Haider S, Hammond M, Holland R, Howe K, Jenkinson A, Johnson N, Kahari A, Keefe D, Keenan S, Kinsella R, Kokocinski F, Kulesha E, Lawson D, Longden I, Megy K, Meidl P, Overduin B, Parker A, Pritchard B, Rios D, Schuster M, Slater G, Smedley D, Spooner W, Spudich G, Trevanion S, Vilella A, Vogel J, White S, Wilder S, Zadissa A, Birney E, Cunningham F, Curwen V, Durbin R, Fernandez-Suarez XM, Herrero J, Kasprzyk A, Proctor G, Smith J, Searle S, Flicek P (2009) Ensembl 2009. Nucleic Acids Res 37:D690–D697

7. Sayers EW, Barrett T, Benson DA, Bryant SH, Canese K, Chetvernin V, Church DM, DiCuccio M, Edgar R, Federhen S, Feolo M, Geer LY, Helmberg W, Kapustin Y, Landsman D, Lipman DJ, Madden TL, Maglott DR, Miller V, Mizrachi I, Ostell J, Pruitt KD, Schuler GD, Sequeira E, Sherry ST, Shumway M, Sirotkin K, Souvorov A, Starchenko G, Tatusova TA, Wagner L, Yaschenko E, Ye J (2009) Database resources of the National Center for Biotechnology Information. Nucleic Acids Res 37:D5–D15

8. Dwinell MR, Worthey EA, Shimoyama M, Bakir-Gungor B, DePons J, Laulederkind S, Lowry T, Nigram R, Petri V, Smith J, Stoddard A, Twigger SN, Jacob HJ (2009) The Rat Genome Database 2009: variation, ontologies and pathways. Nucleic Acids Res 37: D744–D749

9. Curwen V, Eyras E, Andrews TD, Clarke L, Mongin E, Searle SM, Clamp M (2004) The Ensembl automatic gene annotation system. Genome Res 14:942–950

10. Pruitt KD, Tatusova T, Klimke W, Maglott DR (2009) NCBI Reference Sequences: current status, policy and new initiatives. Nucleic Acids Res 37:D32–D36

11. Yeh RF, Lim LP, Burge CB (2001) Computational inference of homologous gene structures in the human genome. Genome Res 11:803–816

12. Smits BM, Cuppen E (2006) Rats go genomic. Genome Biol 7:306

13. Saar K, Beck A, Bihoreau MT, Birney E, Brocklebank D, Chen Y, Cuppen E, Demonchy S, Dopazo J, Flicek P, Foglio M, Fujiyama A, Gut IG, Gauguier D, Guigo R, Guryev V, Heinig M, Hummel O, Jahn N, Klages S, Kren V, Kube M, Kuhl H, Kuramoto T, Kuroki Y, Lechner D, Lee YA, Lopez-Bigas N, Lathrop GM, Mashimo T, Medina I, Mott R, Patone G, Perrier-Cornet JA, Platzer M, Pravenec M, Reinhardt R, Sakaki Y, Schilhabel M, Schulz H, Serikawa T, Shikhagaie M, Tatsumoto S, Taudien S, Toyoda A, Voigt B, Zelenika D, Zimdahl H, Hubner N (2008) SNP and haplotype mapping for genetic analysis in the rat. Nat Genet 40:560–566

14. Hammond MP, Birney E (2004) Genome information resources – developments at Ensembl. Trends Genet 20:268–272

15. Bailey JA, Eichler EE (2006) Primate segmental duplications: crucibles of evolution, diversity and disease. Nat Rev Genet 7:552–564

16. Lin CH, Li LH, Ho SF, Chuang TP, Wu JY, Chen YT, Fann CS (2008) A large-scale survey of genetic copy number variations among Han Chinese residing in Taiwan. BMC Genet 9:92

17. Stranger BE, Forrest MS, Dunning M, Ingle CE, Beazley C, Thorne N, Redon R, Bird CP, de Grassi A, Lee C, Tyler-Smith C, Carter N, Scherer SW, Tavare S, Deloukas P, Hurles ME, Dermitzakis ET (2007) Relative impact of nucleotide and copy number variation on gene expression phenotypes. Science 315: 848–853

18. Cahan P, Li Y, Izumi M, Graubert TA (2009) The impact of copy number variation on local gene expression in mouse hematopoietic stem and progenitor cells. Nat Genet 41:430–437

19. Henrichsen CN, Vinckenbosch N, Zollner S, Chaignat E, Pradervand S, Schutz F, Ruedi

M, Kaessmann H, Reymond A (2009) Segmental copy number variation shapes tissue transcriptomes. Nat Genet 41:424–429

20. Guryev V, Berezikov E, Malik R, Plasterk RH, Cuppen E (2004) Single nucleotide polymorphisms associated with rat expressed sequences. Genome Res 14:1438–1443
21. Stein LD, Mungall C, Shu S, Caudy M, Mangone M, Day A, Nickerson E, Stajich JE, Harris TW, Arva A, Lewis S (2002) The generic genome browser: a building block for a model organism system database. Genome Res 12:1599–1610
22. Kuhn RM, Karolchik D, Zweig AS, Wang T, Smith KE, Rosenbloom KR, Rhead B, Raney BJ, Pohl A, Pheasant M, Meyer L, Hsu F, Hinrichs AS, Harte RA, Giardine B, Fujita P, Diekhans M, Dreszer T, Clawson H, Barber GP, Haussler D, Kent WJ (2009) The UCSC Genome Browser Database: update 2009. Nucleic Acids Res 37:D755–D761
23. Shendure JA, Porreca GJ, Church GM (2008) Overview of DNA sequencing strategies. Curr Protoc Mol Biol. Chapter 7, Unit 7 1
24. Pravenec M, Klir P, Kren V, Zicha J, Kunes J (1989) An analysis of spontaneous hypertension in spontaneously hypertensive rats by means of new recombinant inbred strains. J Hypertens 7:217–221
25. Stein LD (2008) Towards a cyberinfrastructure for the biological sciences: progress, visions and challenges. Nat Rev Genet 9:678–688
26. Stevens R, Goble CA, Bechhofer S (2000) Ontology-based knowledge representation for bioinformatics. Brief Bioinform 1:398–414
27. Shimoyama M, Petri V, Pasko D, Bromberg S, Wu W, Chen J, Nenasheva N, Kwitek A, Twigger S, Jacob H (2005) Using multiple ontologies to integrate complex biological data. Comp Funct Genomics 6:373–378
28. Hill DP, Smith B, McAndrews-Hill MS, Blake JA (2008) Gene Ontology annotations: what they mean and where they come from. BMC Bioinformatics 9(Suppl 5):S2
29. Barrell D, Dimmer E, Huntley RP, Binns D, O'Donovan C, Apweiler R (2009) The GOA database in 2009 – an integrated Gene Ontology Annotation resource. Nucleic Acids Res 37:D396–D403
30. Dowell RD, Jokerst RM, Day A, Eddy SR, Stein L (2001) The distributed annotation system. BMC Bioinformatics 2:7
31. Hubbard T, Andrews D, Caccamo M, Cameron G, Chen Y, Clamp M, Clarke L, Coates G, Cox T, Cunningham F, Curwen V, Cutts T, Down T, Durbin R, Fernandez-Suarez XM, Gilbert J, Hammond M, Herrero J, Hotz H, Howe K, Iyer V, Jekosch K, Kahari A, Kasprzyk A, Keefe D, Keenan S, Kokocinsci F, London D, Longden I, McVicker G, Melsopp C, Meidl P, Potter S, Proctor G, Rae M, Rios D, Schuster M, Searle S, Severin J, Slater G, Smedley D, Smith J, Spooner W, Stabenau A, Stalker J, Storey R, Trevanion S, Ureta-Vidal A, Vogel J, White S, Woodwark C, Birney E (2005) Ensembl 2005. Nucleic Acids Res 33:D447–D453
32. Smedley D, Haider S, Ballester B, Holland R, London D, Thorisson G, Kasprzyk A (2009) BioMart – biological queries made easy. BMC Genomics 10:22

Chapter 4

# Design of Expression Cassettes for the Generation of Transgenic Animals (Including Insulators)

Louis-Marie Houdebine

## Abstract

The use of transgenesis is relatively rare in rats, and this is because of the relative difficulty in adding foreign genes by the conventional methods. Gene knock out and knock in by the conventional techniques of homologous recombination remain difficult in rats. This situation would be less crucial if the gene constructs were more reliable for the expression of foreign genes. The present chapter describes the state of the art in vector design for various genetic modifications in rats.

**Key words:** Insulators, BAC vectors, iPS, 3 UTR, 5 UTR, IRES, shRNAs, NHEJ, Meganucleases

## 1. Introduction

In the early 1980s, the first experiments to generate transgenic mice revealed that transgenes did not often work in the expected manner. In a number of cases, the expression of the transgenes was very weak and not strictly specific of the promoter associated with the foreign gene. In a few cases, it was demonstrated that the ectopic expression of the transgenes was due to the presence of genomic enhancers in the vicinity of the integrated foreign DNA. The frequent transgene silencing was thought to be induced by the integration of the foreign genes near genomic silencers. These putative silencers were rarely identified suggesting that the ectopic transgene expression and their silencing could not be symmetrical phenomena. It was also proved that the level of transgene expression was generally not a function of the integrated copy number. In a number of cases, the expression level appeared even lower when the number of integrated copies was higher. A striking demonstration was given by the experiment in which the

I. Anegon (ed.), *Rat Genomics: Methods and Protocols*, Methods in Molecular Biology, vol. 597
DOI 10.1007/978-1-60327-389-3_4, © Humana Press, a part of Springer Science+Business Media, LLC 2010

human β-globin gene was bordered by two LoxP sequences and integrated in mouse genome as several copies in tandem. The transgene remained silent in these mice but was reactivated in their offspring in which the copy number was reduced to one by the action of the Cre recombinase (1).

After about a decade, it appeared that this was due to chromatin position effects suggesting that the transgenes were recognized as foreign sequences by some unknown cellular mechanisms. One of the most surprising data was that in a genomic DNA sequence containing the whole human β-globin gene including its promoter region, that allowed the gene to be expressed as expected in cultures, red blood cells remained silent in transgenic mice. This discrepancy suggested that the transgene silencing occurred more in vivo than in cultured cells, and that this phenomenon could take place during the early phase of embryo development. A hypothesis was also that the genomic DNA sequence contained the whole β-globin gene and some but not all the transcription regulators. A confrontation of the very low expression level in patients suffering from β-thalassemia and the structure of their DNA in the genomic β-globin gene region revealed that, in some cases, the gene and its promoter were normal but that some remote regions were missing. This suggested that these regions could be the putative regulators that were missing in transgenic mice. An association of these regions with the β-globin gene allowed the later to be highly expressed in transgenic mice. The extensive study of the β-globin gene locus in several species revealed that remote regulatory elements are present on both sides of the locus. These elements bind transcription factors specifically present in differentiated red blood cells, and they form a transcription complex known as a hub in the vicinity of the promoter through a looping process (2). This type of mechanism seems to be common to many, if not all genes in vertebrates, as well as in invertebrates.

These observations may explain, at least in part, why traditionally constructed transgenes are so often poorly active, and they suggest using long genomic DNA fragments to promote transgene expression. The implementation of these tools may be laborious. On the other hand, the failure of transgene expression is because of multiple reasons, which continue to complicate the construction of vectors allowing an efficient and reliable expression of transgenes. In this chapter, the major improvements that could be possible in vectors for transgene expression are described.

## 2. The Overall Composition of the Vectors

A number of observations on the efficiency of the different vectors to express transgenes made it possible to establish a few empiric rules. It is well known that integrated retroviral sequences and transposons are inactivated by a cytosine methylation of the CpG motifs and the local formation of condensed chromatin (heterochromatin) in which histones are deacetylated and methylated in some sites. Transgenes seem to be inactivated by similar mechanisms. Most of the vertebrate genes contain CpG islets in their regulatory regions, which contribute to their expression. The human eF1-α gene is one of those containing a large number of CpG. Some of the CpG motifs belong to the binding site of the transcription factor Sp1, which is present not only in the promoter region of the gene but also in the first part of its first intron. Unexpectedly, vectors based on the use of the human eF1-α gene promoter and the first intron proved poorly efficient to direct the expression of transgenes (3). Several observations led to the conclusion that the CpG motifs have no negative effect per se. Rather it is their exceedingly large number in vectors that induces transgene silencing. The replacement of the eF1-α intron by the second intron from the rabbit β-globin gene which is richer in AT than in GC improved transgene expression. MARs (matrix attached region) are frequently found in the vicinity of genes, and they bind local DNA to the nuclear matrix. MARs are generally AT rich, and they have been added into vectors to tentatively improve transgene expression. This approach has met variable success. The *E. coli* β-galactosidase gene is rich in CpG, and it is also a potent transgene silencer. This silencing potency proved to be reduced as the number of CpG was reduced. The coding sequences of a transgene may thus be obtained by chemical synthesis to replace a part of the CpG rich codons by others without modifying the sequence of the corresponding protein.

## 3. Use of Insulators

In order to improve the expression of transgenes, it is possible to use large genomic DNA fragments (50–250 kb) expected to contain all the regulatory elements of the gene of interest. This approach is being more and more extensively used (4). In practise, unmodified genomic DNA fragments may be used. This is relatively simple as it implies only the isolation and characterization of BACs (bacterial artificial chromosome) from a bank. Linearized BACs may be used to avoid the random cleavage of the gene of interest, or the generation of transgenic animals in which the transgene is not functional. Some experimenters prefer to inject circular BACs to simplify the protocol; they reduce the frequency of the BAC mechanical degradation during its manipulation. The transgenic yield is then relatively high. The transgenic lines must then be examined to indentify those in which the transgene is intact and working as expected. The use of circular BACs may thus be valuable for generating lines of animals to create models but not so useful for identifying the regulatory elements present in a BAC, as the foreign DNA is cleaved randomly before being integrated.

It is worth noting that ICSI (intracytoplasmic sperm injection) is a method that is being increasingly used to generate transgenic animals as it is relatively simple, efficient and compatible with the use of BACs, which remain undegraded during the DNA transfer (5, 6). One drawback of using BACs as vectors is that they often contain several genes. The BACs thus transfer all these genes, potentially generating unknown and unwanted interactions with the animals. If needed, these genes may be inactivated in the BACs by performing short deletions, for example of the cap region, using homologous recombination in bacteria.

An attractive approach consists of using BACs as vectors harbouring the foreign genes. This implies that the BAC structure is well-known, particularly its sequence. Indeed, the foreign DNA sequence must be introduced in the BAC, for example downstream the chosen promoter, using homologous recombination in bacteria. These methods are available and validated, but remain relatively laborious. This homologous recombination may also be used to inactivate the genes which are present in the BAC vectors and which should ideally not be expressed in the transgenic animals.

It is important to note that the transgenes driven by the BACs rarely work in an ideal fashion, if this concept has any real meaning. Plasmid vectors are known to promote the expression of transgenes in a somewhat unreliable manner due to position effects. Long genomic DNA fragments are expected to suppress these effects, but this is not fully the case. Indeed, it is clear that the variegated expression which characterizes the conventional transgenes is less perhaps much less frequent in animals harbouring BAC vectors. A higher proportion of animals expressing the transgenes is generally found with BAC than with plasmid vectors. Yet, the different lines harbouring BAC vectors do not usually express the transgenes at an identical level for a given number of integrated copies. Similarly, it has been rarely reported that the expression of the transgenes was strictly a function of the copy number of the integrated BACs. This means that the BACs provide transgenes with the essential elements for their expression but they do not often remain fully able to suppress the position effects. This disappointing observation may be not surprising from a theoretical point of view. Indeed, a locus has been constructed during evolution to express the genes it contains in an appropriate manner. This implies that these genes are protected against deleterious position effects in their natural chromatin environment. They have no theoretical reasons to be independent of the position effect in all their integration sites. Some BACs may contain all the elements providing the transgenes with a complete independence of the integration site. If not, a BAC vector may still contain enough regulatory elements improving transgene expression significantly to justify its use.

The number of BACs presently validated for use in directing the expression of transgenes is still limited, but it is increasing. It may be considered that in a few years, BACs capable of driving transgene expression in the major cell types in mammals will be available and currently used.

A more sophisticated approach could be the use of vectors containing only the major elements involved in the control of gene and transgene expression and not all the DNA sequence of the BACs. This implies that these regulatory elements have been identified, characterized and introduced into mini BACs or even plasmids. These vectors would facilitate the task of experimenters and would be particularly valuable for gene therapy. The number of studies in this field remains limited as the experiments are not easy to achieve. The regulatory elements are generally spread over a long distance, and some of their effects can be observed only in transgenic animals. It was surprising to discover that some of the regulatory elements of a locus are located within neighbour loci. An example of this is a major regulatory region of the genes present in the mammalian β-globin locus, which is located upstream within the locus of olfactory receptor genes (2). Several essential regulatory regions of the pig WAP gene (whey acidic protein) were found up to 140 kb upstream of the gene, beyond an unrelated gene (7). This genomic DNA fragment greatly favoured the expression of the WAP gene and of reporter genes introduced in the WAP gene, yet with a position effect and not at all as a function of the copy number.

The study of the remote genomic regulatory elements is still in its infancy. Apart from the technical difficulty in studying them is the fact that each gene or gene cluster seems to have used the available genomic sequences to design specific mechanisms allowing a satisfactory expression. Some of these regulatory elements have an unexpected structure. Examples are SINE B2 and Alu sequences or some active tRNA genes, which are essential regulators for neighbour genes (8).

At present, therefore, it is not possible to classify remote regulatory elements precisely. After the discovery of the remote regulatory elements of the β-globin locus, the concept of LCR (locus control region) emerged. This concept supports the idea that related genes in cluster are under the control of common regulators forming an LCR. It is now well established that not only, in vertebrates but also in Drosophila, most of the genes are not regrouped according to their functions or their regulation mechanisms. On the contrary, most of the genes appear to have their own regulatory system with the essential elements located beyond unrelated genes. Hence, the LCR concept cannot be generalized.

In the meantime, the presence of AT rich MARs within the genomic region required for the expression of a gene suggested that these sequences were essential remote regulators. This hypothesis has not been confirmed (9).

The notion of boundary elements and insulators is essential to understand how unrelated genes can be expressed in a specific manner without being dependent on the neighbour gene regulators. This situation is clearly not often encountered for transgenes suggesting that the constructs commonly used do not contain the natural insulators. Insulator activity has been found in the LCR region of the chicken β-globin locus. This activity was identified in a 300 bp region which proved to be able to block in a specific sense the action of an enhancer when added between the said enhancer and a promoter directing the expression of a reporter gene. The enhancer blocker was mediated by the binding of the regulatory protein CTCF to a specific DNA sequence. The CTCF element has now been found in the boundary region of several other genes. These types of elements, known as enhancer-blocking insulators, cannot be assimilated to silencers as the former act only when they are located between an enhancer and a promoter. Moreover, the enhancer-blocking insulators often show unidirectional action.

This 300 bp region of the chicken β-globin locus was found later to contain another sequence known as a chromatin opener. Chromatin openers are regulators capable of maintaining a local euchromatin configuration favouring the expression of the neighbour gene by preventing the local formation of condensed chromatin (heterochromatin) (10). The elements having this function are known as barrier insulators (10). The barrier insulators cannot be assimilated to enhancers as their effect does not occur during transient foreign gene expression in transfected cells. The region locus, known as 5 HS4, containing the 300 bp sequence can improve the expression of a number of unrelated transgenes in mammals (3, 11). However, the potency of the 5′HS4 element remains generally insufficient to express transgenes in a wholly satisfactory manner.

Most vertebrate genes are thus under the control of remote regulatory regions, which are still poorly known, and thus not easily utilisable yet to design compact efficient vectors. These sequences are generally not sufficiently conserved between different genes and species to be easily identified. They often correspond to nuclease hypersensitive sites (HS), having a lower density of nucleosomes. This property may help in their identification. Indeed, HSs generally exist in tissues, in physiological situations, and in the chromatin regions in which the genes of interest are activated. At these particular sites, chromatin is open, and this corresponds to regions in which DNA is not methylated, and histone 3 is hyperacetylated in lysine 9 as well as methylated in lysine

4 following the local recruitment of histone acetyltransferase and histone methyl transferase by some transcription factors. On the contrary, heterochromatin is characterized by local DNA methylation DNA, deacetylated histone 3, methylated histone 3 in lysine 9 and 27, and by the recruitment of the protein HP 1, which itself recruits histone methylases.

## 4. Optimisation of the Transcribed Part

The transcribed region of genes contains as many, if not more, signals as the promoter region. In the simple conventional gene construction, these signals are not often taken into consideration, and the link of a gene fragment to fragments of other genes as a function of the available restriction sites may delete important regulatory regions, create new ones or generate poorly active combinations. More and more frequently, when the expression of a gene must be optimized, the mutations to be done have to be numerous, justifying a chemical synthesis of the cDNA.

### 4.1. 5′-Untranslated Region

The 5′UTR (untranslated region) must be as poor as possible of GC sequences that can stabilize double strand hairpin structures which do not favour ribosome migration to the initiation codon. The AUG initiation codon must preferably be in the Kozak consensus sequence GCCA/GCCAUGG to optimize translation initiation. The natural 5′UTR of the gene of interest may contain sequences regulating translation. It may not be useful then to keep this region, it could be replaced by a short (not less than 80 nucleotides) AT rich 5′UTR region from genes known to be efficiently translated in many cell types or in the targeted cells of the animals.

Some mRNAs encode proteins that are not naturally secreted. Peptide signals may be added to their cDNA. These proteins are thus secreted although other signals may, more or less, target the proteins in other cell compartments. The yield of secreted proteins may then be reduced.

### 4.2. Introns

A transgene must contain at least one intron, which is required to favour the transfer of the mRNA to the cytoplasm. The first intron of many genes contains sites that bind transcription factors. These sequences are true enhancers that prolong the promoter action downstream of the cap site by maintaining a longer open chromatin region. Exogenous enhancers may be added to the introns to increase transcription efficiency. The introns are known to be deleted by a splicing mechanism, which is dependent on several signals comprising consensus sequences in both splicing sites (CAG GUA/GAGUA/UGGG in 5′ and CAG G.....GAA/G.....

GAA/G….in 3′), a CU-rich region immediately upstream of the 3′ splicing site and a BPS site (branched point sequence) U/CNCUGAC at about 30 nucleotides upstream of the 3′splicing site. Additional splicing enhancers may also facilitate intron deletion (12). The second intron of the rabbit β-globin gene is considered to be efficient in expressing transgenes in mammals. The intron(s) must preferably be put before the coding region. If an intron is added after the translated region, the 5′splicing site must be located not more than 50 nucleotides from the termination codon to avoid the activation of the NMD (nonsense mediated decay), which degrades the mRNA (13).

### 4.3. Codon Optimisation

As mentioned above, the vector including the coding region, must contain as few a number of CpG motifs as possible to prevent the silencing of the transgenes. The codons may be chosen from among those that are the most frequently used in the cells in which the transgene must be expressed. This is particularly important if the cDNA belongs to an organism very different from the transgenic animals. The different regions of the vectors must not contain any cryptic 3′or 5′splicing sites so as to avoid interactions with the splicing sites of the intron leading to the elimination of part, or all of the coding sequence. The sequence of the codons and the UTR regions must be modified accordingly. More generally, sequences known to prevent translation, such as those present in some viral genomes, must be eliminated.

### 4.4. 3′ -Untranslated Region

The 3′UTR region of many mRNA contains signals for mRNA translation and stability. A number of mRNAs have an AU rich region with the AUUUA motif in their 3′UTR. These mRNAs have a short half-life controlled by the cell cycle (14). The fortuitous presence of such sequences must be searched and eliminated to prevent a poor transgene expression.

Some mRNAs contain translation regulators that act by the binding of proteins favouring the recycling of ribosomes by binding to the 5′UTR.

CU rich regions in the 3′UTR enhance the stability of the mRNAs, and they may be added in the vectors downstream of the cDNAs. Stabilizing sequences can be taken from the 3′UTR of the human or bovine growth hormone genes and of the α-globin gene, which also contain efficient transcription terminators (15).

Some proteins are anchored to the plasma membrane by a GPI structure (glycophosphatidylinositol). A protein normally not anchored in this way acquires this property by adding the peptide allowing the addition of GPI in the 3′end of the cDNA.

MicroRNAs (miRNA), the role of which was recently discovered, specifically inhibit the translation of an mRNA after forming a hybrid with its 3′UTR. The presence of a target sequence for an miRNA may unduly inhibit the expression of a transgene. This target sequence should then be deleted.

### *4.5. Simultaneous Expression of Several Cistrons*

It is sometimes necessary to express two or even three genes in the same transgenic animals. The co-injection of several independent vectors makes it possible for the generation of up to 80% of the animals harbouring the two or three genes, which are co-integrated at the same site. Vectors containing the two or three independent genes can be constructed, but this may be a difficult task.

An alternative consists of using IRES (internal ribosome entry site). Such sequences exist in the 5′UTR of many mRNAs, the translation of which is controlled by these sequences that bind specific cellular inducible proteins. Such sequences may be added between two cistrons that allow their simultaneous translation from a single vector. The mechanisms of action of IRES are not fully understood, and they may be multiple. The addition of the second cistron 80 nucleotides after the termination codon of the first cistron may contribute to favour the expression of the second cistron (16).

## 5. Vectors to Inhibit the Expression of Endogenous Genes

### *5.1. Gene Knockout*

A gene may be inhibited at the genomic level by a specific knock out based on the replacement of the targeted gene by an inactive version using homologous recombination. This phenomenon is a rare event, and it must be achieved in cells used to generate animals harbouring the transmissible knock out. Pluripotent cells injected into early embryos can generate chimaeric animals capable of transmitting the mutation. ES (embryonic stem) cells are efficient tools, but so far they have been limited essentially to mice, although genuine rat ES cell lines have been established recently (this issue). Recent data indicate that EG (embryonic gonad) cell lines established in chicken are multipotent cells able to participate to chicken gonads development and to generate transgenic animals (17).

Recent experiments have shown that the transfer of three of the genes required for the maintenance of ES cell pluripotency into somatic cells, induced a dedifferentiation of these cells, which became pluripotent. These new pluripotent cells are known as iPS cells (induced puripotent) (18). This protocol, designed originally in mice, has been extended successfully to humans and is in course in pigs and cow. It offers the unprecedented possibility of obtaining pluripotent cells to be used for gene knock out or knock in in a number of vertebrates.

An alternative approach to the use of pluripotent cells is the cloning technique using nuclear transfer. The gene modifications (gene addition or replacement by homologous recombination) may then be achieved in somatic cells that are further used as nuclear donor cells to generate transgenic clones (19).

Gene knock out should ideally be performed not only in the early embryonic stage, but also in only one cell type and at a chosen moment. This was made possible years ago by adding LoxP sites on both ends of the DNA sequence added into cells. This sequence may then be deleted to provoke the gene knock out by the addition of Cre recombinase in the cells. In order to knock out the targeted gene more precisely, the Cre recombinase gene may be under the control of promoters that are active only in a given cell type and are potentially controlled by the administration of tetracycline or its analogue doxycycline into the animals (20). An additional level of control can contribute to enhance knock out specificity. A Cre recombinase to which two sequences sensitive to tamoxifen, an anti-oestrogenic substance, have been added becomes active only in the presence of this molecule given to the animals (21). These methods are efficient and reliable but relatively laborious, and they induce an irreversible inhibition of the targeted genes.

### *5.2. Use of RNAi*

Gene expression can be inhibited also at the mRNA level. Antisense RNAs and ribozymes have given satisfactory results but only in some cases. The use of interfering RNA (RNAi), known as gene knockdown, is much more attractive and met with great success in transgenic plants but not yet in transgenic animals. The RNAi may form a hybrid with a complementary mRNA region inducing a specific degradation of the targeted mRNA or an inhibition of its translation if the RNAi recognizes more particularly the 3′UTR region of the mRNA. In plants, long antisenses which are fragmented into multiple small interfering RNAs (siRNA) induce an efficient degradation of the targeted mRNA usually without any side-effects, in contrast to what happens in animals in which long RNA triggers the action of interferon. Vectors expressing short hairpin RNAs (shRNA) must, therefore, be used in transgenic animals. The promoters of small RNAs, particularly those from U6 and H1 RNA genes are highly efficient in transfected cells, but they are much less so in transgenic animals although they are potent and active in all cell types. Simple shRNA genes under the control of U6 or H1 gene promoter can be expressed in transgenic animals on condition that these constructs are introduced into lentiviral vectors (22)or into vectors validated to efficiently express foreign genes under the control of RNA Polymerase II in transgenic animals (Sawafta et al. unpublished data). ShRNAs can be produced in given cell types only when the shRNA genes are introduced into miRNA genes, the expression of which may be under the control of promoters acting with RNA Polymerase II specifically in given cell types. RNAi may, therefore, act reversibly at any time and in any cell type offering a greater flexibility than gene knock out.

These different approaches are efficient in cultured cells, but they have been much less so, in transgenic animals till date. This is due to several problems not yet completely solved. Several programmes based on empirical data indicate the putative optimal shRNA sequences to allow the preferential use of the RNAi strand complementary to the mRNA. A very important point is to choose a target region of the mRNA that is not in double strand structure and thus is accessible to the RNAi. Banks of shRNA genes in lentiviral vectors are available for the mRNAs of different species. The fact remains however, that most of the RNAis do not inhibit the targeted gene to more than 70–80%, which may be insufficient to obtain relevant animal models. It is tempting to use vectors expressing the shRNA genes at a relatively high level. This may lead to no significant increase of the inhibition but to more intense off-targeting that may be detrimental or even lethal for the animals (23). In fact, it seems that a well targeted RNAi can be highly active even at a low concentration. It appears therefore, to be of paramount importance to select the shRNA capable of strongly inhibiting the targeted mRNA even at a low concentration in cell systems before generating transgenic animals.

### 5.3. Use of Transdominant Negative Proteins

Gene expression can be inhibited at the protein level. Such mechanisms are extensively used in nature. In transgenic animals it is possible to express genes coding for antibodies, secreted or not, and able to inhibit given molecules in the organism. Another possibility consists of over expressing an inactive analogue of a protein playing the role of decoy for the ligand of the targeted protein. Mice overexpressing the inactive analogue of the insulin receptor were thus models for the study of type II diabetes. This approach is limited essentially by the availability of the transdominant negative proteins.

## 6. Gene Targeting

Homologous recombination is extensively used for gene knock out and knock in. A more systematic gene targeting would improve the reliability of transgene expression by reducing the frequency and the intensity of position effects. Systems under study offer attractive new possibilities.

The frequency of homologous recombination, which includes genomic DNA repair at the site of recombination, is greatly increased by local double strand break of DNA. This property has been exploited to tentatively increase the efficiency of knock out. These methods are based on the use of meganucleases, which are restriction enzymes found in yeast, or of engineered ZFN (zinc finger nuclease) which recognize long DNA fragments generally not present in mammalian genomes. The target DNA sequence of one of these meganucleases, I-SceI, was introduced in mouse genome at a chosen site. The knock out at this site was greatly increased by the action of the meganucleases added with the recombination vector (24). The same is true for gene knock in, and this means that gene addition can be targeted with high efficiency not only in somatic cells, but also in ES cells and virtually in one cell embryos. Experiments will determine in course of time, if the frequency of the targeted integration is sufficient to be used in one cell embryos using conventional microinjection. This implies the microinjection of the foreign gene with the I-SceI meganuclease in one cell embryos in which the recognition site of the enzyme has been previously integrated.

A similar approach for targeted foreign gene integration can be achieved using the phage recombinase ΦC31, which recognizes natural sites in various animal genomes (25).

Ideally, this protocol would be much more useful if natural regions of the genomic DNA could be recognized by the nucleases. This has been achieved successfully by mutating existing meganucleases and ZFN until they recognize the natural chosen sites of the genome (26, 27). These tools can thus be used in some cases to target gene integration for gene therapy and transgenesis.

Engineered meganucleases may also be used to inactivate a targeted gene in a genome using a mechanism known as NHEJ (non homologous end-joining). The protocol consists of injecting a meganuclease recognizing a genomic site without the recombination vector. The enzyme cleaves the DNA at this site and the cell either repairs the break with fidelity or randomly. In the latter case, the repairing process induces a gene knockout (28).

## 7. Conclusion

Transgenesis in rats using the conventional microinjection in the embryo is less easy than in mice. The new techniques described in this book should render transgenics in rats more popular. Rat ES cell lines are now available. This tool expected for two decades may be completed by the generation of iPS cells. The difficulty of getting ES cell lines from embryos may be alleviated by the generation of iPS cells. Cloning by nuclear transfer in rats remains a challenge, and it is not currently used for transgenesis. ICSI, transposons and lentiviral vectors should facilitate the generation of transgenic rats. The use of RNAi for gene knockdown is possible as it requires only conventional vectors and methods of gene transfer. The possibility of inactivating a gene using NHEJ rather than conventional knock out seems promising in rats as in other species. All these new approaches can be submitted to experimentation in rats. The cassettes for transgene expression described in this chapter are essentially not species specific, and they can be used in rats.

## 8. Notes

*What to do if your transgene does not work well?*

1. Evaluate the efficiency of your construct by transfecting it into cultured cells in which the promoter of your construct is active.
2. Make sure that the sequence of your construct is this you expected.
3. Make sure that a part of the coding sequence of your construct is not deleted after a cryptic splicing. This can be seen by a Northern blot or by RTPCR. If so, suppress (delete or mutate) the cryptic splicing (donor and acceptor) site(s) from your construct.
4. Add at least one intron preferably upstream of the cDNA to avoid NMD (see below). Choose introns having good splicing consensus sequences and splicing enhancers (12). The second intron of the rabbit b-globin gene is recognized as one of the good introns for transgenes.
5. Make sure that the mRNA coded by the transgene is not degraded by a nonsense mediated decay (NMD) mechanism. This occurs when the donor splicing site of the intron located downstream of the translated region is farther than 50 nucleotides from the termination codon (13).
6. Make sure that the 3'UTR does not contain an AU rich region with the AUUUA motif which induces an mRNA degradation in quiescent cells (14).
7. Use short 5'UTR containing not less than 80 nucleotides and being preferably AU rich to avoid the formation of stable GC rich secondary structure. The 5'UTR must not contain initiation codons within the consensus Kozak sequence.
8. Make sure that the initiation codon is within the Kozak consensus sequence GCCA/GCC**AUGG**.
9. Reduce the overall GC content of the construct and particularly the CpG motifs in the region preceding and following the transcription start point (10).
10. Add one or preferably two copies in tandem of the 5'HS4 insulator from chicken b-globin locus upstream of the promoter-enhancer region and optionally after the transcription terminator (10,11).
11. Use a strong transcription terminator, e.g. from rabbit or human b-globin genes or from human or bovine growth hormone genes.

12. Add mRNA stabilizer such as this present in the 3'UTR of a-globin gene (15).
13. Eliminate the sequences of the transcribed region of the construct (mainly in the 3'UTR) with may be recognized by natural miRNAs of the transgenic host.
14. Use as vectors long genomic DNA fragments cloned in BAC (bacterial artificial chromosome) containing the promoter chosen to express the transgene and introduce your construct (without any promoter) or your cDNA into the BAC after the promoter, for example after the first intron (4).
15. In bicistronic mRNA, put preferably the IRES (internal ribosome entry site) 80 nucleotides after the termination codon of the first cistron to favour the expression of the second cistron (16).
16. Optimize codon usage if the cDNA is not a mammal. This modification and others in the construct may require a complete chemical synthesis of the cDNA.

## References


1. Garrick D, Fiering S, Martin DI, Whitelaw E (1998) Repeat-induced gene silencing in mammals. Nat Genet 18:56–59
2. de Laat W, Klous P, Kooren J, Noordermeer D, Palstra RJ, Simonis M, Splinter E, Grosveld F (2008) Three-dimensional organization of gene expression in erythroid cells. Curr Top Dev Biol 82:117–139
3. Taboit-Dameron F, Malassagne B, Viglietta C, Puissant C, Leroux-Coyau M, Chéreau C, Attal J, Weill B, Houdebine LM (1999) Association of the 5′HS4 sequence of the chicken beta-globin locus control region with human EF1-alpha gene promoter induces ubiquitous and high expression of human CD55 and CD59 cDNAs in transgenic rabbits. Transgenic Res 8:223–235
4. Long X, Miano JM (2007) Remote control of gene expression. J Biol Chem 282: 15941–15945
5. Moreira PN, Pozueta J, Pérez-Crespo M, Valdivieso F, Gutiérrez-Adán A, Montoliu L (2007) Improving the generation of genomic-type transgenic mice by ICSI. Transgenic Res 16:163–168
6. Shinohara ET, Kaminski JM, Segal DJ, Pelczar P, Kolhe R, Ryan T, Coates CJ, Fraser MJ, Handler AM, Yanagimachi R, Moisyadi S (2007) Active integration: new strategies for transgenesis. Transgenic Res 16:333–339
7. Saidi S, Rival-Gervier S, Daniel-Carlier N, Thépot D, Morgenthaler C, Viglietta C, Prince S, Passet B, Houdebine LM, Jolivet G (2008) Distal control of the pig whey acidic protein (WAP) locus in transgenic mice. Gene 401:97–107
8. Lunyak VV, Prefontaine GG, Núñez E, Cramer T, Ju BG, Ohgi KA, Hutt K, Roy R, García-Díaz A, Zhu X, Yung Y, Montoliu L, Glass CK, Rosenfeld MG (2007) Developmentally regulated activation of a SINE B2 repeat as a domain boundary in organogenesis. Science 317:248–251
9. Sippel AE, Sauereesig H, Hubler MC, Faust N, Bonifer C (1997) Insulation of transgenes from chromosomal position effects. In: Houdebine LM (ed) Transgenic animals. Generation and use. Harwood Academic Publishers, Amsterdam, pp 257–266
10. Gaszner M, Felsenfeld G (2006) Insulators: exploiting transcriptional and epigenetic mechanisms. Nat Rev Genet 7:703–713
11. Giraldo P, Rival-Gervier S, Houdebine LM, Montoliu L (2003) The potential benefits of

insulators on heterogonous constructs in transgenic. Transgenic Res 12:751–755

12. Mersch B, Gepperth A, Suhai S, Hotz-Wagenblatt A (2008) Automatic detection of exonic splicing enhancers (ESEs) using SVMs. BMCBioinformatics9:369.doi:10.1186/1471-2105-9-369
13. Chang YF, Imam JS, Wilkinson M (2007) The nonsense-mediated decay RNA surveillance pathway. Annu Rev Biochem 76:51–74
14. Beelman CA, Parker R (1995) Degradation of mRNA in eukaryotes. Cell 81:179–183
15. Chkheidze AN, Lyakhov DL, Makeyev AV, Morales J, Kong J, Liebhaber SA (1999) Assembly of the α-globin mRNA stability complex reflects binary interaction between the pyrimidine-rich 3′ untranslated region determinant and poly(C) binding protein αCP. Mol Cell Biol 19:4572–4581
16. Attal J, Theron MC, Taboit F, Cajero-Juarez M, Kann G, Bolifraud P, Houdebine LM (1996) The RU5 ('R') region from human leukaemia viruses (HTLV-1) contains an internal ribosome entry site (IRES)-like sequence. FEBS Lett 392:220–224
17. Van de Lavoir MC, Diamond JH, Leighton PA, Mather-Love C, Heyer BS, Bradshaw R, Kerchner A, Hooi LT, Gessara TM, Swanberg SE, Delany ME, Etches RJ (2006) Germline transmission of genetically modified primordial germ cells. Nature 441:766–769
18. Pera MF, Hasegawa K (2008) Simpler and safer cell reprogramming. Nat Biotechnol 26:59–60
19. Robl JM, Wang Z, Kasinathan P, Kuroiwa Y (2007) Transgenic animal production and animal biotechnology. Theriogenology 67: 127–133
20. Malphettes L, Fussenegger M (2006) Improved transgene expression fine-tuning in mammalian cells using a novel transcription-translation network. J Biotechnol 124: 732–746
21. Metzger D, Chambon P (2001) Site- and time-specific gene targeting in the mouse. Methods 24:71–80
22. Pfeifer A (2006) Lentiviral transgenesis – a versatile tool for basic research and gene therapy. Curr Gene Ther 6:535–542
23. Sioud M (2006) Innate sensing of self and non-self RNAs by Toll-like receptors. Trends Mol Med 12:167–176
24. Cohen-Tannoudji M, Robine S, Choulika A, Pinto D, El Marjou F, Babinet C, Louvard D, Jaisser F (1998) I-SceI-induced gene replacement at a natural locus in embryonic stem cells. Mol Cell Biol 18:1444–1448
25. Allen BG, Weeks DL (2006) Using phiC31 integrase to make transgenic Xenopus laevis embryos. Nat Protoc 1:1248–1257
26. Epinat JC, Arnould S, Chames P, Rochaix P, Desfontaines D, Puzin C, Patin A, Zanghellini A, Pâques F, Lacroix E (2003) A novel engineered meganuclease induces homologous recombination in yeast and mammalian cells. Nucleic Acids Res 31:2952–2962
27. Porteus MF, Carroll D (2005) Gene targeting using zinc finger nucleases. Nat Biotechnol 23:967–973
28. Woods IG, Schier AF (2008) Targeted mutagenesis in zebrafish. Nat Biotechnol 26:650–651

# Chapter 5

# Inducible and Conditional Promoter Systems to Generate Transgenic Animals

## Yoji Hakamata and Eiji Kobayashi

## Abstract

Transgenic animals are very useful models that can be utilized for the analysis of temporal and spatial gene expression in vivo. However, generation of a transgenic animal may become problematic if the presence of the transgene leads to conditions which are toxic or lethal to cell growth. In an effort to delineate the mechanism by which a specific gene contributes to cell growth and viability, an inducible and/or conditional system was established to generate transgenic animals. The systems comprise the following: (1) Selecting a specific promoter, (2) replacing a normal gene with other gene sequences (knock out), (3) promoting destruction of the mRNA (RNAi), (4) inducing and/or conditioning by drugs (Tet on/off), and (5) conditional cell knock out with cell death. The choice of system employed is dependent on the particular aim of the investigation, and may influence the final result. The inducible and conditional promoter system represents a useful experimental approach for the development of transgenic animals and the precise examination of gene function.

**Key words:** Transgenic rat, Gene expression regulation, Cre/LoxP, Tet on/off, RNA interference, Diphtheria toxin

## 1. Introduction

Transgenic mice are widely used as animal models for the analysis of gene function in vivo. Investigations concerning the use of murine embryonic stem (ES) cells were first performed with laboratory animals. The generation of knock out mice using ES cells is a very important tool for biomedical research. However, embryonic lethality in many conventional gene knockouts has impeded the efforts to further our understanding of gene function. Consequently, in an effort to delineate the precise role of a gene in a specific cell type and at a specific stage of development, inducible and conditional technologies that allow for flexible spatio-temporal

I. Anegon (ed.), *Rat Genomics: Methods and Protocols*, Methods in Molecular Biology, vol. 597
DOI 10.1007/978-1-60327-389-3_5, © Humana Press, a part of Springer Science+Business Media, LLC 2010

control of gene expression have been established (1). Use of a specific promoter can induce a precise pattern of gene expression. However, if the presence of the transgene is detrimental to embryonic development, normal animal development may be obstructed and, in some cases, result in death. A strategy to achieve conditional gene expression has been developed using the yeast FLP recombinase or the bacteriophage P1 Cre recombinase. This system can induce spatio-temporal expression of a particular gene in a transgenic animal. A system that induces downregulation of gene expression employs RNAi through the use of double-stranded RNA (dsRNA) and utilizes a targeting cell knockout model using a cell toxicity agent.

The most critical issue pertaining to the generation of a transgenic animal relates to the design of the transgene DNA construct, and in particular, the manner in which gene expression is promoted. Essentially, the problem involves the design and use of a promoter that strictly regulates the expression of a downstream structural gene. In an effort to delineate the function of a gene in a transgenic animal, it is sometimes useful to express the gene ubiquitously in a variety of cell and tissue types, rather than in a tissue-specific manner. The GFP transgenic rats that we established express GFP ubiquitously under control of the cytomegalovirus-enhanced/β-actin (CAGGS) promoter, and the fluorescence can be detected in oocytes and embryos at all stages of development, including after birth (2,3). Rats are more suitable than mice for organ transplantation research because of their larger body size, thus facilitating the development of transgenic rat technology (*see* Chap. 6). Transgenic rats are very useful tool that can be utilized for the analysis of gene expression in the whole body. We have developed many transgenic rats for bio-medical research (Table 5.1).

This review describes based on our experimental data, the theory, application and problems of established transgenic strategies utilizing inducible and conditional promoter systems in the development of transgenic animals.

## 2. Selection of Promoter for Controlling Gene Expression

The functional unit of a gene consists of a domain (structural gene), which may code for a product such as RNA or protein, and a non-translating domain located upstream to this domain and which typically contains promoter regions that regulate expression of the downstream structural gene. The choice of an appropriate promoter is the most important step to consider when contemplating the design and creation of a transgenic animal. The housekeeping cytomegalovirus-enhanced/β-actin promoter

**Table 5.1**
**Transgenic rat list**

| Promoter-transgene | Rat strain | Main expression tissue | Reference |
|---|---|---|---|
| CAG-GFP | Wistar | muscle, pancreas, skin | (2) |
| CAG-GFP | LEW | muscle, pancreas, skin | (5) |
| CAG-LacZ | DA | muscle, pancreas, skin | (27) |
| Rosa-LacZ | LEW | liver, intestine, skin | (5) |
| CAG-floxedDsRed2-GFP | Wistar | muscle, pancreas, skin | (8) |
| CAG-Cre | Wistar | muscle, pancreas, skin | (8) |
| CAG-Blood A transferase | Wistar | intestine, pancreas, skin | (28) |
| CAG-Blood B transferase | Wistar | intestine, pancreas, skin | (28) |
| Rosa26-luciferase | LEW | lung, heart, liver, skin | (4) |
| Alb-DsRed2 | Wistar | liver | (6) |
| Alb-HSVtk | DA | liver | (20) |
| Alb-ATTR | DA | liver | (21) |

All transgenic rats were produced by pronuclear microinjection
*CAG* b-actin promoter and CMV enhancer; *Alb* mouse albumin promoter; *ATTR* Amyloidogenic transthyretin

(CAGGS) has often been used to induce ubiquitous gene expression in a variety of cell and tissue types in vivo. In addition, the CMV and Rosa promoters have also been widely used to induce gene expression in all tissues in vivo. The Rosa promoter can induce gene expression in the intestinal tract and nerve to a much greater extent when compared with the CAGGS promoter (3,4). Expression of a transgene of relatively short length can sometimes be influenced by the integration site on the chromosome. To prevent this possibility, transgenic mice are produced using a transgene within a large-sized vector such as BAC or YAC. When using this approach, the expression in the tissue of transgenic mice is more stable and ubiquitous compared with that observed when small-sized vectors have been used (5).

Conventional transgenic technology that relies on gene overexpression or knock out is invaluable for modeling, given its uses in biomedical research or pinpointing sequence dysfunction. However, other approaches are required in order to reveal more precise information about the role of a gene in a specific cell type at a particular stage of development. Conditional technologies allow for flexible spatio-temporal control of gene expression.

Tissue-specific promoters can be used to induce gene expression in specific tissues and at specific stages of development. Anegon et al. have detailed a list of current transgenic rat lines and associated promoters and transgenes (http://www.ifr26.nantes.inserm.fr/ITERT/transgenese-rat/Documents/Table%20transgenic%20.doc). We have developed an albumin enhancer/promoter-driven Alb-DsRed2 transgenic rat that expresses the red fluorescent protein DsRed2 specifically in the liver (6). The expression of DsRed2 correlated well with the differentiation of embryonic hepatocytes.

## 3. Gene Knockout

The gene knockout strategy can be used to explore the function of a gene more directly compared with transgenic animal models by simply overexpressing the transgene. The gene knockout is performed by homologous recombination in embryonic stem (ES) cells. Many reports have described the use of cell lines derived from a variety of species and which make use of several important features typical of ES cells. In some knockout mice, the homozygous animal with the knockout gene develops broad organ dysfunction and dies during the embryonic stage because the function of the targeting gene is lost as a result of homologous recombination. Consequently, precise gene functions in terms of the spatio-temporal stages of embryonic development fail to be determined. A number of strategies have therefore been developed that rely on conditional gene inactivation or activation. The Cre/LpxP system has been widely used in this respect as a means of providing conditional gene expression. First, the targeting vector containing the gene of interest is floxed by LoxP and constructed for homologous recombination in ES cells. Mice derived from targeted ES cells can develop normally and breed to homozygosity for the floxed allele. The mice are then crossed with other transgenic mice for Cre recombinase under the control of specific promoters that allow for the spatio-temporal-specific deletion of the floxed segment. However, when functional Cre recombinase is present in the embryo, the recombination must be promoted by a virus vector containing the Cre recombinase sequence (7) or the tamoxfen-induced Cre system to prevent embryonic death. We tested the Cre/LoxP system in a rat model to demonstrate conditional chromosomal translocation during both the fertilization and adult stages, spatio-temporal gene control by catheter-based adenoviral gene transfer, and muscular fusion events in the limb transplant (8). Floxed DsRed2 rat with the CAGGS promoter showed ubiquitous fluorescence of DsRed2; after mating with a Cre transgenic rat, GFP expression is induced at an early embryonic stage. The floxed rat is a useful model to ascertain the characteristic properties of promoters bearing Cre recombinase (Fig. 5.1).

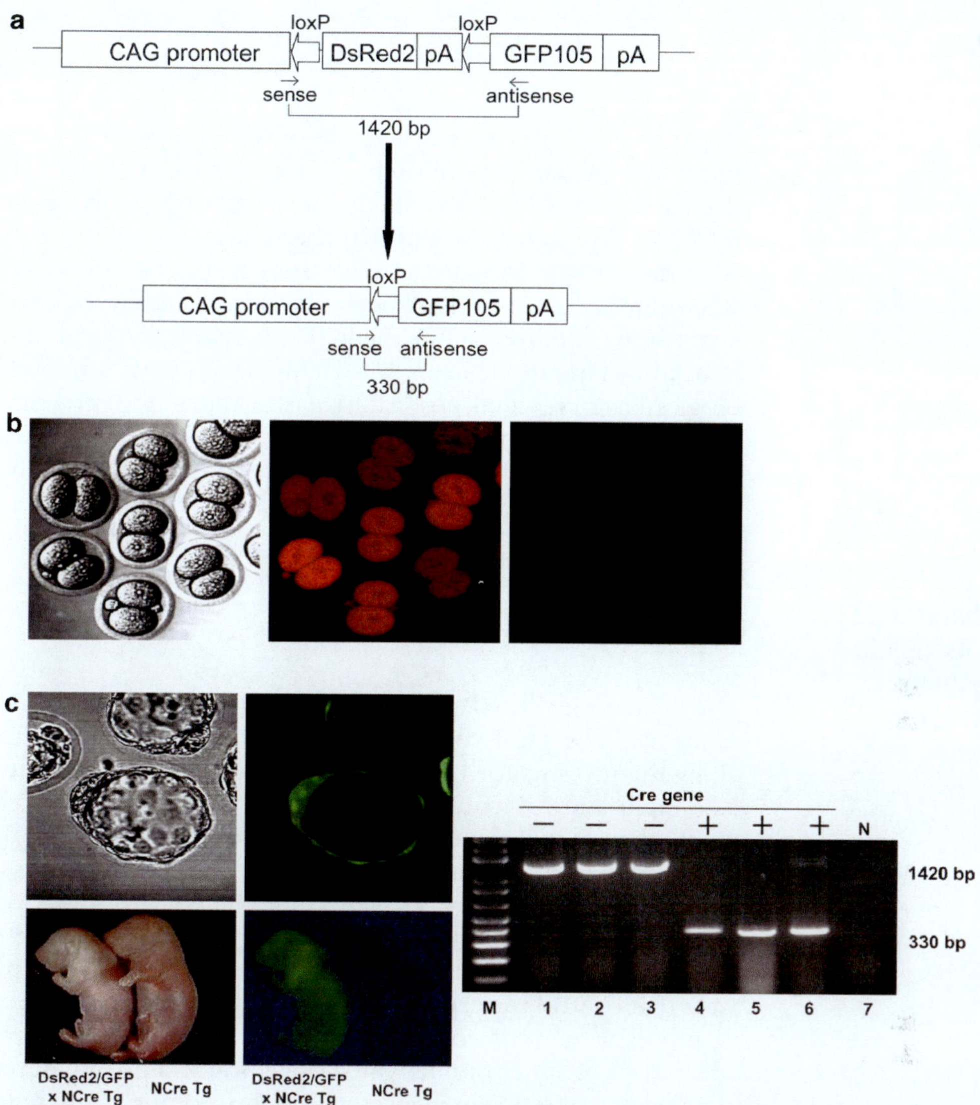


Fig. 5.1. Cre/LoxP system in rat. (**a**) Schematic representation of the DsRed2/GFP double-expression gene and Cre recombinase-mediated LoxP site-specific recombination. (**b**) DsRed2 expression at the 2-cell stage (1.5 embryonic days). DsRed2 expression was observed under 560-nm (*middle*) excitation light using a fluorescence microscope, although GFP expression was not detected under 488-nm excitation light (*right*). (**c**) Cre/LoxP recombination. A homozygous female Cre-expressing transgenic rat was mated with a heterozygous male DsRed2/GFP double-reporter transgenic rat. Blastocysts at 5.5 embryonic days (ED) expressed green fluorescence (GFP), especially in the inner cell mass (*right*). The newborn of double-transgenic rats (DsRed2/GFP_NCre) displayed ubiquitous green fluorescence (*right*). PCR analysis indicated the presence of the DsRed2 stuffer gene in the blastocyst (*lane 4*), the embryo at 15.5 ED (*lane 5*) and the newborn (*lane 6*) of double-reporter transgenic rats. The stuffer was completely excluded by Cre recombinase (*lower band*, 330 bp). PCR amplification from the recombined transgene yielded a 330 bp fragment. However, genes from heterozygous DsRed2/GFP double-reporter transgenic rats showed non-recombination (*upper band*, 1420 bp) in the blastocyst (*lane 1*), the embryo at 15.5 ED (*lane 2*) and the adult (*lane 3*). Negative control neonate of normal Wistar rats (*lane 7*).

## 4. RNAi

RNAi is a reverse genetic technology that allows for the down-regulation of gene expression by introducing short (19- to 21-nt) double-stranded RNA that is complementary to a target gene (9) into cells. We initially confirmed the utility of employing RNAi with rats in the area of transplantation (10). Double-stranded RNA transfected into hepatocytes could specifically block gene expression. Dann et al. succeeded in heritable and stable gene knockdown in rats using an RNAi approach (11). RNAi methodologies have provided powerful tools in the area of conditional gene expression and have impacted many areas of biological research, particularly those areas that require the use of animal models such as rats, without the use of ES cells (11,12).

## 5. Conditional and Inducible Transgene Expression

Use of the Cre/LoxP system irreversibly changes the genome sequence. The ideal conditional gene expression system should allow the investigator to switch transgene expression on and off in a rapid and reversible manner, at any point during development or postnatally, and only in the desired cell types. The "on/off" switching system in controlling gene expression is more suitable for clarifying unknown gene function. Bujard and Gossen established the tetracycline transactivator system as a reliable tool for regulated transgene expression (13,14). This exploits pathways that control the expression of a tetracycline resistance gene in *E. coli.* This gene is constitutively repressed by the tetracycline repressor (tetR) protein, which specifically binds to tetracycline operator (tetO) sequences within the promoter, and renders the gene transcriptionally silent. The repression is relieved by tetracycline, which avidly binds tetR. In this way, tetracycline resistance is controlled in a simple on/off manner by tetracycline itself. Some transgenic rats using this tetracycline system were reported (15,16). Recently, a combinatorial Cre/LoxP and Tet-on/off expression system has been used as an inducible and reversible genetic system (17).

## 6. Conditional Cell Knockout

Another method that can be employed to suppress gene expression involves selective ablation of the targeting cell. Saito et al. generated "toxin-receptor mediated conditional cell knockout"

(TRECK) mice (18). These mice express human heparin binding **epidermal growth factor-like growth factor (hHB-EGF) precursor under the control of the albumin enhancer/promoter. Since hHB-EGF functions as the diphtheria toxin (DT) receptor, the hepatocytes of these mice are selectively ablated following the administration of DT in a dose-dependent manner. Wild-type mice are insensitive to DT as mouse HB-EGF (mHB-EGF) does not function as a DT receptor because of the amino acid substitution within the EGF domain, which disrupts binding to the toxin. This model is a useful tool in investigations concerning liver regeneration (19).

Tissue-specific suicide gene expression in transgenic animals also has been used for cell-specific ablation. To examine the influence of hepatocyte removal, we generated a herpes simplex virus thymidine kinase (HSVtk) transgenic rat, in which the transgene was regulated by an albumin enhancer promoter. Administration of ganciclovir (GCV) into the rat induced infiltration of T cells, macrophages and granulocytes/neutrophils, and hepatocyte cell death (20). Conditional cell knockout can induce spatio-temporal cell death, and subsequently result in conditional targeted gene expression.

## 7. Conclusion

The inducible and conditional promoter system represents a useful experimental approach for examining precise gene function in developing transgenic animals. The genetic background of the strain and species differences are of great importance in conditional transgenic experiments as these may influence the primary pathology and the efficiency of the conditional system. We have developed two different transgenic rats bearing the mouse albumin promoter with the DsRed2 or human amyloidogenic transthyretin V30M (Ueda) genes (21). The expression pattern of the gene in each transgenic rat differs from the pattern observed in transgenic mice with the same promoter (26).

The exploration of gene function is based on genomic manipulations using homologous recombination in ES cells. Recently, some researchers reported that rat ES cells have been established and successfully employed in the development of chimera (23–25). Furthermore, the iPS cell that Nakayama et al. developed could also be a stem cell rather than an ES cell (25). The iPS cell should prove useful in the production of transgenic animals. The transgenic rats referred to in this chapter that we developed are available from the Health Science Research Resources Bank (HSRRB) (http://www.jhsf.or.jp/bank/intro.html), National Bio Resource Project (NBRP) for the Rat in Japan (http://www.anim.med.kyoto-u.ac.jp/nbr/Default.aspx), or from the Rat Resource Research Center (RRRC) in USA (http://www.nrrrc.missouri.edu/).

## References


1. Ryding AD, Sharp MG, Mullins JJ (2001) Conditional transgenic technologies. J Endocrinol 171:1–14
2. Hakamata Y, Tahara K, Uchida H, Sakuma Y, Nakamura M, Kume A, Murakami T, Takahashi M, Takahashi R, Hirabayashi M, Ueda M, Miyoshi I, Kasai N, Kobayashi E (2001) Green fluorescent protein-transgenic rat: a tool for organ transplantation research. Biochem Biophys Res Commun 286:779–785
3. Inoue H, Ohsawa I, Murakami T, Kimura A, Hakamata Y, Sato Y, Kaneko T, Takahashi M, Okada T, Ozawa K, Francis J, Leone P, Kobayashi E (2005) Development of new inbred transgenic strains of rats with LacZ or GFP. Biochem Biophys Res Commun 329: 288–295
4. Hakamata Y, Murakami T, Kobayashi E (2006) "Firefly rats" as an organ/cellular source for long-term in vivo bioluminescent imaging. Transplantation 81:1179–1184
5. Giel-Moloney M, Krause DS, Chen G, Van Etten RA, Leiter AB (2007) Ubiquitous and uniform in vivo fluorescence in ROSA26-EGFP BAC transgenic mice. Genesis 45:83–89
6. Sato Y, Igarashi Y, Hakamata Y, Murakami T, Kaneko T, Takahashi M, Seo N, Kobayashi E (2003) Establishment of Alb-DsRed2 transgenic rat for liver regeneration research. Biochem Biophys Res Commun 311:478–481
7. Ueda S, Fukamachi K, Matsuoka Y, Takasuka N, Takeshita F, Naito A, Iigo M, Alexander DB, Moore MA, Saito I, Ochiya T, Tsuda H (2006) Ductal origin of pancreatic adenocarcinomas induced by conditional activation of a human Ha-ras oncogene in rat pancreas. Carcinogenesis 27:2497–2510
8. Sato Y, Endo H, Ajiki T, Hakamata Y, Okada T, Murakami T, Kobayashi E (2004) Establishment of Cre/LoxP recombination system in transgenic rats. Biochem Biophys Res Commun 319:1197–1202
9. Dillon CP, Sandy P, Nencioni A, Kissler S, Rubinson DA, Van Parijs L (2005) Rnai as an experimental and therapeutic tool to study and regulate physiological and disease processes. Annu Rev Physiol 67:147–173
10. Sato Y, Ajiki T, Inoue S, Fujishiro J, Yoshino H, Igarashi Y, Hakamata Y, Kaneko T, Murakamid T, Kobayashi E (2005) Gene silencing in rat-liver and limb grafts by rapid injection of small interference RNA. Transplantation 79:240–243
11. Dann CT, Alvarado AL, Hammer RE, Garbers DL (2006) Heritable and stable gene knockdown in rats. Proc Natl Acad Sci U S A 103:11246–11251
12. Dann CT (2007) New technology for an old favorite: lentiviral transgenesis and RNAi in rats. Transgenic Res 16:571–580
13. Gossen M, Bujard H (1992) Tight control of gene expression in mammalian cells by tetracycline-responsive promoters. Proc Natl Acad Sci U S A 89:5547–5551
14. Gossen M, Freundlieb S, Bender G, Muller G, Hillen W, Bujard H (1995) Transcriptional activation by tetracyclines in mammalian cells. Science 268:1766–1769
15. Braudeau C, Bouchet D, Toquet C, Tesson L, Menoret S, Iyer S, Laboisse C, Willis D, Jarry A, Buelow R, Anegon I, Chauveau C (2003) Generation of heme oxygenase-1-transgenic rats. Exp Biol Med (Maywood) 228:466–471
16. Tesson L, Charreau B, Menoret S, Gilbert E, Soulillou JP, Anegon I (1999) Endothelial expression of Fas ligand in transgenic rats under the temporal control of a tetracycline-inducible system. Transplant Proc 31: 1533–1534
17. Zeng H, Horie K, Madisen L, Pavlova MN, Gragerova G, Rohde AD, Schimpf BA, Liang Y, Ojala E, Kramer F, Roth P, Slobodskaya O, Dolka I, Southon EA, Tessarollo L, Bornfeldt KE, Gragerov A, Pavlakis GN, Gaitanaris GA (2008) An inducible and reversible mouse genetic rescue system. PLoS Genet 4:e1000069
18. Saito M, Iwawaki T, Taya C, Yonekawa H, Noda M, Inui Y, Mekada E, Kimata Y, Tsuru A, Kohno K (2001) Diphtheria toxin receptor-mediated conditional and targeted cell ablation in transgenic mice. Nat Biotechnol 19:746–750
19. Machimoto T, Yasuchika K, Komori J, Ishii T, Kamo N, Shimoda M, Konishi S, Saito M, Kohno K, Uemoto S, Ikai I (2007) Improvement of the survival rate by fetal liver cell transplantation in a mice lethal liver failure model. Transplantation 84:1233–1239
20. Kawasaki M, Fujino M, Li XK, Kitazawa Y, Funeshima N, Takahashi R, Ueda M, Amano T, Hakamata Y, Kobayashi E (2003) Inducible liver injury in the transgenic rat by expressing liver-specific suicide gene. Biochem Biophys Res Commun 311:920–928
21. Ueda M, Ando Y, Hakamata Y, Nakamura M, Yamashita T, Obayashi K, Himeno S, Inoue S, Sato Y, Kaneko T, Takamune N, Misumi S, Shoji S, Uchino M, Kobayashi E (2007)

A transgenic rat with the human ATTR V30M: a novel tool for analyses of ATTR metabolisms. Biochem Biophys Res Commun 352:299–304

22. Buehr M, Meek S, Blair K, Yang J, Ure J, Silva J, McLay R, Hall J, Ying QL, Smith A (2008) Capture of authentic embryonic stem cells from rat blastocysts. Cell 135:1287–1298
23. Li P, Tong C, Mehrian-Shai R, Jia L, Wu N, Yan Y, Maxson RE, Schulze EN, Song H, Hsieh CL, Pera MF, Ying QL (2008) Germline competent embryonic stem cells derived from rat blastocysts. Cell 135:1299–1310
24. Ueda S, Kawamata M, Teratani T, Shimizu T, Tamai Y, Ogawa H, Hayashi K, Tsuda H, Ochiya T (2008) Establishment of rat embryonic stem cells and making of chimera rats. PLoS ONE 3:e2800
25. Okita K, Ichisaka T, Yamanaka S (2007) Generation of germline-competent induced pluripotent stem cells. Nature 448:313–317
26. Arao Y, Hakamata Y, Igarashi Y, Sato Y, Kayama F, Takahashi M, Kobayashi E, Murakami T (2009) Characterization of hepatic sexual dimorphism in Alb-DsRed2 transgenic rats Biochem Biophys Res Commun 382:46–50
27. Takahashi M, Hakamata Y, Murakami T, Takeda S, Kaneko T, Takeuchi K, Takahashi R, Ueda M, Kobayashi E (2003) Establishment of lacZ-transgenic rats: a tool for regenerative research in myocardium. Biochem Biophys Res Commun 305, 904–908
28. Iwamoto S, Kumada M, Kamesaki T, Okuda H, Kajii E, Inagaki T, Saikawa D, Takeuchi K, Ohkawara, S, Takahashi R, Ueda S, Inoue S, Tahara K, Hakamata Y, Kobayashi E (2002) Rat encodes the paralogous gene equivalent of the human histo-blood group ABO gene. Association with antigen expression by overexpression of human ABO transferase. J Biol Chem 277, 46463–46469

# Chapter 6

# Generation of Transgenic Rats by Microinjection of Short DNA Fragments

**Séverine Ménoret, Séverine Remy, Claire Usal, Laurent Tesson, and Ignacio Anegon**

## Abstract

Here we describe an efficient technique to generate transgenic rats by microinjection of short DNA fragments. We have focused on optimal conditions for superovulation of prepubescent females Sprague–Dawley (CD) strains to have good quality embryos, pseudopregnant females, zygotes preparation, optimal conditions for microinjection and embryo transfer into foster mothers.

**Key words:** Transgenic rats, Microinjection, Embryo transfer, Superovulation, Sprague–Dawley rats, Transgenic efficiency

## 1. Introduction

The rat is a well-characterized rodent conveniently intermediate in size between the mouse and larger mammals. It is more accessible than the mouse for microsurgery, multiple blood sampling (in larger volumes), tissue and organ sampling (e.g. central nervous system) and analysis of organ function in vitro (heart perfusion).

Since the development of the first transgenic rat lines in the 1990s (1, 2), approximately 180 different transgenic rat lines have been produced. We have regularly updated a list of all transgenic and genetically modified rat lines published from 1990, which is available at: http://www.ifr26.univ-nantes.fr/ITERT/transgenese-rat/liste_rats.php

In the absence of rat ES cells, the microinjection of short DNA into the male pronucleus of one-cell embryos was the first efficient technique to be described (0.2–3% transgenic rats/ number of

I. Anegon (ed.), *Rat Genomics: Methods and Protocols*, Methods in Molecular Biology, vol. 597
DOI 10.1007/978-1-60327-389-3_6, © Humana Press, a part of Springer Science+Business Media, LLC 2010

transplanted zygotes) (3). Although the efficiency of generating transgenic rats using DNA microinjection is lower than when using lentiviral transgenesis (3) (0.2–3% vs. 6–18% transgenic rats/number of transplanted zygotes, respectively), the generation of lentiviral vectors of good concentration and quality is technically demanding and the size of the transgene needs to be <9 kb, whereas the size of DNA used in microinjection can be as large as several hundred thousand kb. The generation of transgenic rats remains more difficult compared to transgenic mice, since rat one-cell embryos have more flexible plasma and pronuclear membranes making injection more difficult and increasing embryo lysis (4, 5).

The methods used to generate transgenic rats show considerable variations according to the choice of strain (6), the superovulation protocols (7), the media for embryo culture (6), the concentration of DNA (8) and the reimplantation of embryos either directly or after in vitro culture (6). The protocols we described here have been developed with Sprague–Dawley (CD) strains and in our hands result in a transgenic efficiency of 0.2–3% of transgenic rats/number of transplanted zygotes in 3–6 weeks with four microinjection sessions per week.

The sequencing of the rat genome together with the very rapid development of positional cloning in different disease models has resulted in the identification of numerous putative genes. The function and role of these genes will need to be proven by overexpression in vivo. The generation of transgenic rats by DNA microinjection is a proven and reliable method to do this. In this chapter, we describe the generation of transgenic rats using short transgenes (<8 kb). The technique using large DNA fragments is described in Chapter 7.

## 2. Materials

### 2.1. Male Vasectomy

1. A basic anesthetic delivery system with an isoflurane anesthesia setup.
2. Bunsen burner.
3. Fiber optic light source.
4. Electric shaver.
5. Surgical tools: 1 surgical forceps, 2 × Dumont #5 forceps, 1 dissection scissors. All instruments should be cleaned and sterilized with 70% ethanol.
6. Surgical suture no. 3/0.
7. Surgical gauze.

8. 70% Ethanol.
9. 0.9% NaCl.

### 2.2. Superovulation and Pseudopregnant Females

1. Pregnant Mare's Serum Gonadotropin PMSG (Folligon®, Intervet, Angers, France) at 20 IU/female/0.2 ml. Dissolve 2,500 IU in 20 ml 0.9% NaCl to give 100 IU/ml. Store in 1.8 ml aliquots (for 8× females) at –20°C for several months.
2. Human Chorionic Gonadotropin hCG (Chorulon®, Intervet, Angers, France) at 30 IU/female/0.2 ml. Dissolve 1,500 IU in 10 ml 0.9% NaCl with 0.1% BSA to give 150 IU/ml. Store in 1.8 ml aliquots (for 8× females) at 4°C for 2 weeks.
3. 2 ml syringes.
4. Needles.

### 2.3. Zygote Preparations

1. 8× 26–30-day-old mated prepubescent Sprague–Dawley females.
2. 70-day-old Sprague–Dawley fertile males. They should be replaced every 8 months to a year.
3. Under-stage illumination stereomicroscope.
4. 37°C, 5% $CO_2$ incubator.
5. Surgical tools: 1 dissection scissors, 1 surgical scissors, 1 surgical forceps, 2× fine forceps. All instruments should be cleaned and sterilized with 70% ethanol.
6. 35 mm (35 × 10 mm) culture dishes (Nunc).
7. 4-well plates (Nunc).
8. Transfer Pasteur pipettes.
9. Mouthpiece, tubing and microcapillary holder.
10. M16 medium (Sigma) complemented with 0.1% penicillin–streptomycin, 0.1% glutamine and 10% heat-inactivated fetal calf serum.
11. Embryo-tested mineral oil (Sigma).
12. Embryo-tested bovine testis hyaluronidase (H 4272, Sigma) stock: Dissolve 30 mg in 3 ml of 0.9% NaCl to give 10 mg/ml. Store in 50 µl aliquots at –20°C for several months. For use, resuspend the aliquots in 950 µl of Phosphate Buffer Solution and placed in one well of a 4-well plate.

### 2.4. Injecting Zygotes – Microinjection

1. Work station including:
   (a) Inverted microinjection microscope equipped with 10× eyepieces, 5×, 10× and 40× objectives and with a tray heated.
   (b) Microinjector (Narishige).

(c) Two micromanipulators for both holding and microinjection pipettes (Narishige).

(d) Video system.

2. Micropipette puller (PN-30, Narishige).
3. Holding capillary (Glass capillary, Narishige; cat no. G-1).
4. Microinjection capillaries (Glass capillary with filament, Narishige; cat no. GD-1).
5. Microscope slides for injection chamber.
6. Tips for loading capillaries (Microloader, Eppendorf).
7. Purified DNA plasmid at 2–5 ng/μl.

#### *2.5. Embryo Reimplantation*

1. Pseudopregnant Sprague–Dawley females.
2. Vasectomised Sprague–Dawley males.
3. A stereoscopic microscope.
4. A fiber optic light.
5. Surgical tools: 1 dissection scissors, 1 surgical scissors, 1 surgical forceps, 2× Dumont #5 forceps. All instruments should be cleaned and sterilized by 70% ethanol.
6. Surgical suture.
7. Surgical gauze.
8. Embryo transfer Pasteur Pipette and mouth pipettor.
9. Adrenalin.
10. 0.9% NaCl.

## 3. Methods

#### *3.1. Male Vasectomy*

1. Anesthetize male Sprague–Dawley rats of at least 70 days of age by isoflurane (rats should weight at least 250–400 g) (*see* Note 1).
2. Place the rats on their backs to expose the abdomen.
3. Clean the animals thoroughly with 70% ethanol. All surgical procedures should be performed with sterilized instruments.
4. To open the abdominal cavity, make a transversal incision of 2.5 cm of the abdomen skin and of the abdominal wall with the dissection scissors at a point level with the top of the legs.
5. Apply a moderate pressure to push the testes to the scrotal sac into the abdomen. Pull out one left (or right) of the fat pad by using the blunt forceps. The testis, vas deferens and epidydimis will come with it and be exposed; they should be placed in a sterilized compress soaked in 0.9% NaCl.

6. Pull the vas deferens out with a forceps (the vas deferens has a tube-like structure and is situated below the testis and has a blood vessel running along one side).
7. The vas deferens can be interrupted by cauterization, which is rapid and convenient. Cauterization is performed by dissecting a short portion of the vas deferens. Holding the vas deferens in a loop with one pair of forceps, burn the vas deferens loop with the tips of a heated second pair of forceps. The cauterized ends of the vas deferens will not join back together.
8. Repeat this procedure on the other testis.
9. Stitch up the abdominal wall first, followed by the skin.
10. Do not mate the animals for 15 days. After this period they should be tested for effective vasectomy by two successive matings which should not give pups.

### 3.2. Superovulation and Pseudopregnant Females

To produce zygotes for pronuclear injection (refs. 4, 5, 9), superovulate eight prepubescent females Sprague–Dawley rats (26–30 days old) and mate with Sprague–Dawley fertile males (70 days old) (Fig. 6.1).

1. Superovulate prepubescent rats with an i.p. injection of 20 IU of PMSG between twelve and one p.m. on day –2 (*see* Note 2).

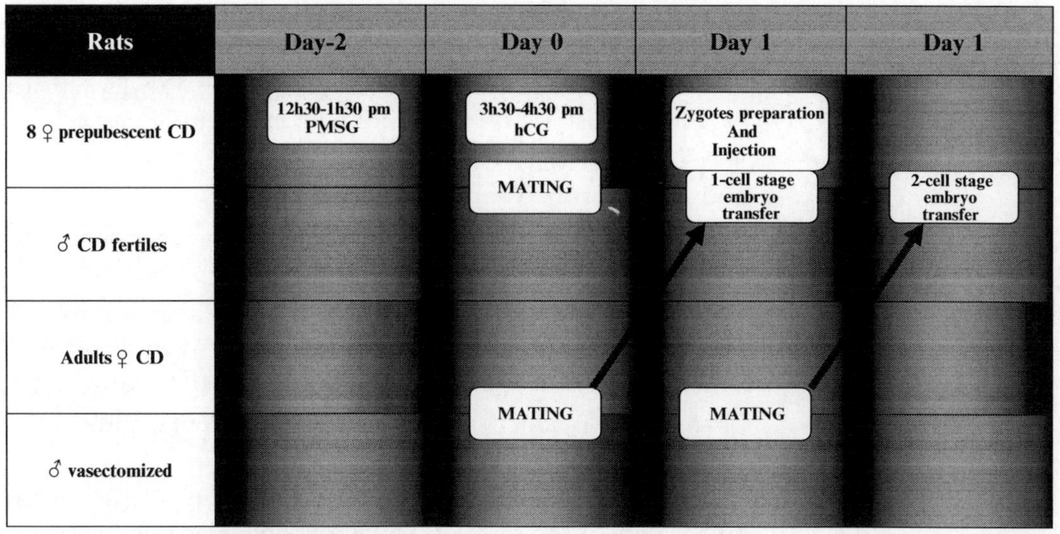


Fig. 6.1. Timelines of PMSG-LHRH treatments and matings. Inject eight females prepubescent 26–30-day-old Sprague–Dawley rats i.p. with PMSG (20 IU) on the afternoon of day –2 between 12 h and 1 p.m. Approximately 48 h after PMSG treatment, at day 0, between 3 and 4 p.m., inject the females with an i.p. injection of 30 IU of hCG and mate with fertile males. At the same time, in order to obtain pseudopregnants females, put one adult female Sprague–Dawley (>70 days old) with one vasectomised male Sprague–Dawley rat at day 0.

2. At day 0, between 3 and 4 p.m., inject the females i.p. with 30 IU of hCG to induce ovulation 10–12 h later.
3. Following the hCG injection, mate females with mature fertile males at a 1:1 ratio in individual cages (with trays and white absorbent paper in order to be able to see the vaginal plugs) (*see* Note 3).
4. To obtain pseudopregnant females, put one adult female Sprague–Dawley (>70 days old) with one vasectomised Sprague–Dawley male at day 0. We recommend a minimum of ten breedings.
5. On the morning of day 1, collect the Sprague–Dawley females with vaginal plugs for zygote preparation and eliminate the females without plugs (*see* Note 4).
6. Similarly, collect the pseudopregnant females with plugs, but leave them in the animal facility until needed later in the afternoon for reimplantation.

### 3.3. Zygote Preparations

1. Prepare a 4-well plate containing PBS in two wells and a solution of hyaluronidase (Sigma) at 1 mg/ml in PBS in a third well and place it for at least 1 h before embryo collection in an incubator at 37°C with 5% $CO_2$.
2. Prepare another 4-well plate with four wells of M16 embryo culture medium and place it for at least 1 h prior to embryo collection in an incubator at 37°C with 5% $CO_2$ (*see* Note 5).
3. Prepare a small 35 mm Petri dish with PBS (to place the oviducts of prepubescent females with vaginal plugs).
4. Between 10 and 11 a.m., kill plugged Sprague–Dawley females by cervical dislocation after anesthetizing by isofluorane.
5. Place females on their backs on absorbent paper and clean thoroughly with 70% ethanol.
6. Pinch the skin at the midline and open up the abdominal cavity by a transversal incision.
7. Push up the gut and take hold of one of the uterine horns with forceps. With a small scissors, make a cut between the oviduct and ovary and then through the uterus near the oviduct.
8. Transfer the oviduct (with some adjacent uterine and ovarian tissue) into a 35 mm dish with PBS.
9. Under a stereomicroscope, locate one-cell stage eggs in the upper part of the oviduct, which is swollen at this time. In a Petri dish containing PBS, tear the oviduct close to where the eggs were located using fine forceps, this releases the clutch of eggs surrounded by cumulus cells.

10. The clutch of eggs is placed in one of the wells with PBS to eliminate cellular debris.
11. The eggs with cumulus are then transferred to the well containing the hyaluronidase and incubated at 37°C for 3–5 min. Eggs can be pipetted up and down to eliminate cumulus cells as quickly as possible (hyaluronidase can be toxic for the eggs).
12. Transfer the eggs to the second wells containing PBS to wash away the hyaluronidase, then place them in the well containing the M16 medium. After that place the dish in the 37°C incubator.

### 3.4. Injecting Zygotes – Microinjection

Before injecting, culture the embryos for one to two hours in M16 medium at 37°C and check that the rat oocytes show visible male pronuclei visible after culture (*see* Fig. 6.2).

1. Place a drop of M16 medium (prewarmed at 37°C) in the center of a clean microscope slide and fix two small plastic coverglass (5–6 mm of wide and 2 mm of thickness) with silicone grease. Place a drop of medium on a small glass coverslip (5 mm of wide). Turn over the coverslip on the medium in the microscope slides. With a Pasteur pipette, place two small drops of mineral oil on the left and right sides of the medium drop to prevent evaporation (*see* Note 6).
2. Place the slide under a stereomicroscope and transfer 20–40 zygotes, depending on the experience of the manipulator, into the upper part of the drop (*see* Note 7).
3. Place the slide onto the stage of a microinjection microscope with a tray heated at 37°C.
4. Place the microcapillary for holding oocytes into the connector piece.

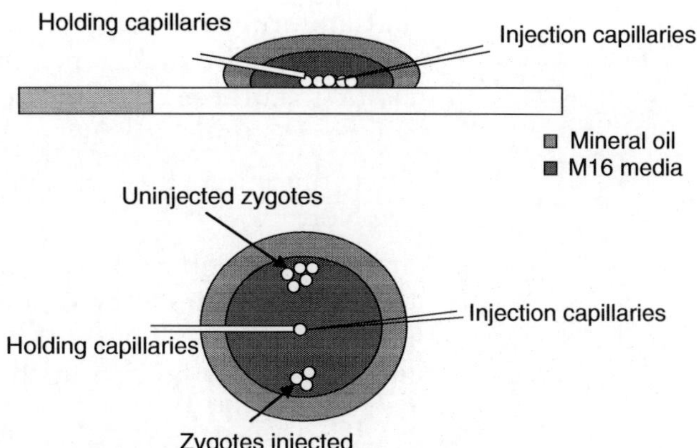


Fig. 6.2. Schematic presentation of the arrangements of zygotes, holding and injections capillaries on injection slides.

5. Fill the injection capillary with an inner filament using a Microloader with 1–2 µl of plasmid DNA. Insert the injection capillary into the connector piece.
6. Adjust the mounting angle of the holding and injection capillary to 35°.
7. First, microinject each new batch of plasmid DNA at 1, 2, 4 and 6 µg/ml (*see* Notes 8 and 9) with 10–20 zygotes for each concentration and determine egg viability and passage to two-cell embryos after overnight culture (*see* Note 10). To generate the transgenic rats, use the highest concentration that preserves reasonable embryo viability. Some decrease in viability is expected when using concentrations that result in the generation of the highest proportion of transgenic animals.
8. To inject a zygote, move the holding pipette to the collected zygotes and apply a minimum pressure.
9. Move one zygote to the center of the slide and position it in the same plane of focus as the opening of the holding pipettes. Examine the oocytes under high power of magnification (×400) and focus on the male pronuclei.
10. Move the injection pipette below the zygote and place the tip of the injection pipette into the holding pipette.
11. To inject, aligned the injection pipette in a horizontal line of the aligned pronuclei (*see* Note 11).
12. Penetrate the male pronucleus with the injection pipette and inject DNA. The male pronucleus swells when the injection is successful (*see* Fig. 6.3).
13. The equipment and procedure for microinjection of rat eggs is basically the same as those used for mouse eggs, although injection of rat eggs is more difficult and time consuming. Rat egg pronuclei are difficult to see, they are less regular and uniform than those of mouse eggs. Furthermore, the plasma and pronuclear membranes of rat eggs are more elastic than those of mouse eggs, making them more difficult to penetrate.
14. Place the injected zygotes in the lower part of the drop (*see* Note 12).
15. Start again with the other zygotes present on the slide. When the injections are successful, take zygotes from the drop and place them in a well of the 4-well Petri dish containing M16 medium. Place the dish in the incubator at 37°C with 5% $CO_2$ (*see* Note 13).
16. Take new zygotes for injection.

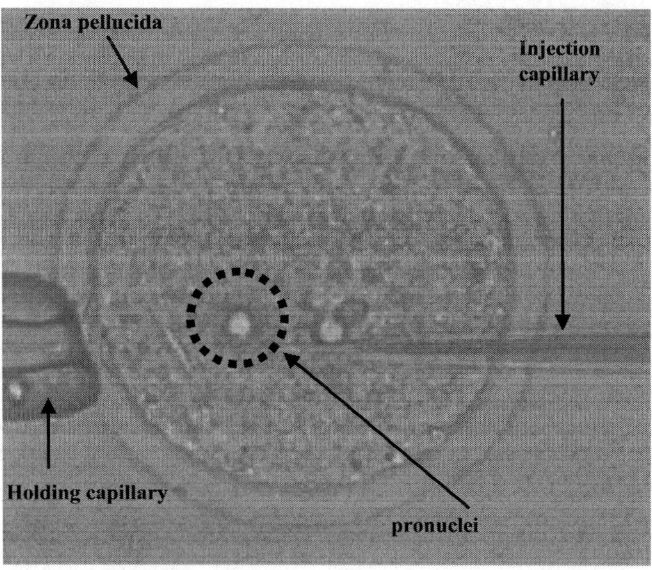


Fig. 6.3. Pronuclear injection.

17. Between 31.5% and 65% of rat eggs are viable after microinjection and are ready to be reimplanted into pseudopregnant females or cultured for 24 h in M16 medium before transfer (*see* Note 14).

### *3.5. Embryo Reimplantation*

Bilateral embryo transfer is used to place eggs into the reproductive tracts of pseudopregnant recipients (*see* Note 15).

1. Anesthetize a pseudopregnant female with isofluorane. Shave a broad area on each side of the rat and clean thoroughly with 70% ethanol. Be sure to perform the surgical procedure with sterile instruments. Place the rat on its right side.
2. Make a 2 cm long incision in the skin, in the middle of the lower back, and the body wall can be penetrated directly over the ovarian fat pad, which is very large in adult rats.
3. Take hold of the fat pad with a forceps and pull the ovary out until the attached oviduct and uterus are clearly visible. Place it in a compress humidified with a solution of 0.9% NaCl.
4. Correct the positioning of these organs under a stereoscopic microscope. The coiled oviduct within the transparent bursa should be clearly visible.
5. Since the bursa (a transparent membrane over the oviduct and the ovary) is more vascular in the rat than in the mice, place two or three drops of adrenalin (acting as a vasoconstrictor) on the top of the ovarian bursa immediately before reimplanting the eggs (*see* Note 16).

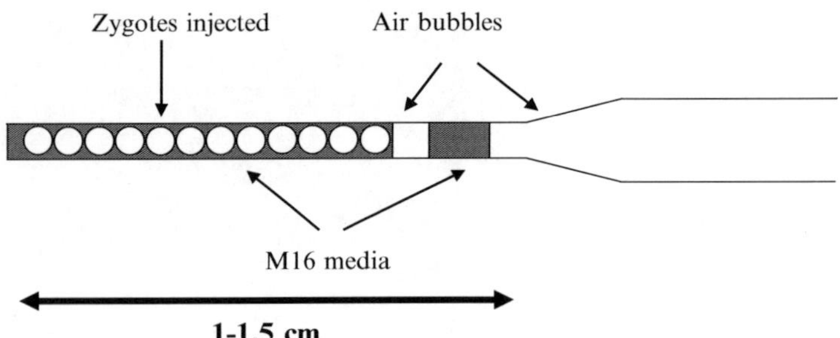


Fig. 6.4. Organization of zygotes injected via the transfer pipette.

6. Make an embryo transfer pipette from a flame-pulled Pasteur pipette and load it with 10–12 zygotes injected in a small quantity of medium (*see* Fig. 6.4) (*see* Note 17).
7. The infundibulum and the swollen ampulla are located underneath the bursa. Use two no. 3 forceps to tear the bursa covering the gap between the ovary and the oviduct.
8. The infundibulum of the rat oviduct is usually oriented horizontally, buried between the oviduct and ovary. It is easier to insert the transfer pipette into the infundibulum if about 4 mm of the tip is bent at a 30° angle. Remove any traces of blood or other liquid with a compress.
9. Carefully widen the cleft that buries the infundibulum with a fine no. 5 forceps without causing any bleeding and insert the reimplantation pipette with the zygotes down the opening of the ampulla. Slowly blow in the infundibulum the zygotes with medium until 2–3 air bubbles become visible in the ampulla, indicating a successful transfer. Wait 30 s before removing the transfer pipette.
10. With blunt forceps, pick up the fat pad and place the uterus, oviduct and ovary back inside the abdominal cavity.
11. Stitch up the abdominal wall and the skin. Repeat the operation of embryo transfer in the other oviduct. Place the reimplanted rats in a cage and transfer them back to the rat facility.

*Notes*: Bilateral reimplantation results in higher rates of pregnancy in transferred females (65–85%) as compared to unilateral transfer (25–50%). The size of litters after bilateral egg transfer varies but it is usually between five and eight newborns (*see* **Note 18**).

#### *3.6. Genotyping*

1. Extract genomic DNA from tail biopsies of the 10-day-old rats by Proteinase K digestion. Briefly, incubate short tail biopsies (~6 mm) in 500 μl of Proteinase K lysis buffer for a night in a water bath at 56°C.

2. Next, centrifuge the tubes at 15,000 g for 10 min. Perform a phenol–chloroform DNA extraction on the supernatant for PCR genotyping.

## 4. Notes

1. Vasectomies have to be performed at least 10–14 days before the first mating (to allow the recovery of males after the surgical procedure). To test the sterility of males, breed those two or three times before use.
2. The timing of injections is critical to obtain a sufficient number of oocytes with clearly visible pronuclei. Another technique can be used for superovulating prepubescent females: osmotic micropumps (model 1003D, Alzet, Palo Alto, CA, USA) containing partially purified follicle-stimulating hormone (FSH, Vetrepharm, London, Ontario, Canada) are implanted subcutaneously or intraperitoneally, using aseptic surgical techniques in prepubescent rats between 8 and 10 a.m. on day –2. These micropumps deliver 1 μl/h, which corresponds to 0.4 IU of LHRH/day. Rats are given an i.p. injection of 1 μg of LHRH on the afternoon of day 0 and are mated with fertile male Sprague–Dawley rats in individual cages.
3. Use dedicated mating cages to check females have loose vaginal plugs after copulation.
4. The superovulation protocol should give 20–25 embryos per females.
5. Stabilize the M16 media in an incubator at 37°C and 5% $CO_2$ for 1 h before use.
6. Use embryo-tested mineral oil.
7. The Pasteur pipette used to transfer the zygote should be slightly larger than a zygote.
8. The technique for DNA purification and quantification is described in the following lines. The fragment is excised from the shuttle plasmid with the restriction enzyme previously chosen to remove a maximum of plasmid sequence. Isolate the fragment from the plasmid backbone by electrophoresis through a 1% agarose gel (SeaKem GTG, FCM Bioproducts, USA), electroelute and purify through an Elutip-d column (Whatmann, Dassel, Germany). The fragment is quantified with a Nanodrop spectrophotometer, stored concentrated (50–100 ng/μl at –20°C) and diluted to a working concentration (2–3 ng/μl in 5 mM Tris–HCl and 0.1 mM EDTA pH 7.4).

9. For short DNA fragments with repeated sequences, heat at 80°C for 3 min.
10. The injection buffer used to dilute the plasmid for microinjection should be filtered to remove all particles.
11. The pronucleus membrane is very flexible.
12. A separation between the non injected zygotes in the upper part of the drop and the injected zygotes prevents mixing up injected and non injected zygotes.
13. Wash injected zygotes 2–3 times in M16 media after microinjection.
14. At least 50–80% of injected zygotes should survive the injection procedure. After injection, oocyte trauma results in cell lysis within 20 min. Low survival rates are frequently caused by too large injection pipettes, nucleus damage or a high concentration of DNA.
15. One-cell embryo reimplantation should be performed 1–2 h after zygotes injection.
16. Rat ovarian bursa is more vascular than in mice, use two or three drops of Adrenalin to reduce bleeding during embryo transfer.
17. Keep the time between loading zygotes into the reimplantation pipette and embryo transfer to a minimum.
18. An average of five pups is born from each foster mother but variability ranges from 0 to 9. Only a few, or even none, are transgenic. If there are no transgenics then reimplantation was successful in this case the injection or the concentration of DNA was too low. Too high a DNA concentration will cause developmental arrest of embryos and results in no pups. Transgenic efficiency is between 1% and 15% of transgenic founders per number of newborns.

## References


1. Mullins JJ, Peters J, Ganten D (1990) Fulminant hypertension in transgenic rats harbouring the mouse Ren-2 gene. Nature 344:541–544
2. Hammer RE, Maika SD, Richardson JA, Tang JP, Taurog JD (1990) Spontaneous inflammatory disease in transgenic rats expressing HLA-B27 and human beta2m: an animal model of HLA-B27 associated human disorders. Cell 63(5):1099–1112
3. Lois C, Hong EJ, Pease S, Brown EJ, Baltimore D (2002) Germline transmission and tissue specific expression of transgenes delivered by lentiviral vectors. Science 295:868–872
4. Charreau B, Tesson L, Soulillou JP, Pourcel C, Anegon I (1996) Transgenic rats: technical aspects and models. Transgenic Res 5:223–234
5. Tesson L, Cozzi J, Ménoret S, Rémy S, Usal C, Fraichard A, Anegon I (2005) Transgenic modifications of the rat genome. Transgenic Res 14:531–546
6. Popova E, Bader M, Krivokharchenko A (2005) Strain differences in superovulatory response, embryo development and efficiency of transgenic rat production. Transgenic Res 14:729–738
7. Filipiak WE, Saunders TL (2006) Advances in transgenic rat production. Transgenic Res 15(6):673–686
8. Hirabayashi M (2008) Technical development for production of gene-modified laboratory rats. J Reprod Dev 54:95–99
9. Nagy A et al (2003) Manipulating the mouse embryo: a laboratory manual, 3rd edn. Cold Spring Harbor Laboratory Press, New York 764 pp

# Chapter 7

# Generation of Transgenic Rats Using YAC and BAC DNA Constructs


## Ri-ichi Takahashi and Masatsugu Ueda



## Abstract

Transgenic rats with a simple plasmid vector smaller than 20 Kb show insufficient expression and tissue specificity of the introduced transgene. Vectors derived from yeast artificial chromosome (YAC) and bacterial artificial chromosome (BAC), consisting of DNA fragments up to ~1 Mb (YAC) and ~200 Kb (BAC), respectively, and containing various endogenous regulatory sequences, were expected to work well and showed expression profiles comparable to their endogenous counterparts in transgenic animals. While attempting to make transgenic rats using YAC and BAC vectors, we faced two problems: how to prepare sufficiently concentrated intact DNA and how to reliably microinject a large DNA fragment into the fragile pronuclear ova of the rat. After solving these problems, we were able to make transgenic rats by introducing YAC/BAC gene constructs (YACs/BACs) into the pronuclear ova. And then we examined the relative transcription rates of these genes in the transgenic rats. In this chapter, we focus on this injection process.



**Key words:** YAC, BAC, Homologous recombination, Microinjection, Rat, Pronuclear ova


## 1. Introduction

Transgenic animals are important tools for in vivo genetic studies. The first generation of transgenic animals was made with conventional simple plasmid vectors smaller than 20 Kb. This consists of a simple promoter and gene. Although these vectors can be handled easily with the standard methods of molecular biology, they are often missing enhancers or other regulatory elements and have drawbacks that may prevent them from achieving sufficient promoter activity and tissue-specific expression (1). The expression profiles of introduced gene in transgenic animals vary among lines, and the expression levels are often lower than those of their endogenous counterparts; moreover, the introduced genes has


I. Anegon (ed.), *Rat Genomics: Methods and Protocols*, Methods in Molecular Biology, vol. 597
DOI 10.1007/978-1-60327-389-3_7, © Humana Press, a part of Springer Science+Business Media, LLC 2010

less accurate tissue specificity The yeast artificial chromosome (YAC) (2) and bacterial artificial chromosome (BAC) vectors contain cloned DNA fragments up to ~1 Mb (YAC) and ~200 Kb (BAC) (3), respectively. A large genomic DNA fragment of YAC/BAC can contain an entire transcription unit with the regulatory regions located upstream or downstream of the gene, and can include a large gene or family gene. Such a fragment can be used for functional studies that cannot be undertaken by using conventional transgenic approaches.

Because YACs have an inherently high frequency of homologous recombinations, it is possible to precisely modify YACs by the deletion or insertion of exogenous DNA (2). The first generation of transgenic animals using large fragments was performed using YAC DNA constructs. Transgenic mice expressed foreign proteins without a position effect with a YAC vector (4–6). A general remark about a method for generating YAC transgenic mice by pronuclear microinjection was reported in 1996 (7). We have also published a series of reports about a human alpha-lactoalbumin chromosomal mapping (8), the production of transgenic rats (9), and application of animal bioreactors (10). We have also reported on the integrity of YAC DNA derived from different purification methods and preparation procedures in the genome of transgenic rats (11) (*see* Note 1–4).

BACs were developed based on *E.coli* F factor and DNA fragments up to 200 Kb were cloned (3). The cloning capacity of BAC was smaller than that of YAC, but cloned DNA fragments can be maintained stably in bacteria without any rearrangements, unlike the case with YAC. The BAC system was employed to make a genomic library in the international genome project for genome mapping and sequencing (12–15). The entire genomic BAC libraries of human, mouse, rat, and others were established, and these BAC clones were made available. For practical usefulness in the study of gene function and regulation, efficient modification methods for BAC vectors have to be developed. Recently, Yang et al. developed the "RecA Method" (16) and Zhang et al. developed the "RedET Method" (17) (*see* Note 5). These methods are based on the homologous recombination. An important platform of the BAC library (cloned fragment and DNA sequence) and manipulation technologies for BACs have been established. Rat genomic BAC clones are readily available, and BAC vectors can be constructed by using recombinant BAC technologies. We describe below the protocols for making transgenic YAC and BAC rats, with special attention paid to problems specific to rats.

## 2. Materials

### 2.1. Preparation of YAC/BAC DNA for Microinjection

1. Low melting temperature agarose: SeaPlaque GTG Agarose (GTG, FMC BioProducts, Philadelphia, PA).
2. TBE buffer (5× stock solution): 54 g Tris base, 24.5 g Boric acid, 20 mL EDTA(0.5 M stock solution; pH 8.0).
3. Pulsed-field gel electrophoresis (PFGE) System: CHEF-DRII (Bio-Rad Laboratories, Hercules, CA).
4. EtBr solution (stored at 4°C; 10 mg/mL; Wako Pure Chemical Industries, Ltd., Osaka, Japan).
5. Microinjection buffer (for YACs/BACs): 10 mM Tris–HCl, pH 7.5, 0.25 mM EDTA, and 100 mM NaCl supplemented with 30 μM spermidine and 70 μM spermin.
6. Lambda ladder marker (stored at 4°C).
7. β-agarose: β-agarase (stored at –20°C; New England Biolabs, Ipswich, MA).
8. Heat block (or incubator): 50, 68, and 42°C.
9. Disposable pipette tips: Genomic tips, 1000G and 180G (Molecular BioProducts, San Diego, CA).
10. Microrefrigerated centrifuge.
11. Filter concentration unit: Ultrafree-MC filter unit (30,000; Millipore, Bedford, MA).

### 2.2. Microinjection of YACs/BACs DNA

1. Image-erected microscope: TMD300 (Nikon, Tokyo, Japan; *see* Fig. 7.1).
2. Micromanipulator: MO-102, MO-130 (Narishige, Tokyo, Japan; *see* Fig. 7.1).
3. Glass capillary tube: G1 (Narishige).

## 3. Methods

Large intact YACs/BACs can be prepared efficiently from yeast or bacteria cells molded in the agarose plug and can be purified for microinjection by pulsed-field electrophoresis separation. There are many good protocols to prepare YACs (7, 18–20)/ BACs (21, 22), but in this section we focus on the procedures to purify a YAC/BAC fragment from an agarose plug, and we describe DNA microinjection into the fragile pronuclear ova of the rat.

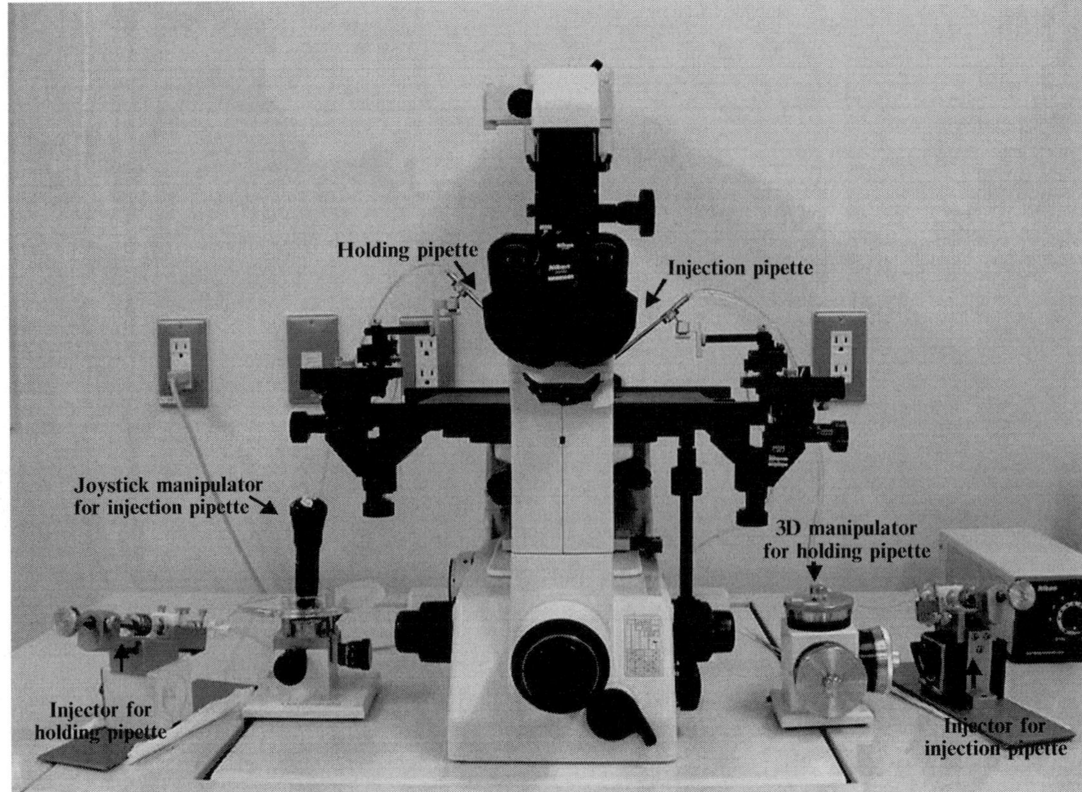


Fig. 7.1. Micromanipulator.

### 3.1. Preparation of YAC/BAC DNA for Microinjection

#### 3.1.1. Band Isolation of YACs/BACs from Agarose Plugs by PEGE (see Fig. 7.2)

1. Melt 3.5 g of Seaplaque GTG agarose in 350 mL of 0.5×TBE buffer by autoclave sterilization. Cool the agarose solution in a water bath (50°C) after autoclaving and keep it at that temperature until use.
2. Also sterilize 3 L of 0.5×TBE electrode buffer in the same way, and cool it in the ice-water bath until use.
3. Tape up the teeth of a 45-well comb except for each end and the center. To load a large number of the plugs, double-ply combs are used to make a preparative lane.
4. Set the gel platform horizontally in a gel stand.
5. Set the comb so that it is not attached to the base of the platform and pour the agarose solution (maintained at 50°C) into it. Leave a little of the solution to cover the plugs and keep it at 50°C. Keep the platform at room temperature for over 20 min until the gel hardens.

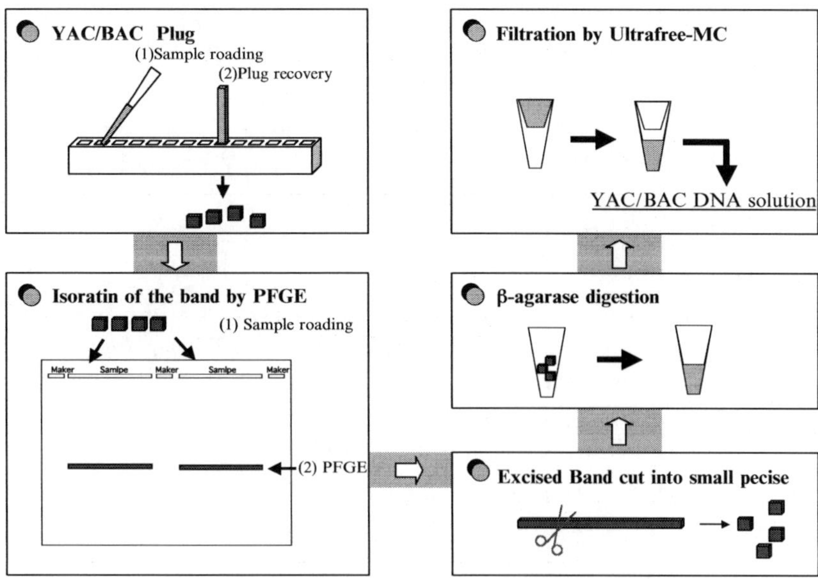


Fig. 7.2. Scheme of purification of YACs/BACs.

6. Pour the electrode buffer into a PFGE bath, then switch on the refrigerator and the circulation pump.
7. After the agarose gel mentioned at step 5 has hardened, pull out the comb to make a preparative lane. To do this successfully, pour a little water on the gel surface.
8. Load the plugs into the preparative lane. Lambda ladder marker plugs are also set in both ends and in the center of the plug lane.
9. Cover the gaps between the preparative lane wall and plugs with agarose solution left mentioned above at step 5 and keep it till the gel hardens.
10. Remove the gel platform from the gel stand and set it into the frame of the PFGE bath.
11. Switch on the drive module and set it to the optimal conditions for separating YAC/BAC DNA (such as 6 V/cm voltage, 40 s switching time, and 20 h of operation time for YAC DNA). The DNA can be concentrated in a 4% agarose plug according to Schedl's method (7).

*3.1.2. Preparation of Isolated YACs/BACs from Agarose Gel (see Fig. 7.2)*

1. After PFGE is finished, switch off the drive modules, remove the gel platform from the equipment, and transfer the gel onto a glass board. Cut off the marker lanes with some of the YAC/BAC lane on both sides and at the center of the gel.

2. Put the agarose gels of the marker lane into EtBr solution (final conc.: 0.1 mg/mL) for DNA staining, with gentle shaking at 25 times/min for 30 min.
3. Measure the length from the top of the gel to the position of YAC/BAC DNA precisely under UV irradiation.
4. Cut off a 2 mm width of the gel at the position of the YAC/BAC DNA according to its position on the marker lanes mentioned above.
5. Cut the separated gel into 1 cm pieces and wash them three times in a 50 mL conical tube with 30 mL of YAC/BAC microinjection buffer by inverting the tube several times. Keep it at 4°C for 30 min and remove the buffer. Repeat washing the gels three times using the same procedure.
6. Place about 0.4 g of each gel piece into a 1.5 mL Eppendorf tube, then spin down the pieces at the bottom and remove the remaining buffer with a micropipette.
7. Melt the agarose gel at 68°C for 10 min in a heat block.
8. Prewarm the Eppendorf tube containing 4 μL of β-agarase solution (1 U/μL) and 20 μL of agarase buffer (1×) at 42°C.
9. Mix the solution once gently by pipetting with Genomic tip 1000 after the agarose gel mentioned at step 7 is melted. Transfer the solution to the β-agarase solution tube mentioned at step 8.
10. Incubate the tube for 2 h at 42°C to digest the agarose.
11. Centrifuge the tube at 18,70 ×$g$ (15,000 rpm) at room temperature for 15 min to sink the undigested matter.
12. Collect each tube's supernatant and transfer it onto an Ultrafree-MC filter unit (30,000) with Genomic tip 180G. Be careful not to transfer undigested matter. Centrifuge the supernatant for 5 min at 4,000×$g$ (7,000 rpm) at room temperature. Be careful to keep about 80 μL of applied volume on the filter, and take care to avoid overcentrifugation.
13. Keep the filter unit for 1 h at room temperature, and recover the upper solution on the filter unit and transfer it in a 1.5 mL Eppendorf tube with Genomic tip 180.
14. Check the integrity of recovered YAC/BAC DNA by PFGE.
15. Calculate the DNA concentration by comparing the EtBr-stain strength of the DNA solution with that of standard DNA solution.

### 3.2. Microinjection of YAC/BAC DNA into Rat Ova

The procedures to microinject YACs/BACs into rat ova are basically the same as those used for mouse ova (23–25). However, rat ovum has an elastic pronucleus membrane and is very fragile against DNA microinjection procedures (26, 27). At several points, the procedures need to be modified and careful attention must be paid (*see* Note 1–5).

*3.2.1. Making a Microinjection Pipette for Rat Ova (see Fig. 7.3)*

1. Pull the capillary tube (Glass capillary tube G1, Narishige) and make a glass pipette with a taper of 1.3–1.5 cm length.
2. To prevent the backflow of HF solution, the pipette must be kept pressurized by using the injection syringe of 25 mL.
   (a) Wash with distilled water.
   (b) Soak the tip of the pipette into HF solution only momentarily.
   (c) Immediately wash the inside of the pipette with distilled water, pressurize it, and confirm that small bubbles emerge from the tip.
3. Bend the pipette head so that the pipette is horizontal to the micromanipulator.
   (a) YAC/BAC DNA is so sticky that the tip of the pipette is often stuffed with it. The proper hole size of the pipette is very important for favorable microinjection procedures. To determine the correct size of the hole for microinjection, confirm the small bubbles emerge from the tip of the pipette when pressurized.
   (b) Another method of opening the tip of the pipette was reported by Filipiak et al. (28). They opened a hole by mechanical destruction to prepare the holding pipette.

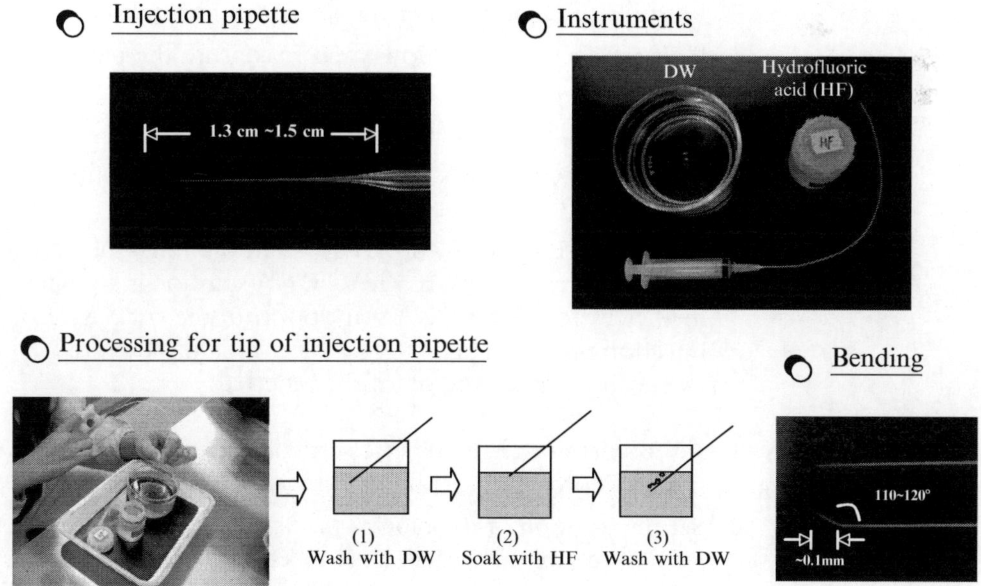


Fig. 7.3. Making the microinjection pipette for rat ova.

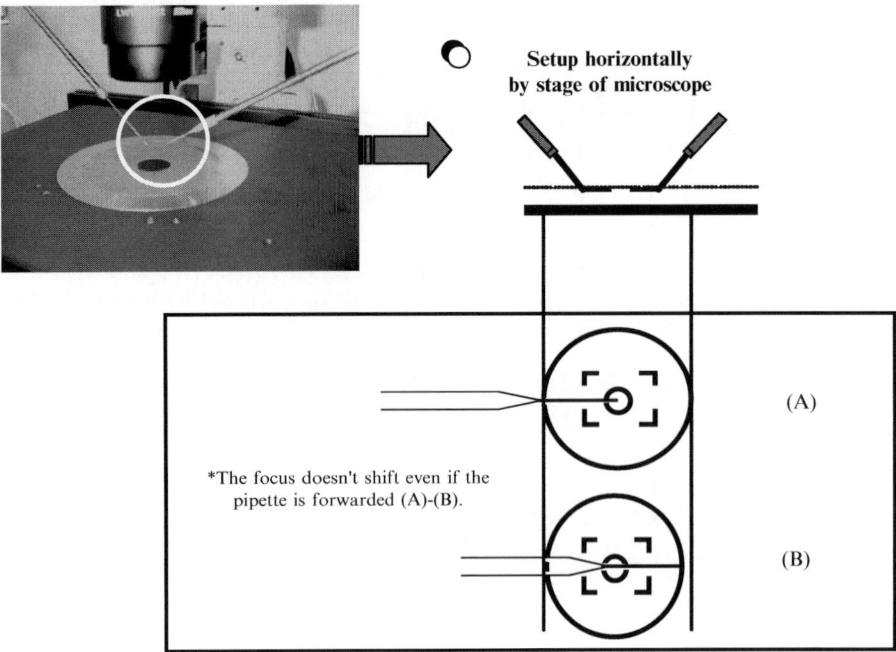


Fig. 7.4. Setting up the injection pipette and holding pipette horizontally.

*3.2.2. Setting Up the Injection Pipette and Holding the Pipette Horizontally (see Fig. 7.4)*

1. Set the pipette angle almost horizontally as judged by the naked eye.
2. Adjust the focus on the tip of the pipette with a magnifying power of 40.
3. Forward the pipette using the micromanipulator.
4. Shift the focus up and down and make sure the pipette is still horizontal.
5. Adjust the pipette angle finely.
6. Repeat procedures (2) to (5) until the focus at points (A) and (B) in Fig. 7.4 is suitable.
7. Regarding the filling of DNA solution into the injection pipette: Because the YAC/BAC DNA solution is so sticky, it is inefficient to fill the DNA solution from the other side of the injection pipette. The DNA solution should therefore be filled by sucking it from the tip of the pipette.

*3.2.3. Improving Pipette Operations (see Fig. 7.5)*

1. Always pressurize the pipette settled in the injector to prevent the backflow of the medium.
2. Set the position of the pipette in such a manner that it has the maximum circumference. This position is very important for the easy introduction of the pipette into a pronucleus (PN) because a PN is soft and its shape is easily changed.

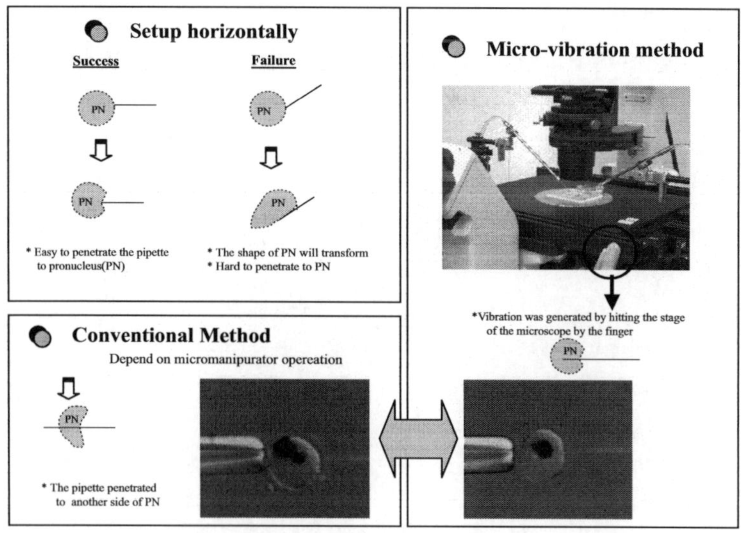


Fig. 7.5. Improvement for pipette operations.

3. Hold the tip of the pipette to a pronucleus membrane and beat the microscope stage lightly to make the tip of the pipette vibrate and thus complete a quick and easy injection. This is called the "microvibration method". Piercing PN quickly with the pipette is important in the "conventional method". Although, it is hard to control not to skewer the ova and the pipette gradually loses its sharpness with the repetition of piercing. With microvibration, the survival rate and handling efficiency can be dramatically improved. The leakage of DNA solution to the cytoplasm can be monitored by injection with black ink.

## 4. Notes

1. *Dilution of purified YACs preparation*
   Our initial study pointed out that the purified YAC preparation contained materials digested by β-agarase. Dilution of the YAC preparation at least three times is recommended, using the buffer for microinjection as shown in Table 7.1.
2. *Efficiency of transgenic rat production with different YAC vectors*

   When a large DNA fragment such as YAC is used to produce transgenic rats, the procedures for purifying the YAC preparation often affect the production efficiency. Transgenic rats were made with YAC preparations purified by different

**Table 7.1**
**Analysis of influent carried over composition from YAC DNA purification process**

| Takahashi and Ueda | Dilution | No. (%) of embryos Injected | Survived | Transferred | Pregant/ (%) recipient | No. (%) of pups transferred |
|---|---|---|---|---|---|---|
| Sham injection | – | 158 | 144 (91.1) | 142 | 6/6 (100) | 73 (51.4) |
| Injection buffer for plasmid[a] | – | 129 | 122 (94.6) | 122 | 6/6 (100) | 64 (52.5) |
| Injection buffer for YAC/ BAC with β-Agarase[b] | 1<br>1/3 | 81<br>81 | 76 (93.8)<br>73 (90.1) | 76<br>73 | 4/4 (100)<br>4/4 (100) | 7 (9.2)<br>20 (27.4) |

[a]10 mM Tris–HCl, pH 7.5; 0.1 mM EDTA-2Na
[b]10 mM Tris–HCl, pH 7.5; 0.25 mM EDTA-2Na; 100 mM NaCl

purification protocols, and the production efficiency was examined. The efficiency was improved by using DNA preparations of YACs with an amplification element (Table 7.2). While a YAC vector with a conditionally amplifiable centromere (29) is a powerful tool to prepare concentrated YACs sufficient for microinjection, the thymidine kinase gene derived from the herpes simplex virus in the YAC vector's amplification element causes sterility in male founders (30), who then cannot transmit their transgene to the next generation. In addition, an endonuclease treatment to remove the amplification element also decreased the efficiency of transgenic rat production. These paradoxical results have not been resolved. Because preparation of a sufficient concentration of YACs for the production of transgenic rats would be difficult without YAC amplification, the transgenesis by YAC vectors has been replaced by BAC vectors that can be easily purified.

3. *Integrity of BAC DNA in the chromosomes of transgenic rats* When a large DNA fragment such as a BAC is purified to make a transgenic rat, the deletion of the introduced DNA fragment in a chromosome of the transgenic rat is a problem that cannot be avoided. Transgenic rats were made with BAC preparations purified by different purification protocols, and the integrity of the BACs in the rat chromosomes was examined (11). A BAC vector can be easily isolated by the same procedures used for conventional plasmid vectors. Results of our initial study on the production of transgenic rats by three different procedures in plasmid vectors are presented in Table 7.3. The three procedures are as follows: linearization by Not I digestion and CsCl gradient centrifugation (Prep.

**Table 7.2**
**Effects of YAC DNA[a] preparation on the production of transgenic rats**

| Amplification element | Restriction enzyme digestion | No. (%) of embryos | | No. (%) of pups transferred | No. (%) of Tg injected | No. (%) of both vector ends present (/Tg) | Overall[b] |
|---|---|---|---|---|---|---|---|
| | | Injected | Transferred | | | | |
| + | – | 551 | 456 | 142 (31.1) | 25 (4.5) | 20 (80.0) | 3.6 |
| + | + | 797 | 713 | 169 (23.7) | 12 (1.5) | 7 (58.3) | 0.9 |
| – | – | 776 | 698 | 154 (22.1) | 3 (0.4) | 2 (66.7) | 0.3 |

[a]Each preparation was the total for the three YAC clones. All YAC DNAs have a 210 kb chain length and are fixed at 1 ng/mL with YAC buffer (10 mM Tris–HClpH 7.5, 0.25 mM EDTA, 100 mM NaCl, 30 μM spermidine, and 70 μM spermin)

[b]Overall ratio of transgenic rats with both vector ends present per microinjected ova

**Table 7.3**
**Effects of BAC DNA[a] preparation on production of transgenic rats.**

| Preparation method | No. (%) of embryos | | | No. (%) of pups (/transferred) | No. (%) of Tg (/transferred) | No. (%) of both vector ends present (/Tg) | Overall[b] |
|---|---|---|---|---|---|---|---|
| | Injected | survived | Transferred | | | | |
| Prep.1 | 206 | 195 (94.7) | 193 | 73 (37.8) | 10 (5.2) | 0 (0.0) | 0.0 |
| Prep.2 | 196 | 177 (90.3) | 175 | 48 (27.4) | 7 (4.0) | 0 (0.0) | 0.0 |
| Prep.3 | 205 | 187 (91.2) | 187 | 47 (25.1) | 12 (6.4) | 4 (33.3) | 2.0 |

[a]BAC DNAs used in this study were prepared with 100-kb chain lengths and were diluted to 5 ng/μL with YAC buffer (Table 7.1)
[b]The overall ratio of transgenic rats with both vector ends present per microinjected ova

1); CsCl gradient centrifugation, linearization by Not I digestion, conventional gel electrophoresis with 0.8% low-melting agarase gel, and β-agarase digestion (Prep. 2); and the same procedures as in Prep. 2 except that PFGE was used for electrophoresis of BAC DNA (Prep. 3). The integration efficiency was similar among the three preparations (5.2, 4.0, and 6.4% respectively), but the integration of the intact transegene was observed only in Prep. 3 (2.0%). It was also reported that the protocol for DNA isolation by PFGE was also superior with regard to the integrity of integrated DNA in BAC transgenic mice (22). The purification procedures for BACs based on those for YACs having PFGE and β-agarose digestion are recommended.

4. *Position-independent expression in transgenic rats with YACs/BACs*

   Position-independent expression with 210 kb human YACs containing the alpha-lactalbumin (α-LA) gene was reported (9). Three female transgenic rat lines were established; they expressed the α-LA gene at similar levels and showed specific expression in the mammary gland. The concentration of human α-LA in their milk was similar to that of human milk (Fig. 7.6).

5. *BAC to the future*

   For BAC vectors to serve as a practical tool for studying gene function and regulation, efficient modification methods for BAC vectors are necessary. Two methods have been developed recently. One is a method with a temperature-sensitive pSV1. RecA shuttle vector carrying the *E.coli* RecA gene developed by Yang et al. (generally called the "RecA Method"). The shuttle vector introduced in the recombination-deficient BAC host of *E.coli* restores the recombination at a desired position in the inserted genomic DNA of BAC (31). This system has been successfully adopted on the NINDS GENSAT BAC Transgenic Project (http://www.ncbi.nlm.nih.gov/projects/gensat/) to make a series of murine BAC transgenic mice with accurate expression of marker genes in their central neurons. The other is the BAC modification protocols that allow recombination at a specific site of BAC DNA regardless of the presence of a suitable restriction site (17). This method is based on homologous recombination mediated by recE (5′-3′ exonuclease) and recT (ssDNA-binding protein), and is generally called the "RedET-cloning method" (32, 33). An important platform of the BAC library (cloned fragment and DNA sequence) and manipulation technologies for BACs have been established. Rat genomic BAC clones are available easily, and BAC vectors can be constructed by the recombinant BAC technologies. These technologies have advanced the applica-

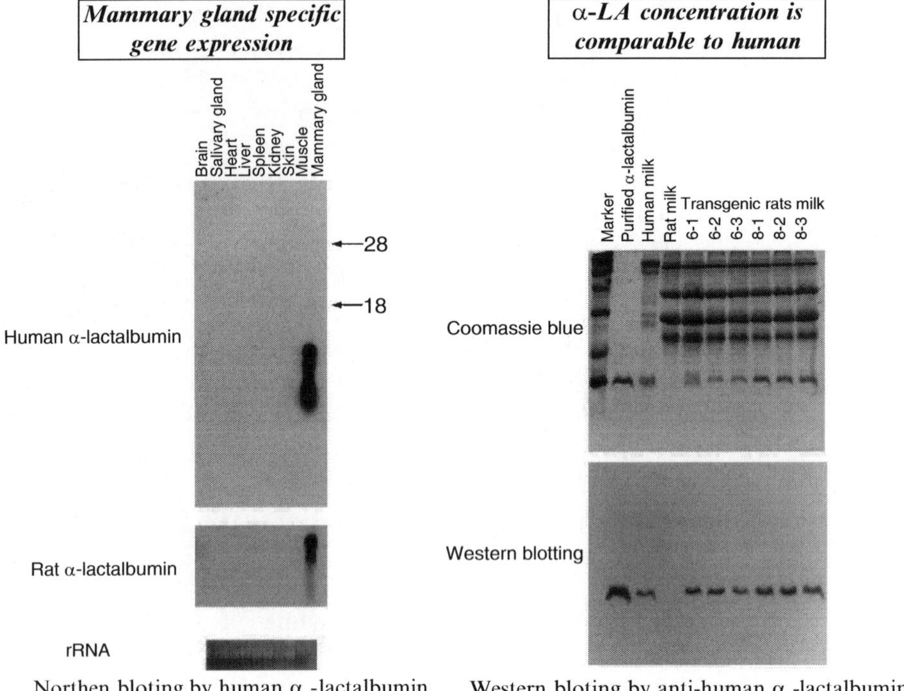


Fig. 7.6. Position independent expression in transgenic rats (YAC).

tion of transgenic rats to the study of life sciences. While the mouse is a suitable species for DNA manipulation such as transgenic and gene targeting, the rat is restricted in gene targeting because it has no reliable ES cells (34). Transgenic rats carrying BAC can expect to have the same expression profiles as the endogenous genes, therefore, relative rates of transcription between transgene and endogenous gene can be compared (We call this the "Semi-knock-in" strategy). The rats have been used in various biological studies for a long time, and many important animal models of various diseases have been developed. BAC transgenic rats are very effective for making various animal disease models and are expected to be applied widely to the study of life sciences.

## Acknowledgments

We would like to thank Dr. Yoshihiro Fujiwara and Dr. Naomi Kashiwazaki for their technical advice in molecular biology and for their helpful discussion.

## References


1. Palmiter RD, Brinster RL (1986) Germ-line transformation of mice. Annu Rev Genet 20:465–499
2. Burke DT, Carle GF, Olson MV (1987) Cloning of large segments of exogenous DNA into yeast by means of artificial chromosome vectors. Science 236:806–812
3. Shizuya H, Birren B, Kim UJ, Mancino V, Slepak T, Tachiiri Y, Simon M (1992) Cloning and stable maintenance of 300-kilobase-pair fragments of human DNA in Escherichia coli using an F-factor-based vector. Proc Natl Acad Sci U S A 89(18):8794–8797
4. Gaensler KM, Kitamura M, Kan YW (1993) Germ-line transmission and developmental regulation of a 150-kb yeast artificial chromosome containing the human beta-globin locus in transgenic mice. Proc Natl Acad Sci U S A 90(23):11381–11385
5. Peterson KR, Clegg CH, Huxley C, Josephson BM, Haugen HS, Furukawa T, Stamatoyannopoulos G (1993) Transgenic mice containing a 248-kb yeast artificial chromosome carrying the human beta-globin locus display proper developmental control of human globin genes. Proc Natl Acad Sci U S A 90(16):7593–7597
6. Schedl A, Montoliu L, Kelsey G, Schüz G (1993) A yeast artificial chromosome covering the tyrosinase gene confers copy number-dependent expression in transgenic mice. Nature 362(6417):258–621
7. Schedl A, Grimes B, Montoliu L (1996) YAC transfer by microinjection. Methods Mol Biol 54:293–306
8. Fujiwara Y, Miwa M, Nogami M, Okumura K, Nobori T, Suzuki T, Ueda M (1997) Genomic organization and chromosomal localization of the human casein gene family. Hum Genet 99(3):368–373
9. Fujiwara Y, Miwa M, Takahashi R, Hirabayashi M, Suzuki T, Ueda M (1997) Position-independent and high-level expression of human alpha-lactalbumin in the milk of transgenic rats carrying a 210-kb YAC DNA. Mol Reprod Dev 47(2):157–163
10. Fujiwara Y, Miwa M, Takahashi R, Kodaira K, Hirabayashi M, Suzuki T, Ueda M (1999) High-level expressing YAC vector for transgenic animal bioreactors. Mol Reprod Dev 52(4):414–420
11. Takahashi R, Ito K, Fujiwara Y, Kodaira K, Kodaira K, Hirabayashi M, Ueda M (2000) Generation of transgenic rats with YACs and BACs: preparation procedures and integrity of microinjected DNA. Exp Anim 49(3): 229–233
12. Mozo T, Dewar K, Dunn P, Ecker JR, Fischer S, Kloska S, Lehrach H, Marra M, Martienssen R, Meier-Ewert S, Altmann T (1999) A complete BAC-based physical map of the Arabidopsis thaliana genome. Nat Genet 22(3):271–275
13. Hoskins RA, Nelson CR, Berman BP, Laverty TR, George RA, Ciesiolka L, Naeemuddin M, Arenson AD, Durbin J, David RG, Tabor PE, Bailey MR, DeShazo DR, Catanese J, Mammoser A, Osoegawa K, de Jong PJ, Celniker SE, Gibbs RA, Rubin GM, Scherer SE (2000) A BAC-based physical map of the major autosomes of Drosophila melanogaster. Science 287(5461):2271–2274
14. Osoegawa K, Tateno M, Woon PY, Frengen E, Mammoser AG, Catanese JJ, Hayashizaki Y, de Jong PJ (2000) Bacterial artificial chromosome libraries for mouse sequencing and functional analysis. Genome Res 10(1): 116–128
15. McPherson JD, Marra M, Hillier L, Waterston RH, Chinwalla A, Wallis J, Sekhon M, Wylie K, Mardis ER, Wilson RK, Fulton R, Kucaba TA, Wagner-McPherson C, Barbazuk WB, Gregory SG, Humphray SJ, French L, Evans RS, Bethel G, Whittaker A, Holden JL, McCann OT, Dunham A, Soderlund C, Scott CE, Bentley DR, Schuler G, Chen HC, Jang W, Green ED, Idol JR, Maduro VV, Montgomery KT, Lee E, Miller A, Emerling S, Kucherlapati, Gibbs R, Scherer S, Gorrell JH, Sodergren E, Clerc-Blankenburg K, Tabor P, Naylor S, Garcia D, de Jong PJ, Catanese JJ, Nowak N, Osoegawa K, Qin S, Rowen L, Madan A, Dors M, Hood L, Trask B, Friedman C, Massa H, Cheung VG, Kirsch IR, Reid T, Yonescu R, Weissenbach J, Bruls T, Heilig R, Branscomb E, Olsen A, Doggett N, Cheng JF, Hawkins T, Myers RM, Shang J, Ramirez L, Schmutz J, Velasquez O, Dixon K, Stone NE, Cox DR, Haussler D, Kent WJ, Furey T, Rogic S, Kennedy S, Jones S, Rosenthal A, Wen G, Schilhabel M, Gloeckner G, Nyakatura G, Siebert R, Schlegelberger B, Korenberg J, Chen XN, Fujiyama A, Hattori M, Toyoda A, Yada T, Park HS, Sakaki Y, Shimizu N, Asakawa S, Kawasaki K, Sasaki T, Shintani A, Shimizu A, Shibuya K, Kudoh J, Minoshima S, Ramser J, Seranski P, Hoff C, Poustka A, Reinhardt R, Lehrach H; International Human Genome Mapping Consortium (2001) A physical map of the human genome. Nature 409(6822):934–941

16. Yang XW, Model P, Heintz N (1997) Homologous recombination based modification in Escherichia coli and germline transmission in transgenic mice of a bacterial artificial chromosome. Nat Biotechnol 15(9):859–865
17. Zhang Y, Buchholz F, Muyrers JP, Stewart AF (1998) A new logic for DNA engineering using recombination in Escherichia coli. Nat Genet 20(2):123–128
18. Shen S, Harmar A, Hastie N (2006) Modification and amplification of yeast artificial chromosomes. Methods Mol Biol 349:67–74
19. MacKenzie A (2006) Production of yeast artificial chromosome transgenic mice by pronuclear injection of one-cell embryos. Methods Mol Biol 349:139–150
20. Peterson KR (2007) Preparation of intact yeast artificial chromosome DNA for transgenesis of mice. Nat Protoc 2(11): 3009–3015
21. Marshall VM, Allison J, Templeton T, Foote SJ (2004) Generation of BAC transgenic mice. Methods Mol Biol 256:159–182
22. Abe K, Hazama M, Katoh H, Yamamura K, Suzuki M (2004) Establishment of an efficient BAC transgenesis protocol and its application to functional characterization of the mouse Brachyury locus. Exp Anim 53(4):311–320
23. Gordon JW, Ruddle FH (1983) Gene transfer into mouse embryos: production of transgenic mice by pronuclear injection. Methods Enzymol 101:411–433
24. Brinster RL, Chen HY, Trumbauer ME, Yagle MK, Palmiter RD (1985) Factors affecting the efficiency of introducing foreign DNA into mice by microinjecting eggs. Proc Natl Acad Sci U S A 82(13):4438–4442
25. Ittner LM, Goz J (2007) Pronuclear injection for the production of transgenic mice. Nat Protoc 2(5):1206–1215
26. Hochi S, Ninomiya T, Honma M, Yuki A (1990) Successful production of transgenic rats. Animal Biotechnology 1:175–184
27. Heideman J (1991) Transgenic rats: a discussion.Biotechnology 16:325–332
28. Filipiak WE, Saunders TL (2006) Advances in transgenic rat production. Transgenic Res 15(6):673–686
29. Smith DR, Smyth AP, Strauss WM, Moir DT (1993) Incorporation of copy-number control elements into yeast artificial chromosomes by targeted homologous recombination. Mamm Genome 4(3):141–147
30. al-Shawi R, Burke J, Wallace H, Jones C, Harrison S, Buxton D, Maley S, Chandley A, Bishop JO (1991) The herpes simplex virus type 1 thymidine kinase is expressed in the testes of transgenic mice under the control of a cryptic promoter. Mol Cell Biol 11(8): 4207–4216
31. Yang Y, Sharan SK (2003) A simple two-step, 'hit and fix' method to generate subtle mutations in BACs using short denatured PCR fragments. Nucleic Acids Res 31(15):e80
32. Muyrers JP, Zhang Y, Testa G, Stewart AF (1999) Rapid modification of bacterial artificial chromosomes by ET-recombination. Nucleic Acids Res 27(6):1555–1557
33. Zhang Y, Muyrers JP, Testa G, Stewart AF (2000) DNA cloning by homologous recombination in Escherichia coli. Nat Biotechnol 18(12):1314–1317
34. Tesson L, Cozzi J, Méoret S, Rémy S, Usal C, Fraichard A, Anegon I (2005) Transgenic modifications of the rat genome. Transgenic Res 14(5):531–546

# Chapter 8

# The Use of Lentiviral Vectors to Obtain Transgenic Rats

**Séverine Remy, Tuan Huy Nguyen, Séverine Ménoret, Laurent Tesson, Claire Usal, and Ignacio Anegon**

## Abstract

Lentiviral vectors are now well recognized as good vehicles for gene delivery. This is because they can efficiently transduce both dividing and post-mitotic cells, and stably integrate into the host genome allowing for long-term expression of the transgene. Their potential utility for the generation of transgenic animals has been recognized as an attractive and promising alternative to the conventional DNA-microinjection method which lacks efficiency. The initial success of lentiviral transgenesis in mice considerably broadened its use in other species, in which classical transgenic techniques are difficult, such as in the rat.

In this chapter, we describe detailed procedures for both the production of human immunodeficiency virus-1 (HIV-1)-derived lentiviral vectors and for the generation of transgenic rats by injection of these vectors into the perivitelline space of fertilized one-cell eggs.

**Key words:** Lentiviral vectors, Transgenic rat, Lentiviral transgenesis

## 1. Introduction

The growing interest in developing new tools and approaches to modify the genome of animals has led to the emergence of transgenic technologies. The first transgenic mice were generated in 1976 by Jaenish, after the infection of mouse embryos with Moloney leukemia retroviruses (1). Despite the successful integration and germ line transmission of exogenous DNA of retroviral origin, the absence of transgene expression in the founder animals due to retroviral DNA methylation considerably hampered the use of oncoretroviruses in animal transgenesis. Subsequently, alternative methods were developed to generate transgenic animals, with particular focus on the pronuclear microinjection of DNA. Since its development by Gordon et al. in 1980 (2), this is the

I. Anegon (ed.), *Rat Genomics: Methods and Protocols*, Methods in Molecular Biology, vol. 597
DOI 10.1007/978-1-60327-389-3_8, 

most commonly used method to generate small (2–4) and large transgenic animals (5). Unfortunately, the lack of efficiency of this method in certain species other than mice, associated with its high cost in larger animals, has led scientists to develop new strategies to overcome these problems. The idea of using lentiviral vectors to generate transgenic animals emerged several years ago, as a result of their ability to efficiently integrate into the host genome of both dividing and non-dividing cells (6). Lois et al. (7) and Pfeifer's group (8) were the first to generate transgenic mice using lentiviral vectors with a very high efficiency (80% of founder mice). Interestingly, transgene expression in these animals was not only observed in the founder animals but also in the offspring, and thus with lentiviral vectors they did not observe the phenomenon of transgene expression-silencing, previously described with traditional oncoretroviral vectors. These encouraging data subsequently led the scientific community to focus on the use of lentiviral vectors as a promising tool to generate other transgenic animal species, in particular rats (9, 10) and also non-human primates (11). Indeed, among the different gene transfer vectors, lentiviral vectors have unique properties: (i) they allow for efficient gene transfer into non-dividing cells, also including cells resting in a G0 quiescent stage (12), (ii) they integrate the new genetic material into the host cell genome and preferentially into transcriptionally-active regions of the genome, and (iii) they have a large cloning capacity, 8 kb (although up to 18 kb has been reported (13)), allowing for insertion of additional genetic elements, such as WPRE or a locus control region from the human beta-globin gene, to improve transgene expression, and avoid gene silencing, respectively (14, 15). Notably, although viral titer is influenced by insert type, lentiviral vectors with an insert size of less than 5 kb have been shown to have the best titer (13).

In this chapter, we describe the production of HIV-1-derived recombinant lentivirus pseudotyped with the vesicular stomatitis virus (VSV) G envelope protein using the so-called second generation packaging system. In the second generation packaging system, the *gag*, *pol*, *tat*, and *rev* genes are expressed in one plasmid (16). The described method is therefore suitable for production of any lentiviral vector. The VSV G envelope protein is routinely used, because it is highly stable, allowing for the concentration and cryopreservation of the recombinant lentivirus, and because it confers a large tropism to the virus. In our experience, there is no generation of replication-competent lentivirus (RCL) using the second generation packaging system (a single plasmid coding for gag-pol, rev, and tat HIV-1 proteins) and the plasmids used in the described method.

In the second part of this chapter, we describe the generation of transgenic rats by microinjection of lentiviral vectors into one-cell rat embryos.

## 2. Materials

### 2.1. Materials for Lentiviral Vector Production

#### 2.1.1. Cell Culture

1. HEK 293 T (293 T/17, ATCC, CRL-11268) and Hela (ATCC, CCL-2) cells.
2. DMEM culture medium:
   Dulbecco's Modified Eagle Medium (DMEM), (4,500 mg/L D-Glucose), 110 mg/L Sodium Pyruvate (Invitrogen) supplemented with 10% fetal bovine serum, 2 mM L-glutamine (Invitrogen), and Penicillin/streptomycin solution (100 IU/ml penicillin G sodium and 100 μg/ml streptomycin sulfate, Invitrogen).
3. Solution of trypsin (0.25%) and ethylenediamine tetraacetic acid (EDTA) (1 mM) (Invitrogen).
4. D-PBS pH 7.4 (Invitrogen).
5. Trypan blue, final concentration of 0.08% in D-PBS (Invitrogen).
6. Malassez cell counter.
7. EPISERF serum-free medium (Invitrogen).
8. 10 cm tissue culture plates and 50 ml sterile conical centrifuge tubes.
9. Humidified incubators at 37°C and 5% $CO_2$.
10. Cell culture vacuum pump.

#### 2.1.2. Transfection

1. Plasmids were available from Didier Trono's Laboratory (Lab of virology and genetics, EPFL, Lausanne, Switzerland, http://tronolab.epfl.ch/):
   - psPAX2 (encoding HIV-1 Gag, Pol, Tat and Rev proteins)
   - pMD2G (encoding the VSV G envelope protein)
   - pRRLSIN.cPPT.PGK-GFP.WPRE (lentiviral vector)

Plasmids (in TE Buffer, pH 8.0) need to be cleaned, i.e., purified using a column-based maxiprep (Macherey-Nagel, Invitrogen, Sigma). The 260/280 ratio should be 1.8–2.0. Store at –20°C.

2. Buffered water (H20-Hepes):
   Add 125 μl of 1 M Hepes, pH 7.2–7.5 (Invitrogen) to 50 ml of sterile $H_2O$, pH 7. Store at +4°C.
3. 2X HeBS solution:
   Dissolve 16.4 g of NaCl (Sigma, MW 58.44, ref. S7653) and 0.213 g of $Na_2HPO_4$, anhydrous (Sigma, MW 142, ref. S7907) (1.5 mM final concentration) in 50 ml of 1 M Hepes solution pH 7.2–7.5 (Invitrogen) (0.05 M final) then add 800 ml of $H_2O$. Adjust the pH to 7.05 with 5 N NaOH, then add water to a final volume of 1 L. Filter the solution through a 0.22 μm nitrocellulose filter. Store at 4°C for up to 1 month, or at –20°C for longer periods.

4. $CaCl_2$ 0.5 M solution:
   Dissolve 36.7 g of $CaCl_2$, $2H_2O$ (Sigma, MW 147, ref. C5080) in 500 ml of $H_2O$. Filter the solution through a 0.22 μm nitrocellulose filter. Store at 4°C for up to 1 month, or at –20°C for longer periods.
5. 50 ml sterile conical centrifuge tubes.
6. Vortex.

*2.1.3. Vector Collection and Concentration*

1. 50 ml sterile conical centrifuge tubes.
2. Refrigerated table-top centrifuge.
3. Stericup-HV, 150 ml, 0.45 μm, PDVF (Millipore, Ref. SCHVU01RE).
4. Vacuum pump for cell culture.
5. Ultracentrifuge with SW28 rotor (swinging bucket rotor) and bucket (Beckman-Coulter).
6. 30 ml-Beckman Konical™ open-top tubes (25 × 89 mm) and SW28 tube adaptaters (Beckman-Coulter, Ref. 358126).
7. 20% sucrose:
   Dissolve 40 g of sucrose (Sigma, Ref. S7903) in 200 ml of H20. Filter through a 0.22-μm nitrocellulose filter. Store at +4°C.
8. Sterile 1.5-ml microcentrifuge tubes.
9. Chromatography purification kit:
   Fast-Trap Lentivirus Purification and Concentration kit (Millipore, Ref. FTLV00003, 3 purifications) or vivapure LentSELECT 40 kit (Sartorius stedim, ref. VS-LVPQ040, 4 purifications).

*2.1.4. Vector Titration*

1. 6-well cell culture plates available from many manufacturers.
2. Trypan blue, final concentration 0.08% in D-PBS (Invitrogen).
3. Malassez cell counter.
4. Genomic DNA purification (DNeasy Qiagen kit).
5. Filter tips.
6. SYBR Green PCR 2X master mix, which includes SYBR Green dye, Taq Polymerase, ROX, and dNTP, available from many manufacturers (Eurogentec, Invitrogen, etc … ).
7. Primers:
   GAG-F: 5′-GGAGCTAGAACGATTCGCAGTTA-3′ (vector amplification)
   GAG-R: 5′-GGTTGTAGCTGTCCCAGTATTTGTC-3′ (vector amplification)
   HB2-F: 5′-TCCGTGTGGATCGGCGGCTCCA-3′ (beta-actin amplification)

HB2-R: 5′-CTGCTTGCTGATCCACATCTG-3′ (beta-actin amplification)

8. 96-well optical reaction plate and optical adhesive covers for qPCR (Applied Biosystems).
9. Real-time qPCR thermocycler (ABI7700 Sequence detector).
10. HIV p24 ELISA (NEN Life Science Products, ref. NEK050B).

#### *2.1.5. Safety Guidelines*

Experiments to produce recombinant lentiviruses are performed in accordance with a biosafety Level 2 or 3, depending on whether the vectors express a gene that modifies the cell cycle, such as an oncogene. All manipulations should be approved by national competent authorities. Since the lentiviral vectors are aliquoted in small volumes (<10 μl), individual aliquots with a relatively low amount of total particles can be manipulated in an L2 laboratory, even if they contain dangerous transgenes. All lentivirus-containing liquids and disposable materials have to be inactivated/cleaned with ethanol, bleach, or detergent solutions before manipulation outside the culture hood.

### ***2.2. Materials and Animals for Production of Transgenic Rats***

#### *2.2.1. Superovulation and Fertilized Embryo Collection*

1. Immature females (4–5 weeks old and around 75–100 grams in weight).
2. Fertile males (2 months old) for mating with the immature females. They should be replaced every 8 months to a year.
3. Pregnant Mare's Serum Gonadotrophin (PMSG; Intervet Laboratories Ltd, Cambridge, UK).
   Working solution: 125 IU/ml made up with 0.9% (w/v) NaCl. Store frozen at –20°C in 1 ml aliquots.
4. Human Chorionic Gonadotrophin (hCG; Chorulon, Intervet). Working solution: 150 IU/ml made up with 0.9% (w/v) NaCl. Store at 4°C (up to 2 weeks) in 1 ml aliquots.
5. Hyaluronidase (Sigma, cat.no.H4272). Stock solution: 10 mg/ml in M16 medium (Sigma; cat.no.M7292). Store at –20°C (stable for several months) in 1 ml aliquots.
6. Embryo culture medium: M16 medium supplemented with 10% fetal bovine serum, 2 mM L-glutamine, 100 IU/ml penicillin, and 100 μg/ml streptomycin.
7. Embryo-tested mineral oil (Sigma; cat.no.M8410).
8. 35-mm Petri dishes.
9. Egg transfer pipette (Pasteur Pipette, *see* Note 1) assembled into a mouth-operated system made up of a mouthpiece, rubber tube (~40 cm) and a pipette holder.
10. Stereomicroscope with under-stage illumination.
11. Humidified incubator at 37°C and 5% $CO_2$.

*2.2.2. Microinjection of One-Cell Embryos*

1. Inverted microscope equipped with a 10X lens for low-magnification work and a 40X objective for microinjection.
2. Two micromanipulators for both holding and microinjection pipettes (Narishige).
3. Microinjector (Narishige).
4. Micropipette puller (PN-30, Narishige).
5. Holding pipette (Glass capillary, Narishige; cat.no. G-1) (*see* Note 2).
6. Microinjection pipette (Glass capillary with filament, Narishige; cat.no. GD-1) (*see* Note 3).
7. Tips for loading capillaries with virus (Microloader, Eppendorf).
8. Injection chamber (*see* Note 4).

*2.2.3. Transfer of Microinjected Embryos Into Recipient Females*

1. Recipient females (8–16 weeks of age) that have successfully had at least one litter (virgins often eat pups).
2. Vasectomized males are needed to engender pseudopregnancy in the recipient females. They have to be replaced every 6–8 months.
3. Surgical microscope and fiber optic illumination.
4. Adrenalin (1 mg/ml).
5. Embryo Transfer Pipette assembled into a mouthpiece equipped with a membrane filter (0.02 µm, Whatman, cat. no. 6809-1102, Anotop 10) for biosafety.

## 3. Methods

### *3.1. Lentivirus Production Protocol*

Production of recombinant lentiviruses relies on co-transfecting a minimum of three plasmids into 293 T cells: the transfer vector plasmid, the packaging plasmid encoding viral structural and enzymatic proteins, and the envelope protein-coding plasmid. Here we describe the production of a 100 ml crude vector supernatant. This volume of production is sufficient for generating transgenic animals using most lentiviral vectors, providing that that they have a viral titer $\geq 1 \times 10^6$ Hela Transducing units per ml ($^{Hela}$TU/ml) before vector concentration. High-titer lentiviral vector stocks are then produced, usually by ultracentrifugation given the stability of the VSV G protein. We further describe the purification of lentiviral particles using chromatography. Both concentration/purification methods enable the production of vector stock suitable for transgenesis. However, chromatography purification generates vector stocks with a higher purity as compared to ultracentrifugation, which may be important for some vector preparations.

1. Maintain the 293 T cells in 10 ml of DMEM culture medium in 10-cm tissue culture plates. Keep the cells growing exponentially, and do not let them reach confluence for more than a day. At cell confluence, split the cells into new plates at a ratio of 1:10. New cell confluence should be reached after 3 days of culture, otherwise start a new 293 T cell culture from a frozen stock (usually every 3 months).
2. The day before the transfection, trypsin three confluent 293 T cells plates and count the cells in tryptan blue solution. Seed $2 \times 10^6$ 293 T cells in a new set of 20 10-cm plates in a final volume of 10 ml of DMEM culture medium. Incubate the cells for 16–24 h.
3. Observe some of the seeded plates, rapidly and at least 2 h before transfection. Cell confluence should be at least 50–70% and cells should be evenly distributed across the surface of the plate, otherwise throw away the cells and repeat the previous step. Keep the cells in the incubator until the transfection mixture is ready to be added.
4. Pre-warm the transfection solution at 37°C (H20-Hepes, 2X HeBS, 0.5 M $CaCl_2$).
   In one 50 ml conical tube, mix the plasmids in a final volume of 2.5 ml H20-Hepes:
   Lentiviral vector plasmid: 200 μg
   Packaging plasmid, PsPAX2: 150 μg
   Envelope plasmid, pMD2G: 100 μg
   Vortex the tube.
   Prepare a second tube in the same way.
5. Add 5 ml of 2X HeBS to each 50 ml tube and vortex.
6. While vortexing vigorously (3/4 of maximum speed), add 2.5 ml of 0.5 M $CaCl_2$ to each 50-ml tube in a drop-wise manner.
7. Incubate at room temperature for 15 min to allow the formation of DNA precipitates. Vortex gently, then add the formed DNA precipitates to the cells within 15 min.
8. Remove ten plates of the seeded 293 T plates (Subheading 3.1, step 2) from the incubator and gently add 1 ml of the DNA precipitate solution to each plate in a drop-wise manner. Gently move the plates back and forth to evenly distribute the DNA precipitate solution. Proceed within 5 min to avoid pH changes in the medium. Incubate the cells for 16–24 h in the incubator. Repeat the same procedure for the last ten 293 T plates.
9. Observe the cells. If transgene expression is not toxic for the cells, they should be confluent. Aspirate the culture medium, gently (cells are loosely attached) add 5 ml of pre-warmed D-PBS and move the plates back and forth. Aspirate and

gently add 5 ml of pre-warmed serum-free EPISERF medium. Incubate for 16–24 h.

If the lentiviral vector expresses GFP (or another visible marker) under a ubiquitous promoter, transfection efficiency could can be assessed under a fluorescent microscope. Transfection efficiency should be ≥50%.

10. Collect and pool the virus-containing medium (95–100 ml) in 50 ml tubes. Centrifuge at 1,000 × g for 5 min at 4°C in a table-top centrifuge to pellet cell debris. Pool and filter the virus supernatant through a 0.45 μm Stericup filter to further clear supernatant of cell debris. Notably, filtered supernatants can be stored at 4°C for up to 3 days before vector concentration.

### 3.2. Concentration of Recombinant Lentivirus

#### 3.2.1. Ultracentrifugation

1. Centrifuge 90 ml of filtered virus supernatant (Subheading 3.1, step 10) in an SW28 rotor at 26,000 rpm (or 50,000 × g) for 90 min at 4°C in three 30 ml-Beckman Konical™ tubes. For some vector preparations, it may be important to centrifuge virus supernatant through a sucrose cushion to obtain a higher removal of protein contaminants. For this, create a sucrose cushion by slowly layering 25 ml of filtered virus supernatant onto 5 ml of 20% sucrose in four 30 ml-Beckman Konical™ tubes. In addition, a virus ultrafiltration step may be performed to pre-purify and/or pre-concentrate filtered virus supernatant before virus ultracentrifugation. For this purpose, centrifuge the 90 ml of filtered virus supernatant in Amicon ultra 15–100,000 MWCO (15 ml per unit) (Millipore, ref UFC9 100 08) or Vivaspin 20–100,000 MWCO (20 ml per unit) (Sartorius-Stedim, ref VS2042) centrifugal units in a table-top centrifuge with 50 ml tube adapters, at 3,000 rpm, 4°C for 10–30 min until the vector supernatant volume is reduced to 2–3 ml in the upper chamber. Add cold D-PBS (12–13 ml) to the upper chamber, mix by gentle pipeting, and centrifuge as described above. Collect all vector supernatants in the upper chambers, mix and ultracentrifuge the ultrafiltered virus supernatant in a 30 ml-Beckman Konical™ tube.
2. Pour off the supernatant.
3. Let the pellet dry by inverting briefly for 2 min on a paper towel and aspirate the remaining liquid droplets.
4. Add 30 μl of EPISERF medium (10,000X volume concentration) to each viral pellet and incubate at room temperature for 30 min or overnight at 4°C.
5. Resuspend the pellet by pipeting up and down several times, avoiding frothing.
6. Transfer the resuspended virus solution to a 1.5 ml microfuge tube, vortex, and centrifuge at low speed (5,000 rpm) for 2 min in a bench-top centrifuge.

7. Transfer the supernatant to a fresh 1.5 ml tube and make 10 μl aliquots of vector stock. For titering (see below), take 2 μl of the vector stock and mix with 198 μl DMEM culture medium (dilution of 100X).
8. Freeze and store the vector stock, including the 100X-diluted vector stock, at –80°C.

*3.2.2. Chromatography*

During virus production, 293 T cells may release a certain quantity of genomic DNA, cellular and subcellular debris along with lentivirus particles. The amounts of these cell-released contaminants depend on the vector design and/or the transgene, and their expression can be toxic to the 293 T producer cells. A high amount of cell-released contaminants may be detrimental to transgenesis efficiency. Since ultracentrifugation co-pellets cell-released contaminants with virus particles, it may be crucial to prepare some VSV G-pseudotyped lentiviral vector stocks by using chromatography.

Two commercial lentivirus chromatography-based purification kits are now available: "Fast-Trap Lentivirus Purification and Concentration" kit and "LentiSELECT 40" kit. For both kits, a chromatography device is loaded with 40–45 ml of filtered virus supernatant (Subheading 3.1, step 10), and final volume-fold concentration is about 50–100X. Therefore, as compared to ultracentrifugation, the vector stocks obtained using these kits are less concentrated, but have a higher purity.

## 3.3. Vector Titration

### 3.3.1. "Biological Titer" Measurement by Cell Transduction and Real-Time Quantitative PCR

#### 3.3.1.1. Transduction of Target Hela Cells

This procedure is used to determine the titer of any infectious virus, provided that the primers amplify a region in the vector sequence.

1. Seed $5 \times 10^4$ Hela cells in each well of a 6-well culture plate in a final volume of 2 ml DMEM culture medium.
2. The next morning, observe the cells. Cells should be 50% confluent and evenly distributed across the surface of the well. Remove media and add 2 ml of fresh DMEM culture medium before transduction.
3. Rapidly thaw the frozen 100X-diluted vector stock (Subheading 3.2.1, step 7) at 37°C. Make a tenfold serial dilution of this 100X-diluted vector stock (Subheading 3.2.1, step 7) to a $10^{-4}$ dilution.
4. Add to 100 μl of seeded Hela cells, 10 μl of the 100X-diluted vector stock and 10 μl of the $10^{-3}$- and $10^{-4}$-diluted vector. For some vector preparations, transduce Hela cells with 5–10 μl of undiluted vector stock.
5. After 24 h, wash the cells with D-PBS and grow the cells for another 4 days.

6. Lyse the cells directly in the well to extract genomic DNA by using a genomic DNA extraction kit. Genomic DNA concentration is 50–100 ng/μl. Store genomic DNA at –20°C until use. All procedures should be performed with filter tips.

#### 3.3.1.2. Real-Time Quantitative PCR

All procedures have to be performed with filter tips.

1. A standard curve has to be run in order to relatively quantify the Ct threshold values of the unknown DNA sample to determine the number of vector copies per cell.
   Prepare a standard curve in which the lentiviral vector plasmid (harboring the sequence amplified by the PCR primers) is present at 100, 10, 1, 0.1, 0.01, and 0 copies in Hela genomic DNA.
   We usually prepare a standard curve with the pRRLPGKGFP-WPRE vector (7,384 bp), as follows:
   (i) Prepare a solution of pRRLPGKGFPWPRE plasmid at 1.612 ng/μl, which contains $2 \times 10^8$ vector copies/μl according to this formula:
   Number of copies/μl = (concentration ($g.\mu l^{-1}$) X $6.02 \times 10^{23}$ ($mol^{-1}$))/(660 ($g.mol^{-1}$) x plasmid size (bp)), where Avogadro number is $6.02 \times 10^{23}$ and bp weight is 660 $g.mol^{-1}$.
   (ii) Prepare a Hela genomic DNA solution at 64 ng/μl, which contains $1 \times 10^4$ cells/μl, by assuming that the human diploid genome size is $5.89 \times 10^9$ bp, and therefore that the DNA content of a human cell weighs 6.38 pg (=$5.82 \times 10^9$ X 660/$6.02 \times 10^{23}$).
   (iii) Mix 5 μl of the 1.612 ng/μl plasmid solution ($1 \times 10^9$ vector copies) with 1 ml of the 64 ng/μl Hela genomic DNA ($1 \times 10^7$ cell) to make a solution containing 100 vector copies per cell. Make a tenfold serial dilution by diluting the 100 vector copies/cell solution in the 64 ng/μl Hela genomic DNA, starting from 100 to 0.01 vector copies/cell.
2. Mix the GAG-F with GAG-R primer (vector amplification) and the HB2-F with HB2-R (beta-actin gene amplification) primer to a final concentration of 10 μM in water. For example, add 10 μl of 100 μM GAG-F and 10 μl of 100 μM of GAG-R to 80 μl of water.
3. Prepare PCR master mix for vector detection (GAG mix):

| | **1 run** | **50 runs** |
|---|---|---|
| 2x SYBR Green mix Reaction mix | 10 μl | 500 μl |
| GAG Forward + Reverse primer (10 μM) | 0.4 μl | 20 μl |
| H2O | 7.6 μl | 380 μl |

Prepare the HB2 mix (HB2 primers) in the same way.

Calculate GAG and HB2 volume mix for analyzing samples in triplicate. A blank is made with water instead of the DNA sample. Samples for generating a standard curve are prepared at 100, 10, 1, 0.1, 0.01, and 0 copies of vector per cell (step 3.3.B.1).

4. Load 18 µl of the GAG mix into the wells of a 96-well optical reaction plate, then distribute 18 µl of the HB2 mix in another well.
5. Add 2 µl of DNA sample to each well.
6. Place the optical adhesive cover over the plate, taking care not to touch the cover.
7. Centrifuge the plate briefly at 2,000 RPM for 1 min before placing it in the real-time qPCR thermocycler. PCR conditions are the following: 1 cycle of (95°C for 10 min) (activation) and 40 cycles of (95°C for 15 s and 60°C for 60 s).
8. Set the Ct values: they should be set in the exponential phase of the log graph of the amplification plots for both the GAG and HB2 amplification curves.
9. Export the Ct results to a Microsoft Excel sheet.
10. In Microsoft excel, open the Ct values file. For each DNA sample, including the standard curve:
    (i) Calculate the mean GAG ($n=3$) and HB2 Ct values ($n=3$)
    (ii) Calculate the delta Ct (dCt) mean value (GAG Ct mean minus HB2 Ct mean)
    (iii) From the dCt mean values of the standard curve, display a graph of dCt mean value (X-axis) against the vector copy number/cell (Y-axis, logarithmic scale). The $R^2$ correlation value should be 0.99. Display the equation formula of the exponential curve.
    (iv) Using the formula, calculate the vector copy number/cell for unknown DNA samples
    (v) The titer of the lentiviral vector stock can be then calculated as follows:

Titer ($^{Hela}$TU/ml) = ($1 \times 10^5$ (Number of Hela cells at the time of transduction) x number of vector copies per cell (qPCR result)/volume of supernatant (ml)) x (vector dilution).

#### 3.3.1.3. Determination of the "Physical Titer" by Anti-p24 Immunoassay

Determining the number of physical virus particles in the vector stock is important to calculate the index infectivity, i.e., the number of TU per physical particle, or per p24 content. Indeed, a high index infectivity is an indication of an efficient vector production/concentration process. Its quantification is based on an HIV-1 p24 antigen capture assay. The physical titer can be measured by using a commercial HIV p24 ELISA. By this assay, both infectious and noninfectious virus particles (virus devoid of viral genomic RNA or of envelope protein, or non mature virus) are quantified. The index infectivity of VSV G-pseudotyped vector should be ≥1 $^{Hela}$TU/pg of p24; below that value some trouble-shooting may be necessary for transgenesis experiments. Note that if biological titration is performed on target cells other than Hela cells, the index infectivity is different and not comparable.

##### 3.3.1.4. Use of Vector Stock

1. Rapidly thaw the frozen vector stock at 37°C.
2. Centrifuge at low speed (5,000 rpm) for 2 min in a bench-top centrifuge.
3. Transfer the supernatant to a fresh 1.5 ml microcentrifuge tube.
4. Vector stock is ready to be used.

Minimize freeze/thaw cycles of virus (<3). A virus will lose approximately 50% of its activity with each freeze thaw cycle.

### *3.4. Superovulation and Fertilized Embryo Collection*

Female rats are administered gonadotropins prior to mating to increase the number of released eggs (superovulation). This technique can yield around 30–40 eggs per donor female.

1. Prepubescent female rats are injected intraperitonally with 25 IU of PMSG between 12 am and 1 pm on day -2, followed by 30 IU of hCG between 3 and 4 pm on day 0.
2. Individually mate each hormone-treated female with one fertile male overnight. On the morning of day 1, check the females for copulation plugs.
3. Sacrifice female rats by cervical dislocation around 10 a.m. on the morning of day 1.
4. Excise the oviducts and transfer them to a dish containing M16 medium at room temperature. Embryos, enclosed by cumulus mass cells, can be released from the swollen ampullae (the upper portion of the oviduct) by gently tugging and opening the walls of the ampullae with fine forceps (*see* Note 5).
5. Transfer the embryos using an egg transfer pipette to a dish containing a pre-warmed hyaluronidase solution (500 μg/ml in M16 medium), which enzymatically digests the cumulus cells thus releasing the embryos. A few minutes of treatment is sufficient (longer incubation can be toxic for embryos); gentle up and down pipeting can facilitate the process.
6. Transfer the embryos to fresh pre-warmed M16 medium to wash off the hyaluronidase solution and preserve their viability.
7. Finally, transfer the embryos in a microdrop of embryo culture medium (20–30 embryos per drop) overlaid with mineral oil in a humidified 37°C incubator under 5% $CO_2$ until needed.

### 3.5. Delivery of Lentiviruses to One-Cell Embryos

Lentiviruses are delivered to the fertilized oocytes on the same day of collection, targeting only one-cell embryos to minimize mosaicism. Two methods are available to deliver the lentiviruses to the embryos.

#### 3.5.1. Microinjection of Lentiviruses into the Perivitelline Space of Single-Cell Embryos

1. In an L2 laboratory, thaw a newly frozen aliquot (<10 µl) of lentivirus at room temperature and centrifuge at 10,000 rpm for 2 min to pellet aggregates of cellular debris.
2. Under a biosafety laminar flow hood, load approximately 3 µl of the viral solution into the microinjection pipette with a microloader tip.
3. Transfer one-cell embryos (in batches of ~30) in a microdrop of embryo culture medium in the injection chamber and cover with mineral oil to prevent evaporation and maintain osmolarity.
4. Mount the chamber on the stage of an inverted microscope, and monitor the injection procedure under 400X magnification.
5. Hold fertilized embryos (pronuclei visible) in place against the holding pipette using gentle negative pressure. Hold the micropipette loaded with the virus in place onto a micromanipulator.
6. Using the micromanipulator to guide the pipette, push the tip through the zona pellucida into the perivitelline space (region between the zona pellucida and the oocyte cell membrane). Using gentle positive pressure, the viral solution flows continuously from the pipette. The embryo should be slightly displaced within the zona pellucida by the pressure of the liquid (*see* Note 6). At the end of the injection, the embryo should bounce back into place. Leave the micropipette in the perivitelline space for 10–20 s before withdrawing it from the zygote.
7. After the injection, wash the surviving embryos twice in a microdrop of embryo culture medium equilibrated at 37°C, 5% $CO_2$, and then keep them in a 37°C humidified incubator under 5% $CO_2$ until implantation (*see* Note 7).
8. Clean all materials contaminated with lentiviral solution with bleach or detergent solutions.

#### 3.5.2. Co-Incubation of Denuded One-Cell Embryos with Lentiviruses

Although, the delivery of the lentiviral vectors by injection into the perivitelline space is relatively easy and highly efficient to generate transgenic rats, it requires expensive equipment and it is very difficult to control the volume of viruses delivered (*see* Note 8), leading from one to multiple proviral insertions in the genome (7). Lois's group reported on an alternative approach to generate transgenic mice, which consists in co-incubating zona-free embryos with a lentiviral suspension until they reach the morula/ blastocyst stage (7). However, they found that the development of denuded embryos was delayed and that the rate of implantation was reduced. Unfortunately, even though this technique is logistically and technically easier, it can be difficult to apply to some species other than mice, for which culture of denuded embryos is currently difficult. For example, many previous attempts to culture denuded rat embryos have failed due to the loss of embryo integrity (blastomere separation) following cell division (unpublished data from our lab and from Dann's group (cited in the following review (17))).

### 3.6. Transfer of Embryos into Recipient Females

1. Obtain pseudopregnant females needed to host the microinjected embryos by mating sexually mature females in estrus with vasectomized mature males (*see* Note 9), the night before the day of implantation. Confirm mating the next morning by checking for a plug.
2. Transfer embryos injected or infected with lentivirus into the oviduct of host females by using an embryo transfer pipette (*see* Note 10), preferably the same day as the microinjection to increase the rate of implantation. In general, no more than 30 embryos are transferred bilaterally into the uterus.
3. Pups should be born 21 days after transfer. The birth rate is around 30%.

### 3.7. Genotyping Analysis of Founders

The resulting offspring are screened for the presence of the transgene in their genomes by PCR, and the proviral copy number is determined by Southern Blot analysis.

In contrast to DNA pronuclear injection, which often leads to the insertion of multiple transgene copies in only one site in the host genome, transgenic animals generated by lentiviral perivitelline injection carry one transgene copy in multiple integration sites (*see* Table 8.1). Bryda et al. recently described an LM-PCR strategy to exactly identify the number and chromosomal location of the transgene (18) (*see* in this book, chapter "PCR techniques to determine sites of transgene insertion into the genome" by EC Bryda and BA Bauer).

The expected number of transgenic rats obtained by lentiviral perivitelline microinjection depends on two crucial parameters; the lentiviral vector purification method and its titer (*see* Table 8.1).

## 4. Notes

1. Embryo transfer pipettes are prepared according to the detailed procedure in the article by Si-Hoe et al. (19).
2. Holding pipettes are prepared according to the detailed procedure described in this book, in the chapter entitled "Generation of transgenic rats by ooplasmic injection of sperm cells exposed to exogenous DNA" by M Hirabayashi and S Hochi.

**Table 8.1**
**Importance of the purification method and the lentiviral suspension titer on the efficiency of lentiviral perivitelline injection (data obtained in our lab)**

| LV vector | Purification method | Titer (TU/ml) | Egg donor female strain | Recipient female strain | Injection mode[a] | % of surviving eggs | % of pups/no. implanted eggs | % of transgenic | F0 vector insertion site number % of transgenic/no. pups |
|---|---|---|---|---|---|---|---|---|---|
| LV 1 | ultracentrifugation | $1.8 \times 10^{10}$ | SD | SD | SI of 90s | 87 | 11 | 60 | 1–3 |
| LV 2 | ultracentrifugation | $2.4 \times 10^{9}$ | SD | SD | bp[b] | 57 | 25 | 19 | 2–5 |
| LV 3 | ultracentrifugation | $1.98 \times 10^{9}$ | SD | SD | bp | 70 | 35.5 | 18 | ND |
| | | | BN | SD | bp | 80 | 7.5 | 33 | 1–4 |
| LV 4 | ultracentrifugation | $1.2 \times 10^{9}$ | SD | SD | bp | 71 | 27 | 12.5 | ND |
| LV 5 | chromatography | $2.5 \times 10^{8}$ | SD | SD | bp | 85 | 25 | 86 | 1–6 |

*SD* Sprague–Dawley, *BN* Brown Norway, *SI* single injection, *ND* not determined
[a]Continuous injection
[b]bp, back pressure is the mechanism of the automatic injector system which prevents the pipette filling with medium between injections due to capillary attraction

3. Microinjection pipettes are prepared by pulling glass capillaries on a pipette puller. The pipettes should have an outer diameter of between 10 and 15 μm.
4. The injection chamber is prepared as described in this book, in the chapter entitled "Generation of transgenic rats by microinjection of short DNA fragments" by S. Menoret et al.
5. The procedure for removal of the oviduct and the position of eggs in the swollen ampulla is well illustrated in the article by Si-Hoe et al. (19).
6. If the viral solution stops flowing, apply some more positive pressure. If the flow does not resume then it is likely that the pipette has clogged with some large cellular debris. Apply positive pressure until the debris is expelled from the micropipette. If the debris cannot be moved, discard the pipette in a flask containing bleach.
7. Preliminary experiments are advised to determine optimal conditions (pressure, injection time), by injecting 10–20 embryos and culturing them in vitro. At least 80% of the injected embryos should reach the first cell division.
8. The volume injected is dependent on the pressure applied, the injection time and the size of the hole in the injection pipette.
9. Males are vasectomized by cauterizing the vas deferens at two separate locations. The efficiency of the vasectomy is assessed by mating males with 2–3 females (19).
10. The procedure for transfer of microinjected eggs is well detailed and illustrated in the article by Si-Hoe et al. (19).

## References

1. Jaenisch R (1976) Germ line integration and Mendelian transmission of the exogenous Moloney leukemia virus. Proc Natl Acad Sci U S A 73:1260–1264
2. Gordon JW, Scangos GA, Plotkin DJ, Barbosa JA, Ruddle FH (1980) Genetic transformation of mouse embryos by microinjection of purified DNA. Proc Natl Acad Sci U S A 77:7380–7384
3. Filipiak WE, Saunders TL (2006) Advances in transgenic rat production. Transgenic Res 15:673–686
4. Tesson L, Cozzi J, Menoret S, Remy S, Usal C, Fraichard A, Anegon I (2005) Transgenic modifications of the rat genome. Transgenic Res 14:531–546
5. Hammer RE, Pursel VG, Rexroad CE Jr, Wall RJ, Bolt DJ, Ebert KM, Palmiter RD, Brinster RL (1985) Production of transgenic rabbits, sheep and pigs by microinjection. Nature 315:680–683
6. Naldini L, Blomer U, Gallay P, Ory D, Mulligan R, Gage FH, Verma IM, Trono D (1996) In vivo gene delivery and stable transduction of nondividing cells by a lentiviral vector. Science 272:263–267
7. Lois C, Hong EJ, Pease S, Brown EJ, Baltimore D (2002) Germline transmission and tissue-specific expression of transgenes delivered by lentiviral vectors. Science 295:868–872
8. Pfeifer A, Ikawa M, Dayn Y, Verma IM (2002) Transgenesis by lentiviral vectors: lack of gene silencing in mammalian embryonic stem cells and preimplantation embryos. Proc Natl Acad Sci U S A 99:2140–2145
9. van den Brandt J, Wang D, Kwon SH, Heinkelein M, Reichardt HM (2004) Lentivirally generated eGFP-transgenic rats allow efficient cell tracking in vivo. Genesis 39:94–99

10. Michalkiewicz M, Michalkiewicz T, Geurts AM, Roman RJ, Slocum GR, Singer O, Weihrauch D, Greene AS, Kaldunski M, Verma IM, Jacob HJ, Cowley AW Jr (2007) Efficient transgenic rat production by a lentiviral vector. Am J Physiol Heart Circ Physiol 293:H881–H894
11. Pfeifer A (2006) Lentiviral transgenesis-a versatile tool for basic research and gene therapy. Curr Gene Ther 6:535–542
12. Frecha C, Costa C, Negre D, Gauthier E, Russell SJ, Cosset FL, Verhoeyen E (2008) Stable transduction of quiescent T cells without induction of cycle progression by a novel lentiviral vector pseudotyped with measles virus glycoproteins. Blood 112:4843–4852
13. Kumar M, Keller B, Makalou N, Sutton RE (2001) Systematic determination of the packaging limit of lentiviral vectors. Hum Gene Ther 12:1893–1905
14. May C, Rivella S, Callegari J, Heller G, Gaensler KM, Luzzatto L, Sadelain M (2000) Therapeutic haemoglobin synthesis in beta-thalassaemic mice expressing lentivirus-encoded human beta-globin. Nature 406:82–6
15. Zufferey R, Donello JE, Trono D, Hope TJ (1999) Woodchuck hepatitis virus posttranscriptional regulatory element enhances expression of transgenes delivered by retroviral vectors. J Virol 73:2886–2892
16. Zufferey R, Nagy D, Mandel RJ, Naldini L, Trono D (1997) Multiply attenuated lentiviral vector achieves efficient gene delivery in vivo. Nat Biotechnol 15:871–875
17. Dann CT (2007) New technology for an old favorite: lentiviral transgenesis and RNAi in rats. Transgenic Res 16:571–580
18. Bryda EC, Pearson M, Agca Y, Bauer BA (2006) Method for detection and identification of multiple chromosomal integration sites in transgenic animals created with lentivirus. Biotechniques 41:715–719
19. Si-Hoe SL, Wells S, Murphy D (2001) Production of transgenic rodents by the microinjection of cloned DNA into fertilized one-cell eggs. Mol Biotechnol 17:151–182

# Chapter 9

# Generation of Transgenic Rats by Ooplasmic Injection of Sperm Cells Exposed to Exogenous DNA

**Masumi Hirabayashi and Shinichi Hochi**

## Abstract

Intracytoplasmic sperm injection (ICSI) has been successfully achieved in mice and rats using a piezo-driven injection pipette, with the offspring rate of >30%. The ICSI technique was applied not only to rescue infertile male strains but also to produce transgenic rodents. The ICSI-mediated DNA transfer, that the sperm heads and exogenous DNA solution are mixed and co-injected into ooplasm, has been equally effective to the conventional pronuclear DNA microinjection. Production efficiency of transgenic founders by the ICSI-mediated DNA transfer was comparable between mice and rats, while the optimal DNA concentration was lower in rats than mice. The production efficiency was improved when membrane structure of sperm heads was partially disrupted by detergent or ultrasonic treatment before exposure to the exogenous DNA solution. Exogenous DNAs with various chain lengths were stably integrated into the rodent genomes of various genetic backgrounds by this method. The ICSI-mediated DNA transfer in which the preparation of pronuclear-stage fertilized zygotes is not required would be alternative to conventional pronuclear DNA microinjection.

**Key words:** DNA transfer, ICSI, Sperm sonication, Transgenic rat

## 1. Introduction

Transgenic technology in mammals is increasingly important in the design and implementation of biological and physiological studies. The pronuclear microinjection of exogenous DNA is the most convenient and reproducible technique to produce transgenic animals, and the efficiency of producing transgenic mice and rats is generally 1–5% of the total injected zygotes or 10% of the newborn offspring. Sperm-mediated DNA transfer into mouse genomes was first reported by Lavitrano et al. (1) who demonstrated that the oocytes fertilized in vitro with exogenous DNA-bound sperm cells could develop into transgenic mice, with the maximum efficiency as 30% of 250 newborn offspring.

I. Anegon (ed.), *Rat Genomics: Methods and Protocols*, Methods in Molecular Biology, vol. 597
DOI 10.1007/978-1-60327-389-3_9, © Humana Press, a part of Springer Science+Business Media, LLC 2010

This approach for production of transgenic mice using sperm cells as vectors of exogenous DNA appeared to be more effective and less laborious, but the numerous numbers of experiments by the other laboratories failed to reproduce their results (2). Perry et al. (3) reported that transgenic mice were successfully produced by intracytoplasmic sperm injection (ICSI) of mouse oocytes using sperm heads co-incubated with exogenous DNA solution. The ICSI-mediated DNA transfer was found to be a reproducible technique in mice (4, 5) and rats (6, 7). The ICSI-mediated DNA transfer is equally effective as the conventional pronuclear DNA microinjection in terms of the efficiency of producing transgenic rats, the stability of transgenes, and the applicable size of exogenous DNA (7–9). In the present manuscript, successful protocol for rat transgenesis via ICSI is described.

## 2. Materials

### 2.1. Media

Modified rat 1-cell embryo culture medium (modified-R1ECM) is used for in vitro culture of ICSI zygotes up to 24 h and transfer of the zygotes to surrogate mothers, and for suspension, sonication and cryopreservation of epididymal rat spermatozoa. Hepes-buffered R1ECM (hepes-R1ECM) is used for collection of oocytes, temporal incubation of the oocytes in air, and the ICSI procedure.

#### 2.1.1. Modified-R1ECM

1. Two stock solutions (A and B; 100 mL each) are prepared according to the reagent list shown in the Table 9.1, and stored at 4°C after filtration through a 0.22-μm filter unit (*see* Note 1).
2. All components of modified-R1ECM (10), except bovine serum albumin (BSA), are mixed. Osmolarity of the medium is approximately 310 mOsm. The BSA-free modified R1ECM is sterilized by the filtration, and stored at 4°C.
3. On the day of ICSI, an aliquot of the medium (approximately 10 mL) is placed in a conical tube and BSA is added (no agitation). The modified-R1ECM is sterilized by the filtration, and incubated at 37°C in 5% $CO_2$ and 95% air until use.

#### 2.1.2. Hepes-R1ECM

1. According to the reagent list (*see* Table 9.1), all components of hepes-R1ECM except polyvinyl alcohol (PVA) are mixed. Since the PVA is not readily dissolved in the mixture, the hepes-R1ECM supplemented with the PVA is kept in a refrigerator (+4°C) for several days until the PVA is completely dissolved.
2. The volume of hepes-R1ECM is adjusted to 100 mL, and stored at 4°C after the filtration. An aliquot of the medium is placed at room temperature on the day of ICSI.

**Table 9.1**
**Composition of modified-R1ECM medium and Hepes-R1ECM medium**

| Reagents | Modified-R1ECM | Hepes-R1ECM |
|---|---|---|
| Stock solution A<br>( NaCl 6.428 g<br>KCl 0.239 g<br>Glucose 1.352 g<br>Penicillin G 0.075 g<br>Streptomycin 0.050 g<br>Na lactate (60%) 1.9 mL<br>in 100 mL distilled water (Sigma W1503)) | 10 mL | 10 mL |
| Stock solution B<br>(CaCl ·2H O 0.294 g<br>MgCl ·6H O 0.102 g<br>in 100 mL distilled water (Sigma W1503)) | 10 mL | 10 mL |
| $NaHCO_3$ | 0.2100 g | 0.0337 g |
| Na-pyruvate (Sigma P2256) | 0.0055 g | 0.0055 g |
| MEM essential amino acids solution (×50; Gibco 11130-051) | 2.0 mL | 2.0 mL |
| MEM non-essential amino acids solution (×100; Gibco 11130-050) | 1.0 mL | 1.0 mL |
| L-glutamine (Sigma G3126) | 0.0146 g | 0.0146 g |
| Hepes buffer (1 M; Gibco 15630-080) | | 2.2 mL |
| Bovine serum albumin (Sigma A9647) | 0.4000 g | |
| Poly vinyl alcohol (Sigma P8136) | | 0.0100 g |
| Distilled water (Sigma W1503) | Mess up to 100 mL | Mess up to 100 mL |

### 2.2. Microtool

Proceeding to the ICSI, glass microtools (pipettes for sperm injection and oocyte holding) should be prepared by processing glass capillaries (#B100-75-10, Sutter Instrument Co., Novato, CA) with equipments such as mechanical puller (PN-3, Narishige Scientific Instrument Laboratory, Tokyo, Japan) and microforge (MF-900, Narishige) (*see* Note 2).

#### 2.2.1. Sperm Injecting Pipette

1. Platinum filament of the mechanical puller in the adjustable region of the heater is molded using forceps and precision tools. A glass capillary is set as the central part of the capillary is enclosed by the filament of the heater. Scales for the heater and magnet to make pulled capillaries with 1-cm shank region (*see* Fig. 9.1a) may vary according to the puller model, temperature, and humidity.

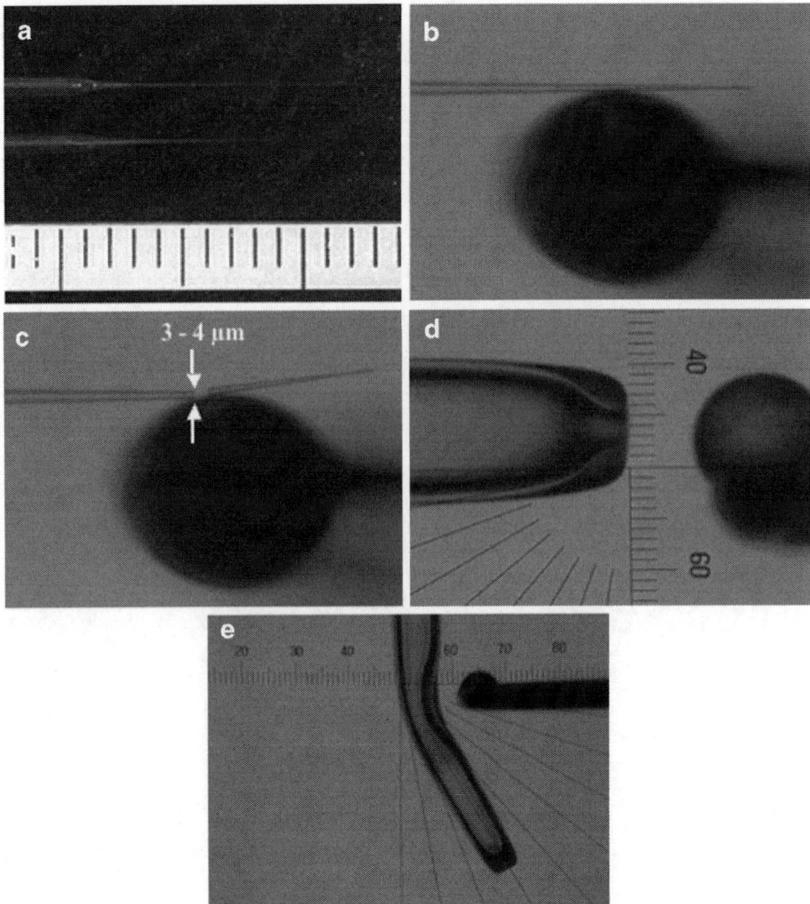


Fig. 9.1. Microtool processing for rat piezo-ICSI. (**a**) A glass capillary is heat-pulled by mechanical puller, to give the shank region of approximately 1-cm length. The extent of tapering off is dependent of the puller model (*upper*, Narishige PN-3, *bottom*; Sutter P-97/IVF). (**b**) In processing sperm injecting pipette, unheated glass bead of microforge is contacted with the pulled capillary at the outer diameter of 3–4 µm. (**c**) The heater is turned on/off quickly, so that the tip of the capillary is vertically cut off. (**d**) As for oocyte holding pipette, blunt end of pulled/cut capillary is processed by the heated bead, to give the inner diameter of 20–30 µm. (**e**) The holding pipette is bent by the heated bead to give an angle of approximately 150°.

2. The capillary extended by the puller is horizontally attached to the microforge where a glass bead is equipped at the tip of the platinum heater (*see* Fig. 9.1b). The outer diameter of the capillary contacting with the glass bead should be 3–4 µm, as determined macroscopically.
3. As soon as a fusion of the capillary with the glass bead by the heating starts, the power-supply of the heater is turned off, by which the tip of the capillary is vertically cut off (*see* Fig. 9.1c).

*2.2.2. Oocyte Holding Pipette*

1. A capillary extended by the puller is scratched with an ampoule-cutter at a site with an outer diameter of 90–120 μm in order to cut the tip vertically.
2. The capillary is horizontally attached to the microforge, and the tip is rounded by heating of the glass bead until the inner diameter of the tip decreases to 20–30 μm (*see* Fig. 9.1d).
3. The capillary is positioned vertically to the glass bead, and then bent at approximately 150° using the heated glass bead (*see* Fig. 9.1e).

## 3. Methods

### *3.1. Oocyte Recovery*

Depending on the experimental design or client requests, rat strains for oocyte donor females as well as sperm donor males must be carefully chosen. Superovulation treatment to females at immature or younger age is recommended because gonadotropin can be administered regardless of their estrus cycles. Forward rat oocytes to ICSI immediately once they were released from the bodies, or the oocytes would activate spontaneously under in vitro conditions (11).

1. Young female rats at 5–7 weeks old, which have not yet exhibited stable estrus cycles, are superovulated by intraperitoneal administrations of 300 IU/kg eCG and 300 IU/kg hCG at an interval of 48 h (12).
2. Hyaluronidase (H3506, Sigma) is dissolved in modified-R1ECM to yield a final concentration of 1 mg/mL and the solution is sterilized through a 0.22-μm filter unit. One 100 μL-microdrop of the hyaluronidase solution along with three 100 μL-microdrops of modified R1ECM are overlaid with the mineral oil (M8410, Sigma) in a 35-mm Petri dish (Falcon® 1008, Becton Dickinson, Franklin Lakes, NJ) and equilibrated at least for a few hours at 37°C in 5% $CO_2$ and 95% air.
3. Rats are sacrificed by cervical dislocation 14 h after the hCG administration. The oviducts are excised and placed in the mineral oil near the drop containing hyaluronidase (*see* Fig. 9.2a). Then, cumulus-oocyte complexes (COCs) are released from the ampullae of the oviducts and pulled into the hyaluronidase drop using a 26G × 1/2 syringe needle connected with 1 mL-syringe under a stereoscopic microscope (*see* Fig. 9.2b).
4. Five minutes later, oocytes are freed from cumulus cells with an aid of gentle pipetting (inner diameter of sterile capillary; 100 μm). The oocytes are washed three times by changing the modified-R1ECM drops, and incubated at 37°C in 5% $CO_2$ and 95% air.

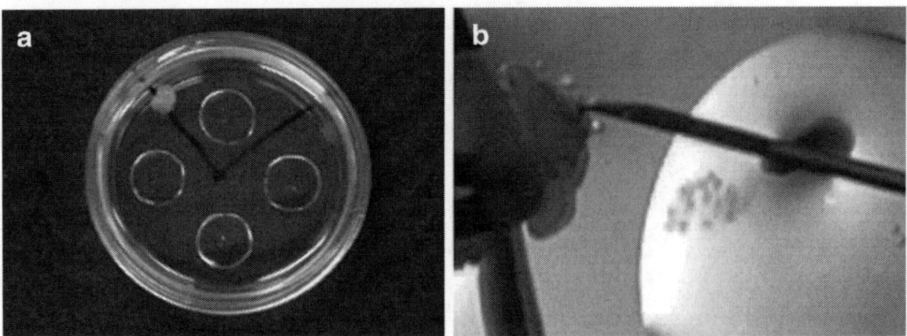


Fig. 9.2. Oocyte recovery from rat oviductal ampullae. (**a**) Excised ampullae are placed near a microdrop containing hyaluronidase (*top*: surrounded by black line). The other three drops of modified R1ECM are used for washing of denuded oocytes. (**b**) A mass of cumulus-oocyte complexes are released from the oviductal ampulla and brought into the hyaluronidase drop by a 27G sterile needle.

### 3.2. Preparation of DNA-Bound Sperm

Exogenous DNAs with various chain lengths including plasmids, YAC and BAC (up to a few hundreds kbp) can be used for the ICSI-mediated DNA transfer in rats (7, 9). Epididymal rat spermatozoa are sonicated to dissociate the heads from the midpieces and tails, and the sperm heads can be cryopreserved in the absence of protective additives before they are co-incubated with exogenous DNA.

1. A mature male rat at 3–6 months old is sacrificed by cervical dislocation, and the cauda epididymides are excised and placed on a sterile filter paper. After cutting the seminiferous duct with a surgical blade, the masses of epididymal spermatozoa are squeezed out and suspended in 2 mL of modified-R1ECM.
2. The spermatozoa are allowed to swim-up during the incubation for 30 min at 37°C in 5% $CO_2$ and 95% air. The motile spermatozoa in the supernatant (1.0 mL) are transferred into a 5-mL conical tube.
3. The sperm suspension is sonicated for 10 s using a 10% power output from an ultrasonic cell disruptor (Sonifire250, Branson, Danbury, CT) to dissociate sperm heads from midpieces and tails. They are divided into several cryotubes (Nunc366656; Thermo Fisher Scientific, Roskilde, Denmark), and stored in liquid nitrogen tank.
4. After being thawed in 20–25°C waterbath, 9-μL of the sperm suspension are mixed with 1 μL of exogenous DNA solution, and kept for 1 min at room temperature. The concentration of exogenous DNA solution should be adjusted to 1 ng/μL (*see* Note 3).

### 3.3. ICSI Procedure

Piezo-driven micromanipulator (PMM-150FU; Prime Tech, Ibaraki, Japan) with a pulse controller (PMASCT150; Prime Tech) is used for rat ICSI (13). Due to the large and unique shape of rat sperm heads, the sperm head is hung on (rather than aspirated into) the tip of much fine injecting pipettes and expelled into oocytes with a limited amount of the accompanying medium.

1. Besides the hepes-R1ECM, hepes-R1ECM supplemented with 12% (w/v) polyvinylpyrrolidone (PVP 360 kDa; ICN Pharmaceuticals, Costa Mesa, CA) should be prepared. The hepes-R1ECM/PVP medium is also filtered and stored at –20°C until use. Using the lid of a 35-mm plastic dish, two 6-μL microdrops of hepes-R1ECM (drop-[I] and drop-[II]) and two 6-μL microdrops of hepes-R1ECM/PVP (drop-[III] and drop-[IV]) overlaid with mineral oil are prepared (*see* Fig. 9.3a).
2. Oocytes (10–15) and sperm heads (1-μL suspension) are transferred into drop-[I] and drop-[III], respectively. First, the inner surface of injecting pipette is washed by repeated blow-off of mercury and aspiration of hepes-R1ECM/PVP in the drop-[IV]. Then, a single sperm head is hung on the tip of injecting pipette in drop-[III], and stands by for ICSI in drop-[I]. The outer surface of the pipette is washed in

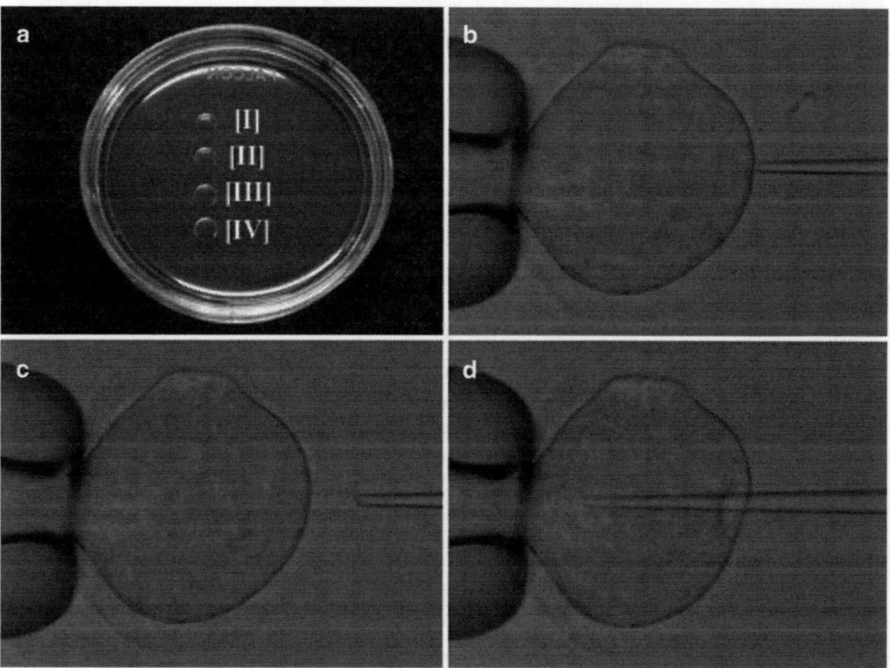


Fig. 9.3. Protocol for rat piezo-ICSI. (**a**) The ICSI chamber contains two 6-μL microdrops of hepes-R1ECM ([I] and [II]) and two 6-μL microdrops of hepes-R1ECM/PVP ([III] and [IV]). Oocytes and sperm heads are placed in drop-[I] and -[III], respectively. (**b**) Zona pellucida is drilled using several piezo-pulses (intensity 2, speed 1). Loading position of mercury in the sperm injecting pipette influences the "piezo effect". (**c**) Sperm head is hung on the blunt end of the injecting pipette. (**d**) The oolemma is punctured by one piezo-pulse (intensity 1, speed 1) with a light positive pressure, so the sperm head is injected into ooplasm with a minimum amount of the accompanying medium.

PVP-free drop-[II] when the pipette is transferred from the drop-[III] to the drop-[I].

3. An oocyte is held by the holding pipette as the second metaphase plate is located at 12:00 or 6:00 o'clock position. The sperm head is released near the tip of the injecting pipette, and the zona pellucida is drilled by several piezo-pulses (intensity 2, speed 1) (*see* Fig. 9.3b).

4. Again, the sperm head is hung on the injecting pipette (*see* Fig. 9.3c). The pipette is then advanced mechanically deep into the oocyte, stretching the oolemma extensively. Upon application of light positive pressure and one piezo-pulse (intensity 1, speed 1), the oolemma is punctured at the pipette tip. At the same time, the sperm head jumps out into the ooplasm with a minimum amount of the accompanying medium (*see* Fig. 9.3d). The injecting pipette is quickly withdrawn.

5. The ICSI oocytes are collected in hepes-R1ECM, and kept at room temperature for 10 min. Then, the oocytes are washed three times with modified-R1ECM for subsequent culture.

### 3.4. Oviductal Transfer of ICSI Zygotes

Rat embryos at early cleaving stage and blastocyst stage can be transferred into oviducts and uteri of surrogate mothers, respectively. The former is recommended; Rat one-cell zygotes are capable of developing into blastocysts if modified-R1ECM medium with lowered osmolarity is applied (14), but genetic background of donor and recipient rats as well as extent of synchrony between embryos and uteri are among the factors influencing in vivo survival of the blastocysts (15).

1. ICSI zygotes are cultured in 100-μL microdrops of modified-R1ECM overlaid with mineral oil at 37°C in 5% $CO_2$, 95% air. At 6 h of the culture, oocytes with both two pronuclei and second polar body are considered normally fertilized. At 24 h of the culture, oocytes cleaved to 2-cell stage are subjected to subsequent oviductal transfer.

2. Mature female rats (>10 weeks old) are mated with a vasectomized male rat, and embryo transfer is performed on the day that the vaginal plug is detected. The recipient rats are anesthetized by an intraperitoneal administration of 40 mg/kg nembutal® (Dainippon Sumitomo Pharma, Osaka, Japan).

3. Under the full anesthesia, the ovary is exposed on sterile gauze through a lateral incision. Before the bursa is opened with thermo-knife (Gemini™ Cautery System; Bio Research Center, Nagoya, Japan) and watchmakers forceps, the oviduct is inspected for the presence of a swollen ampulla, indicating the recent occurrence of ovulations. Unless one or two drops of epinephrine (Bosumin®, Daiichi Pharmaceutical, Tokyo, Japan) are dripped on the ovarian bursa to constrict blood

vessels, opening of the ovarian bursa with a slit parallel to the direction of the clearly visible vessels may cause some bleeding (*see* Note 4).

4. When the opening of the infundibulum becomes visible, a glass capillary is gently inserted into this opening. The 2-cell embryos (usually 5–10) aspirated in the capillary with an outer diameter of 140–160 μm can be blown into the oviduct. It is recommended to place an air bubble into the capillary to mark the position of the last embryo to be introduced.
5. The peritoneal wall (muscle layer) is closed using suture-needle and suture, and then the skin is closed by autoclip (MikRon® AUTOCLIP; Becton Dickinson, Franklin Lakes, NJ). The recipient is allowed to recover on 37°C warm plate. The resultant offspring are examined for the presence of exogenous DNA in their genomes, by the conventional PCR analysis (*see* Note 5).

## 4. Notes

1. Both stock solutions (A and B) should be used within 3 months of storage at +4°C. The BSA-containing media (modified-R1ECM and hepes-R1ECM) should be used within 1 month after preparation. Hepes-R1ECM/PVP medium can be stored at –20°C, but should not be kept beyond 3 months. The hepes-R1ECM/PVP medium is thawed by placing at room temperature, and should be used within 1 week (ideally within the day of ICSI).
2. Borosilicate capillaries are kept overnight in 3% HCl suspended in 99% ethanol solution, and are rinsed twice with Milli-Q water. The capillaries are dried using the hot air sterilizer at 60°C, and then sterilized at 200°C for 2 h. Glass capillaries from other commercial sources (e.g., G-1, Narishige) are also available, as well as mechanical puller (e.g., P-97/IVF, Sutter) and microforge (e.g., F-1000, H. SAUR, Reutlingen, Germany) from different suppliers.
3. Linearized plasmid DNA is separated by 1.2% agarose gel electrophoresis, and extracted from the gel using the QIAquick Kit (Qiagen, Valencia, CA). The DNA is dissolved in 10 m*M* Tris-HCl (pH7.6)/0.1 mM EDTA to yield a final concentration of 1 μg/mL and stored at +4°C. Immediately before use, the DNA solution is centrifuged at 12,100*g* for 10 min at +4°C. Some web sites (http://med.stanford.edu/transgenic/, http://www.med.umich.edu/tamc/tgoutline.html) are available at preparing DNA solution for production of transgenic animals.

4. Iridectomy scissors, instead of thermo-knife, are available to open the ovarian bursa. Repeated removal of cotton-sticks from the site of bursa opening can stop the bleeding, therefore the use of cotton-sticks is helpful to facilitate the operation even when the epinephrine drops are applied. The skin incision can be closed by suture in case autoclip is not available.
5. Our data are presented: In case that Sprague-Dawley rats are used for the ICSI-mediated transfer of plasmid DNA (3.0–5.0 kb), integration efficiency of the exogenous DNA into offspring rat genomes range from 1.3 to 8.2% of the total injected oocytes (>80% survival rate, 30–50% offspring rate) [7, 9]. Application of the larger DNA (186–208 kb) to the ICSI-mediated transgenesis resulted in the integration efficiency of 0.9–2.4% [7].

## References

1. Lavitrano M, Camaioni A, Fazio VM, Dolci S, Farace MG, Spadafora C (1989) Sperm cells as vectors for introducing foreign DNA into eggs: genetic transformation of mice. Cell 57:717–723
2. Brinster RL, Sandgren EP, Behringer RR, Palmiter RD (1989) No simple solution for making transgenic mice. Cell 59:239–241
3. Perry ACF, Wakayama T, Kishikawa H, Kasai T, Okabe M, Toyoda Y, Yanagimachi R (1999) Mammalian transgenesis by intracytoplasmic sperm injection. Science 284:1180–1183
4. Szczygiel MA, Moisyadi S, Ward WS (2003) Expression of foreign DNA is associated with paternal chromosome degradation in intracytoplasmic sperm injection-mediated transgenesis in the mouse. Biol Reprod 68: 1903–1910
5. Kaneko T, Moisyadi S, Suganuma R, Hohn B, Yanagimachi R, Pelczar P (2005) Recombinase-mediated mouse transgenesis by intracytoplasmic sperm injection. Theriogenology 64: 1704–1715
6. Kato M, Ishikawa A, Kaneko R, Yagi T, Hochi S, Hirabayashi M (2004) Production of transgenic rats by ooplasmic injection of spermatogenic cells exposed to exogenous DNA: a preliminary study. Mol Reprod Dev 69: 153–158
7. Hirabayashi M, Kato M, Ishikawa A, Kaneko R, Yagi T, Hochi S (2005) Factors affecting production of transgenic rats by ICSI-mediated DNA transfer: effects of sonication and freeze-thawing of spermatozoa, rat strains for sperm and oocyte donors, and different constructs of exogenous DNA. Mol Reprod Dev 70:422–428
8. Hirabayashi M, Kato M, Hochi S (2006) Mini review; transgenesis via intracytoplasmic sperm injection (ICSI) in rodents. J Mammal Ova Res 23:86–90
9. Hirabayashi M, Kato M, Amemiya K, Hochi S (2008) Direct comparison between ICSI-mediated DNA transfer and pronuclear DNA microinjection for producing transgenic rats. Exp Anim 57:145–148
10. Oh SH, Miyoshi K, Funahashi H (1998) Rat oocytes fertilized in modified rat 1-cell embryo culture medium containing a high sodium chloride concentration and bovine serum albumin maintain developmental ability to the blastocyst stage. Biol Reprod 59:884–889
11. Hirabayashi M, Ito K, Sekimoto A, Hochi S, Ueda M (2001) Production of transgenic rats using young Sprague-Dawley females treated with PMSG and hCG. Exp Anim 50:365–369
12. Keefer CL, Schuetz AW (1982) Spontaneous activation of ovulated rat oocytes during in vitro culture. J Exp Zool 224:371–377
13. Hirabayashi M, Kato M, Aoto T, Sekimoto A, Ueda M, Miyoshi I, Kasai N, Hochi S (2002) Offspring derived from intracytoplasmic injection of transgenic rat sperm. Transgenic Res 11:221–228
14. Miyoshi K, Abeydeera LR, Okuda K, Niwa K (1995) Effects of osmolarity and amino acids in a chemically defined medium on development of rat one-cell embryos. J Reprod Fertil 103:27–32
15. Kato M, Ishikawa A, Hochi S, Hirabayashi M (2004) Donor and recipient rat strains affect full-term development of one-cell zygotes cultured to morulae/blastocysts. J Reprod Dev 50:191–195

# Chapter 10

# Procedures for Somatic Cell Nuclear Transfer in the Rat

**Jean Cozzi, Eryao Wang, Christelle Jacquet, Alexandre Fraichard, Yacine Cherifi, and Qi Zhou**

## Abstract

Somatic cell nuclear transfer (SCNT) is a powerful tool for the investigation of the mechanisms of nuclear remodeling. In addition, SCNT may offer the possibility of introducing targeted mutations by homologous recombination in species for which ES cell technology is not available. The rat specific features of the oocyte have long impeded the development of SCNT. We detail here the procedures developed and optimized during the last several years for the optimization of rat cloning.

**Key words:** Cloning, Nuclear transfer, Sprague-Dawley, MG132, Spontaneous activation, Fetal fibroblast

## 1. Introduction

Somatic cell nuclear transfer (SCNT) offers the possibility to clone animals using in vitro cultured cells. Since the pioneering work of Wilmut et al. in 1997 in the sheep (1), SCNT techniques have subsequently developed rapidly leading to the cloning of several other mammalian species. Nuclear transfer constitutes a relevant tool to study the mechanisms of nuclear reprogramming and gene expression regulation (2). On the other hand, cloning may also provide an alternative strategy for the introduction of targeted gene mutation into the genome of species in which embryonic stem cell technology is not available. This strategy has proved successful in the pig and sheep with the generation of cloned knock-out animals (3–6). Because of its relevance to human physiology, the laboratory rat (R. norvegicus) remains one of the best systems for modeling human disease. The absence of germ line competent ES cells in the rat has led to tremendous efforts from the scientific community for the development of alternative

I. Anegon (ed.), *Rat Genomics: Methods and Protocols*, Methods in Molecular Biology, vol. 597
DOI 10.1007/978-1-60327-389-3_10, © Humana Press, a part of Springer Science+Business Media, LLC 2010

transgenic technologies (7). Despite worldwide efforts, rat cloning has proved exceptionally difficult. The poor outcome of the cloning technology in the rat has been mainly attributed to the remarkable sensitivity of rat oocyte to its environment with rapid and abortive in vitro spontaneous activation upon recovery attempt. A major breakthrough was achieved by Zhou et al. in 2003 (8) with the obtention of the first alive cloned pups from fetal fibroblast nuclei after regulation of oocyte activation using cell cycle inhibitors. Hereafter, procedures that were established to maximize the development capacity of rat oocytes after nuclear transfer, are detailed. One should keep in mind that aside from the technical skill of the experimenter and the procedures aimed at maximizing cloned rat embryo development, rat cloning limitations mainly relate to the type and quality of nucleus donor cells.

## 2. Materials

### 2.1. Oocytes Production

#### 2.1.1. Superovulation

1. 60–70 g immature female rats (*see* Note 1).
2. Isoflurane anesthesia setup (*see* Note 2).
3. Isoflurane (anesthetic).
4. Buprenorphine (analgesic).
5. Surgical tools: fine scissors, curved forceps, blunt tip forceps, wound clips and applier. All instruments should be clean and sterilized with 70% ethanol or heat sterilized prior to use.
6. Osmotic minipumps (1003D, Alzet).
7. Follicle Stimulation Hormone (FSH) (Folltropin®-V, Vetrepharm) (*see* Note 3). Dissolve 400 mg NIH (corresponding to the content of 1 tube) in 4.9 ml of provided diluent (Bacteriostatic Sodium Chloride Injection USP). Add 105 μλ of a 400 UI hCG solution. Store aliquots at –20°C for up to 2 months.
8. Human Chorionic Gonadrotrophin (hCG) (Chorulon, Intervet). Dissolve 1,500 UI in 3.75 ml PBS, 0.1% BSA. Store aliquots at –20°C for up to 2 months.
9. Luteinising hormone releasing hormone (LHRH) (Sigma). Dissolve 1 mg in 5 ml PBS, 0.1% BSA. Store aliquots at –20°C for up to 2 months.
10. 1-ml Syringe.
11. Non powedered gloves.
12. Ethanol 70%.

#### 2.1.2. Oocytes Collection

1. Superovulated SD rat females.
2. Under-stage illumination stereomicroscope equipped with heated stage.

3. Surgical tools: surgical scissors, fine forceps, iris forceps and iris scissors.
4. 37°C, 5% $CO_2$ incubator.
5. 35 mm (35 × 10 mm) culture dishes.
6. Mouthpiece, tubing and microcapillary holder.
7. Transfer pipettes: borosilicate glass capillaries pulled by hand over a Bunsen burner and broken at an adequate internal diameter using a diamond cutter.
8. MG132 stock solution. Dissolve 5 mg MG132 powder (SIGMA, C2211) in 525 μl DMSO (final 20 mM). Store 10 μl aliquots at –20°C for up to 6 months.
9. M2 (Sigma) medium containing 10 μM MG132.
10. mR1ECM-BSA (*see* Table 10.1) containing 10 μM MG132.
11. Mineral oil, embryo tested (*see* Note 4).

**Table 10.1**
**Formulation and preparation of mR1ECM-BSA and mR1ECM-PVA media**

| Ingredient | mR1ECM-BSA | mR1ECM-PVA |
|---|---|---|
| NaCl | 110 mM | 76.7 mM |
| KCl | 3.2 mM | 3.2 mM |
| $CaCl^2$ | 2.0 mM | 2.0 mM |
| $MgCl^2$ | 0.5 mM | 0.5 mM |
| $NaHCO^3$ | 25.0 mM | 25.0 mM |
| Sodium lactate | 10.0 mM | 10.0 mM |
| Sodium pyruvate | 0.5 mM | 0.5 mM |
| Glucose | 7.5 mM | 7.5 mM |
| BSA[a] | 4.0 mg | – |
| PVA[d] | – | 1.0 mg/ml |
| Glutamine | 0.1 mM | 0.1 mM |
| EAA[b] | 2% (v/v) | 2% (v/v) |
| NEAA[c] | 1% (v/v) | 1% (v/v) |
| Osmolarity | 300 mOsM | 246 mOsM |

Reconstitute the media in double distilled or higher purity water and sterilize using a 0.22 μm sterile-filter. Store in 5-ml polypropylene bottles at 4°C for up to 1 month. Use a fresh bottle of medium on each experimental day

[a]Embryo-tested fraction V powder (Sigma)

[b]Minimal essential medium (MEM) amino acid solution

[c]MEM non essential amino acid solution

[d]Polyvinylalcohol 2% (v/v)

12. Embryo tested bovine testis hyaluronidase (Sigma) stock containing MG132. Dissolve 30 mg in 3 ml M2 containing 10 μm MG132. Store in 50 μl aliquots at –20°C for several months. Use at a final concentration of 300 μg/ml.

### 2.2. Nucleus Donor Cells Preparation

1. Primary rat embryonic fibroblasts at early passage.
2. 100-mm sterile plastic tissue culture dishes.
3. DMEM supplemented with 10% newborn calf serum (FBS).
4. Colcemid (GIBCO, 15212-046).

### 2.3. Nuclear Transfer

1. Vibration-damped table.
2. Work station including:
   (a) Inverted microinjection microscope (Leica DM IRB or equivalent) with Hoffmann optics equipped with 10× eyepieces and 5×, 20× and 40× objectives.
   (b) Heated stage.
   (c) Manual microinjector for holding oocytes (Cell Tram oil, Eppendorf).
   (d) 2× micromanipulator (TransferMan NK2, Eppendorf).
   (e) Piezo-actuated micromanipulator, such as the PMAS-CT150 (Prime Tech) or the PiezoDrill (Burleigh).
3. Holding pipette (outer diameter 90 μm, inner diameter 10 μm) and blunt end pipette microinjection capillaries (*see* Note 5). Borosilicate glass capillaries are pulled using a pipette puller (e.g., Flaming/brown P97/IVF, Sutter instrument Co.). Holding and microinjection capillaries are further modified using a microforge (manufacturers include Narishige and De Fonbrune) to adjust the internal diameter.
4. Culture dishes (Falcon 1008, Falcon).
5. Fluorinert® FC-77 (SIGMA).

### 2.4. Activation of Embryos and Culture

1. Butyrolactone I (BLI) stock solution. Dissolve 1 mg powder of BLI (BIOMOL, CC210) in 94 μl DMSO. (final 37.5 mM). Store 2 μλ aliquots at –20°C.
2. mR1ECM-BSA medium (*see* Table 10.1).
3. mR1ECM-BSA medium containing 150 μM BLI.
4. mR1ECM-PVA medium (*see* Table 10.1).

### 2.5. Embryo Transfer

1. 8–12 week old females (*see* Note 6).
2. Sprague Dawley vasectomized rats (*see* Note 7).
3. Isoflurane anesthesia setup and isoflurane gas.
4. Buprenorphine.

5. Surgical tools: 2× Dumont #5 forceps, surgical forceps, iris forceps, iris scissors, serafine clamp, vessel clamp, wound clips and applier. All instruments should be clean and sterilized with 70% ethanol or heat sterilized prior to use.
6. Surgical suture.
7. Surgical gauze or Kimwipe tissues.
8. Embryo transfer capillary and mouth pipetor.
9. Fiber optic light source.
10. Warming pad or red lamp.

## 3. Methods

Timelines of the whole procedure from superovulation to in vivo reimplantation of reconstructed activated embryos are presented in Fig. 10.1.

### *3.1. Oocyte Production*

#### *3.1.1. Superovulation*

Oocyte production is a critical step in the overall rat nuclear transfer procedure. Oocyte production efficiency is dependent on various factors such as the rat strain, the age and weight of animals, the timing of hormone delivery and oocyte collection, the quality of hormonal preparation and the hormones doses, the housing condition of the animals (light/dark cycle duration) and the seasonal effect. The rate of developing cloned embryos being very low in the rat, a large number of embryos should be reconstructed per experiment for obtention of live fetuses after in vivo reimplantation. Reliable procedures that yield large numbers of viable oocytes exhibiting normal developmental potential are not easy to set up in the rat. PMSG has been widely used to induce ovulation stimulation in rodents including the rat because

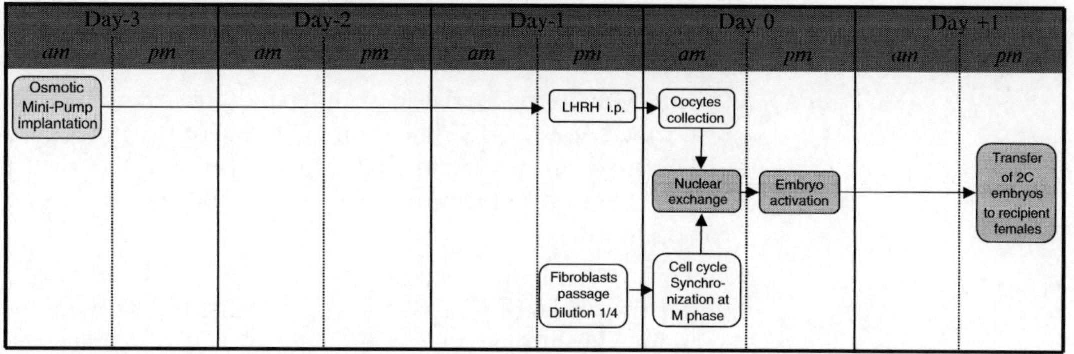


Fig. 10.1. Timelines of the nuclear transfer procedure from estrus stimulation and synchronization to in vivo reimplantation of reconstructed and activated embryos.

it is easy to setup and cost effective. Although viable oocytes can be obtained after induction of ovulation with low doses of gonadotrophin, high doses are associated with variability in ovulatory response, decreased developmental capacity and abnormal embryo development. We achieved the best results using immature females and continuous diffusion of FSH by Alzet osmotic minipumps implanted sub-cutaneously. The timing of experiments indicated here are valid for animals kept under a 12:12-h light-dark cycle.

1. Osmotic minipumps are filled with the FSH preparation and immediately implanted subcutaneously under anesthesia, on morning of day –3, between 8 and 10 a.m. (*see* Note 8).
2. Inject females i.p. with 0.2 ml of LH solution on afternoon of day –1, between 6 and 7 p.m. for synchronization of ovulation.

#### 3.1.2. Oocyte Collection and Preparation

The main steps of the procedure for oocyte collection and preparation for SCNT are similar to those used in the mouse. However, special care should be taken all along the procedure to avoid spontaneous activation. With that objective, rat oocytes should be transferred as rapidly as possible upon female sacrifice into the medium containing the proteasome inhibitor (MG132). This condition not being sufficient to preserve the developmental capacity of the oocytes, one should also avoid to expose oocytes to heat shocks throughout the collection and further on during the nuclear transfer procedure and up to in vivo reimplantation. The failure in this procedure leads to poor SCNT outcome. The occurrence of spontaneous activation before nuclear transfer may be detected by the appearance of a marked cytoplasmic protrusion (Fig. 10.3) and/or expulsion of the second polar body (*see* Note 9).

1. Prepare the dishes required for oocyte collection, preparation and incubation (Fig. 10.2) (*see* Note 10).
2. Sacrifice superovulated females using $CO_2$. 12–14 h after LH injection (*see* Note 11).
3. Place the rat on its back on absorbent paper and rinse thoroughly with 70% ethanol. Incise the peritoneum to expose the inner organs.
4. Collect oviducts on both sides and rapidly transfer them in pre-warmed M2-MG132 medium. The rest of the procedure is performed on a warming plate using pre-warmed media (37°C). A transfer pipette is used to move oocytes from one drop to another.
5. Under a stereoscopic microscope (15–20× magnification) transfer the oviducts in a 200 μl drop of M2-MG132 medium containing hyaluronidase at 300 mg/ml and dilacerate the ampulla (swollen part of the oviduct) using a pair of fine forceps. Gently squeeze the oviduct to help the oocyte-cumulus masses to flow out.

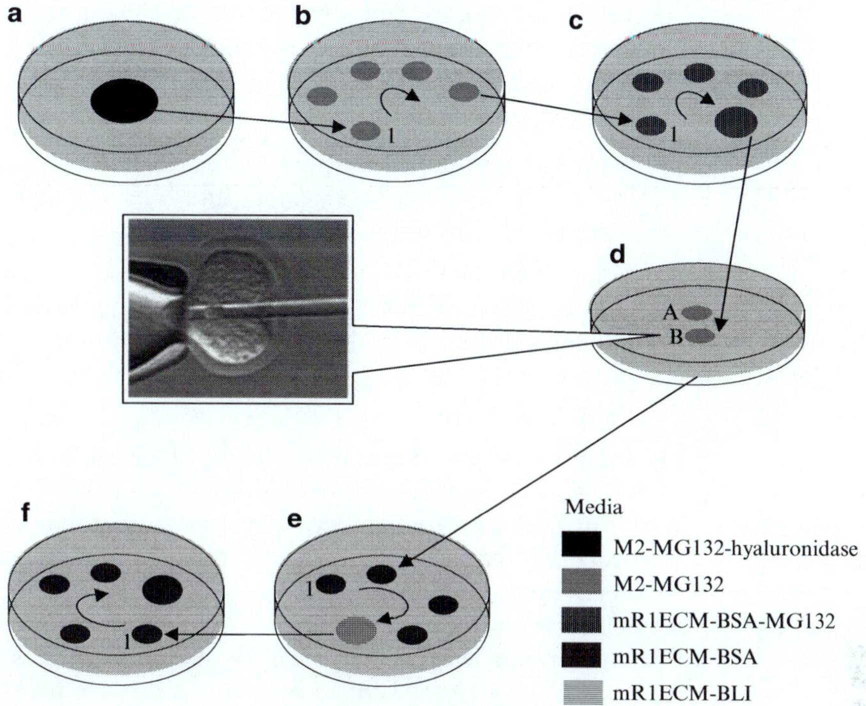


Fig. 10.2. Scheme representing oocytes preparation, incubation and injection setups with the different associated media. All media are covered by embryo tested mineral oil. Oocytes are freed from cumulus cells in a 200 μl drop of M2 containing hyaluronidase (**a**). Free cumulus oocytes are rinsed in five 10 μl drops of M2 containing MG132 under oil (**b**). Oocytes are rinsed in four drops of 10 μl mR1ECM containing MG132 and transferred in a 50 μl drop of the same medium before nuclear transfer (**c**). Oocytes are micromanipulated in a 10 μl M2-MG132 drop on a heating stage (**d**). Reconstructed oocytes are rinsed in 10 μl mR1ECM drops and incubated in 50 μl of the same medium containing the activation drug butyrolactone I (**e**). Activated oocytes are rinsed 4 times in 10 μl mR1ECM drops and transferred in 50 μl of the same medium for in vitro culture to the two-cell stage (**f**).

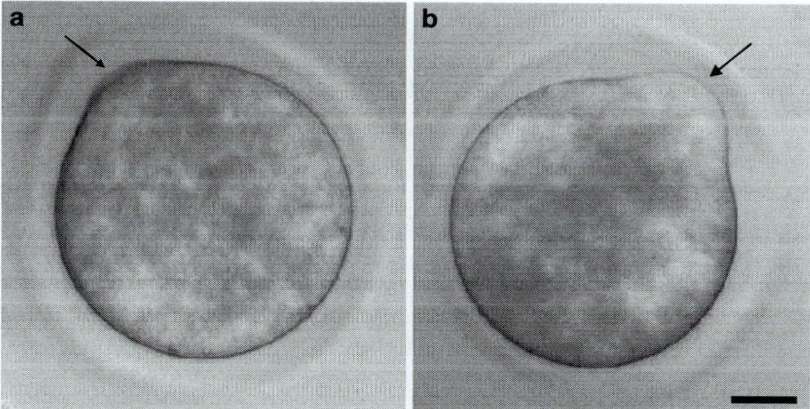


Fig. 10.3. Detection of spontaneous activation in rat oocytes through observation with microscope before SCNT. Non activated rat oocyte showing a slight protrusion caused by the presence of MII below (*arrow*). This oocyte is suitable for micromanipulation (b) Rat oocyte showing a clearly visible bulge of the plasma membrane. This oocyte is spontaneously activated and should be discarded. Scale bar = 20 μm.

6. Leave the oocyte-cumulus masses in the same drop until complete dispersion of the cumulus (about 5 min).
7. Rinse oocytes by serial transfer in five 10 μl drops of fresh M2-MG132 medium.
8. Cumulus granulosa cells that remain after hyaluronidase treatment can be stripped by pipetting eggs through transfer pipettes 90–100 μm in diameter.
9. Transfer all collected oocytes to a 50 μl drop of equilibrated culture medium mR1ECM-BSA containing MG132 under mineral oil after 3–4 rinses in the same medium.
10. Place the dish containing the oocytes in a 37°C, 5% $CO_2$, humidified incubator.
11. Incubate oocytes for at least 30 min before SCNT.

### *3.2. Nuclei Donor Cell Preparation*

In our studies, post-implantation development has been observed after reconstruction of embryos with stem cells (skin and neural stem cells) and embryonic fibroblasts synchronized at M phase of the cell cycle (*see* Note 12). We have observed that M phase donor nuclei always provided with higher developmental rates than G1 nuclei. The protocol described hereafter is suitable for the use of embryonic fibroblasts as donor cell nuclei. Primary rat fibroblasts can be easily isolated from fetuses obtained from females at day 12.5 of pregnancy according to the same protocol as described for mouse fetal fibroblasts (9).

1. Confluent cultures are passaged the day before and diluted 4 times with DMEM-FBS medium in 100 mm dishes.
2. On the morning of nuclear transfer, demecolcin is added to fibroblast dishes (final concentration: 0.05 μg/ml). After 2 h of incubation the mitotic cells loosely attached to the dish are recovered in the medium after shaking.
3. Wash the cells twice in DMEM-FBS free of drug and resuspend in low volume of medium to obtain high cell concentration (*see* Note 15).
4. Keep the cell suspension on ice before using in nuclear transfer experiment.
5. Use only those cells which show regular round shape.

### *3.3. Somatic Cell Nuclear Transfer*

To preserve the developmental ability of the cloned embryo, it is crucial that the nuclear exchange is achieved within the shortest time possible at 37°C upon transfer of the oocytes to the microscope stage. With that objective, a one step nuclear exchange micromanipulation procedure (Fig. 10.4) is used with prewarmed M2-MG132 medium and a heating microscope stage.

1. Prepare the injection chamber as described and place it under the stereoscopic microscope.

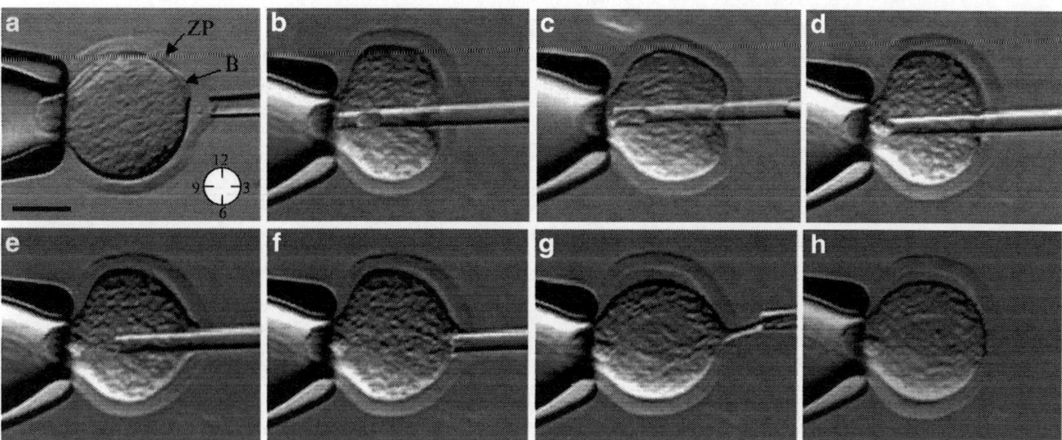


Fig. 10.4. One step nuclear transfer procedure. Successive micromanipulation steps (**a–h**) are described in the text. The oocyte's MII is located under the cytoplasmic bulge (B) formed by the oocyte plasma membrane (*arrow*). The bulge is positionned at 2 o'clock to facilitate the injection and enucleation steps. The zona pellucida (ZP) is drilled using several piezo pulses at high intensity (**a**). The injection pipet containing the donor cell nucleus is inserted through the hole created in the ZP and pushed through the ooplasm till it reaches the opposite side of the oocyte thus creating a deep invagination in the oocyte plasma membrane (**b**). A slight positive pressure is maintained within the pipette while the plasma membrane is ruptured by one or two piezo pulses at low intensity (**c**). The donor cell nucleus is expulsed into the cytoplasm with a miminimun amount of injection medium (**d**). The pipette is then rapidly moved backward before complete relaxation of the plasma membrane (**e**) until its tip is located below the MII (**f**). The MII is aspirated into the pipette with the two ends of the ruptured plasma membrane while the pipette is slowly pulled off the cytoplasm. The pipette is moved backward till the plasma membrane closes (**g**) thus allowing survival of the reconstructed oocyte (**h**). Scale bar = 40 μm.

2. Place the microcapillary for holding oocytes into the connector piece.
3. Fill the injection micropipette with Fluorinert (*see* Note 13) using a fine glass capillary and insert it into the connector piece. Expel any remaining air and some Fluorinert into droplet A (Fig. 10.3).
4. Transfer 20 to 30 oocytes (*see* Note 14) and 1–3 μl of donor cell suspension (*see* Note 15) to the microinjection chamber in droplet B (Fig. 10.3).
5. Position the holding and injection microcapillaries in droplet B on either side of a selected oocyte at the center of the field. Move the oocyte to the center of the droplet.
6. Rotate the oocyte using the microinjection capillary so that the metaphase II plate is positionned at 2 o'clock or 4 o'clock (Fig. 10.4) (*see* Note 16).
7. This orientation is crucial to avoid damaging the MII plate during injection of the exogenous nucleus and to allow its successful removal while retrieving the capillary from the oocyte's cytoplasm.
8. Move the oocyte to the center of the droplet.

9. Apply few strong piezo pulses to drill the zona (*see* Note 17). Retrieve the pipette and select an appropriate donor cell. Remove the donor nucleus by gently sucking the cell in and out the pipette until a maximum of cytoplasm material is removed.
10. Push the nucleus until the near tip of the pipette and advance the pipette through the breach previously created in the zona until it almost reaches the opposite side of the oocyte's cortex (Fig. 10.4b). Apply one or two weak piezo pulses (*see* Note 18 and Fig. 10.4c) to break the oolemma. Expel the donor nucleus with a minimal amount of medium (Fig. 10.4d). Quickly withdraw the injection pipette until the tip is closed to the MII plate (Fig. 10.4e, f). Aspirate the MII plate while slowly pulling back the pipette (Fig. 10.4f, g) (*see* Note 19). The removal of the MII plate allows the plasma membrane to close (Fig. 10.4h). The complete injection-enucleation procedure should be completed within about 5 s (*see* Note 20).
11. Move the injected oocyte to the bottom of the drop in order to avoid a mix-up of reconstructed and intact oocytes.
12. Repeat the procedure until all oocytes are injected.
13. Rinse the reconstructed oocytes 4 times in equilibrated mR1ECM-BSA medium to remove HEPES and MG132. Incubate the reconstructed oocytes for 30 min in the same medium before transfer into the activation medium (*see* Subheading 10.3.4).

### 3.4. Activation of Oocytes and Culture of Reconstructured Embryos

1. Transfer and culture each group of reconstructed oocytes into drops of mR1ECM-BSA containing BLI for 2 h in the incubator (*see* Note 21).
2. Wash the activated reconstructed oocytes 4 times in equilibrated mR1ECM.
3. Culture the activated oocytes into the same medium till cleavage to the 2-cell stage.
4. To obtain in vitro development beyond the 2-cell stage, transfer the embryos a few hours after cleavage into drops of mR1ECM-PVA medium (10).

### 3.5. Reimplantation of Reconstructed Embryos

The embryo reimplantation procedure is close to that in the mouse except for the administration of anesthesia, which is preferentially achieved using anesthetic gasses such as isoflurane (*see* Note 22). Outbred strain females exhibiting very good reproductive performance should be preferentially used as surrogate mothers for the reconstructed eggs (e.g., Sprague Dawley or Wistar). Bilateral embryo transfer is performed. Reimplantation of 40 to 50 cloned embryos per recipient is commonly achieved on day 2, at the two-cell stage in a 0.5-dpc pseudopregnant females.

This timing allows observation of cleavage of cloned embryos and selection of viable activated embryos for transfer.

1. Place the animal in the induction chamber.
2. For induction anesthesia: adjust the flowmeter to 1–1.5 l/m (oxygen as carrier gas) and the isoflurane vaporizer to 5%.
3. Inject buprenorphine subcutaneously about 2 min before surgery. Buprenorphine ensures analgesia during the peri- and early postoperative periods.
4. Maintain anesthesia with 3% isoflurane by placing the nose of the pre-anesthetized rat in the nose-cone connected to the breathing circuit.
5. To verify surgical depth of anesthesia before surgery, pinch the foot of the narcotized rat; absence of retraction reflex should be observed.
6. Isoflurane will be administered for approximately 12–15 min; 2 min prior and until the end of the surgical procedure.
7. When the animal is properly anesthetized (lack of spontaneous movement to toe pinch), operate the embryo transfer as previously described (9).

## 4. Notes

1. This corresponds to 3–4 weeks of age in the Sprague-Dawley strain.
2. It is assumed that prolonged and repetitive exposure to anesthetic gasses can be toxic to procedure area personnel. Ensure that all individuals responsible for anesthesia are properly trained. Maintain equipment in good working order, and have it certified yearly to ensure optimal performance. Work in a well-ventilated area, ideally under a fume hood or hard-ducted biosafety cabinet.
3. The quality of FSH preparation is crucial for the outcome of superovulation. Highly purified FSH preparations with low LH activity should be used.
4. Mineral oil may contain contaminants deleterious to embryonic development. Use embryo tested mineral oil preferentially from Sigma.
5. Injection pipette diameter should be adapted to the donor nucleus cell type and cell cycle phase at which the cells are used. Inner diameter is about 15–20 µm for fetal fibroblasts at M phase.

6. Females tend to lose the vaginal plug a few hours after copulation. It is thus advisable to check for plugs early in the morning (7–8 a.m.) or use dedicated mating cages that enable plug recovery thereby greatly facilitating identification of mated females (e.g., "E" type cut-out base section complete cage for rat, Charles River Lab).
7. Vasectomized rats can be bought from animal suppliers or prepared in house using 2–3-month-old males. Keep track of plugging performance for best results. Vasectomized males should be replaced approximately every 10 months.
8. Details of s.c. minipump implantation procedure is available at "WEB site Alzet." Take care not to introduce air bubbles in the minipump. Manipulation of the minipump should be done with gloves to avoid bacterian contamination and obstructing the pump pores with skin grease.
9. In the rat, the first polar body (PBI) rapidly degenerates after ovulation. By consequence, freshly collected rat oocytes rarely show any PB in the perivitelline space (PVS). Presence of a well-formed PB in the PVS after egg collection thus frequently relates to spontaenous activation with expulsion of the second polar body (PBII).
10. It is crucial that the temperature of media is at 37°C at the time of oocytes incubation.
11. It is necessary to reduce the time between the sacrifice of the female and the time of transfer of oviducts to MG132-containing pre-warmed medium as much as possible in order to limit spontaneous activation of oocytes. Considering this constraint we strongly advise to sacrifice of one female at a time.
12. Genetic background or number of passages of cell lines will affect the success rate. Favor hybrid genetic background and early passage cells. We routinely use fetal fibroblasts from passage 1 to 5.
13. Fluorinert increases the penetration capacity of the microinjection pipette by enhancing piezo pulse's impact power.
14. The number of oocytes per injection session depends on the experimenter's skill. The injection session duration should be 15–20 min max.
15. Cell suspension should be very concentrated in order for the experimenter to spend some time finding an appropriate donor cell.
16. In the rat oocyte, the metaphase II spindle is located below the cytoplasmic bulge.
17. The pipette should drill the zona in a few seconds removing a cylindric plug thus creating a hole that could easily be found at the time of nucleus injection.

18. Strong pulses of the piezo at the injection step can induce damage to the cell nucleus and dramatically decrease the embryo development capacity.
19. The MII plate is not aspirated into the pipette but should be kept strongly adherent to the tip of the pipette thanks to the negative pressure applied. For successful enucleation and healing of the oocyte oolema the MII plate is then gently pulled out of the oocyte's cortex before the plasma membrane has completely relaxed.
20. A well trained experimenter should obtain a survival rate of reconstructed oocytes of 90–100%. Special care should be paid to aspirating a minimum of cytoplasmic material at enucleation stage.
21. Expulsion of the pseudo polar body should occur in the majority (80% at least) of the reconstructed oocytes within 45–60 min after exposure to BLI containing medium. Inadequate donor cell quality, damage caused to the oocyte during nuclear exchange or deficient activation procedure, are most of the time responsible for a too low rate of pseudo polar body expulsion.
22. Anesthesia through inhalation of halogenated gasses cause significant body heat loss. Prevent heat loss until the animal recovers using red lamp or heating pad devices. The animal should recover within 5 min postoperatively.

## Acknowledgments

This work was supported by grants from ANR "ANR-05-PRIB-026" and the EuraTool project (European Rat Tools for Functional Genomics).

The author would like to thank staff from the embryology and molecular biology department for their technical help in writing the document.

### References

1. Wilmut I, Schnieke AE, McWhir J, Kind AJ, Campbell KH (1997) Viable offspring derived from fetal and adult mammalian cells. Nature 385:810–813
2. Yang X, Smith SL, Tian XC, Lewin HA, Renard JP, Wakayama T (2007) Nuclear reprogramming of cloned embryos and its implications for therapeutic cloning. Nat Genet 39:295–302
3. Denning C, Burl S, Ainslie A, Bracken J, Dinnyes A, Fletcher J, King T, Ritchie M, Ritchie WA, Rollo M et al (2001) Deletion of the alpha(1, 3)galactosyl transferase (GGTA1) gene and the prion protein (PrP) gene in sheep. Nat Biotechnol 19:559–562
4. Schnieke AE, Kind AJ, Ritchie WA, Mycock K, Scott AR, Ritchie M, Wilmut I, Colman A, Campbell KH (1997) Human factor IX transgenic sheep produced by transfer of nuclei from transfected fetal fibroblasts. Science 278:2130–2133

5. Phelps CJ, Koike C, Vaught TD, Boone J, Wells KD, Chen SH, Ball S, Specht SM, Polejaeva IA, Monahan JA et al (2003) Production of alpha 1, 3-galactosyltransferase-deficient pigs. Science 299:411–414
6. Dai Y, Vaught TD, Boone J, Chen SH, Phelps CJ, Ball S, Monahan JA, Jobst PM, McCreath KJ, Lamborn AE et al (2002) Targeted disruption of the alpha1, 3-galactosyltransferase gene in cloned pigs. Nat Biotechnol 20:251–255
7. Cozzi J, Fraichard A, Thiam K (2008) Use of genetically modified rat models for translational medicine. Drug Discov Today 13:488–494
8. Zhou Q, Renard JP, Le Friec G, Brochard V, Beaujean N, Cherifi Y, Fraichard A, Cozzi J (2003) Generation of fertile cloned rats by regulating oocyte activation. Science 302:1179
9. Nagy A, e.a. (2003) Manipulating the mouse embryo: a laboratory manual, 3rd edn. Cold Spring Harbor Laboratory Press, 764 pp
10. Zhou Y, Galat V, Garton R, Taborn G, Niwa K, Iannaccone P (2003) Two-phase chemically defined culture system for preimplantation rat embryos. Genesis 36:129–133

# Chapter 11

# ENU Mutagenesis to Generate Genetically Modified Rat Models

**Ruben van Boxtel, Michael N. Gould, Edwin Cuppen, and Bart M.G. Smits**

## Abstract

The rat is one of the most preferred model organisms in biomedical research and has been extremely useful for linking physiology and pathology to the genome. However, approaches to genetically modify specific genes in the rat germ line remain relatively scarce. To date, the most efficient approach for generating genetically modified rats has been the target-selected *N*-ethyl-*N*-nitrosourea (ENU) mutagenesis-based technology. Here, we describe the detailed protocols for ENU mutagenesis and mutant retrieval in the rat model organism.

**Key words:** ENU mutagenesis, Rat strains, Mutation discovery, Mutation frequency, Mutant, Knockout models

## 1. Introduction

The availability of technology that allows for introducing targeted genetic modifications in model organisms has greatly contributed to our understanding of specific gene function. In the mouse, homologous recombination in embryonic stem (ES) cells has proven to be a powerful tool for generating genetic knockouts (1). Although the laboratory rat (*Rattus norvegicus*) is one of the preferred model organisms in physiological and pharmacological research, gene knockout technology using homologous recombination approaches is still not available, due to the lack of pluripotent ES cells (2). A successful alternative approach for the production of knockout rats is using an *N*-ethyl-*N*-nitrosourea (ENU) mutagenesis-based technology. An advantage of this approach is that the created mutants are not "transgenic" in nature, since no artificial

I. Anegon (ed.), *Rat Genomics: Methods and Protocols*, Methods in Molecular Biology, vol. 597
DOI 10.1007/978-1-60327-389-3_11, © Humana Press, a part of Springer Science+Business Media, LLC 2010

DNA construct is integrated into the genome. And the technology does not require special (ES) cell lines and/or advanced oocyte or embryo manipulation, since it is based on the mutagenic property of the germ line mutagen ENU applied in vivo (3).

The mutagenicity of ENU results from the ability to transfer its ethyl group to oxygen or nitrogen radicals present in the DNA. Subsequent replication cycles of this damaged DNA can result in mispairing followed by single base pair substitutions. Treatment of male rats with ENU will cause DNA alkylation and subsequently point mutations in the DNA of cells with a high turnover rate, as in the spermatogonial stem cells. Because of the randomness of the mutagenesis, every affected cell will contain a unique set of induced mutations, which will result in a genetically heterogeneous sperm cell population. To transfer the induced mutations through the germ line, the treated males are crossed with untreated females. All individuals of the F1 population are screened for induced heterozygous mutations in preselected genes of interest. Occasionally, knockout-like alleles are generated by introducing a mutation in the open reading frame (ORF) or a splice donor/acceptor site of a gene of interest that will lead to a premature stopcodon and an absent or truncated protein product. Additionally, hypomorphic and hypermorphic alleles of the same gene could be generated by nonsynonymous mutations, which enable the study of gene-dosage effects and of amino acid residues important for protein–protein interaction or catalytic activity. Animals carrying interesting heterozygous mutations are identified using high-throughput mutation discovery methods, outcrossed and eventually bred to homozygosity.

The effectiveness of the technique depends on the efficiency of the mutagenesis and the mutation discovery methodology. The first knockout rats were identified using a yeast-based screening assay that specifically identifies mutations interfering with translation of the protein (4). Although it has been shown that this assay is highly effective for identifying knockout-like alleles, it neglects mutations that potentially results in interesting amino acids substitutions. Two other methods have been successfully applied to discover induced mutations in rat ENU mutagenesis experiments, namely CELI cleavage-mediated (5) and Mu transposase-based heteroduplex identification (6). These methods rely on the discovery of mismatches that arise after denaturing and reannealing a DNA fragment containing a heterozygous mutation. Both methods have the potency to identify all mutation types; however, the observed moderate mutation rates suggest false negatives in the discovery. These problems are overcome by resequencing the genes of interest of animals of the F1 population (7). Resequencing is considered to be the golden standard for mutation discovery. However, it is a relatively expensive method. In the near future, the emerging massive parallel sequencing platforms may offer novel possibilities for resequencing complete ORFs, which would increase the efficiency of the ENU mutagenesis-based knockout technology considerably.

Several rat knockout models for genes involved in human inherited diseases have already been made using this method. Characterization of the rat knockout models for the tumor suppressor genes *BRCA1, BRCA2, APC,* and *MSH6* demonstrated that the rat complements the equivalent mouse models for studying specific aspects of tumorigenesis, especially owing to its larger size, prolonged viability, and ability to bear larger tumors (4, 8–10). The serotonin transporter (*SERT*) knockout rat model has been shown to exhibit a disturbed serotonin homeostasis (11). Serotonin is evolutionarily the most ancient neurotransmitter. It is involved in a wide array of biological functions, such as emotion, motivation, and cognition, for which the rat is the preferred model organism to study. These examples illustrate that rat knockout models are a valuable addition to the toolkit to study the genetic basis of specific aspects of human health and disease.

The ENU mutagenesis-based knockout technology protocol follows a simple outline: ENU mutagenesis of male founder rats, F1 library generation by mating with untreated females, mutation discovery in genes of interest, and isolation of induced mutations into the desired genetic background (Fig.11.1). Here, we describe detailed protocols for each of these steps, including protocols for two mutation discovery strategies: high-throughput resequencing, developed by the Cuppen lab (Hubrecht Institute, Utrecht, The Netherlands), and the yeast-based screening assay, developed by the Gould lab (McArdle Lab for Cancer Research, Madison, WI).

## 2. Materials

### 2.1. Animals and ENU Mutagenesis

1. Safety wear for protection during ENU solution preparation and animals treatment: lab coat, gloves, mouth mask, and goggles.
2. One-way absorption paper and 0.1 M sodium hydroxide (NaOH) solution.
3. Syringes of 10 ml and 50 ml, needles of 21G, filters (∅ 0.2 μm) and disposable cuvettes and parafilm.
4. Ethanol (100%).
5. Phosphate citrate buffer: 0.1 M $NaH_2PO_4$, 0.05 M citric acid, and pH set at 5.0 with phosphoric acid. The buffer is filter sterilized.
6. Weigh scale.

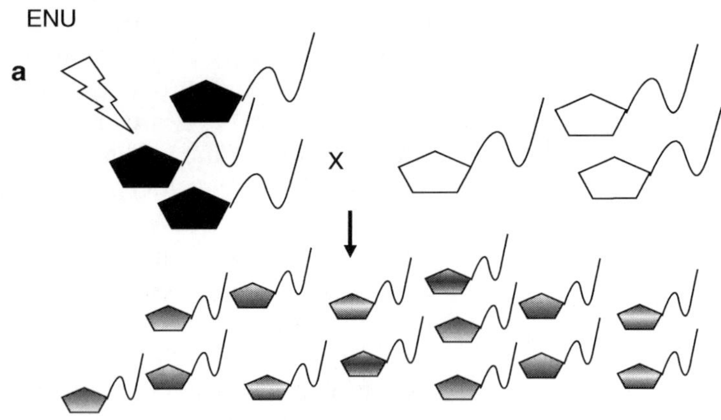


Generate F1 and isolate DNA

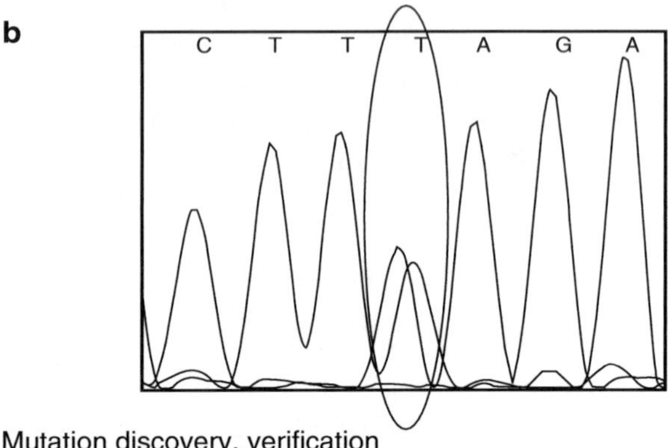


Mutation discovery, verification

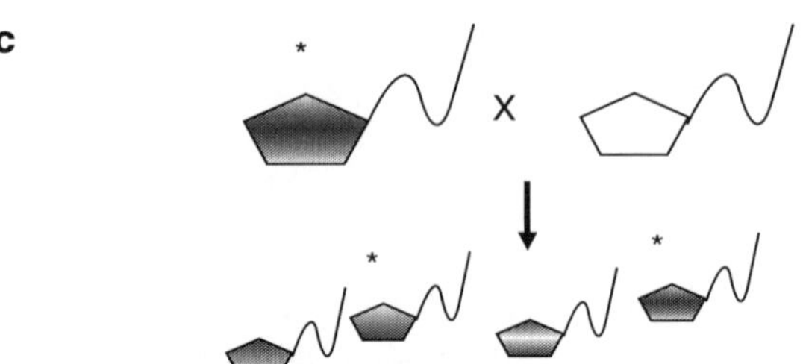


Outcrossing and genotyping

Fig. 11.1. Outline of the target-selected ENU mutagenesis-based technology. (**a**) Male rats (*black*) are mutagenized with ENU at 11–12 weeks of age. These males are crossed with untreated females (*white*) to generate an F1 population. Each of the animals of the F1 population harbors a unique set of heterozygous mutations (illustrated by variable *white*, *gray*, and *black* shading). From a tissue sample of all F1 animals generated atleast 10 weeks after the last injection, a genomic DNA sample is isolated. (**b**) The DNA is screened for ENU-induced heterozygous mutations in preselected genes of interest. The mutations of interest (i.e., stopcodons or nonsynonymous mutations) are verified in independent PCR and sequencing reactions. (**c**) The corresponding mutant rats are kept and outcrossed to the parental genetic background to eliminate putative confounding background mutations. In each backcross, heterozygous mutants are identified by genotyping and could be intercrossed to generate homozygous mutants.

7. Spectrophotometer.
8. Animals: Male rats 11–12 weeks of age and untreated fertile females for test breeding.

### 2.2. Library Generation and Genomic DNA Isolation

1. Sufficient untreated females for F1 progeny generation.
2. Instruments and tubes/deep-well plates for tissue collection.
3. Tissue lysis buffer: 100 mM Tris–HCl (pH 8.5), 200 mM NaCl, 0.2% SDS (w/v), 5 mM EDTA, and 100 μg/ml of freshly added Proteinase K.
4. Phenol:chloroform (1:1).
5. Isopropanol.
6. 70% ethanol.
7. Sterile 10 mM Tris–HCl (pH 8.0).
8. Tube microfuge and plate centrifuge.

### 2.3. Mutation Discovery

#### 2.3.1. Automated Resequencing

1. Primers designed for a nested PCR amplifying exons of genes of interest.
2. Taq polymerase.
3. dNTPs.
4. 5× PCR buffer: 25 mM of tricine, 7.0% glycerol (w/v), 1.6% DMSO (w/v), 2 mM $MgCl_2$ and 85 mM $NH_4Ac$, pH 8.7 with 25% ammonia (w/v).
5. PCR machine with 96- and/or 384-well blocks.
6. Standard gel electrophoresis unit.
7. Agarose.
8. Ethidium bromide.
9. BigDye (v3.1; Applied Biosystems). BigDye mix should be kept at –20°C and in the dark until use.
10. Sanger BigDye Dilution Buffer version 2 (SBDDv2; Applied Biosystems).
11. Ice-cold 80% ethanol.
12. Precipitation mix: 80% ethanol (w/v) supplemented with 40 mM NaAc.
13. OPTIONAL: liquid handling station.

#### 2.3.2. The Yeast-Based Screening Assay

1. Universal gap repair vector (12), available upon request.
2. Maxiprep kit (Qiagen).
3. PCR purification kit (Qiagen).
4. *Sma*I restriction enzyme and corresponding reaction buffer.
5. PCR primers having 5′ tails that match the universal gap repair vector.

6. PCR machine of choice.
7. Yeast strain yIG397 (MATa *ade*2-1 *leu*2-3,112 *trp*1-1 *his*3-11,15 *can*1-100 *ura*3-1 URA3 3xRGC::pCYC1::ADE2).
8. Yeast culture medium YPDA++, containing per liter: 50 g Difco broth (0428-17), 0.2 g filter-sterile adenine (added after autoclaving the broth).
9. Selective yeast culture plates, containing per liter (~50 plates): 6.7 g yeast nitrogen base (with ammonium sulfate, without amino acids), 20 g glucose, and 25 g agar. After autoclaving and cooling, the medium is supplemented with 5 mg of adenine, 20 mg tryptophan, 20 mg histidine, 20 mg arginine, 20 mg methionine, 30 mg isoleucine, 50 mg phenylalanine, 100 mg glutamine, 100 mg aspartic acid, 150 mg valine, 200 mg threonine, and 400 mg serine.
10. LiOAc/TE: 0.1 M lithium acetate, 10 mM Tris–HCl, pH 8.0, 1 mM EDTA.
11. LiOAc/TE/PEG: LiOAc/TE, 40% PEG (w/v).
12. Shaking incubator at 30°C, and incubator or water bath at 42°C.
13. 96-well flat bottom polystyrene microtiter plates.
14. OPTIONAL: automated plate counter (ProtoCOL/Microbiology International).

### 2.4. Outcrossing

1. Animals with the same genetic background as the mutagenized males.
2. A genotyping system of choice.

## 3. Methods

### 3.1. ENU Mutagenesis

Theoretically, treatment with high doses of ENU will yield a high molecular mutation frequency and therefore a high chance of generating modified alleles of genes of interest. However, higher doses of ENU will negatively affect fertility and even viability of the treated animals. Hence, the optimal dose is largely determined by the tolerance to the toxic effects of the mutagen. In general, the optimal dose will have to result in more than 25% fertile males to be able to efficiently generate a large F1 population. It has been demonstrated in the mouse that two or three weekly administrations of low doses of ENU will yield higher mutagenic efficiency than a single high dose (13). The optimal dose is strongly strain-dependent (Table 11.1), which could be a major consideration when selecting a strain to initiate an ENU mutagenesis

**Table 11.1**
**Differences in strain-dependent ENU tolerance**

| Strain/line | Optimal dose (mg ENU/kg bodyweight)[a] | Sterility dose (mg ENU/kg bodyweight)[a] | Mutation rate | Reference |
|---|---|---|---|---|
| BN | 3×20 | 3×40 | 1 in $2.91\times10^6$ bp | (7) |
| F344/Crl | 3×40 | 3×60 | 1 in $1.76\times10^6$ bp | (7) |
| F344/NHsd | 2×60 | 2×75 | 1 in 29 pups[b] | (4) |
| LEW | Unknown | 3×20 | Unknown | (7) |
| WF | Unknown | 2×50 | Unknown | (4) |
| WKy | Unknown | 2×50 | Unknown | (4) |
| Hsd:SD | 2×60 | 2×100 | 1 in 64 pups[b] | (4) |
| Wistar | 3×40 | 3×60 | 1 in $1.24\times10^6$ bp | (7) |
| *Msh6*[1Hubr] [d] | 3×30 | 3×35 | 1 in $5.85\times10^5$ bp[c] | (15) |

[a]Doses are listed as number of injections in weekly intervals
[b]For these strains, the mutation rate is shown as the rate of appearance of phenotypically aberrant pups
[c]Mutation frequency was shown to decline in F1 progeny that was generated more than 14 weeks after the last ENU
[d]This mutant rat line is Wistar-derived

experiment (*see* Note 1). Notably, outbred strains seem to have higher tolerance to the toxic effects of ENU when compared with inbred strains and yield higher mutation frequencies. The highest mutation frequency has been obtained using a Wistar-derived line, *Msh6*[1Hubr] (Table 11.1). This rat line lacks the MSH6 component of the mismatch repair (MMR) machinery and specifically fails to recognize single nucleotide mismatches (10). In rodents, it has been shown that the MMR system accounts at least partially for the repair of ENU-induced mutations (14, 15). Interestingly, when males of this rat line were mutagenized, not only did the mutation frequency increase more than twofold, but even the mutation spectrum was changed, resulting in an increased chance of introducing a premature stopcodon (15).

*3.1.1. ENU Stock Solution Preparation*

1. It is strongly advised to put on safety wear. ENU solutions should be handled with great caution, since it is a potent mutagen. ENU solutions should always be transported in a closed vial. A 0.1 M NaOH solution should be within reach to quickly neutralize any spillages (*see* Note 2).
2. Roughly 1 h prior to the planned injections, the ENU bottle is unpacked and the metal lid is removed (*see* Note 2).
3. Using a syringe, 5 ml of 100% ethanol is injected, and the solution is shaken vigorously (*see* Note 3).

4. 95 ml of the phosphate citrate buffer is slowly added, allowing the air to flow into the syringe for depressurization, followed by shaking vigorously for approximately 5 min to dissolve the ENU.
5. The ENU solution is filter sterilized (∅ 0.2 μm).
6. 100 μl of the ENU solution is mixed with 900 μl of the phosphate citrate buffer to measure the optical density (OD) using a spectrophotometer at 395 nm wavelength. The concentration is calculated by assuming that 1 OD unit equals a concentration of approximately 1 mg/ml (*see* Note 4).

*3.1.2. ENU Administration to Male Rats*

1. The volume of the ENU solution to be injected should be calculated for each male rat and follows the equation (optimal strain-dependent split dose of ENU × kg bodyweight)/(10 × measured OD). If the calculated volume of the ENU solution is too low for accurate injection, dilute the solution with phosphate-citrate buffer.
2. Prepare a room for the ENU treatment by covering the floor with the one-way absorption paper to absorb any spilled ENU.
3. Male rats of 11–12 weeks of age are mutagenized by intraperitoneally (IP) injecting a precalculated volume of the ENU solution (*see* Note 5).
4. The ENU treatment is repeated weekly until the optimal dose is reached.
5. It is recommended to monitor the animals' health at least once a week after the last ENU injection by determining their bodyweight and general status. Temporal stabilization of body weight is normal, but a decrease in body weight after 1 week after injection could be a measure for taking that animal out of the study.
6. The fertility rate is determined in the first 10 weeks after the last injection as a surrogate for the efficiency of mutagenesis. Three weeks after the last injection, mutagenized males are crossed with untreated females, and the progeny are counted. A reduction and regain of the fertility rate, as well as smaller litter sizes, in the first 10 weeks are indicative of an effective mutagenesis (16), as ENU is capable of killing the mature sperm population. F1 pups sired in the first 10 weeks after mutagenesis are not suitable for screening (*see* Note 6).

### 3.2. Library Generation

After the mutagenesis procedure, the ENU-treated males are crossed with untreated females to generate F1 progeny. Depending on the available animal space, there are different approaches to generate a library. If space is not limited, a large living animal repository can be generated as fast as possible, which can be repeatedly screened for mutations in genes of interest. In contrast, a rolling-cycle model significantly reduces the amount of animal space needed. In this model, the F1 animals are screened for interesting mutations in a preselected set of genes before weaning, and only animals harboring interesting mutations are retained. Here, we describe the general protocol for producing a genomic DNA (gDNA) library from an F1 population.

1. An F1 population is generated by crossing the mutagenized males with untreated females. 10 weeks after the last ENU injection, the F1 animals are retained for tissue collection (*see* Note 7).
2. From each F1 animal, a tissue fragment (e.g., toe or tail material) is collected for DNA extraction.
3. The tissue sample is incubated overnight at 55°C in 400 µl of lysis buffer, preferably under shaking or rotating conditions.
4. If tissue debris is present (which is most likely from a tail clip), the samples are spun at maximal speed for 1 min. The supernatant is transferred to a fresh tube.
5. 400 µl of phenol/chloroform (1:1, v/v) is added, and the mixture is vortexed vigorously for 2 min followed by centrifugation for 3 min at maximal speed. The aqueous layer is transferred to a new tube, and the phenol/chloroform extraction is repeated. After centrifugation, 300 µl of the aqueous layer is transferred to a new tube.
6. The gDNA is precipitated by adding 300 µl of isopropanol and by inverting the tube ten times. The sample is centrifuged at 12,000 *g*, at 4°C for 20 min. The supernatant is removed, and 100 µl of 70% ethanol is added to wash the pellet. Again, the sample is centrifuged at 14,000 rpm, at 4°C for 5 min, and the supernatant is removed. The sample is spun another minute, and all remaining ethanol is carefully removed using a pipette. The DNA is dissolved by adding 500 µl 10 mM Tris–HCl (pH 8.0) and incubating at 55°C for 10 min with occasional vortexing.
7. The DNA can be stored at –20°C.

### 3.3. Mutation Discovery Protocols

As described in the introduction, different mutation discovery platforms can be used to screen the DNA of the F1 progeny for induced mutations in exons of interest. Here, we will give detailed descriptions of the two procedures of our choice, namely high-throughput resequencing, developed by the Cuppen lab, and the yeast-based screening assay, developed by the Gould lab.

#### 3.3.1. High-Throughput Resequencing

This assay is based on amplifying the preselected genetic fragments using a nested-PCR setup, followed by dideoxy sequencing. Depending on the strategy that is used for the F1 library generation, two different procedures can be followed. If a large living animal repository was generated, the DNA of the F1 animals could be sampled in 96-well format and screened. This approach allows for multiplexing primers of up to eight amplicons in the first PCR. In contrast, if the rolling-cycle model was applied, the DNA of the F1 animals could be separately screened for all preselected amplicons (*see* Note 8). The setup described here can be fully automated using robotics and is well suited for scaling.

1. These instructions assume the use of a GeneAmp® PCR system 9700 (Applied Biosystems), although other brands are expected to perform equally.
2. The first PCR reaction contains 5–10 ng of template DNA, 2 μl of 5× PCR buffer, 0.1 mM of each dNTP (*see* Note 9), 0.2 μM of each primer (*see* Note 10), and 0.2 units of Taq polymerase in a total volume of 10 μl and is carried out using a touchdown thermocycling program (94°C for 60 s; 15 cycles of 92°C for 30 s, 65°C for 30 s with a decrement of 0.2°C per cycle and 72°C for 60 s; followed by 30 cycles 92°C for 30 s, 58°C for 30 s and 72°C for 60 s; 72°C for 180 s).
3. After thermocycling, the first PCR reaction is diluted with 20 μl of Milli-Q water, and 1 μl is hatched into the second PCR mix, which contains 1 μl of 5× PCR buffer, 0.1 mM of each dNTP, 0.2 μM of each primer, and 0.1 units of Taq polymerase in a total volume of 4 μl. The second PCR is carried out using a standard thermocycling program (94°C for 60 s; 35 cycles of 92°C for 20 s, 58°C for 30 s and 72°C for 60 s; 72°C for 180 s).
4. Several samples of the second PCR are checked on a 1% agarose gel containing ethidium bromide for the presence of the correct amplification product after thermocycling.
5. The second PCR mix is then diluted with 20 μl of Milli-Q, and 1 μl is hatched into the sequencing mix, which contains 1.9 μl of sequencing buffer, 0.1 μl of BigDye, and 0.4 μM of sequencing primer (*see* Note 11) in a total volume of 4 μl. The sequencing reaction is carried out using a specially designed thermocycling program (40 cycles of 92°C for 10 s, 50°C for 5 s, and 60°C for 120 s).
6. The sequencing fragments need to be purified. To each well, 30 μl of precipitation mix is added, the plate is sealed, the mixtures are vortexed vigorously, and are spun at maximal speed, at 4°C for 40 min in a cooled plate centrifuge. The supernatant is discarded, and 30 μl of ice-cold 80% ethanol is added. The samples are spun for an additional 10 min. The supernatant is discarded, and the plates are air-dried for approximately 15 min, preferably protected from light. The precipitate is dissolved in 10 μl of Milli-Q.

7. The plates are analyzed on the 96-capillary 3730XL DNA analyzer (Applied Biosystems) using the standard RapidSeq protocol on a 36 cm array.
8. The sequencing reads that belong to one amplicon are aligned, which facilitates the discovery of heterozygous point mutations (*see* Note 12).
9. All candidate mutations are verified in an independent PCR and sequencing reaction.

#### *3.3.2. Yeast-Based Screening Assay*

The yeast-based screening assay developed by the Gould lab can be applied to any coding exon of interest, as it makes use of a previously designed universal gap repair vector (12). The vector is available upon request. The vector contains a small unique gap repair sequence cassette that can be cleaved by a *Sma*I digestion. By using PCR primers with 3′ tails that match the cleaved ends of the unique gap repair cassette, any PCR product of interest will be cloned in vivo into the gap repair vector upon cotransformation of the PCR product and the linearized vector into competent yeast cells. Once cloned, the PCR fragment will be located behind the yeast *ADH1* promoter and in frame with the *ADE2* reporter gene, which could give rise to a functional fusion protein. The ADE2-negative yeast cells that receive a gap-repaired vector will have restored ADE2 function and will grow efficiently into large white colonies on selective media. However, if the DNA donor rat has a truncating mutation in the amplified fragment, the fusion protein will be nonfunctional and the cells will grow poorly into small red colonies. Since the F1 animals are heterozygous for putative induced truncating mutations, a plate derived from such animals will have half white and half red colonies, which is easily recognizable by eye.

The procedure consists of the following steps: Vector linearization, template generation, PCR, yeast cotransformation, and colony screening.

1. For a successful screening assay, the PCR primers need to be designed carefully. The gene-specific part of the primers should be designed such that the amplified coding fragment of an exon runs in frame with the *ADE2* gene upon in vivo cloning (*see* Note 13). For robustness of the reaction, it is advised that the gene-specific primers amplify an exonic piece of no more than 2 kb. The universal adapter sequences are 5′-GGCCATCGATAGCTCGATGTAACGTG′, and 5′-GGCCTACTAACAGATACGCTATGCAGGACTCTGGATTGCCC-3′. These should be added to the 5′ end of the upstream and downstream gene-specific primers respectively.
2. The circular universal gap repair vector is isolated in large quantity, using the maxiprep system (Qiagen). The vector is linearized using a standard *Sma*I digestion and cleaned up

using the PCR purification kit (Qiagen). Linearization is verified by standard gel electrophoresis, and the concentration is adjusted to 100 ng/µl.

3. To generate gDNA templates for PCR, the extracted gDNA samples are diluted to the desired concentration. The proper dilution factor must be determined empirically and may differ between primer combinations. Determination of the dilution factor is done using the PCR conditions of choice, on a series of template dilutions.

4. The PCR could be done using reaction conditions of choice. The conditions of our choice are as follows: 1× Herculase Hotstart buffer (Stratagene), 0.2 mM dNTPs, 25 ng/µl of each primer, and 0.15 units/µl of Hotstart High-Fidelity DNA polymerase (Stratagene) in a total volume of 10 µl. The cycling conditions of our choice are 94°C for 2 min, then 35 cycles of 94°C for 45 s, 61°C for 45 s, and 72°C for 1–3 min, with a final elongation step of 72°C for 10 min. To verify successful amplification, a fraction of each reaction (~10%) should be analyzed using standard agarose gel electrophoresis. The proper template dilution and PCR conditions should yield a solid PCR product in the vast majority of DNA samples tested.

5. The transformation of the yIG397 yeast cells with the linearized universal gap-repair vector and the PCR product requires preparation of a fresh yeast culture. About 48 h in advance, a preculture is started by transferring a single colony to 10 ml of YPDA++ media, which is incubated overnight at 30°C in a shaking incubator. The next morning, the preculture is removed from the shaking incubator and rested at room temperature. At the end of the day, the appropriate amount of large overnight cultures is started by diluting 75–90 µl the preculture into 500 ml of YPDA++ in large flasks. For each transformation, 2.5 ml of culture is needed. The cultures are incubated ~16 h at 30°C in a shaking incubator until the $OD_{600nm}$ reaches 0.6. The yeast cells are pelleted by centrifugation for 10 min at 2,500 × *g*. The supernatant is removed by decanting the bucket. To wash, the pellet is resuspended in 1/4 of the original volume of $dH_2O$. The cells are pelleted again and resuspended in the appropriate amount of freshly prepared LiOAc/TE (8 µl/ml of grown culture). The transformation is performed in a 96-well flat bottom polystyrene microtiter plate. The final transformation mixture contains 30 µl yeast cells, 0.1 µl of the linearized vector (10 ng), 2.5 µl of the salmon sperm DNA carrier (25 µg), 150 µl of the LiOAc/PEG/TE, and 2–5 µl of the unpurified PCR product. It is required that the yeast cells remain in suspension. To mix the transformation reactions, the 96-well plate is mildly shaken (~1,000 rpm) on a plate shaker for 10 min. The mixture is

incubated for 30 min at 30°C, followed by a heat-shock of 15 min at 42°C. The cells are pelleted by centrifugation of the plate for 5 min at 2,000 rpm. The supernatant is removed from the wells by decanting and gently flicking the plate above a sink. To the pelleted cells, 60 μl of sterile $dH_2O$ is added. The plate can now be stored up to 24 h until ready for plating.

6. The final step is the preparation and screening of the colony plates. The cells are plated on prewarmed synthetic minimal medium lacking leucine and supplemented with 5 μg/ml of adenine. The transformants are grown for 3 days at 30°C. After 3 days, the plates are scanned by eye or analyzed using an automated colony counter. The background level of red colonies derived from a wild type animal is generally very low (0.5–1%), but may be increased depending on the size and sequence content of the amplicon.
7. To definitively verify the nature of the mutation, an independent PCR and sequencing reaction must be performed on the original gDNA sample of the putative heterozygous knockout animal.

### 3.4. Outcrossing

Once F1 animals harboring interesting mutations are identified, it is important to cross them with untreated animals of the parental strain. This outcross yields more heterozygous animals, which can be used to cross the mutation to homozygosity to study its phenotypic effect. Additionally, however, to eliminate as many background mutations as possible, more backcrosses to the parental strain genetic background should be initiated. Generally, six to ten backcrosses are considered necessary to vanish the vast majority of induced random background mutations. However, background mutations that are close to the site of the mutation of interest could be coinherited. Such linked mutations are hard to evade (17). However, the change that such a linked mutation landed in a genetic feature that attenuates the phenotype under study is small. Nevertheless, once homozygous animals are generated that show a phenotype, nonhomozygous littermates should always be included as controls for phenotypic analysis. There should always be a 1:1 relationship between genotype and phenotype. If this is not the case, a linked, confounding background mutation might be present. The ultimate control would be the generation of a second allele of the same gene. Homozygous mutants of the second allele will not carry the confounding background mutation and should verify the phenotype.

1. Animals of the F1 population that harbor interesting mutations are crossed with untreated animals of the parental strain.
2. The backcross progeny are genotyped to determine if they inherited the induced mutation (*see* Note 14). The heterozygous animals are retained.

3. Further outcrossing of the mutation to the parental strain genetic background is performed by backcrossing heterozygous animals of each subsequent generation to the parental strain. In every outcrossing stage, homozygous animals can be generated by intercrossing two heterozygotes. This procedure also yields nonhomozygous littermates that could serve as controls in experiments.

## 4. Notes

1. Before starting any ENU mutagenesis experiment in the rat, it is important to select a strain and its corresponding optimal dose of ENU. As the ENU tolerance and subsequent induced mutation rate are strongly strain-dependent (Table 11.1), not every strain is equally suitable for mutagenesis. If the genetic background is not of utmost importance, using an outbred strain (Wistar, Sprague Dawley) is recommended. If an isogenic genetic background is desired, the Fisher inbred strain (F344) is recommended. Using other inbred strains, such as Lewis or Brown Norway, can be problematic due to their limited ENU tolerance and/or reproductive capabilities. An ENU mutagenesis experiment in such inbred strains requires a carefully designed mutagenesis, breeding, and mutation discovery plan. This plan might involve more males to be mutagenized and more base pairs to be screened for induced mutations, as the mutation rate could be moderate. Alternatively, a more suitable strain could be chosen for the mutagenesis, such as an outbred strain or the *Msh6–/–* line. A recovered mutation could be transferred to any genetic background by outcrossing. Similar considerations should be taken into account as with normal outcrossing to the parental genetic background (*see* Subheading 11.3.4).
2. We recommend preparing the ENU solution 1 h prior to treatment. Because the half-life of ENU is less than a minute at high pH, alkaline solutions (0.1–1 M NaOH) should be kept closely to inactivate any spillage and to clean equipment.
3. It is recommended to thaw the ENU isopac before solution preparation. This greatly facilitates dissolving the ENU.
4. The final concentration of the ENU stock differs between batch numbers and should always be determined by measuring the $OD_{395nm}$. Typically, this concentration varies between 6 and 8.5 mg/ml. In ENU mutagenesis experiments in rodents, the concentration of ENU is generally measured at a wavelength of 395 nm; however, in other ENU experiments, a wavelength of 238 nm might be used (e.g., zebrafish ENU mutagenesis). For consistency purposes, it is recommended to use $OD_{395nm}$.

5. For safety reasons, two persons should perform the ENU injections. The first person will restrain the animal, and the other person will perform the injection of the ENU solution.
6. Mutation discovery in F1 animals generated in the first 10 weeks after ENU mutagenesis might result in the identification of chimeric animals. It is thought that such chimeras are the result of ethyl adducts in the fertilized oocyte, which originate from mutagenized postmeiotic sperm cells. This could result in heterogeneous mutation fixation in different cell lineages during embryonic development.
7. It is recommended to produce the F1 library as quickly as possible. Not only does ENU affect the viability of the founders, but there are also indications that time negatively influences the mutation rate in the germ line of the founders (especially if the MSH6-deficient line is used). This is thought to be the result of selective repopulation of the testis by the most viable stem cells presumably containing the lowest amount of genotoxic damage after initial depletion of the population (15). This not only decreases the apparent mutation frequency, but also potentially increases the chance of clonal progeny.

   Recently, an attractive advance to the library generation has been developed. Male F1 animals could now be utilized for generating an archive of cryopreserved sperm. In this way, the library could be screened exhaustively, and interesting mutants could be recovered by intracytoplasmic sperm injection (6).
8. In the rolling-cycle model, the animals are screened before weaning. In this case, it is recommended to use PCR plates with pregridded primers to increase the throughput. Management of mutation discovery screening projects and the design of large batches of primers for groups of selected genes can be done using LIMSTILL (http://limstill.niob.knaw.nl).
9. For multiplexing, it is necessary to increase the amount of each dNTP by fourfold in the first PCR.
10. The setup described here assumes the use of primers that were designed with an optimal melting temperature of 58°C.
11. The sequence reaction can be carried out using the forward primer of the second PCR reaction or by using a universal M13 primer if the second PCR primers were designed with universal M13 tails on their 5′ ends.
12. For aligning and analyzing the sequence, reads PolyPhred (18) can be used. This computer program automatically detects the presence of heterozygous single nucleotide substitutions.
13. The yeast-based screening assay is most efficiently employed if the exon of interest is above 1.5 kb. However, not all genes harbor such a large exon. To bypass any exon–intron structures, the yeast-based screening assay comes in a second version that

uses a cDNA sample derived from the mRNA as the template for the PCR. There are three major drawbacks associated with this version: a high false positive rate, the possibility of nonsense-mediated decay of the aberrant mRNA, and the mRNA of interest may not be expressed at all in the collected tissue sample. Despite these disadvantages, the *BRCA1* knockout rat was identified using the cDNA version of the yeast-based screening assay (4). The protocol differs on two points: along with gDNA, total RNA is extracted from a second tissue sample of the same F1 animal, and a reverse transcriptase (RT) step using mRNA-specific primers is incorporated to generate cDNA as a PCR template.

14. Genotyping could be performed by resequencing, or any other cost-reducing strategy (e.g., RFLP). If the MMR-deficient strain is used as the strain of choice, the first outcross is also used to eliminate the *Msh6–/–* allele, as MSH6-deficiency would result in an instable genetic background. Hence, heterozygous progeny from the first outcross should also be genotyped to check for absence of the *Msh6–/–* allele.

## References


1. Capecchi MR (2005) Gene targeting in mice: functional analysis of the mammalian genome for the twenty-first century. Nat Rev Genet 6:507–512
2. Aitman TJ, Critser JK, Cuppen E, Dominiczak A, Fernandez-Suarez XM, Flint J, Gauguier D, Geurts AM, Gould M, Harris PC, Holmdahl R, Hubner N, Izsvak Z, Jacob HJ, Kuramoto T, Kwitek AE, Marrone A, Mashimo T, Moreno C, Mullins J, Mullins L, Olsson T, Pravenec M, Riley L, Saar K, Serikawa T, Shull JD, Szpirer C, Twigger SN, Voigt B, Worley K (2008) Progress and prospects in rat genetics: a community view. Nat Genet 40:516–522
3. Russell WL, Kelly EM, Hunsicker PR, Bangham JW, Maddux SC, Phipps EL (1979) Specific-locus test shows ethylnitrosourea to be the most potent mutagen in the mouse. Proc Natl Acad Sci U S A 76:5818–5819
4. Zan Y, Haag JD, Chen KS, Shepel LA, Wigington D, Wang YR, Hu R, Lopez-Guajardo CC, Brose HL, Porter KI, Leonard RA, Hitt AA, Schommer SL, Elegbede AF, Gould MN (2003) Production of knockout rats using ENU mutagenesis and a yeast-based screening assay. Nat Biotechnol 21:645–651
5. Smits BM, Mudde J, Plasterk RH, Cuppen E (2004) Target-selected mutagenesis of the rat. Genomics 83:332–334
6. Mashimo T, Yanagihara K, Tokuda S, Voigt B, Takizawa A, Nakajima R, Kato M, Hirabayashi M, Kuramoto T, Serikawa T (2008) An ENU-induced mutant archive for gene targeting in rats. Nat Genet 40:514–515
7. Smits BM, Mudde JB, van de Belt J, Verheul M, Olivier J, Homberg J, Guryev V, Cools AR, Ellenbroek BA, Plasterk RH, Cuppen E (2006) Generation of gene knockouts and mutant models in the laboratory rat by ENU-driven target-selected mutagenesis. Pharmacogenet Genomics 16:159–169
8. Amos-Landgraf JM, Kwong LN, Kendziorski CM, Reichelderfer M, Torrealba J, Weichert J, Haag JD, Chen KS, Waller JL, Gould MN, Dove WF (2007) A target-selected Apc-mutant rat kindred enhances the modeling of familial human colon cancer. Proc Natl Acad Sci U S A 104:4036–4041
9. Cotroneo MS, Haag JD, Zan Y, Lopez CC, Thuwajit P, Petukhova GV, Camerini-Otero RD, Gendron-Fitzpatrick A, Griep AE, Murphy CJ, Dubielzig RR, Gould MN (2007) Characterizing a rat Brca2 knockout model. Oncogene 26:1626–1635
10. van Boxtel R, Toonen PW, van Roekel HS, Verheul M, Smits BM, Korving J, de Bruin A, Cuppen E (2008) Lack of DNA mismatch repair protein MSH6 in the rat results in

hereditary non-polyposis colorectal cancer-like tumorigenesis. Carcinogenesis 29:1290–1297

11. Homberg JR, Olivier JD, Smits BM, Mul JD, Mudde J, Verheul M, Nieuwenhuizen OF, Cools AR, Ronken E, Cremers T, Schoffelmeer AN, Ellenbroek BA, Cuppen E (2007) Characterization of the serotonin transporter knockout rat: a selective change in the functioning of the serotonergic system. Neuroscience 146:1662–1676
12. Chen KS, Gould MN (2004) Development of a universal gap repair vector for yeast-based screening of knockout rodents. Biotechniques 37:383–388
13. Justice MJ, Carpenter DA, Favor J, Neuhauser-Klaus A, Hrabe de Angelis M, Soewarto D, Moser A, Cordes S, Miller D, Chapman V, Weber JS, Rinchik EM, Hunsicker PR, Russell WL, Bode VC (2000) Effects of ENU dosage on mouse strains. Mamm Genome 11:484–488
14. Claij N, van der Wal A, Dekker M, Jansen L, Te Riele H (2003) DNA mismatch repair deficiency stimulates N-ethyl-N-nitrosourea-induced mutagenesis and lymphomagenesis. Cancer Res 63:2062–2066
15. van Boxtel R, Toonen PW, Verheul M, van Roekel HS, Nijman IJ, Guryev V, Cuppen E (2008) Improved generation of rat gene knockouts by target-selected mutagenesis in mismatch repair-deficient animals. BMC Genomics 9:460
16. Smits BM, Haag JD, Cuppen E, Gould MN (2007) Rat knockout and mutant models. In: Conn PM (ed) Sourcebook of models for biomedical research. Humana, Totowa, NJ, pp 171–178
17. Keays DA, Clark TG, Flint J (2006) Estimating the number of coding mutations in genotypic- and phenotypic-driven N-ethyl-N-nitrosourea (ENU) screens. Mamm Genome 17:230–238
18. Nickerson DA, Tobe VO, Taylor SL (1997) PolyPhred: automating the detection and genotyping of single nucleotide substitutions using fluorescence-based resequencing. Nucleic Acids Res 25:2745–2751

# Chapter 12

# Establishment of Embryonic Stem Cells from Rat Blastocysts

**Masaki Kawamata and Takahiro Ochiya**

## Abstract

Rats have important advantages over mice as an experimental system for physiological and pharmacological investigations. Their embryonic stem (ES) cells, after differentiation into each tissue or organ, are applied in regenerative medicine, which enables examination of the effects of drugs for various diseases. Knockout rats will also provide a suitable model system for many human diseases and a great amount of new insights into gene functions, which have not been revealed by knockout mice. In 2008, we experienced the world's first success in establishing rat ES cells with chimeric contribution. Following on the heels of our report, others reported the establishment of rat ES cells that could complete a germline transmission. Recent studies on rat as well as mouse ES cells suggest that modifications of signal inhibitors and serum in the medium are critical for the maintenance of the pluripotency of ES cells. In this chapter, we discuss techniques for the successful establishment and maintenance of rat ES cells.

**Key words:** ES cell, Signal inhibitor, Pluripotency, Chimera rat, Knockout rat

## 1. Introduction

Embryonic stem (ES) cells are derived from the inner cell mass (ICM) of preimplantation blastocysts (1). ES cells have been routinely derived from mice since 1981 (2, 3). These cells have a stable developmental potential to form derivatives of all three embryonic germ layers even after prolonged culture (4), and have been used to study the mechanism of cell differentiation. Moreover, they are capable of generating germline chimeras following injection into the blastocyst (5) (Fig. 12.1).

The laboratory rat was the first mammalian species domesticated for scientific research, and it has been used as an animal model in physiology, toxicology, nutrition, behavior, immunology, and neoplasia for over 150 years (6). Despite this history, rats

I. Anegon (ed.), *Rat Genomics: Methods and Protocols*, Methods in Molecular Biology, vol. 597
DOI 10.1007/978-1-60327-389-3_12, © Humana Press, a part of Springer Science+Business Media, LLC 2010

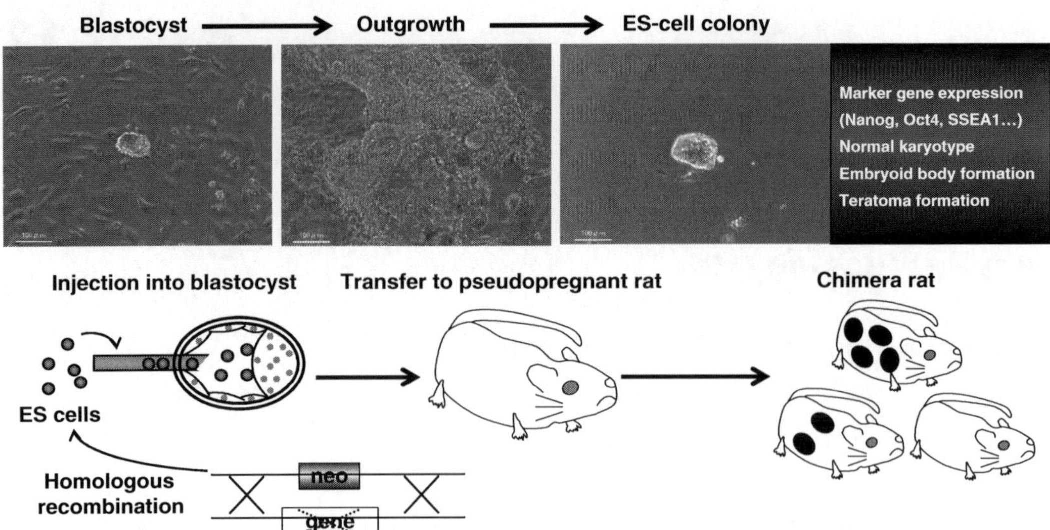


Fig. 12.1. Establishment of rat ES cells and strategy to generate gene-targeting animals. Signal inhibitors enable outgrowth of ICM of rat blastocysts and colony formation from single cells. The rat ES cells show typical features of mouse ES cells. To generate gene-targeting rats, homologous recombination is performed in the ES cell. After donor ES cells are injected into host blastocysts, they are transferred to uteri of pseudopregnant rats. Heterozygous rats are obtained from germline chimera mated with wild-type rats.

lag far behind mice in functional genetic studies and generation of knockout animals reflecting human disease models because of the absence of germline-competent rat ES cells, which are vital in a reverse genetics approach (7). After the first mouse ES cell lines were derived 28 years ago (2, 3), many efforts were made to establish rat ES cells. However, no one was able to succeed in establishing authentic ones. Although we established new lines of rat ES cells with chimeric contribution, they could not complete germline transmission (8). Soon after our report, other groups succeeded in establishing rat ES cells with germline transmission (9, 10). The common technique for maintaining the pluripotency of rat ES cells was to reduce the content of fetal bovine serum (FBS) in the culture medium (8–10). Two groups overcame the difficulty of generation of germline-competent rat ES cells by using signal inhibitors (9, 10). Recent studies suggest that signal inhibitors have a critical role in the maintenance of rat ES cells (9, 10) as well as rat induced pluripotent stem (iPS) cells (11).

## 2. Problems in Current Methods to Maintain Self-Renewal of ES Cells

Derivation and maintenance of the undifferentiated state of mouse ES cells originally relied on cocultivation with feeder cells, usually mitotically inactivated mouse embryonic fibroblasts

(MEFs), and the presence of serum. Later, it was shown that leukemia inhibitory factor (LIF) is the key cytokine secreted by feeders in supporting mouse ES cell self-renewal (12, 13). Ying et al. demonstrated that bone morphogenetic proteins (BMPs) could replace serum and act together with LIF to maintain mouse ES cell self-renewal (14). Although a culture medium composed of either serum and LIF, or BMP and LIF has been thought to be available in all species, several groups failed to establish rat ES cells under similar conditions (15–19). We have already succeeded in cloning complete rat LIF cDNA and demonstrated that rat LIF had an effect on rat ES cells for the maintenance of a stem-cell phenotype (20). However, their self-renewal potential was temporal in early passages, which may be due to the similar culture medium of mouse ES cells.

## 3. Establishment of Rat ES Cells by Reduction of Fetal Bovine Serum (FBS)

Recently, we succeeded in establishing rat ES cells with chimeric contribution by using a devised culture medium and passaging method (8). While a general culture medium for mouse ES cells contains 15 or 20% FBS, our culture medium contains only 3% FBS. The method for rat ES cells is briefly mentioned below.

The rat ES-cell culture medium consisted of DMEM/F12 supplemented with 3% FBS, 0.1 mM 2-mercaptoethanol, 1% non-essential amino acid, 2 mM L-glutamine, 1 mM sodium pyruvate, antibiotic antimycotic, and nucleoside solution. Frozen embryos of Wistar rats obtained at 4.5 days postcoitum (d.p.c.) were used for establishment of ES cells. After removal of zona pellucida by treating with tyrode's solution, seven to ten embryos were placed on a plate preseeded with mitotically inactivated MEFs by a treatment with mytomycin C. After 2 or 3 days, ICM-derived cells were dissociated into clumps mechanically or by exposure to 0.05% collagenase type IV before transfer onto a new feeder layer. The propagated cells were routinely passaged every 3–4 days up to 5 passages in the culture medium in the presence of rat LIF at 1,000 U/ml, and then they were cultured in a medium supplemented with 0, 250, 500 or 1,000 U/ml rat LIF.

These rat ES cells showed marker gene expression of ES cells such as Oct4, Nanog, SSEA1, and so on, and formed embryoid bodies (EBs) after the ES-cell colonies were dissociated by treating with collagenase IV. Teratomas were formed by subcutaneous, intratesticular, or intraperitoneal injection of rat ES cells into scid mice. Finally, chimeric rats were generated from embryos in which the rat ES cells, cultured in the presence of rat LIF, were injected (8).

Germline transmission could not occur in the chimeric rats. The reasons might be due to chromosomal instability, inadequate expression levels of ES cell marker genes, or inability of colony formation from single cells. Chromosomal abnormality is one of the major causes for the loss of germline competence of mouse ES cells (21). Chimeric mice or germline transmitters were obtained from ES cells exhibiting over 28 or 38% of normal karyotype, respectively (22). Although the rat ES cells could sustain approximately 40% of the normal karyotype, a higher rate may be necessary to complete germline transmission.

The marker genes of the ES cells were detected in the rat ES cells. However, the genuine expression levels of the marker genes were obscure because authentic rat ES cells did not exist as a positive control. We, however, recently succeeded in establishing new rat ES cell lines by a further devised culture medium containing signal inhibitors, which showed quite higher expression levels of the ES-cell marker genes (Fig. 12.1).

The rat ES cells scarcely formed a colony from a single cell through enzymatic treatment. When the ES cells were injected into blastocysts, most of the cells were manipulated as single cells. The injected rat ES cells might have difficulty proliferating and developing into various tissues or organs with high rates during the embryogenesis, resulting in a failure of germline transmission.

## 4. Maintenance of Pluripotency and Self-Renewal by Signal Inhibitors

Recent reports suggest that small molecules, which inhibit glycogen synthase kinase-3 (GSK3), fibroblast growth factor-4 (FGF4) through the mitogen-activated protein kinase (MAPk) pathway, or Rho-associated kinase (ROCK) signaling, have effects on ES cells for the maintenance of pluripotency and self-renewal. GSK3 is a central node for negative modulation of a range of anabolic processes and generally acts to suppress cellular biosynthetic capacity (23). GSK3 is inhibited by phosphorylation downstream of growth factors that activate phosphatidyl inositol 3 kinase and Akt. GSK3 is also a key component of the β-catenin destruction complex, and pharmacological inhibition of GSK3 increases cytoplasmic and nuclear β-catenin, mimicking canonical Wnt signaling (24). The Wnt pathway was assumed to maintain self-renewal of ES cells because the main components of the canonical Wnt pathway were detected in undifferentiated human ES cells (25). Actually, the Wnt pathway activation by 6-bromoindirubin-3′-oxime (BIO), a specific pharmacological inhibitor of GSK3 (26), maintained an undifferentiated phenotype in mouse and human ES cells and sustained expression of the pluripotent state-specific transcription factors Oct4, Rex1, and Nanog even in the absence of LIF and MEF (27). However, BIO is not highly selective and cross-reacts with cyclin-dependent kinases and other kinases, while CHIR99021 was defined as a more selective inhibitor of GSK3 (28–30). Ying et al. found that the activity of mouse ES cells was reduced by BIO but not by CHIR99021 (31). In a report relating to the Wnt pathway, a high-throughput cell-based assay showed that a small molecule IQ-1 allowed for long-term expansion of mouse ES cells and inhibited spontaneous differentiation to prevent β-catenin from switching coactivator usage from CBP to p300 (32). These reports suggest that the addition of GSK3 inhibitor or Wnt recombinants in the ES culture medium might be a useful method to continuously propagate undifferentiated ES cells.

FGF signaling is a conserved initiator of vertebrate neural development (33–37). Activation of FGF receptors (FGFRs) can initiate transduction via three major intracellular pathways: classical MAPk, phospatidylinositol 3 -OH kinase (PI3K), and phospholipase C gamma (PLCγ), of which the last two can activate protein kinase C proteins (PKCs), which can in turn stimulate ERK1/2 signaling (38). All three pathways were implicated in effects on vertebrate embryos that induce differentiation into neural tissue (34, 39, 40). A high-throughput chemical screen with a library of 50,000 compounds revealed that the compound SC-1 dually inhibited RasGAP and ERK1, which propagate mouse ES cells in an undifferentiated, pluripotent state even in the absence of MEF, serum, and LIF (41). Treatment of ES cells with the specific inhibitor for MEK, PD098059 (42), ERK, PD184352, or FGFR, PD173074 and SU5402 also suppressed differentiation of ES cells (43, 44). Furthermore, the majority of Fgf null (Fgf−/−) ES cells (45) or Erk2−/− ES cells were able to retain expression of Oct4 under a differentiation condition without LIF (43). Since Fgf4 mRNA is expressed specifically in ES cells of various animals, FGF4 has been considered as a marker gene of ES cells. On the other hand, these reports suggest that an autoinductive stimulation of the MAPk by FGF4 enhances differentiation of ES cell, especially into neural cells. Thus, the MAPk inhibition might be a key method for suppressing differentiation of ES cells.

Recently, Watanabe et al. found that a ROCK inhibitor, Y-27632, caused human ES cells to block apoptosis after dissociation into single cells by enzymatic treatment. It is characteristic of human ES cells that they need to be subcultured by the bulk-passage method since single ES cells form scant colonies. The propagated ES cells cultured by Y-27632 were positive for alkaline phosphatase (ALP), marker genes such as E-cadherin, Oct4, and SSEA4, and the number of chromosomes was normally kept during a long-term culture (46). Although the mechanism that allows Y-27632 to form human ES-cell colony with undifferentiated state is unknown, the compound is recently used for a single-cell-passaging method. Actually, we have succeeded in propagating rat ES cells that maintain pluripotency by the addition of Y-27632 in a devised ES culture medium (Fig.12.1, unpublished observation). Thus, the use of Y-27632 might be able to maintain a self-renewal and undifferentiated state of ES cells in various species.

## 5. Generation of Rat ES Cells by Combinations of Signal Inhibitors

Ying et al. succeeded in producing germline-competent mouse ES cells by a combination of the three inhibitors (3i), CHIR99021, PD184352, and SU5402 (Fig. 12.2) in a serum-free N2B27 medium (31). The report was striking because it is contrary to dogma in that the mouse ES cell self-renewal neither required activation signals from the LIF/STAT3 nor BMP/SMAD pathways, but was maintained simply by being shielded from inductive differentiation cues.

Ying et al. eventually succeeded in establishing germline-competent rat ES cells by the same 3i medium (10). Interestingly, the rat ES cells formed domed colonies like those of the mouse and had a quite similar expression pattern to marker genes of mouse. In particular, the gene expression analysis of the cell surface marker showed that, similar to mouse ES cells, the rat ES cells expressed SSEA-1, but not SSEA-4 or GCTM-2 (TRA-1-60), while human ES cells adversely expressed SSEA-4 and GCTM-2, but not SSEA-1. Concurrently with this report, Buehr

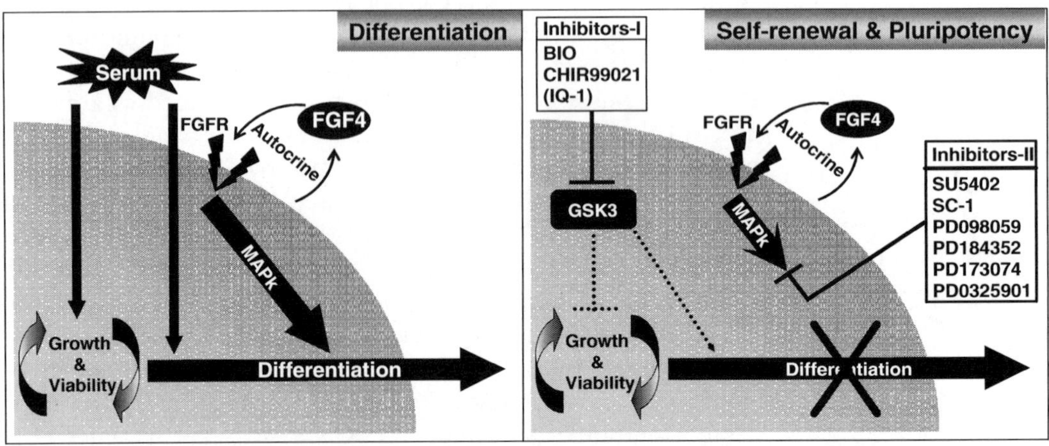


Fig. 12.2. Mechanism of maintenance of self-renewal and pluripotency in rat ES cells. Undefined factors in serum and autoinductive stimulation of FGF4 enhance differentiation of rat ES cells (*left*). Removal of serum from culture medium and inhibition of GSK and/or FGF4/MAPk pathway by the small molecules such as Inhibitors-I and -II sustain self-renewal and pluripotency (*right*). The combinations of CHIR99021, SU5402, and PD184352 (3i) or CHIR99021 and PD0325901 (2i) can establish authentic rat ES cells with germline transmission.

et al. also reported establishment of germline-competent rat ES cells by a combination of the two inhibitors (2i), CHIR99021 and PD0325901 (9). PD0325901 is known to be a more potent MEK inhibitor (28), and it can replace both PD184352 and SU5402. The 2i medium was found to improve the viability and growth of rat ES cells (9, 10).

## 6. Future Prospects

The rat ES cells generated by Stem Cell Sciences of Cambridge, UK were able to proliferate and form colonies from single cells after dissociation by enzymatic treatment, which might result in generating high-content chimeric rats, followed by overcoming the difficulty of germline transmission. However, the incidence of germline transmission was low because only one ES-cell line could complete germline transmission in each group. Currently, researchers led by Sheng Ding of Scripps Research Institute in La Jolla, California, Hongkui Deng of Peking University created iPS cells from cultured rat liver progenitor cells by using serum-free medium with a combination of 2i (CHIR99021, PD0325901) and an inhibitor of the type 1 TGFβ receptor, ALK5 (A-83-01). Although their iPS cells contributed chimera, they could not overcome the difficulty of germline transmission (11).

The genetic backgrounds of the host embryos and the donor ES cells are critical for germline transmission in mice (47). This was also likely to be true in the rat. Thus, a systematic screening of different donor and the host strain combinations should give a preferred combination for the production of germline chimeras in rats (9, 10). The karyotype of the rat ES cells was reasonably stable at early passage numbers, but chromosomal abnormalities increased with higher passages, which might cause the failure of germline transmission (10). In order to establish more potent rat ES cells, we need a further devised culture medium in which the ES cells can maintain self-renewal and pluripotency with normal chromosomal numbers even in long-term passages. We believe that availability of rat ES cells as well as iPS cells should open up promising tools for generating knockout rats lacking the gene of interest.

## References


1. Brook FA, Gardner RL (1997) The origin and efficient derivation of embryonic stem cells in the mouse. Proc Natl Acad Sci USA 94:5709–5712
2. Evans MJ, Kaufman MH (1981) Establishment in culture of pluripotential cells from mouse embryos. Nature 292:154–156
3. Martin GR (1981) Isolation of a pluripotent cell line from early mouse embryos cultured in medium conditioned by teratocarcinoma stem cells. Proc Natl Acad Sci USA 78:7634–7638
4. Thomson JA, Marshall VS (1998) Primate embryonic stem cells. Curr Top Dev Biol 38:133–165

5. Bradley A, Evans M, Kaufman MH, Robertson E (1984) Formation of germ-line chimaeras from embryo-derived teratocarcinoma cell lines. Nature 309:255–256
6. Jacob HJ (1999) Functional genomics and rat models. Genome Res 9:1013–1016
7. Liao J, Cui C, Chen S, Ren J, Chen J, Gao Y, Li H, Jia N, Cheng L, Xiao H, Xiao L (2009) Generation of induced pluripotent stem cell lines from adult rat cells. Cell Stem Cell 4:11–15
8. Ueda S, Kawamata M, Teratani T, Shimizu T, Tamai Y, Ogawa H, Hayashi K, Tsuda H, Ochiya T (2008) Establishment of rat embryonic stem cells and making of chimera rats. PLoS ONE 3:e2800
9. Buehr M, Meek S, Blair K, Yang J, Ure J, Silva J, McLay R, Hall J, Ying QL, Smith A (2008) Capture of authentic embryonic stem cells from rat blastocysts. Cell 135:1287–1298
10. Li P, Tong C, Mehrian-Shai R, Jia L, Wu N, Yan Y, Maxson RE, Schulze EN, Song H, Hsieh CL, Pera MF, Ying QL (2008) Germline competent embryonic stem cells derived from rat blastocysts. Cell 135:1299–1310
11. Li W, Wei W, Zhu S, Zhu J, Shi Y, Lin T, Hao E, Hayek A, Deng H, Ding S (2009) Generation of rat and human induced pluripotent stem cells by combining genetic reprogramming and chemical inhibitors. Cell Stem Cell 4:16–19
12. Smith AG, Heath JK, Donaldson DD, Wong GG, Moreau J, Stahl M, Rogers D (1988) Inhibition of pluripotential embryonic stem cell differentiation by purified polypeptides. Nature 336:688–690
13. Williams RL, Hilton DJ, Pease S, Willson TA, Stewart CL, Gearing DP, Wagner EF, Metcalf D, Nicola NA, Gough NM (1988) Myeloid leukaemia inhibitory factor maintains the developmental potential of embryonic stem cells. Nature 336:684–687
14. Ying QL, Nichols J, Chambers I, Smith A (2003) BMP induction of Id proteins suppresses differentiation and sustains embryonic stem cell self-renewal in collaboration with STAT3. Cell 115:281–292
15. Brenin D, Look J, Bader M, Hubner N, Levan G, Iannaccone P (1997) Rat embryonic stem cells: a progress report. Transplant Proc 29:1761–1765
16. Buehr M, Nichols J, Stenhouse F, Mountford P, Greenhalgh CJ, Kantachuvesiri S, Brooker G, Mullins J, Smith AG (2003) Rapid loss of Oct-4 and pluripotency in cultured rodent blastocysts and derivative cell lines. Biol Reprod 68:222–229
17. Demers SP, Yoo JG, Lian L, Therrien J, Smith LC (2007) Rat embryonic stem-like (ES-like) cells can contribute to extraembryonic tissues in vivo. Cloning Stem Cells 9:512–522
18. Fandrich F, Lin X, Chai GX, Schulze M, Ganten D, Bader M, Holle J, Huang DS, Parwaresch R, Zavazava N, Binas B (2002) Preimplantation-stage stem cells induce long-term allogeneic graft acceptance without supplementary host conditioning. Nat Med 8:171–178
19. Vassilieva S, Guan K, Pich U, Wobus AM (2000) Establishment of SSEA-1- and Oct-4-expressing rat embryonic stem-like cell lines and effects of cytokines of the IL-6 family on clonal growth. Exp Cell Res 258:361–373
20. Takahama Y, Ochiya T, Sasaki H, Baba-Toriyama H, Konishi H, Nakano H, Terada M (1998) Molecular cloning and functional analysis of cDNA encoding a rat leukemia inhibitory factor: towards generation of pluripotent rat embryonic stem cells. Oncogene 16:3189–3196
21. Liu X, Wu H, Loring J, Hormuzdi S, Disteche CM, Bornstein P, Jaenisch R (1997) Trisomy eight in ES cells is a common potential problem in gene targeting and interferes with germ line transmission. Dev Dyn 209:85–91
22. Suzuki H, Kamada N, Ueda O, Jishage K, Kurihara Y, Kurihara H, Terauchi Y, Azuma S, Kadowaki T, Kodama T, Yazaki Y, Toyoda Y (1997) Germ-line contribution of embryonic stem cells in chimeric mice: influence of karyotype and in vitro differentiation ability. Exp Anim 46:17–23
23. Frame S, Cohen P (2001) GSK3 takes centre stage more than 20 years after its discovery. Biochem J 359:1–16
24. Ding VW, Chen RH, McCormick F (2000) Differential regulation of glycogen synthase kinase 3beta by insulin and Wnt signaling. J Biol Chem 275:32475–32481
25. Sato N, Sanjuan IM, Heke M, Uchida M, Naef F, Brivanlou AH (2003) Molecular signature of human embryonic stem cells and its comparison with the mouse. Dev Biol 260: 404–413
26. Meijer L, Skaltsounis AL, Magiatis P, Polychronopoulos P, Knockaert M, Leost M, Ryan XP, Vonica CA, Brivanlou A, Dajani R, Crovace C, Tarricone C, Musacchio A, Roe SM, Pearl L, Greengard P (2003) GSK-3 selective inhibitors derived from Tyrian purple indirubins. Chem Biol 10:1255–1266
27. Sato N, Meijer L, Skaltsounis L, Greengard P, Brivanlou AH (2004) Maintenance of pluripotency in human and mouse embryonic stem

cells through activation of Wnt signaling by a pharmacological GSK-3-specific inhibitor. Nature Med 10:55–63

28. Bain J, Plater L, Elliott M, Shpiro N, Hastie CJ, McLauchlan H, Klevernic I, Arthur JS, Alessi DR, Cohen P (2007) The selectivity of protein kinase inhibitors: a further update. Biochem J 408:297–315
29. Murray JT, Campbell DG, Morrice N, Auld GC, Shpiro N, Marquez R, Peggie M, Bain J, Bloomberg GB, Grahammer F, Lang F, Wulff P, Kuhl D, Cohen P (2004) Exploitation of KESTREL to identify NDRG family members as physiological substrates for SGK1 and GSK3. Biochem J 384:477–488
30. Zhen Y, Sørensen V, Jin Y, Suo Z, Wiedłocha A (2007) Indirubin-3 -monoxime inhibits autophosphorylation of FGFR1 and stimulates ERK1/2 activity via p38 MAPK. Oncogene 26:6372–6385
31. Ying QL, Wray J, Nichols J, Batlle-Morera L, Doble B, Woodgett J, Cohen P, Smith A (2008) The ground state of embryonic stem cell self-renewal. Nature 453:519–523
32. Miyabayashi T, Teo JL, Yamamoto M, McMillan M, Nguyen C, Kahn M (2007) Wnt/betacatenin/CBP signaling maintains long-term murine embryonic stem cell pluripotency. Proc Natl Acad Sci USA 104:5668–5673
33. Bertrand V, Hudson C, Caillol D, Popovici C, Lemaire P (2003) Neural tissue in ascidian embryos is induced by FGF9/16/20, acting via a combination of maternal GATA and Ets transcription factors. Cell 115:615–627
34. Delaune E, Lemaire P, Kodjabachian L (2005) Neural induction in Xenopus requires early FGF signalling in addition to BMP inhibition. Development 132:299–310
35. Launay C, Fromentoux V, Shi DL, Boucaut JC (1996) A truncated FGF receptor blocks neural induction by endogenous Xenopus inducers. Development 122:869–880
36. Streit A, Berliner AJ, Papanayotou C, Sirulnik A, Stern CD (2000) Initiation of neural induction by FGF signalling before gastrulation. Nature 406:74–78
37. Wilson SI, Graziano E, Harland R, Jessell TM, Edlund T (2000) An early requirement for FGF signalling in the acquisition of neural cell fate in the chick embryo. Curr Biol 10:421–429
38. Schonwasser DC, Marais RM, Marshall CJ, Parker PJ (1998) Activation of the mitogen-activated protein kinase/extracellular signal-regulated kinase pathway by conventional, novel, and atypical protein kinase C isotypes. Mol Cell Biol 18:790–798
39. Ribisi S Jr, Mariani FV, Aamar E, Lamb TM, Frank D, Harland RM (2000) Ras-mediated FGF signaling is required for the formation of posterior but not anterior neural tissue in Xenopus laevis. Dev Biol 227:183–196
40. Otte AP, Koster CH, Snoek GT, Durston AJ (1988) Protein kinase C mediates neural induction in Xenopus laevis. Nature 334:618–620
41. Chen S, Do JT, Zhang Q, Yao S, Yan F, Peters EC, Schöler HR, Schultz PG, Ding S (2006) Self-renewal of embryonic stem cells by a small molecule. Proc Natl Acad Sci USA 103:17266–17271
42. Burdon T, Stracey C, Chambers I, Nichols J, Smith A (1999) Suppression of SHP-2 and ERK signalling promotes self-renewal of mouse embryonic stem cells. Dev Biol 210:30–43
43. Kunath T, Saba-El-Leil MK, Almousailleakh M, Wray J, Meloche S, Smith A (2007) FGF stimulation of the Erk1/2 signalling cascade triggers transition of pluripotent embryonic stem cells from self-renewal to lineage commitment. Development 134:2895–2902
44. Stavridis MP, Lunn JS, Collins BJ, Storey KG (2007) A discrete period of FGF-induced Erk1/2 signalling is required for vertebrate neural specification. Development 134: 2889–2894
45. Wilder PJ, Kelly D, Brigman K, Peterson CL, Nowling T, Gao QS, McComb RD, Capecchi MR, Rizzino A (1997) Inactivation of the FGF-4 gene in embryonic stem cells alters the growth and/or the survival of their early differentiated progeny. Dev Biol 192:614–629
46. Watanabe K, Ueno M, Kamiya D, Nishiyama A, Matsumura M, Wataya T, Takahashi JB, Nishikawa S, Nishikawa S, Muguruma K, Sasai Y (2007) A ROCK inhibitor permits survival of dissociated human embryonic stem cells. Nat Biotechnol 25:681–686
47. Schwartzberg PL, Goff SP, Robertson EJ (1989) Germ-line transmission of a c-abl mutation produced by targeted gene disruption in ES cells. Science 246:799–803

# Chapter 13

# Derivation, Culture, and In vivo Developmental Capacity of Embryonic Cell Lines from Rat Blastocysts

**Simon-Pierre Demers and Lawrence C. Smith**

## Abstract

Embryonic stem (ES) cells have been used extensively for site-specific gene targeting in the mouse. The resulting knock-out and knock-in mouse models generated so far have demonstrated their usefulness in biomedical research. However, for many diseases and fields of study, the rat still represents a superior model. The derivation and culture of germline-competent ES cells in the rat would allow the application of site-specific gene targeting technologies to this species of indisputable importance to biomedical research. We have recently shown the derivation, culture, and for the first time, in vivo contribution of rat ES-like cells to developing tissues. This represents an important step forward in making gene targeting technologies available to the rat research community, via development of rat ES cells. Here, we describe the materials, methods and techniques that have been used to obtain rat blastocysts, derive and culture embryonic cell lines from these, and assess the developmental capacity of the cells in vivo.

**Key words:** Rat, Blastocyst, Inner cell mass (ICM), Outgrowth, Derivation, Embryonic stem cells, Long-term culture, ES-like, Developmental capacity

## 1. Introduction

The ability to create gene-targeted knock-out rats for manipulation of the rat genome and the generation of rat models of human disease is long awaited. At present, there are no published reports of germline transmission of gene-targeted rat ES cells (1–3). Before this powerful technology becomes available in the rat, it will be necessary to show that genetic modifications resulting from site-specific gene targeting in rat ES cells can be carried efficiently through the germline. We have previously demonstrated the derivation and full characterization of rat ES-like cells, and most importantly the in vivo contribution of these cells to devel-

I. Anegon (ed.), *Rat Genomics: Methods and Protocols*, Methods in Molecular Biology, vol. 597
DOI 10.1007/978-1-60327-389-3_13, © Humana Press, a part of Springer Science+Business Media, LLC 2009

oping extraembryonic tissues (4). This was the first demonstration of in vivo contribution of embryonic cell lines in rats, and represents an important step forward in the development of germline-competent rat ES cells. Using the method described below, embryonic cell lines can be efficiently and reproducibly derived from rat blastocysts. In addition to their capacity to contribute to tissues in vivo, these cell lines maintain a robust and stable phenotype over long-term culture in vitro, all the while retaining pluripotent cell characteristics such as *Oct4* mRNA and protein expression, alkaline phosphatase (AP) activity, and stage-specific embryonic antigen-1 (SSEA-1) cell surface antigen expression.

Much of what is known in the literature concerning the derivation of ES cell lines comes from studies using the mouse (5, 6). The process of derivation per se and the culture conditions required for rat ES cells is poorly described in the literature and has proven to be challenging over the past decade. It appears as though one of the major difficulties has been the maintenance of a constant and stable cell line phenotype over long-term culture (4). Few groups have shown longitudinal studies of culture over long time periods. Some factors involved in the derivation process, such as feeder layers composed of MEFs, are believed to facilitate derivation and may function as both an attachment matrix and a source of secreted factors favouring self-renewal (7). It is also possible that the embryos obtained from natural in vivo reproduction are more suitable for cell line derivation than those obtained from superovulation regimes (8). Another source of difficulty may be the selection of correct cell types obtained following initial disaggregation of the blastocyst outgrowth (5, 6). However, a major shortcoming of past unsuccessful attempts in obtaining rat ES cells has been the rare assessment of their in vivo developmental capacity and their ability to contribute to the germline.

Here we provide a detailed description of a protocol for the derivation, culture, and assessment of the developmental capacity of stem cell lines from rat embryos. The major procedures are: Animals and set-up of pairings, feeder layers and culture media, embryo collection, initial culture, disaggregation of blastocyst outgrowths, long-term culture and maintenance of cell lines, and finally the assessment of in vivo developmental capacity.

## 2. Materials

### 2.1. Animals and Cells

1. Rat blastocysts (4.5 dpc) (*see* Note 1).
2. Mouse embryonic fibroblasts (MEFs) (*see* Note 2).

### 2.2. Culture Media and Materials

Wherever possible, use embryo-tested or at least tissue-culture grade reagents and sterile technique (*see* Note 3).

1. MilliQ water.
2. Dulbecco's Modified Eagle's Medium (DMEM) (Gibco, Burlington, ON, Canada).
3. Fetal Bovine Serum (FBS), Qualified, Heat-inactivated (Gibco). (*see* Note 4).
4. DMEM-10% medium: Combine 90 mL DMEM with 10 mL FBS, and add Penicillin-Streptomycin (Gibco). Penicillin (50 units/mL final) and Streptomycin (50 μg/mL final).
5. Phosphate Buffered Saline (PBS) without calcium, without magnesium (Gibco).
6. Mitomycin-C (Sigma, Oakville, ON, Canada): For a 10 μg/mL final working solution, dissolve the contents of one ampoule (2 mg) in 5 mL of PBS, and then add to 195 mL of DMEM-5% medium. Aliquot into 10 mL tubes, and store at −20°C.
7. Gelatin (Sigma): 0.1% final working solution in MilliQ water, then autoclave.
8. Charcoal/dextran-treated FBS: 750 mg Charcoal (Sigma), 75 mg dextran (Sigma), and 100 mL DMEM. Mix at 4°C for 6–14 h. Add 20 mL of this solution to 100 mL of FBS. Mix at 4°C for at least 1 h. Centrifuge at 4°C and 2060 g for 15 min. Filter first through 0.8 μm (Corning Incorporated Life Sciences, Lowell, MA, USA) and then 0.22 μm filter units. Aliquot and store at −20°C (*see* Note 5).
9. Non-essential amino acids (Gibco): Aliquot and store at 4°C.
10. L-Glutamine (Gibco): Aliquot and store at −20°C.
11. β-mercaptoethanol (Sigma): Dilute 7.2 μL in 10 mL PBS. Mix and store at 4°C for 1 week.
12. Nucleosides 100× stock solution: Dissolve the following into 100 mL water at 37°C: 80 mg Adenosine (Sigma), 85 mg Guanosine (Sigma), 73 mg Cytidine (Sigma), 73 mg Uridine (Sigma), 24 mg Thymidine (Sigma). Aliquot and store at 4°C.
13. Penicillin-Streptomycin (Gibco): Aliquot and store at −20°C.
14. Rat ES culture medium: Combine the following: 9 mL Knockout DMEM (Gibco), 1 mL Knockout Serum Replacement (Gibco), 1.25 mL Charcoal/dextran-treated FBS (*see* Subheading 2.2, Item 8), 125 μL Non-essential amino acids (*see* Subheading 2.2, Item 9), 125 μL L-Glutamine (*see* Subheading 2.2, Item 10), 125 μL β-mercaptoethanol (*see* Subheading 2.2, Item 11), 125 μL Nucleosides (*see* Subheading 2.2, Item 12), 50 μL Penicillin-Streptomycin (*see* Subheading 2.2, Item 13). Mix and filter, then add 2 μL/mL

ESGRO just before use (Chemicon International, Temecula, CA, USA).

15. Puck's Saline (Sigma): For 1 L, add 200 mg ethylenediaminetetraacetic acid (EDTA) (Sigma) and adjust the pH to 7.2. Store at 4°C.
16. Trypsin-EDTA: For a working solution of 0.25%, dilute 10 mL Trypsin (Gibco) in 90 mL Puck's Saline, aliquot and store at –20°C.

### 2.3. Embryo Collection

1. Glass slides (Fisher Scientific, Ottawa, ON, Canada).
2. 70% Ethanol (v/v).
3. Flush medium: Combine 80 mL DMEM, 10 mL Charcoal/dextran-treated FBS, and 10 mL of 250 mM HEPES solution (Sigma), aliquot and store at 4°C.
4. 3 cc syringe (Becton Dickinson, Oakville, ON, Canada) fitted to a 25G 5/8 hypodermic needle (Becton Dickinson).
5. Dissection instruments including large and fine forceps, Mayo scissors (CDMV, St-Hyacinthe, QC, Canada), and fine scissors (CDMV). Sterilise by autoclaving.
6. Pulled Pasteur pipettes (VWR International, Ville Mont-Royal, QC, Canada) (*see* Note 6).
7. Four-well plates (Nalge Nunc International, Rochester, NY, USA), 35 mm dishes (Nalge Nunc International) and 60 mm dishes (Nalge Nunc International).
8. Dissection stereomicroscope with transmitted light.

## 3. Methods

While attempting to carry out cell line derivation, it is important to keep the final objective in mind; to successfully derive a cell line that retains maximum developmental capacity over long-term culture. Starting in the reproductive tract of the pregnant dam, and all the way through to culture conditions and reagents, every effort should be made to minimise any form of stress to the embryo, as this is the sole source of starting material for the eventual cell line. Embryonic stem cell culture is not trivial and requires meticulous attention to detail at all steps and strict adherence to sterile technique to avoid contamination. In addition, all the procedures using animals should be approved by, and performed according to, the appropriate institutional animal care and use committee (IACUC). Follow the protocol below according to the timeline in Fig. 13.1 to derive embryonic cell lines from rat embryos:

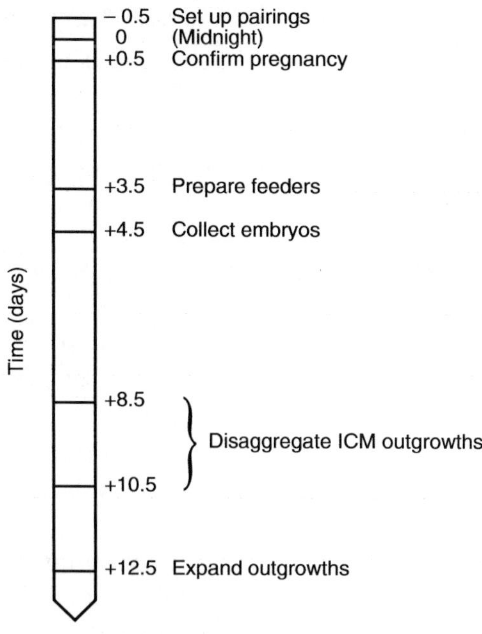


Fig. 13.1 Timeline of major procedures involved in the derivation process

### 3.1. Animals and Setup of Pairings

1. Identify cycling female rats by performing vaginal smears. Using an eyedropper with a few drops of clean lukewarm water, gently flush the vagina and examine the resulting smear on a standard glass slide under an upright microscope (*see* Note 7).
2. Mate the appropriate number of cycling females found to be in the proestrous or early estrous stage with fertile males (*see* Note 8).
3. The next morning, confirm that fertilization has occurred by looking for spermatozoa in the vaginal smears, and identify the females found to be positive. These will be sacrificed four days later to recover the embryos (*see* Note 9).

### 3.2. Feeder Layers and Culture Media

1. Thaw and expand MEFs in T75 flasks to confluency, ensuring that sufficient numbers of cells are maintained in culture to provide a continuous source of feeder layers. The day prior to embryo collection, remove and discard the medium from the MEFs and replace it with 10 mL of mitomycin-C solution (10 μg/mL) for 3 h at 37°C to induce mitotic arrest.
2. Remove and discard the mitomycin-C solution and wash the MEFs with PBS three times.
3. Add 5 mL of 0.25% trypsin-EDTA, incubate at 37°C for 3–5 min, then collect cells into a 50 mL tube containing 10 mL of DMEM-10%, simultaneously neutralizing the trypsin.

4. Centrifuge at 200 g for 10 min.
5. Discard the supernatant and resuspend the cell pellet with DMEM-10%. Count cells and adjust the concentration to 100,000 cells/mL (*see* Note 10).
6. Seed the cells onto pre-gelatinised four-well culture plates (0.5 mL per well) (*see* Note 11).

### 3.3. Embryo Collection

1. Prepare the laminar flow hood with the materials listed in subheading 13.2.3.
2. Humanely sacrifice the previously identified pregnant females according to the IACUC protocols and soak the abdomen with 70% ethanol.
3. Open the abdominal cavity with the large forceps and scissors, pushing aside the large intestine to reveal the reproductive tract.
4. Grasp the uterus immediately above the cervix with fine forceps and make a single cut crosswise. Still holding on to the uterus, cut all the way up along the length of the mesometrium with fine scissors, and cut across the ovaries to liberate the entire uterus from the body wall. Place the uterus in a 35 mm dish containing 1 mL of pre-warmed flush medium.
5. Remove all excess fat and tissue from the uterus and cut across the uterus just below the oviduct, at the level of the utero-tubal junction. In a new 60 mm dish, insert the flushing needle into the uterus at the cervical extremity and flush each uterine horn with approximately 1 mL of flush medium.
6. Under a stereomicroscope, and using a pulled Pasteur pipette, pick the blastocysts out of the flush medium in the 60 mm dish and transfer them to one well of a four-well plate containing pre-warmed flush medium. Wash the collected blastocysts through the three other wells to remove all traces of debris.

### 3.4. Initial Culture

1. Distribute the collected blastocysts individually into separate wells of four-well dishes containing inactivated MEFs and rat ES culture medium (*see* Note 12).
2. Allow the blastocysts to attach and expand for four to six days (Fig. 13.1) (*see* Note 13).

### 3.5. Disaggregation of Blastocyst Outgrowths

1. When the morphology of the cultured blastocyst resembles Fig. 13.2a, prepare several 30 μL drops of PBS in 35 mm dishes. Also prepare the equivalent number of 30 μL drops of 0.25% trypsin-EDTA in 35 mm dishes.
2. Isolate the inner cell mass (ICM)-derived cells from the underlying trophoblast cells using a flame-pulled Pasteur pipette (*see* Notes 6 and 14).

3. Transfer the dislodged ICMs individually into separate 50 μL drops of PBS to wash briefly, then into 50 μL drops of 0.25% trypsin-EDTA for 3–5 min at 37°C.
4. Gently aspirate and expel the trypsinized ICMs a few times, until they begin to break up into several small clumps. Immediately transfer the trypsinized ICMs into pre-warmed four well plates containing inactivated MEFs and rat ES culture medium as in step 3.4.1 and Note 12. (*see* Note 15).
5. Within two days, primary colonies of various types will appear. Do not passage large flat trophectoderm or epithelial cells. Rather, look for tightly packed aggregates of small cells with a high nucleus-to-cytoplasm ratio (Fig. 13.2b) (*see* Note 16).

### 3.6. Long-Term Culture and Maintenance

1. To maintain the derived cell lines (*see* Fig. 13.2c), passage the aggregates mechanically at a ratio of 1:2 to 1:3 every 3 days using a pulled Pasteur pipette (*see* Notes 17 and 18).
2. Remember to practice careful record-keeping. Note the cell line, date and passage number directly on the culture dish.

### 3.7. Assessment of In vivo Developmental Capacity

Once cell lines are successfully established, they should be characterized for expression of ES cell markers and characteristics in vitro. More importantly, their in vivo developmental capacity should be assessed. The gold standard for determining the in vivo developmental capacity of ES cell lines is re-introduction of the cells into host embryos by blastocyst injection and embryo transfer, and analysis of the extent of contribution of the cells to fetal, neonatal and adult tissues (*see* Note 19).

1. Isolate cells by performing a passage like that described for long-term culture and maintenance in subheading 13.3.6.

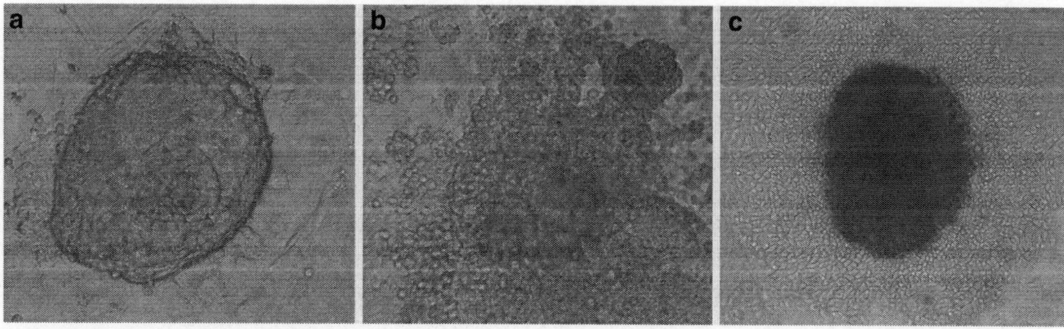


Fig. 13.2. (**a**) Blastocyst outgrowth after four days of culture. Note the tight ICM and clear border, as well as the underlying trophoblast cells. (**b**) Morphology of primary outgrowth colonies following disaggregation of blastocyst outgrowths. The tightly packed overlying cells should be passaged and expanded. (**c**) Morphology of established rat ES-like cell line. The central aggregate contains the undifferentiated cells from which the underlying epithelial cells originate. For long-term culture, select and passage the aggregate.

2. Inject 10–15 cells into each host blastocyst using a Piezo-equipped micromanipulator and a flat tip microinjection pipette with an internal diameter of 15 to 20 μm.
3. Allow the injected blastocysts to re-expand for 2–3 h.
4. Transfer the injected embryos to the uteri of pseudopregnant recipient females (*see* Note 20).

## 4. Notes

1. See your local commercial supplier to obtain rats.
2. Can be obtained commercially from various suppliers (i.e. Millipore, Billerica, MA, USA). Lower passage stocks are preferable, amplify and freeze cells at low passage numbers.
3. For culture media, store at 4°C, protect from exposure to light and use within 1–2 weeks. Unless otherwise noted, all solutions and culture media are filter-sterilized using 0.22 μm syringe-driven filter units (Millipore).
4. It is advisable to test the suitability of each batch and arrange to reserve multiple bottles of the same batch with the supplier.
5. This treatment of FBS is reported to reduce serum concentrations of steroid hormones and is often used to minimise lot-to-lot variability (9).
6. To prepare pulled Pasteur pipettes, hold the pipette briefly over a Bunsen flame and quickly pull the two ends apart. The internal diameter of the pipette should be slightly larger than the blastocysts. Becoming comfortable with embryo handling takes plenty of patience and practice.
7. Use rats aged between 60 days and 6 months.
8. The proestrous stage is characterized by a uniform abundance of aggregates of large, nucleated epithelial cells resembling bunches of grapes.
9. We use a 14-h light 10-h dark regime, with noon on the day after mating considered to be 0.5 days post-coitum (dpc), as fertilization normally occurs at the middle of the dark phase. Look for numerous loose or aggregated filaments (i.e. sperm tails) in the vaginal smear.
10. We have found that it is also possible to use feeder layers prepared a few days earlier if more convenient, with similar results.
11. The 0.1% gelatin solution is simply added to culture dishes to cover the entire surface, removed, and the surface allowed to dry completely. Alternatively, the inactivated MEFs can be seeded directly onto the surface immediately after removing excess gelatine.

12. Pre-warm wells with the rat ES culture medium at 37°C before adding the blastocysts.
13. Monitor the morphology of the blastocyst outgrowths, because they all do not develop at the same rate.
14. Pick the compact blastocyst outgrowths that have a tight border and few rounded endodermal cells. Flat, heterogeneous blastocyst outgrowths without clearly defined borders around the ICM do not give rise to cell lines as efficiently. Keep the individual blastocyst outgrowths and subsequent manipulations separate from this point on.
15. Do not trypsinize the ICMs to single cells, as isolated ICM cells will quickly differentiate.
16. It is not recommended to trypsinize the whole well to passage its contents. In our experience, this results in extensive differentiation, clearly seen morphologically by an abundance of epithelial cells.
17. Squeeze the aggregate against the bottom of the dish with a pipette to split it into equal pieces. Alternatively, trituration can also be used to split the cells.
18. The colour of the medium will vary according to the growth rate and total cell number. Ideally, it should be reddish to slightly orange (but not yellow) at the time of passage. Epithelial, cobblestone-like cells will be constantly generated from the aggregates, and although this is a good sign that the morphology of the cell line is stable, the epithelial cells are considered differentiated and should not be passaged (*see* Fig. 13.2c).
19. Karyotype, marker gene expression by reverse transcription polymerase chain reaction (RT-PCR), sex determination by PCR for a Y-chromosome specific gene, and immunostaining and flow cytometry analyses should be carried out according to standard procedures. However, in vitro characterization alone is insufficient to assess the full developmental capacity.
20. Appropriate care should be used in determining precise conditions for any in vivo procedure using animals, and since these vary by institution and country, it is advised to work closely with your IACUC to ensure that the regulations are followed.
21. Recently a report was published describing the establishment of putative rat ES cells that are capable of embryonic contribution in chimeras (3). Cells that had been transduced with lentivirus at a multiplicity of infection ranging from 250 to 500 were tested for their developmental capacity by blastocyst injection, teratoma formation, and in vitro differentiation. It is unknown whether normal untransduced cells are

capable of generating chimeras. Moreover, germline contribution was not demonstrated. These cells were reported to be AP negative, contrary to mouse and human ES cells and to the rat ES-like cells described here. There are some differences in cell line establishment that may explain the discrepancy in results, such as the use of frozen rat embryos instead of fresh embryos to grow blastocyst outgrowths. Further, the primary outgrowths from different embryos (both male and female embryos) were mixed and combined, giving a genotypically heterogeneous population of cells. Perhaps the high cell density obtained through this approach allowed the retention of a wide developmental capacity.

## Acknowledgments

The work described here was funded by the Natural Sciences and Engineering Research Council (NSERC) of Canada, (Industrial Post-Graduate Scholarship to S.-P. D.) and by Clonagen inc. The authors thank Carmen Léveillée and Joëlle A. Desmarais for assistance in preparation of this manuscript.

### References

1. Aitman TJ, Critser JK, Cuppen E, Dominiczak A, Fernandez-Suarez XM, Flint J et al (2008) Progress and prospects in rat genetics: a community view. Nat Genet 40:516–522
2. Tesson L, Cozzi J, Menoret S, Remy S, Usal C, Fraichard A et al (2005) Transgenic modifications of the rat genome. Transgenic Res 14:531–546
3. Ueda S, Kawamata M, Teratani T, Shimizu T, Tamai Y, Ogawa H et al (2008) Establishment of rat embryonic stem cells and making of chimera rats. PLoS ONE 3:e2800
4. Demers SP, Yoo JG, Lian L, Therrien J, Smith LC (2007) Rat Embryonic Stem-Like (ES-Like) Cells Can Contribute to Extraembryonic Tissues In Vivo. Cloning Stem Cells 9: 512–522
5. Hogan B (1994) Manipulating the mouse embryo : a laboratory manual. Cold Spring Harbor Laboratory Press, Plainview, N.Y
6. Nagy A (2003) Manipulating the mouse embryo : a laboratory manual. Cold Spring Harbor Laboratory Press, Cold Spring Harbor, N.Y
7. Hoffman LM, Carpenter MK (2005) Characterization and culture of human embryonic stem cells. Nat Biotechnol 23:699–708
8. Van der Auwera I, D'Hooghe T (2001) Superovulation of female mice delays embryonic and fetal development. Hum Reprod 16:1237–1243
9. Dang ZC, Lowik CW (2005) Removal of serum factors by charcoal treatment promotes adipogenesis via a MAPK-dependent pathway. Mol Cell Biochem 268:159–167

# Chapter 14

# Interference RNA for In vivo Knock-Down of Gene Expression or Genome-Wide Screening Using shRNA

**Silvère Petit and Kader Thiam**

## Abstract

With the lack of tools available to manipulate the rat genome, alternative technologies have been investigated to generate loss-of-function rat models by gene invalidation. The recent demonstration that RNA interference (RNAi)-mediated gene silencing occurs in rodents has opened new opportunities for rat functional genetics. In this chapter, we provide some practical guidelines for RNAi working in rat, based on the recent design and development of mice and rat Knock down models.

**Key words:** Transgenic rats, RNAi, shRNA, Knock down, Cre/loxP system, Tet system, BAC

## 1. Introduction

The development of high throughput methods for the sequencing and annotation of cDNA and whole genomes have led to the expansion of reverse genetics, in which access to gene sequences makes it possible to elucidate their biological function. Although this approach has greatly extended our knowledge of simple organisms (1), it is still of limited use for investigating more complex mammalian biological systems. Therefore, it appears crucial to further elucidate gene function in vivo in suitable animal models.

For this purpose, the generation of hypomorphic mutants using mouse knock out technology by homologous recombination into embryonic stem (ES) cells provides a powerful way of elucidating gene function in vivo. To date, published knock out mouse models exist for about 10% of mouse genes.

In addition, characteristics of suitable animal models should include physiological parameters closely matching those of

I. Anegon (ed.), *Rat Genomics: Methods and Protocols*, Methods in Molecular Biology, vol. 597
DOI 10.1007/978-1-60327-389-3_14, © Humana Press, a part of Springer Science+Business Media, LLC 2009

humans as well as proper model design in order to reproduce disease and enable target validation. In this respect, the rat may constitute a model of choice in many areas of biomedical research because of its relevance to human physiology. Since the development of the first transgenic rat lines in 1990 (2, 3), transgenic rats have been used in many research programs investigating mammalian physiology and the molecular basis of human genetic disorders (4, 5).

Nevertheless, the lack of tools available to manipulate the rat genome has dramatically slowed down the use of the rat model for investigating gene function. Up to recently, all attempts to isolate pluripotent rat ES cells displaying the property to colonize blastocysts and contribute to germline transmission have failed strongly limiting the process of genome engineering in rat. As a consequence, only a few rat lines have been produced with the aim of achieving partial or total loss-of-function (5): these rats lines were mainly developed either by spontaneous mutations or by ENU mutagenesis (ENU mutagenesis to obtain genetically modified rats, R. Van Boxtel).

In the absence of a robust ES cell technology in rat, alternative technologies have been investigated to generate loss-of-function models by gene invalidation. In particular, the recent discovery of RNA interference (RNAi) in cultured mammalian cells (6) and the proof that RNAi-mediated gene silencing can occur in rodents (7) have opened opportunities for rat functional genetics. In addition, the recent demonstration by De Sauza et al. (8) that the molecular profile and phenotype were comparable between Knock out mice and the corresponding Knock down mice has validated the utility of RNAi approach in vivo.

RNAi may thus provide a powerful alternative approach in rat for the development of relevant gene loss-of-function models.

The aim of this chapter is to provide some practical guidelines for RNAi working in rat, based on the recent design and development of mice and rat knock down models.

This article will therefore focus on the practical aspects of the use of RNAi mediated gene Knock Down in rat emphasizing the design, delivery and functional validation of small hairpin RNA (shRNA) targeting molecules. Key parameters and methods fulfilling major requirements for successful use of RNAi in rat will then be discussed. These methods combine the lessons learned from (1) in vitro and in vivo mouse and rat studies, using RNAi-mediated gene knock down and (2) features and limitations of transgenic models. Finally, the advantages and limitations of the different transgenic approaches for efficient endogenous gene knock down in rat will be addressed.

## 2. Prerequisites for Successful Use of RNAi in Rat: Preliminary In vitro Demonstration of shRNA Effectiveness

### *2.1. Lessons Learned from the RNAi Technology*

RNAi is a sequence-specific posttranscriptional gene silencing (PTGS) mechanism. It is triggered by double stranded RNA (dsRNA), an important signaling molecule in eukaryotic cells enabling the break down of mRNAs that are homologous in sequence to itself. Mammalian cells were until recently not amenable to RNAi since the use of in vitro transcribed, long dsRNA (>30 bp) led to sequence-unspecific response. It has been then reported that chemically synthesized duplexes of 21 nt RNAs with 2 nt 3′ overhang, known as siRNA (small interfering RNA), can specifically interfere with gene expression and bypass the sequence-independent response of mammalian cells to long dsRNAs (9, 10). More recently, it was demonstrated that RNAi-mediated gene silencing can also be obtained in cultured mammalian cells by endogenous expression of shRNA harboring a fold-back stem-loop structure (11–15).

Among the others technologies available for in vitro functional genomics (ribozyme or antisens techniques), RNAi-mediated gene silencing has been demonstrated to be a technology of choice. Indeed, its high gene knock down efficiency and high specificity make the use of siRNA or shRNA molecules attractive as these characteristics are difficult to assess in other in vitro methods.

However, both in vitro and in vivo experiments require caution, as it is uncertain whether long-term expression of shRNA may trigger side effects including the induction of off-target knockdown silencing by cross hybridation of shRNA partially homologous to other transcripts (1). In this case, the observed phenotype may be difficult to interpret. Expression of long-term shRNA could also have negative effects by competing with endogenously expressed micro RNA (miRNA) (16). Another major concern with in vivo RNAi-mediated gene silencing is the possibility of activating the interferon defense system (17–19).

Given the major concern about specificity of RNAi-mediated silencing, working at the lowest possible concentration of shRNA became one of the major rules for effective RNAi assays (1, 16). All the limitations associated with the RNAi technology in general might be of importance for interpreting RNAi effects in cell culture as well as in vivo rat experiments.

### *2.2. Design of shRNA Against Rat Genes – In vitro Validation of shRNA Specificity and Functionality*

As described above, one of the main factors for successful RNAi-mediated knock down in rat is the high selectivity of the shRNA molecule since RNAi targeting has to be exclusively limited to the targeted rat gene. Cross hybridisation of the antisens shRNA strand with other transcripts displaying full or partial identity with the targeted mRNA may lead to off-targeted invalidation. In this case, the observed rat phenotype will be difficult to interpret since it will represent a mixture of several genes invalidation events.

For this reason, before the development of knock down transgenic rat models, it should be of prime importance to closely design the shRNA sequence and to validate the efficiency and the specificity of the designed shRNA into in vitro assays.

### 2.3. shRNA Design and Target Selection

To achieve efficient gene silencing, Elbashir et al. have first suggested the design of shRNA, to select 19 nucleotide long sequences in the coding region flanked by AA at the 5′ and TT at the 3′ end. In addition, 30–70% GC content has been shown to be required for the internal stability of the shRNA (9, 10). The better understanding of the RNAi mechanism at the molecular level as well as the statistical analysis of hundreds of functional shRNA molecules are continuously contributing to the refinement of shRNA design rules (such as thermodynamic and preferred nucleotides positions rules).

To date, several algorithms for the design of shRNA and incorporation of newly established rules have been developed and are freely accessible (*see* some examples in Table 14.1).

After the design of several shRNA candidates against the same target, it is particularly important to perform a homology search (for example a BLAST search using the option "search for short, nearly exact matches") of the predicted targeting sequence against the mRNA and EST libraries databases. This in silico quality control is crucial to reducing the chance of off-target effects. It is advised to select shRNA sequences showing at least two mismatches to any gene other than the targeted one and to select sequences that are identical in different species (e.g., human, mouse and rat) (20, 21).

**Table. 14.1**
**Web sites proposing algorithms for the design of effective shRNA**

| Supplier or Institute | Web site URL |
|---|---|
| Ambion | http://www.ambion.com/techlib/misc/siRNA_finder.html |
| Dharmacon | http://www.dharmacon.com/DesignCenter/DesignCenterPage.aspx?WT.mc_id=sidesign_homepage |
| Invitrogen | http://rnaidesigner.invitrogen.com/rnaiexpress/index.jsp |
| Hannon Lab | http://katahdin.cshl.org:9331/homepage/siRNA/RNAi.cgi?type=siRNA |
| Whitehead institute | http://jura.wi.mit.edu/bioc/siRNAext/ |
| Institut Pasteur | http://mobyle.pasteur.fr/cgi-bin/MobylePortal/portal.py?form=sirna |
| University of Minnesota | http://sirecords.umn.edu/siDRM/ |

Note that the quality of the original sequence used for shRNA design is crucial as a single mismatch between the shRNA sequence and the target sequence is sufficient to prevent silencing effectiveness. In addition, known polymorphic regions may be avoided while designing shRNA sequences (20).

### 2.4. Synthesis of shRNA Expression Plasmids for In vitro Validation Assays

When the goal is a long-lasting gene silencing as it is in knock down rat model, the most effective approach for in vitro validation of the shRNA sequences effectiveness is the use of shRNA expressed from plasmid or, in the case of cells difficult to transfect, from lentivirus- or retrovirus-based vectors.

Commercially, expression vectors are widely proposed by different suppliers such as Ambion (pSilencer™), Clontech (pSiren™), Oligoengine (pSuper™), Imgenex (pSuppressor™), and so on.

These shRNA expression vectors principally differ by the type of RNA polymerase III promoters used to drive the expression of the shRNA and by the antibiotic selectable marker (neomycin, hygromycin…).

Most shRNA expression vectors display a RNA polymerase III promoter for driving the expression of shRNA. These promoters include the well-characterized human/mouse U6 promoters and the human H1 promoter.

To construct these shRNA expression vectors, two 50–70-nucleotide long oligonucleotides encoding the desired shRNA sequence together with the $(T)_5$ transcription STOP sequence are chemically synthesized, annealed and then ligated into the vector downstream of the RNA polymerase III promoter.

### 2.5. In vitro Functional Validation of shRNA Effectiveness

In order to evaluate the shRNA functionality, shRNA expression vectors are transfected or transduced into rat cells (such as PC12 or FR rat cells), and gene knock down is then monitored at mRNA or at protein level. It is strongly recommended that the evaluation into in vitro silencing effectiveness of several shRNA sequences candidates (at least more than three against the same target) and that the use of the best shRNA candidates should be done for further in vivo studies (20).

Whenever an antibody is available, detection of the in vitro Knock down at the endogenous protein level is recommended. Alternatively, the target gene coding sequence could be cloned in frame with a small antigen epitope TAG (His, Myc…), and silencing effect is monitored on the over expressed protein using an antibody against the TAG.

One other option is to over-express the target cDNA cloned downstream of the translational STOP codon of a fluorescent reporter gene (ex: EGFP) or enzymatic reporter gene (ex: luciferase, LacZ). This construct produces a chimeric mRNA from which only the reporter gene is expressed. Efficient silencing by

RNAi of the target gene will lead to the degradation of the chimeric mRNA, and consequently, no reporter is expressed (10). Such useful and efficient approach is commercially proposed by suppliers such as Promega (psiCheck™) or Invitrogen (Block-It™). The coexpression of a second reporter allows normalization for transfection efficiency.

Finally, note that transfection or transduction optimisation experiments may be required in order to be sure to achieve high transfection efficiency, especially when primary cells are used for analysis.

### 2.6. In vitro Functional Validation of shRNA Specificity

Once having demonstrated the effectiveness of the Knock down of the target gene, it is then crucial to closely validate the specificity of the Knock down. As described in Subheading 14.2.1, cross-hybridation of shRNA partially homologous to others transcripts could indeed induce off-target silencing. Therefore, appropriated controls should be added while performing the in vitro functional validation assays for assessing shRNA specificity.

A first simple control is to show that the use of an irrelevant and scrambled shRNA (or the tested shRNA sequence displaying one or two mismatches) does not exert the same phenotype as the tested shRNA.

A more widely used control experiment is to Knock down the target gene by using one or two additional shRNAs that target distinct sequence of the same mRNA and to show that same phenotype is obtained. As complementary control, experiment showing that expression of an irrelevant but functional shRNA (against GFP as instance) lead to the absence of phenotype could be done. Nevertheless, such controls are not appropriate to definitely conclude on the absence of off-target effect.

The gold standard control experiment to validate the specificity of the silencing is to reintroduce the targeted gene in a form that is resistant to the shRNA used (22). This rescue approach is performed by transfection or transduction of a plasmid overexpressing a mutated form (silent mutation) of the target gene into the knock down cells. Once the mutated form of the target sequence is reintroduced, the observed phenotype should return to the wild-type state.

## 3. Shifting from In vitro Rat Gene Knock Down to In vivo Rat Gene Knock Down

Having demonstrated in vitro, the effectiveness and the specificity of shRNA sequences against a specific rat target gene, how to successfully develop a Knock down rat model?

Indeed, the proper delivery of shRNA into much more complex mammalian models has been and remains a subject of intense investigation.

The lack of tool to manipulate the rat genome and the absence of a robust ES cell technology in the rat have slowed the development of Knock down rat models in contrast to the mice for which several Knock down approaches have already been tested and evaluated.

As a consequence, for developing relevant Knock down rat lines, it may be of great interest to learn from the different methods experienced for the development of knock down mouse lines and, especially from the advantages and limitations of these approaches.

### 3.1. Lessons Learned for the Development of Transgenic Knock Down Mice Models

One obvious way of inducing prolonged knock down effect is to transgenically supply shRNA-expressing constructs into the genome. Three main transgenic RNAi strategies, leading to the random integration of shRNA-expressing constructs into the genome, have been tested and evaluated in mice: oocytes pronucleus microinjection, recombinant viral delivery methods and embryonic stem (ES) cell approaches.

Microinjection of transgenic constructs into fertilized oocytes is the method of choice for the rapid and cost–effective generation of transgenic rodents. Hasuwa and coworkers demonstrated that transgenically supplied shRNA induced strong inhibition of the EGFP fluorescent reporter gene in mice and rats (23). Immunoblot quantification indicated that EGFP level of expression was reduced by about 80% in day 10.5 whole embryos. This strong reduction in fluorescence was further confirmed in newborn mice in a large panel of organs, including brain, heart, liver, pancreas, kidney and skin, suggesting that the silencing effect was widespread throughout the body. Moreover, the knockdown phenotype observed in the founder's embryos or adult mice was efficiently transmitted to the germline.

The second method is the use of lentiviral or retroviral vectors enabling the generation of transgenic animals by in vitro transduction of fertilized eggs at different preimplantation stages (24). This method has been successfully applied to transduce shRNA-expressing construct in mice (25–27). As reported by Tiscornia et al., after the transduction of fertilized eggs from GFP-positive transgenic mice with lentivirus expressing shRNA targeted against GFP, markedly reduced fluorescence was observed in blastocysts. Pups from F1 progeny also presented a sharp decrease in fluorescence, indicating germline transmission of the silencing effect. One major drawback of this type of approach is that mosaïsism due to transcriptional shut-off mechanism can be observed.

Another illustration that RNAi can be used to create transgenic mice came from the work of Carmell and coworkers (28). Eighty percent gene knockdown was observed in stable integrants after ES cell transfection with the shRNA-expressing constructs targeted towards *Neil1*, a gene involved in DNA repair following ionizating radiation. Two independent ES cell clones were injected into blastocysts, and germline transmission of the shRNA-expressing construct was observed in about half the F1 progeny of both generated transgenic cell lines.

Several interesting features are associated with the ES cell electroporation method. If the target gene is expressed in ES cells, it is possible to monitor the efficiency of the knockdown effect of shRNA constructs. ES cell lines can then be preselected, based on either the knockdown effect or the copy number of integrated shRNA constructs. Finally, a high number of founders, with controlled shRNA copy number, can be generated.

These three transgenic mice approaches reviewed above demonstrate that RNAi can be an efficient alternative for the study of gene loss-of-function phenotypes in rodent. However, important drawbacks are associated with random integration of shRNA-expressing constructs into the genome of transgenic animals.

Not only the power but also the limitations of random integration of transgenes into the genome are now well understood and characterized, as the first transgenic mice were generated two decades ago. The application of these techniques to generate transgenic shRNA-expressing animals presents intrinsic limitations similar to those that exist for the generation of "classical" transgenic animals. First, the number of integrated shRNA-expressing constructs cannot be pre-selected. In the case of integration of a large number of copies, the F1 generation phenotype may be highly variable, due to segregation of the mutant alleles followed by alteration of the final level of expressed shRNA molecules. The well-known positional variegation effect, namely the integration of the shRNA-expressing construct into chromatin regions with different transcriptional activities, can also lead to great variation in the level of expression of the shRNA of interest. Another consequence of the random integration of the shRNA-expressing construct is that comparative studies between different shRNA designs cannot be performed. In order to perform significant in vivo analysis, several independent transgenic lines (3–5) must be analysed to overcome the positional variegation effect.

### 3.2. First Proof of Concept of Knock Down Rat Models

The feasibility of the modulation of gene expression by RNAi in rat was first demonstrated by Hasuwa and coworkers (23). In this study, a shRNA driven by a Pol III human H1 promoter allowed the complete or partial silencing of a reporter EGFP construct in a wide range of organs. The efficiency of RNA interference in depleting a specific endogenous gene product in the rat was recently demonstrated by Dann et al. (29), who targeted the expression of the germ cell-specific mRNA, dazl, to create a transgenic rat model with male sterility. In all cases, gene silencing was shown to be heritable, the progeny showing the same knockdown phenotype as the founders.

In vivo RNAi has been achieved in the rat by DNA pronuclear injection (23) and transfection of embryos by lentiviruses ((29, 30), The use of lentiviral vectors to obtain transgenic rats, S. Remy). As described above, these techniques have important limitations since they do not enable control over the number of integrated copies or site of insertion of the transgene. Consequently, the expression levels of interfering RNA molecules can vary greatly within the transgenic founders and their progeny. This effect will be particularly important in the case of multiple transgene insertions with segregation of the mutant alleles in successive generations. The generation of several lines is thus required in order to validate the shRNA effect and/or target specificity. The analysis of the phenotype of several transgenic lines can be labor-intensive and may seem contradictory with the short development time often associated with RNAi projects.

### 3.3. Bypassing the Limitations Associated to Random Integration: The Control of the Integration Site and the Number of Inserted Copies

All reports describing the generation of Knock down mice and rats suggest that the key parameters for successful in vivo RNAi effects seem to be the necessity to control the site of integration of the shRNA expression constructs and to limit the number of the integrated copies. In fact, the integration of a high copy number is often associated with loss of RNAi specificity, off-target gene inhibition and interference with embryo development (16, 31).

### 3.4. Lessons Learned for the Development of Targeted Knock Down Mice Models

The limitations described above are found in any transgenic models based on transgene or shRNA random integration. Gene targeting through homologous recombination allows for modifying the genome in a localized and precise manner. Knock in gene targeting refers to the replacement of one particular genome sequence with another. This approach constitutes the best solution for overcoming the drawbacks linked to random integration of the shRNA-expressing construct. It is based on the insertion through homologous recombination of the shRNA-expressing construct in a ubiquitously expressed, permissive locus, such as HPRT (hypoxanthine phosphoribosyl transferase) and ROSA26 (32). These loci have been described as "neutral loci", with no enhancer activity, and are positioned in constantly active regions of the genome that have an open chromatin structure and allow permanent access to transcription factors.

This approach has became a common way for the successful generation of Knock down mice models.

The first proof of concept demonstrating that inserting the shRNA construct into a well-defined neutral locus was reported by Zheng et al. (33). In the study, a shRNA cassette against the Enhanced Green Fluorescent Protein (EGFP) was integrated into the HPRT locus in a site specific manner. Reproducibility and uniformity of the suppressive effect occurred in individual ES cell clones before and after differentiation.

Following this approach, Yu et al. demonstrated that the insertion of shRNA expression cassette into the ROSA26 locus leads to reproducibility and to the high efficient Knock down in mice of exogenous gene such as the Yellow Fluorescent Protein (YFP) gene or endogenous genes such as *Smo* and *Ptch1* genes involved in the Hedgehog signalling pathway (34).

Other have developed ROSA26 targeted Knock down mice models and have demonstrated that this type of approach is an efficient way to get reproducible and predictable knock down in rodent (35–37).

Inserting shRNA-expressing constructs in such neutral locus offers several major advantages. Firstly, the site of integration as well as number of copy of the shRNA construct is mastered and can be easily preselected by screening of ES cell clones. The level of expression of the shRNA-expressing construct is controlled, as the HPRT and ROSA26 loci exhibit a high accessibility to transcriptional machinery. This approach enables in vivo analysis on a single transgenic lineage, and the transgenic animals can be obtained directly with inbred genetic backgrounds. Most importantly, comparative analysis of the in vivo efficiency of different shRNA constructs becomes possible.

### 3.5. Bypassing the Limitations Associated to Random Integration in Rat

As described above, unpredictable or unfaithful Knock-down patterns often reported in transgenic mouse models were succesfully overcome using specific neutral locus in which the shRNA expression cassette is inserted. In the absence of a robust ES cell technology in the rat enabling the targeted insertion of the shRNA expression construct, an alternative strategy would be the use of artificial chromosome-type vectors.

Thanks to their size, these vectors have the capacity to harbor genomic regions comprising genes together with their regulatory sequences. Moreover, the size of the surrounding flanking regions can act as insulators protecting the gene of interest from potential interference from the host genome (38). The choice of such a vector is mainly driven by the size of the genomic region to be inserted: transgenes up to 100 kb can be hosted by PACs (Phage Artificial Chromosome), whereas BACs (Bacterial Artificial Chromosome) are capable of mobilizing genomic fragments 100–300 kb in length. Bigger fragments (up to 1 Mb) can be cloned into YACs (Yeast Artificial Chromosome).

These vectors have been successfully used in the mouse using pronuclear microinjection of large exogenous DNA fragments (39). Gene over-expression using artificial chromosome vectors has been used successfully in the rat in order to secure faithful transgene expression patterns (40, 41). For example, the expression of the human α-lactalbumin in the milk of transgenic rats was achieved after microinjection of a YAC vector (41). The 210 kb YAC DNA vector harbouring the entire human α-lactalbumin gene allowed position-independent and specific transgene expression.

In the majority of transgenic mice generated using this technology, the transgene expression was found comparable to the endogenous levels. Furthermore, contrary to transgenic animals generated using conventional small vectors, most transgenic animals generated with artificial chromosome-type vectors display single or few (<5) copies of the integrated transgene. Therefore, artificial chromosome-type vectors present several advantages that make them very attractive for shRNA transgenesis in rats.

Indeed, introduction of this type of construct into the rat genome makes possible the expression of shRNA in conditions approximating the HPRT or ROSA26 neutral locus genomic context that is shown to be convenient locus for predictable shRNA expression. Their cloning capacity, ranging from less than 100 kb to more than 1 Mb, allows the inclusion of most regulatory sequences required to preserve the particular properties of such neutral locus. With this strategy, it would possible to "mimic" the targeted mice approach in which shRNA constructs are integrated into the HPRT or ROSA26 permissive locus. In addition, since one of the key parameters for successful RNAi application in rat is the insertion of a low number of shRNA construct, this method may be adapted to generate transgenic rats displaying single or few copies of the integrated shRNA constructs.

The overall transgenesis efficiency achieved following pronuclear microinjection of large DNA molecules such as artificial chromosome vectors is usually low, and the generation of several transgenic lines is labour-intensive. These bottlenecks are associated with the difficulties of purifying and injecting large DNA fragments into single cell embryos (39). Nevertheless, the development of alternative technologies based on sperm as a DNA carrier such as ICSI-mediated transgenesis has alleviated the technical constraints associated with the use of large DNA constructs (Sperm cryopreservation and intracytoplasmic sperm injection into rat embryo for the generation of transgenic rats, M. Hirabayashi).

## 4. Future Development for In vivo Rat Gene Knock Down

The conventional knock down strategies described above show that a specific gene can be silenced with high efficiency in rodent models. One important improvement to these approaches is the development of inducible and tissue-specific shRNA expression systems. Indeed, the stable loss of function of genes through RNAi may cause compensatory responses or embryonic lethal phenotype. Moreover, a constitutive expression of shRNA molecules may induce nonphysiological responses and pleiotropic effects in multiple tissues. Thus, phenotypic analysis can be complex and difficult to interpret. Therefore, being able to obtain the expression of the desired shRNA in a spatio-temporal manner is of great interest.

To date, most shRNA expression vectors have been designed to drive shRNA transcription from U6 or H1 RNA polymerase III promoters. These promoters are suitable for efficient ubiquitous and constitutive expression of small hairpin RNA. Nevertheless, one major limitation of RNA polymerase III promoters is their lack of tissue specificity. Therefore, alternative strategies were developed to express shRNA in a conditional manner.

### 4.1. Lessons Learned from Development of Conditional Knock Down Mice Models

Two main systems enabling the conditional expression of shRNA have been tested and evaluated in mice: the tetracycline system and the Cre/loxP system.

### 4.2. Control of the shRNA Expression Using the Tetracycline System

The development of conditional vectors using the tetracycline operator/repressor interaction system was the first description of an efficient inducible polymerase III promoter driving shRNA transcription (42, 43). This system is characterized by the use of modified H1 or U6 polymerase III promoter harbouring the operator region of the bacterial tetracycline operon sequence (tetO) (44) close to the initiation site of transcription. The tet repressor (tetR) binds to the tetO sequence and represses any expression of shRNA. In contrast, addition of tetracycline or doxycycline prevents tetR binding to tetO sequence, thereby allowing for shRNA transcription (*see* Fig. 14.1).

A few mice models were developed, and they demonstrated the feasibility of this approach for in vivo conditional gene Knock down (35, 45). Seibler et al. reported the generation of a mouse model of reversible insulin resistance by expression of an insulin receptor specific shRNA. For this purpose, the shRNA driven by the tetO-modified promoter and the tetR expression cassette were coinserted into the ROSA26 locus. Temporal control of ubiquitous shRNA expression was obtained by administration of doxycycline. Upon induction, mice develop severe hyperglycemia within seven days. The phenotype returns to baseline shortly after the withdrawal of the doxycycline inductor (35).

These findings are consistent with the literature reports on the in vitro use of tetracycline inducible system. Nevertheless, in vivo tetracycline control systems have already been described as displaying substantial leakiness in many transgenic approaches (45). It is worth noting that this type of methodology requires the insertion of two transgenes: the tetR repressor expression cassette and the shRNA driven by the tetO-modified promoter.

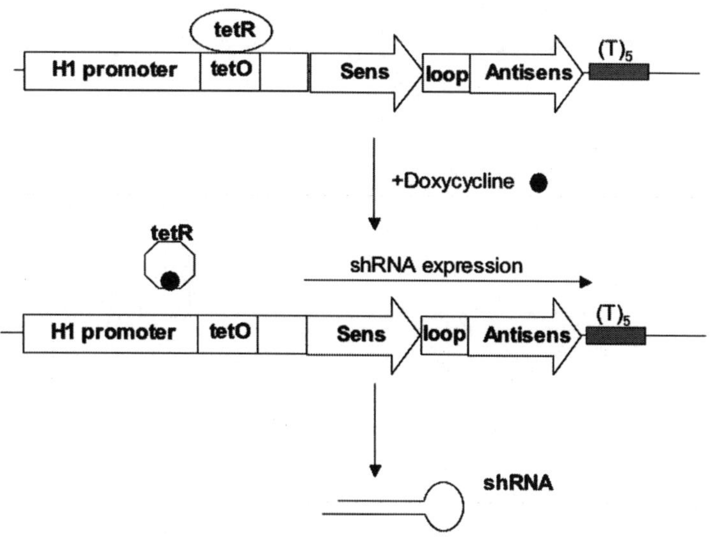


Fig. 14.1. Doxycycline based strategy for regulation of shRNA expression. *Sens and Antisens* represent the sens strand and the antisens strand of the shRNA molecule, respectively. *Loop* represents the fold-back stem-loop structure of the shRNA molecule. The $(T)_5$ *gray box* represents the polymerase III promoter transcription "stop" sequence. *tetR* tetracycline repressor, *tetO* tetracycline operator sequence contained into the modified polymerase III promoter.

Finally, the lack of tissue specific tetR transgenic mice makes this approach useful only for time specific induction of shRNA expression.

### 4.3. Control of the shRNA Expression Using the Cre/loxP System

Over the last decade, the manipulation of the mouse genome using site-specific recombinase strategies, such as the Cre/loxP system, has proven to be a very useful tool for activating or inactivating specific genes in spatially and temporally manners.

Thus, it is not surprising that this approach was investigated for conditional gene knock-down in mice as the Cre/loxP system offers all the desired characteristics for tight in vivo control of shRNA expression. To date, several Cre/loxP systems were developed and tested for the conditional expression of shRNA.

#### *4.3.1. Insertion of a Transcription "Stop" Cassette in the Loop of the shRNA*

The first system is based on the insertion of a loxP flanked transcription "Stop" cassette in the loop of shRNA driven by a H1 or U6 polymerase III promoter (36, 37, 46, 47). The loxP flanked transcription "Stop" cassette contains a termination site recognized by the polymerase pol III. In the absence of Cre-recombinase, incomplete and hence nonfunctional shRNA is transcribed. In the presence of Cre recombinase, the loxP flanked transcription "Stop" cassette is removed, enabling the correct transcription of a functional shRNA (*see* Fig. 14.2).

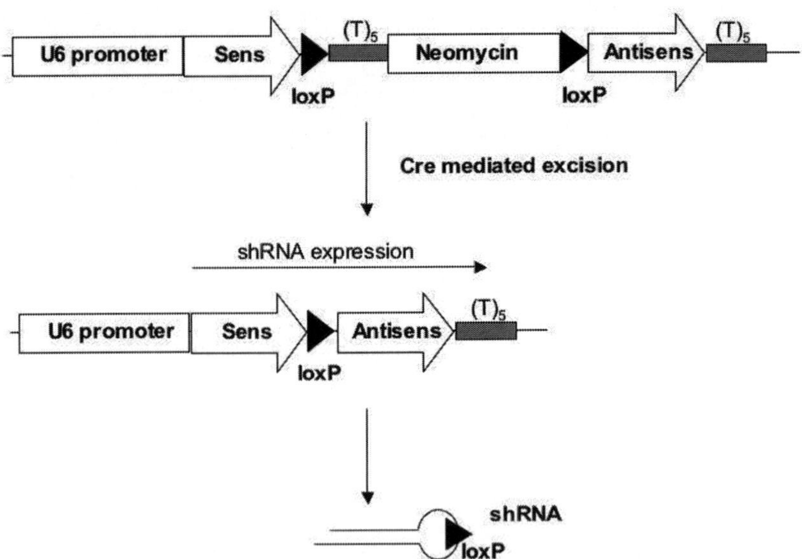


Fig. 14.2. Cre/loxP system based strategy for regulation of shRNA expression. Insertion of the transcription "stop" cassette in the loop of the shRNA. *Sens and Antisens* represent the sens strand and the antisens strand of the shRNA molecule, respectively. *Black arrows* represent the loxP sites. The loxP flanked "stop" cassette contains the $(T)_5$ transcription "stop" sequence and the neomycin positive selection cassette. *$U_6$ promoter* human/mouse $U_6$ polymerase III promoter.

Using this approach, Hitz et al. demonstrated the efficiency of this system in mice for silencing three genes of the MAPK signaling pathway (*Braf*, *Mek1* and *Mek2*) in a time and tissue dependent manner (36).

#### 4.3.2. Insertion of a Transcription "Stop" Cassette Within the Polymerase III Promoter

The second system for generating Cre-mediated shRNA expression cassette is based on the insertion of the loxP flanked transcription "Stop" cassette within the polymerase III promoter (25, 34, 48, 49) (*see* Fig. 14.3).

Two groups detailed the generation of Knock down mice in which the transcription "Stop" cassette is placed in between the distal sequence element (DSE) and the proximal sequence element (PSE) of the promoter (34, 48). Yu et al. demonstrated the usefulness of this approach for conditional expression of shRNA against exogenous gene, such as the Yellow Fluorescent Protein (YFP) gene or endogenous genes such as Smo and Ptch1 genes (34). This methodology was shown to be effective for driven the *Fgfr2* shRNA expression in a Cre recombinase-dependent manner (48).

Two others groups developed a strategy consisting of a shRNA construct carrying a modifed mouse U6 promoter including a hybrid between a loxP site and a TATA box. This latest approach has also been shown to be effective for in vivo Cre-mediated control of shRNA expression and thus for Cre-mediated silencing of reporter and endogenous genes (49).

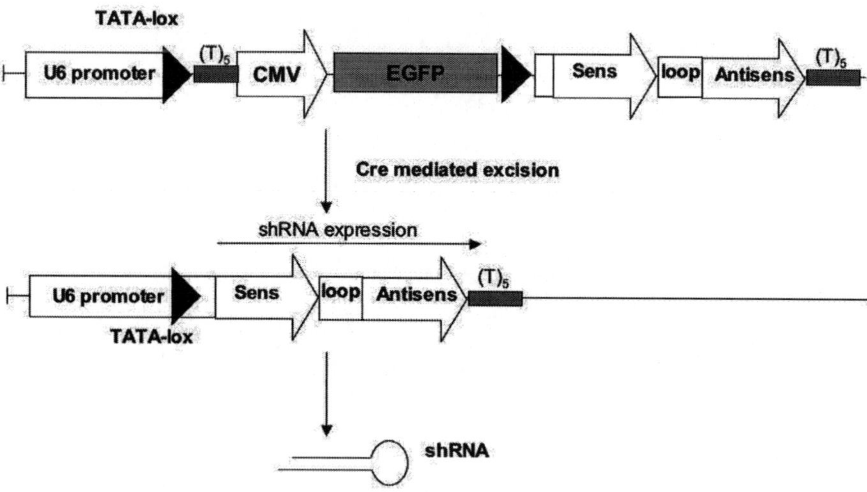


Fig. 14.3. Cre/loxP system based strategy for regulation of shRNA expression. Insertion of the transcription "stop" cassette within the polymerase III promoter. *Sens and Antisens* represent the sens strand and the antisens strand of the shRNA molecule, respectively. *Loop* represents the fold-back stem-loop structure of the shRNA molecule. *Black arrows* represent the loxP sites. The loxP flanked "stop" cassette contains the $(T)_5$ transcription "stop" sequence and the Enhanced Green Fluorescent Protein (EGFP) reporter cassette. *$U_6$ promoter* human/mouse $U_6$ polymerase III promoter containing the a modified a TATA-loxP sequence.

#### *4.3.3. Insertion of a Transcription "Stop" Cassette in Between the Polymerase III Promoter and the shRNA Element*

The third system for generating Cre-mediated shRNA expression cassette is based on the insertion of the loxP flanked transcription "Stop" cassette in between the polymerase III promoter and the shRNA element (50, 51) (*see* Fig. 14.4).

To investigate the role of thymosin ß4 (*Tß4*) during heart development, we generated in coolaboration with Smart et al. a model enabling the generation of mouse embryos with heart specific *Tß4* deficiency using this Cre/loxP strategy. By crossing the mice harbouring the foxed *Tß4* shRNA construct with cardiomyocites specific Cre expessing mice, the resultant mutant embryos were shown to display numerous cardiac defects. Impressively, the severity of phenotype correlated with the degree of *Tß4* knock down.

All these reports using the Cre/loxP dependent shRNA expression vectors demonstrated that this system could be very promising for efficient and tight control of shRNA expression in vivo and may provide the highest security with regard to shRNA expression in vivo.

While developing such conditional shRNA vectors using loxP flanked transcription "Stop" cassette, it is strongly advised to control the functionality of these conditional shRNA vectors before moving to in vivo applications. Indeed, the shRNA vector should be activated upon Cre-mediated excision of the transcription "Stop" cassette. Since a single 34 bp loxP site remains in the Cre-excised vector, preliminary in vitro tests may be performed to

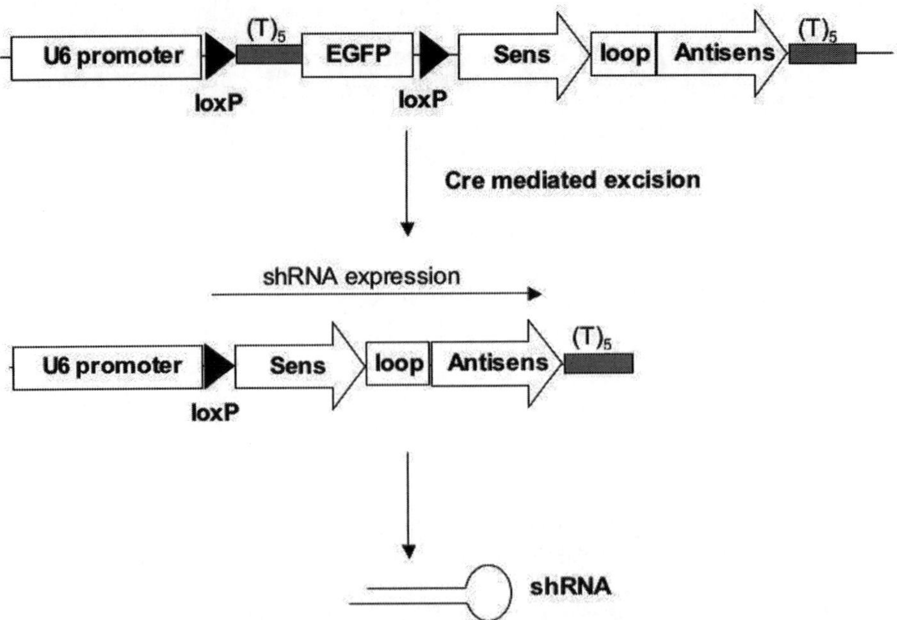


Fig. 14.4. Cre/loxP system based strategy for regulation of shRNA expression. Insertion of the transcription "stop" cassette in between the polymerase III promoter and the shRNA element. *Sens and Antisens* represent the sens strand and the antisens strand of the shRNA molecule, respectively. *Loop* represents the fold-back stem-loop structure of the shRNA molecule. *Black arrows* represent the loxP sites. The loxP flanked "stop" cassette contains the $(T)_5$ transcription "stop" sequence and the Enhanced Green Fluorescent Protein (EGFP) reporter cassette. *$U_6$ promoter* human/mouse $U_6$ polymerase III promoter.

check whether this remaining sequence disturbs RNAi functionality and efficiency. In addition, it may be worth testing both H1 and U6 promoter in order to define which is the best polIII promoter candidate for conditional shRNA expression.

### 4.4. Perspectives for the Development of Conditional Knock Down Rat Models

As described above, several useful and efficient approaches were developed for the generation of mice models expressing the shRNA in a time and tissue specific manner. Nevertheless, despite the fact that either the tetracycline or the Cre/loxP systems started to be investigated for the conditional over-expression of transgenes in rat, the full functionalities of such approaches for conditional Knock down in rat remain to be investigated (Inducible and conditional promoter systems to generate transgenic animals, E. Kobayashi).

#### 4.4.1. Control of shRNA Expression Using the Tetracycline System

In the rat, the application of the tetracycline system has been restricted to systems including single transgenes under the control of tetracycline, thus allowing temporal control of expression only (52, 53). The functionality of such systems in vivo has been hampered by insufficient levels of induction, leakiness and lack of tissue specificity. These problems are partly related to the insertion of the tet-responsive element in an inappropriate genomic environment. The tightness and range of regulation of the Tet system have been improved in mice by integration of the transgene under the control of the tetO-promoter into a BAC vector containing genomic sequences of a specific locus that have been shown to ensure the tight tetracycline regulation of transgene expression (54). This strategy should be applicable for the generation of knock down rat, thanks to the recent advances achieved in the generation of transgenic animals using artificial chromosome vectors and ICSI.

#### 4.4.2. Control of the shRNA Expression Using the Cre/loxP System

The Cre/loxP system has been described only recently in the rat (55, 56). Ajiki et al. (57) used a DsRed2/GFP double-reporter transgenic rat and a Cre transgenic rat. These rats exhibited the unique characteristic of changing from red fluorescence to green fluorescence by Cre/loxP recombination when cell-to-cell fusion occurred between the two transgenic lines.

Most of Cre lines are generated by DNA pronuclear microinjection of Cre-expressing plasmid. Even though tissue-specific promoters are often used, the resulting expression pattern is highly dependent upon position effects associated with the integration site. This unreliable Cre expression pattern may lead to artefacts because of, for example, induction of transgene expression in unexpected tissues or at the wrong time during development. Recently, several studies have reported the use of Cre-expressing lines generated using recombinant BAC vectors for the efficient, targeted expression of recombinase in specific tissues (58). As ICSI increases the efficiency of large DNA fragment transgenesis in the rat, one would anticipate that reliable rat Cre-expressing lines could be generated using BAC transgenesis. The development of relevant Cre-expressing rat lines will enable the use of the cre/loxP system for conditional expression of shRNA in rat.

## 5. Conclusion

The proof of concept of the use of RNAi technology for rapid and efficient gene silencing has been well established in vitro and more recently in mice and rat models. Because of the impossibility to produce gene knock out through homologous recombination in rat, RNAi holds great promise for investigating gene function or therapeutic gene target validation in rat.

Before any project related to in vivo rat RNAi can be undertaken, a first important step is the careful design and study of the efficiency and specificity into in vitro assays of a given shRNA construct.

The lessons learned from the development of knock down mice enable the definition of some of the key parameters for successful in vivo RNAi applications. For instance, the control of the number of shRNA expression constructs and the integration of the shRNA into specific genomic locus has been shown to be particularly important to secure shRNA expression and effectiveness. Therefore, since rat transgenesis rely mostly on random insertion strategies, alternative approaches may be developed to avoid the major limitations associated with random integration of shRNA construct. For instance, the use of artificial chromosome vectors, allowing the inclusion of the shRNA construct into genomic environments favoring the expression of the shRNA cassette while enabling the generation of transgenic animals displaying low copies of shRNA, is tempting.

The ability to regulate transgene expression in spatio-temporal manner using the tetracycline based or the Cre/loxP based systems in rat brings further improvements to the regulation of shRNA expression.

The opportunity to generate knock down rat in which the shRNA construct is targeted into a well-defined locus, thus securing its expression, is the next major step to be investigated. This may be accomplished in the future through the set up of new technologies, allowing the generation of rats from gene targeted somatic or germ cells after nuclear transfer or cell transplantation. In this regard, the recent paper of Buehr et al. (59) opens new opportunity to translate the shRNA Knock in approaches used into mouse to the rat genetics. However, it is mandatory to ensure the robustness of rat ES cell lines available before using this technology for the validation of target by gene Knock down.

Finally, the recent demonstration that RNAi could occur in rodent through the specific expression of miRNA has also open new opportunities for the generation of relevant knock down rats (60, 61). One important advantage of this approach is that the expression of miRNA could be directly driven by ubiquitous or tissue-specific polymerase II promoters, thus allowing the silencing of target gene in a tissue-specific manner without the use of a conditional expression system.

## References

1. Hannon GJ, Rossi JJ (2004) Unlocking the potential of the human genome with RNA interference. Nature 431:371–378
2. Mullins JJ, Peters J, Ganten D (1990) Fulminant hypertension in transgenic rats harbouringthe mouse Ren-2 gene. Nature 344(6266):541–544
3. Hammer RE, Maika SD, Richardson JA, Tang JP, Taurog JD (1990) Spontaneous inflammatory disease in transgenic rats expressing HLA-B27 and human beta 2 m: an animal model of HLA-B27-associated human disorders. Cell 63(5):1099–1112
4. Gill TJ, Smith GJ, Wissler RW, Kunz HW (1989) The rat as an experimental animal. Science 245(4915):269–276
5. Tesson L, Cozzi J, Ménoret S, Rémy S, Usal C, Fraichard A, Anegon I (2005) Transgenic modifications of the rat genome. Transgenic Res 14(5):531–546

6. Hammond SM, Boettcher S, Caudy AA, Kobayashi R, Hannon GJ (2001) Argonaute2, a link between genetic and biochemical analyses of RNAi. Science 293(5532):1146–1150
7. Lewis DL, Hagstrom JE, Loomis AG, Wolff JA, Herweijer H (2002) Efficient delivery of siRNA for inhibition of gene expression in postnatal mice. Nat Genet 32(1):107–108
8. De Souza AT, Dai X, Spencer AG, Reppen T, Menzie A, Roesch PL, He Y, Caguyong MJ, Bloomer S, Herweijer H, Wolff JA, Hagstrom JE, Lewis DL, Linsley PS, Ulrich RG (2006) Transcriptional and phenotypic comparisons of Ppara knockout and siRNA knockdown mice. Nucleic Acids Res 34(16):4486–4494
9. Tuschl T (2001) RNA interference and small interfering RNAs. Chembiochem 2(4): 239–245
10. Elbashir SM, Harborth J, Lendeckel W, Yalcin A, Weber K, Tuschl T (2001) Duplexes of 21-nucleotide RNAs mediate RNA interference in cultured mammalian cells. Nature 411(6836):494–498
11. Paddison PJ, Caudy AA, Bernstein E, Hannon GJ, Conklin DS (2002) Short hairpin RNAs (shRNAs) induce sequence-specific silencing in mammalian cells. Genes Dev 16(8): 948–958
12. Miyagishi M, Taira K (2002) Development and application of siRNA expression vector. Nucleic Acids Res Suppl (2):113–114
13. Miyagishi M, Taira K (2002) U6 promoter-driven siRNAs with four uridine 3′ overhangs efficiently suppress targeted gene expression in mammalian cells. Nat Biotechnol 20(5): 497–500
14. Miyagishi M, Taira K (2004) RNAi expression vectors in mammalian cells. Methods Mol Biol 252:483–491
15. Brummelkamp TR, Bernards R, Agami R (2002) A system for stable expression of short interfering RNAs in mammalian cells. Science 296(5567):550–553
16. Grimm D, Streetz KL, Jopling CL, Storm TA, Pandey K, Davis CR, Marion P, Salazar F, Kay MA (2006) Fatality in mice due to oversaturation of cellular microRNA/short hairpin RNA pathways. Nature 441(7092):537–541
17. Sledz CA, Holko M, de Veer MJ, Silverman RH, Williams BR (2003) Activation of the interferon system by short-interfering RNAs. Nat Cell Biol 5(9):834–839
18. Persengiev SP, Zhu X, Green MR (2004) Nonspecific, concentration-dependent stimulation and repression of mammalian gene expression by small interfering RNAs (siRNAs). RNA 10(1):12–18
19. Pebernard S, Iggo RD (2004) Determinants of interferon-stimulated gene induction by RNAi vectors. Differentiation 72(2–3):103–111
20. Sandy P, Ventura A, Jacks T (2005) Mammalian RNAi: a practical guide. Biotechniques 39(2):215–224
21. Cullen BR (2005) Induction of stable RNA interference in mammalian cells. Gene Ther 13(6):503–508
22. Whither RNAi? (2003) Nat Cell Biol 5(6):489–490
23. Hasuwa H, Kaseda K, Einarsdottir T, Okabe M (2002) Small interfering RNA and gene silencing in transgenic mice and rats. FEBS Lett 532(1–2):227–230
24. Pfeifer A, Brandon EP, Kootstra N, Gage FH, Verma IM (2001) Delivery of the Cre recombinase by a self-deleting lentiviral vector: efficient gene targeting in vivo. Proc Natl Acad Sci USA 98(20):11450–11455
25. Tiscornia G, Singer O, Ikawa M, Verma IM (2003) A general method for gene knockdown in mice by using lentiviral vectors expressing small interfering RNA. Proc Natl Acad Sci USA 100(4):1844–1848
26. Hemann MT, Fridman JS, Zilfou JT, Hernando E, Paddison PJ, Cordon-Cardo C, Hannon GJ, Lowe SW (2003) An epi-allelic series of p53 hypomorphs created by stable RNAi produces distinct tumor phenotypes in vivo. Nat Genet 33(3):396–400
27. Rubinson DA, Dillon CP, Kwiatkowski AV, Sievers C, Yang L, Kopinja J, Rooney DL, Ihrig MM, McManus MT, Gertler FB, Scott ML, Van Parijs L (2003) A lentivirus-based system to functionally silence genes in primary mammalian cells, stem cells and transgenic mice by RNA interference. Nat Genet 33(3):401–406
28. Carmell MA, Zhang L, Conklin DS, Hannon GJ, Rosenquist TA (2003) Germline transmission of RNAi in mice. Nat Struct Biol 10(2):91–92
29. Dann CT, Alvarado A, Hammer R, Garbers D (2006) Heritable and stable gene knockdown in rats. Proc Natl Acad Sci USA 103(30):11246–11251
30. Dann CT (2007) New technology for an old favorite: lentiviral transgenesis and RNAi in rats. Transgenic Res 16(5):571–580
31. Cao W, Hunter R, Strnatka D, McQueen CA, Erickson RP (2005) DNA constructs designed to produce short hairpin, interfering RNAs in transgenic mice sometimes show early lethality and an interferon response. J Appl Genet 46(2):217–225

32. Zambrowicz BP, Imamoto A, Fiering S, Herzenberg LA, Kerr WG, Soriano P (1997) Disruption of overlapping transcripts in the ROSA beta geo 26 gene trap strain leads to widespread expression of beta-galactosidase in mouse embryos and hematopoietic cells. Proc Natl Acad Sci USA 94(8):3789–3794
33. Zheng GD, Hidaka K, Morisaki T (2005) Stable and uniform gene suppression by site-specific integration of siRNA expression cassette in murine embryonic stem cells. Stem Cells 23(8):1028–1034
34. Yu J, McMahon AP (2006) Reproducible and inducible knockdown of gene expression in mice. Genesis 44(5):252–261
35. Seibler J, Kleinridders A, Küter-Luks B, Niehaves S, Brüning JC, Schwenk F (2007) Reversible gene knockdown in mice using a tight, inducible shRNA expression system. Nucleic Acids Res 35(7):e54
36. Hitz C, Wurst W, Kühn R (2007) Conditional brain-specific knockdown of MAPK using Cre/loxP regulated RNA interference. Nucleic Acids Res 35(12):e90
37. Steuber-Buchberger P, Wurst W, Kühn R (2008) Simultaneous Cre-mediated conditional knockdown of two genes in mice. Genesis 46(3):144–151
38. West AG, Gaszner M, Felsenfeld G (2002) Insulators: many functions, many mechanisms. Genes Dev 16(3):271–288
39. Schedl A, Montoliu L, Kelsey G, Schütz G (1993) A yeast artificial chromosome covering the tyrosinase gene confers copy number-dependent expression in transgenic mice. Nature 362(6417):258–261
40. Takahashi R, Ito K, Fujiwara Y, Kodaira K, Kodaira K, Hirabayashi M, Ueda M (2000) Generation of transgenic rats with YACs and BACs: preparation procedures and integrity of microinjected DNA. Exp Anim 49(3): 229–233
41. Fujiwara Y, Miwa M, Takahashi R, Hirabayashi M, Suzuki T, Ueda M (1997) Position-independent and high-level expression of human alpha-lactalbumin in the milk of transgenic rats carrying a 210-kb YAC DNA. Mol Reprod Dev 47(2):157–163
42. Van de Wetering M, Oving I, Muncan V, Pon Fong MT, Brantjes H, van Leenen D, Holstege FC, Brummelkamp TR, Agami R, Clevers H (2003) Specific inhibition of gene expression using a stably integrated, inducible small-interfering-RNA vector. EMBO Rep 4(6):609–615
43. Czauderna F, Santel A, Hinz M, Fechtner M, Durieux B, Fisch G, Leenders F, Arnold W, Giese K, Klippel A, Kaufmann J (2003) Inducible shRNA expression for application in a prostate cancer mouse model. Nucleic Acids Res 31(21):e127
44. Yao F, Svensjo T, Winkler T, Lu M, Eriksson C, Eriksson E (1998) Tetracycline repressor, tetR, rather than the tetR-mammalian cell transcription factor fusion derivatives, regulates inducible gene expression in mammalian cells. Hum Gene Ther 9(13):1939–1950
45. Szulc J, Wiznerowicz M, Sauvain MO, Trono D, Aebischer P (2006) A versatile tool for conditional gene expression and knockdown. Nat Methods 3(2):109–116
46. Fritsch L, Martinez LA, Sekhri R, Naguibneva I, Gerard M, Vandromme M, Schaeffer L, Harel-Bellan A (2004) Conditional gene knock-down by CRE-dependent short interfering RNAs. EMBO Rep 5(2):178–182
47. Kasim V, Miyagishi M, Taira K (2004) Control of siRNA expression using the Cre-loxP recombination system. Nucleic Acids Res 32(7):e66
48. Coumoul X, Li W, Wang RH, Deng C (2004) Inducible suppression of Fgfr2 and Survivin in ES cells using a combination of the RNA interference (RNAi) and the Cre-LoxP system. Nucleic Acids Res 32(10):e85
49. Ventura A, Meissner A, Dillon CP, McManus M, Sharp PA, Van Parijs L, Jaenisch R, Jacks T (2004) Cre-lox-regulated conditional RNA interference from transgenes. Proc Natl Acad Sci USA 101(28):10380–10385
50. Chang HS, Lin CH, Chen YC, Yu WC (2004) Using siRNA technique to generate transgenic animals with spatiotemporal and conditional gene knockdown. Am J Pathol 165(5):1535–1541
51. Smart N, Risebro CA, Melville AA, Moses K, Schwartz RJ, Chien KR, Riley PR (2007) Thymosin beta4 induces adult epicardial progenitor mobilization and neovascularization. Nature 445(7124):177–182
52. Braudeau C, Bouchet D, Toquet C, Tesson L, Ménoret S, Iyer S, Laboisse C, Willis D, Jarry A, Buelow R, Anegon I, Chauveau C (2003) Generation of heme oxygenase-1-transgenic rats. Exp Biol Med (Maywood) 228(5): 466–471
53. Tesson L, Charreau B, Ménoret S, Gilbert E, Soulillou JP, Anegon I (1999) Endothelial expression of Fas ligand in transgenic rats under the temporal control of a tetracycline-inducible system. Transplant Proc 31(3): 1533–1534
54. Schönig K, Schwenk F, Rajewsky K, Bujard H (2002) Stringent doxycycline dependent control of CRE recombinase in vivo. Nucleic Acids Res 30(23):e134

55. Sato Y, Endo H, Ajiki T, Hakamata Y, Okada T, Murakami T, Kobayashi E (2004) Establishment of Cre/LoxP recombination system in transgenic rats. Biochem Biophys Res Commun 319(4):1197–1202
56. Ueda S, Fukamachi K, Matsuoka Y, Takasuka N, Takeshita F, Naito A, Iigo M, Alexander DB, Moore MA, Saito I, Ochiya T, Tsuda H (2006) Ductal origin of pancreatic adenocarcinomas induced by conditional activation of a human Ha-ras oncogene in rat pancreas. Carcinogenesis 27(12):2497–2510
57. Ajiki T, Kimura A, Sato Y, Murakami T, Hakamata Y, Kariya Y, Hoshino Y, Kobayashi E (2005) Composite tissue transplantation in rats: fusion of donor muscle to the recipient site. Transplant Proc 37(1):208–209
58. Gong S, Doughty M, Harbaugh CR, Cummins A, Hatten ME, Heintz N, Gerfen CR (2007) Targeting Cre recombinase to specific neuron populations with bacterial artificial chromosome constructs. J Neurosci 27(37):9817–9823
59. Buehr M, Meek S, Blair K, Yang J, Ure J, Silva J, McLay R, Hall J, Ying QL, Smith A (2008) Capture of authentic embryonic stem cells from rat blastocysts. Cell 135: 1287–1298
60. Matsuda T, Cepko CL (2007) Controlled expression of transgenes introduced by in vivo electroporation. Proc Natl Acad Sci USA 104(3):27–32
61. Dickins RA, McJunkin K, Hernando E, Premsrirut PK, Krizhanovsky V, Burgess DJ, Kim SY, Cordon-Cardo C, Zender L, Hannon GJ, Lowe SW (2007) Tissue-specific and reversible RNA interference in transgenic mice. Nat Genet 39(7):914–921

# Chapter 15

# Generation of Gene-Specific Mutated Rats Using Zinc-Finger Nucleases

**Aron M. Geurts, Gregory J. Cost, Séverine Rémy, Xiaoxia Cui, Laurent Tesson, Claire Usal, Séverine Ménoret, Howard J. Jacob, Ignacio Anegon, and Roland Buelow**

## Abstract

The genetic dissection of physiological and pathological traits in laboratory model organisms is accelerated by the ability to engineer loss-of-function mutations at investigator-specified loci. This chapter describes the use of zinc-finger nucleases (ZFNs) for the targeted disruption of endogenous rat genes directly in the embryo. ZFNs can specifically disrupt target genes in cultured rat cells and in embryos from inbred and outbred strains, leading to permanently genetically modified animals. This technology allows for the rapid, targeted modification of the rat genome.

**Key words:** Zinc-finger nuclease, ZFN, Knockout, Rat model, QTL, Gene targeting, Congenic, NHEJ

## 1. Introduction

Zinc-finger nucleases (ZFNs) are engineered synthetic proteins which combine the highly sequence-specific DNA binding ability of the zinc-finger protein motif with the endonuclease activity of the restriction enzyme *Fok*I (1). The plasticity of the zinc-finger domain allows for the design of ZFNs which can bind to a broad range of sequences (2, 3). When introduced into a cell, two ZFNs interact as a heterodimer around a target sequence and introduce a site-specific, double-strand break (DSB) in the chromosome (4, 5), which is subsequently repaired by the highly conserved homologous repair (HR) or nonhomologous end joining (NHEJ) DNA repair pathways (6). NHEJ-mediated repair can occasionally result in the loss or addition of sequence information, resulting in a mutation (7).

I. Anegon (ed.), *Rat Genomics: Methods and Protocols*, Methods in Molecular Biology, vol. 597
DOI 10.1007/978-1-60327-389-3_15, © Humana Press, a part of Springer Science+Business Media, LLC 2009

Beginning with pioneering work in the fruit fly (8), engineered ZFNs have been used to generate site-specific mutations in a variety of cells from several species, including embryos (9).

Here we describe the use of zinc-finger nucleases to produce heritable, targeted mutations in the rat by combining *in vitro* transcription of ZFN-encoding nucleic acids, testing in cultured rat cells, and delivery to the one-cell embryo via standard transgenic microinjection techniques. Action of the ZFNs during the earliest cell divisions leads to a high percentage of modified chromosomes among the resulting offspring (10). Modified alleles are transmitted through the germline and can be backcrossed to establish multiple strains with unique mutant alleles.

## 2. Materials

### 2.1. Materials for Production and Gel Visualization of ZFN Messenger RNA

#### 2.1.1. ZFN mRNA Production

1. 20 μg of plasmids encoding ZFNs.
2. Restriction enzyme XbaI or XhoI.
3. Phenol/Chloroform, pH 8.0 (Sigma, St. Louis, MO).
4. 3 M Sodium Acetate.
5. Ethanol, 70% ethanol.
6. 0.1× TE (1 mM Tris-HCl pH 8.0, 0.1 mM ethylenediamine tetraacetic acid (EDTA)).
7. MessageMax T7 ARCA-Capped Message Transcription Kit (Epicentre Biotechnologies, Madison, WI) or equivalent.
8. A-Plus Poly (A) Polymerase tailing kit (Epicentre Biotechnologies) or equivalent.
9. MegaClear kit (Ambion, Austin, TX) or equivalent.
10. Formamide loading buffer (0.05% xylene cyanol, 0.05% bromophenol blue in formamide).
11. 1× TBE buffer (90 mM Tris base, 90 mM boric acid, 2 mM EDTA pH 8.0).
12. Agarose.
13. Ethidium bromide.

#### 2.1.2. Plasmid DNA Preparation for Microinjection

1. GenElute HP midiprep kit (Sigma).
2. DNA Injection Buffer (10 mM Tris-HCl, 0.1 mM EDTA, pH7.5, sterile filtered).

#### 2.1.3. Screening of ZFNs in Rat Cells

1. C6 cells (ATCC; cat no.CCL-107).
2. F-12 medium supplemented with 5% fetal bovine serum and 15% horse serum, Trypsin-EDTA (0.25%), 1× phosphate buffered saline (PBS) (Invitrogen, Carlsbad, CA).

3. Amaxa nucleofection kit V (Lonza, Cologne, Germany) or equivalent.
4. Masterpure DNA Preparation kit (Epicentre Biotechnologies) or equivalent.

### 2.2. Materials and Animals for Producing Gene Knockout Rats

#### 2.2.1. Superovulation and Fertilized Embryo Collection

1. Immature females (4–5 weeks old and around 75–100 g in weight).
2. Fertile males (2 months old) for mating with the immature females. They should be replaced every 8 months to a year.
3. Pregnant Mare's Serum Gonadotrophin (PMSG; Intervet Laboratories Ltd, Cambridge, UK). Working solution: 125 IU/mL made up with 0.9% (w/v) NaCl. Store frozen at –20°C in 1 mL aliquots.
4. Human Chorionic Gonadotrophin (hCG; Chorulon, Intervet). Working solution: 150 IU/mL made up with 0.9% (w/v) NaCl. Store at 4°C (up to 2 weeks) in 1 mL aliquots.
5. Hyaluronidase (Sigma; cat.no. H4272). Stock solution: 10 mg/mL in M16 medium (Sigma; cat.no.M7292). Store at –20°C (stable for several months) in 1 mL aliquots.
6. Embryo culture medium: M16 medium supplemented with 10% fetal bovine serum, 2 mM L-glutamine, 100 IU/mL penicillin and 100 mg/mL streptomycin.
7. Embryo-tested mineral oil (Sigma; cat.no. M8410).
8. 35-mm Petri dishes.
9. Egg transfer pipette (Pasteur Pipette, *see* Note 1) assembled into a mouth-operated system made up of a mouthpiece, rubber tube (~40 cm) and a pipette holder.
10. Stereomicroscope with under-stage illumination.
11. Humidified incubator at 37°C and 5% $CO_2$.

#### 2.2.2. Microinjection of One-Cell Embryos

1. Inverted microscope equipped with a 10× lens for low-magnification work and a 40× objective for microinjection.
2. Two micromanipulators for both holding and microinjection pipettes (Narishige, London, UK).
3. Microinjector (Narishige).
4. Micropipette puller (PN-30, Narishige).
5. Holding pipette (Glass capillary, Narishige; cat.no. G-1) (*see* Note 2).
6. Microinjection pipette (Glass capillary with filament, Narishige; cat.no. GD-1) (*see* Note 3).
7. Tips for loading capillaries with mRNAs solution (Microloader, Eppendorf).
8. Injection chamber (*see* Note 4).

*2.2.3. Transfer of Microinjected Embryos into Recipient Females*

1. Recipient females (8–16 weeks of age) that have successfully had at least one litter (virgins often eat pups).
2. Vasectomized males are needed to engender pseudopregnancy in the recipient females. They have to be replaced every 6–8 months.
3. Surgical microscope and fiber optic illumination.
4. Adrenalin (1 mg/mL).
5. Embryo transfer pipette assembled into a mouthpiece.

### 2.3. Materials for Assaying ZFN-Mediated Genome Editing

*2.3.1. Preparation of DNA from Rat Tissue*

1. Tissue digestion buffer: 100 mM Tris-HCl pH 8.5, 10 mM EDTA, 0.2% (w/v) SDS, 200 mM NaCl, 200 μg/mL Proteinase K. Add Proteinase K immediately before use from a freshly made stock solution.
2. Isopropanol.
3. 70% ethanol.
4. TE: 10 mM Tris-HCl pH 8.5, 1 mM EDTA.

*2.3.2. Analysis of Rat Tissue for ZFN Modification*

1. Accuprime HiFi polymerase and 10× PCR buffer (Invitrogen) or equivalent PCR reagents. Use a polymerase that adds untemplated adenosines to the 3′ end of the PCR product.
2. Oligonucleotides specific to the target locus.
3. Surveyor Nuclease (Transgenomic, Omaha, NE).
4. Polyacrylamide gel, 10% acrylamide, (Bio-Rad Laboratories, Hercules, CA) or equivalent gel.
5. 6× Surveyor nuclease stop buffer: 0.25% Orange G (Sigma).
6. 30% (v/v) glycerol.
7. 10 mM Tris-HCl pH 8.5.
8. 60 mM EDTA.

*2.3.3. Genotyping of ZFN-Modified Rats*

1. Topo-TA cloning kit (Invitrogen).
2. Sequencing primer, e.g.: T7 (5′-TAATACGACTCACTATA-GGG-3′).

## 3. Methods

ZFNs can be injected in either plasmid DNA or mRNA form. ZFN mRNA synthesis is performed using commercially available kits and protocols with the additional notations below. Gel electrophoresis of the mRNA and validation in cultured rat cells is crucial if working with previously untested transcription vectors or reagents. The embryo microinjection procedures describe the pronuclear and intracytoplasmic injection of ZFN mRNA as both methods can lead to ZFN-mediated genome editing in rat embryos.

Detection of ZFN activity is achieved using the Surveyor nuclease assay, based on the endonuclease activity of the Surveyor enzyme at sites of heteroduplex (mismatched) DNA. ZFN action on one or both chromosomes followed by PCR of the locus of interest will create a population of wild-type and mutated PCR products. Melting and reassortment by annealing of the strands creates heteroduplex DNA which can be cleaved by the Surveyor nuclease and resolved by gel electrophoresis.

ZFN-modified founder animals should be bred to assay for germline modification and to create uniformly modified F1 rats. If the mutated region is in coding sequence, about one-third of mutations will result in an in-frame deletion or insertion that may not functionally disable the gene. For this reason, we recommend genotyping Surveyor nuclease-positive rats by sequencing to identify desirable alleles prior to attempting germline transmission.

### 3.1. In Vitro Transcription of ZFN mRNA

#### 3.1.1. Linearization of the Template DNA

All ZFN plasmids that have been built with Sangamo's vector backbone have a T7 promoter upstream of the ZFN open reading frame (ORF) and unique XbaI and XhoI sites immediately downstream of the ZFN ORF, either of which can be used to linearize the plasmid.

1. Digest 20 μg of ZFN expression plasmid DNA in 100 μL reaction containing 1× buffer, 1× BSA, and 80 units of XbaI or XhoI, at 37°C for 2 h.
2. Extract the reactions with 100 μL of phenol/chloroform, pH 8.0, and centrifuge at 20,000 × g for 10 min.
3. Transfer the aqueous phase to a clean tube and precipitate with 10 μL 3 M NaOAc and 250 μL 100% ethanol and centrifuge at top speed for 25 min at room temperature.
4. Decant supernatant and wash pellet with 300 μL 70% ethanol and centrifugation at top speed for 3 min.
5. Air dry the pellet for 5 min and resuspend in 20 μL of 0.1× TE.

#### 3.1.2. Capped In Vitro Transcription

Use an *in vitro* transcription kit which utilizes the T7 RNA polymerase and incorporates a 7-methylguanosine cap. Ambion's MessageMax T7 ARCA-capped message kit uses the Anti-reverse cap analog (ARCA) and improves the yield of capped mRNA. Follow manufacturer's instructions (*see* Note 5).

#### 3.1.3. Polyadenosine-Tailing Reaction

Immediately following *in vitro* transcription, use a polyadenosine (polyA)-tailing kit such as the A-plus Poly (A) Polymerase tailing kit. Follow manufacturer's instructions precisely (*see* Note 6).

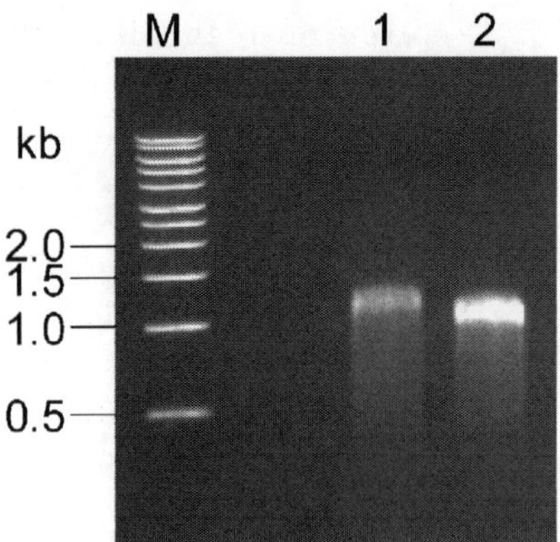


Fig. 15.1. Example gel visualization of in vitro transcribed ZFN mRNA. Lane M – Sigma's DirectLoad 1 kb DNA ladder; Lane 1 – 0.27 mg ZFN mRNA 1; Lane 2 – 0.4 mg ZFN mRNA 2. The transcribed mRNA migrates near 1.0 kb. The smear is normally visible.

#### *3.1.4. Purification and Gel Visualization of mRNA*

1. Immediately following the polyA-tailing reaction, use a kit such as the Ambion MegaClear kit to purify the mRNA and elute in an RNAse-free solution. Follow manufacturer's instructions precisely.
2. Measure RNA concentration and OD260/280 ratio (*see* Note 7).
3. Mix 1 μL of the eluted ZFN mRNA and 1 μL of formamide loading buffer (0.05% xylene cyanol, 0.05% bromophenol blue in formamide).
4. Heat to 70°C for 3 min and place on ice.
5. Load to 1% TBE agarose gel and run in 1× TBE buffer at 12 V/cm for 20 min (*see* Note 8) with a proper size marker. ZFN mRNAs runs at ~1 kb as one defined band with minor smearing below the major band (*see* Fig. 15.1).
6. Dilute mRNA to 10 ng/μL and store at –80°C until use.

#### *3.1.5. Preparation of Plasmid ZFN DNA for Microinjection*

ZFNs can be delivered to embryos as supercoiled plasmid DNA as an alternative to mRNA (*see* Note 9).

1. Prepare plasmid DNA using the GenElute HP midiprep kit following the manufacturer's instructions.
2. Resuspend in DNA Injection Buffer at 10 ng/μL and store at –20°C until use.

### ***3.2. Validating ZFN Activity Assay in Cultured Rat Cells***

Different transfection methods can be used to deliver the ZFN DNA or mRNA into cultured cells, including nucleofection, electroporation, and lipid-based transfection reagents and should be determined empirically for your cell line. Messenger RNA should be handled with extra care to prevent degradation during transfection. At least two transfections should be performed: one with

ZFNs, the other with a GFP (or equivalent) expressing plasmid to estimate transfection efficiency (*see* Note 10) and serve as a negative control for ZFN activity. We give an example of validating ZFN activity in cultured rat C6 cells. Importantly, activity in rat embryos has proven to be generally higher than in C6 cells. Any detectable signal generated by C6 cell transfection is sufficient grounds for proceeding with embryo injection.

1. Transfect ~2 million C6 cells with 1 μg each ZFN plasmid DNA or with 5 μg each ZFN mRNA according to the manufacturer's instructions. We use the Amaxa nucleofection kit V (*see* Note 11).
2. Two days post-transfection, harvest cells via trypsinization.
3. Prepare chromosomal DNA using the Masterpure DNA preparation kit according to the manufacturer's instructions. Alternately, use the protocol for rat tail DNA preparation in Subheading 15.3.4.1 below.
4. Analyze the chromosomal DNA for ZFN modification as in Subheading 15.3.4.2 below.

### *3.3. Production of ZFN-Mediated Knockout Rats*

#### *3.3.1. Selection of Rat Strain*

As some laboratory rats are incompletely inbred, they may contain an appreciable level of single-nucleotide polymorphisms (SNPs). Since an SNP will create a signal in the Surveyor nuclease assay, their presence in the founder animal is undesirable. We recommend that the Surveyor nuclease assay can be performed on the genomic DNA of potential founder animals and those scoring positive for a SNP be excluded from further use. (*See* Subheading 15.3.4.2 for details of the Surveyor nuclease assay.)

#### *3.3.2. Superovulation and Fertilized Embryo Collection*

Female rats are administered gonadotropins prior so as to mating to increase the number of released eggs (superovulation). This technique can yield around 30–40 eggs per donor female for some strains.

1. Inject the prepubescent female rats intraperitonally with 25 IU of PMSG between 12 am and 1 pm on day –2, followed by 30 IU of hCG between 3 and 4 pm on day 0.
2. Individually mate each hormone-treated female with one fertile male overnight. On the morning of day 1, check the females for copulation plugs.
3. Sacrifice female rats by cervical dislocation around 10 a.m. on the morning of day 1.
4. Excise the oviducts and transfer them to a dish containing M16 medium at room temperature. Embryos, enclosed by cumulus mass cells, can be released from the swollen ampullae (the upper portion of the oviduct) by gently tugging and opening the walls of the ampullae with fine forceps (*see* Note 12).

5. Transfer the embryos using an egg transfer pipette to a dish containing a pre-warmed hyaluronidase solution (500 μg/mL in M16 medium), which enzymatically digests the cumulus cells thus releasing the embryos. A few minutes of treatment is sufficient (longer incubation can be toxic for embryos); gentle up and down pipeting can facilitate the process.
6. Transfer the embryos to fresh pre-warmed M16 medium to wash off the hyaluronidase solution and preserve their viability.
7. Finally, transfer the embryos in a microdrop of embryo culture medium (20–30 embryos per drop) overlaid with mineral oil in a humidified 37°C incubator under 5% $CO_2$ until needed.

#### *3.3.3. Delivery of ZFNs to One-Cell Embryos*

The ZFNs are delivered either into the pronucleus (as supercoiled plasmid DNA or mRNA) or into the cytoplasm (mRNA) of single-cell embryos.

1. Thaw a newly frozen aliquot (~10 μL) of ZFN mRNAs solution (*see* Note 13) at room temperature and centrifuge at 10,000 rpm for 2 min. Keep the mRNAs solution on ice during the microinjection. Plasmid DNA ZFNs should be diluted to the desired concentration in DNA Injection Buffer (*see* Note 13).
2. Load approximately 1–2 μL of nucleic acid solution into the microinjection pipette with a microloader tip.
3. The loaded micropipette is held in place onto a micromanipulator connected to the $N_2$ gas-operated pressure injector.
4. Transfer one-cell fertilized embryos (in batches of 20–30) in a microdrop of embryo culture medium in the injection chamber, and cover with mineral oil to prevent evaporation and maintain osmolarity.
5. Mount the chamber on the stage of an inverted microscope, and monitor the injection procedure under 400× magnification.
6. Hold fertilized embryos (pronuclei visible) in place against the holding pipette using gentle negative pressure. Hold the micropipette loaded in place onto a micromanipulator.
7. Using the micromanipulator to guide the pipette, push the tip through the zona pellucida into the cytoplasm or into the pronucleus (*see* Note 14).
8. Using gentle positive pressure, the solution flows continuously from the pipette (*see* Note 15). If it is delivered into the cytoplasm, the injected solution spreads to form a drop having a diameter approximately similar to the pronuclei's one. The delivery into the pronucleus is successful when the pronucleus swells to double its original volume.
9. The microinjection pipette should be changed every 25–30 embryos.

10. After the injection, transfer the surviving embryos in a microdrop of embryo culture medium equilibrated at 37°C, 5% $CO_2$ and then keep them in a 37°C humidified incubator under 5% $CO_2$ until implantation.

*3.3.4. Transfer of Embryos into Recipient Females*

1. Obtain pseudopregnant females needed to host the microinjected embryos by mating sexually mature females in estrus with vasectomized mature males (*see* Note 16), the night before the day of implantation. Confirm mating the next morning by checking for a plug.
2. Transfer injected embryos into the oviduct of host females by using an embryo transfer pipette (*see* Note 17), preferably the same day as the microinjection to increase the rate of implantation. In general, no more than 30 embryos are transferred bilaterally into the uterus. Embryo survival decreased with pronuclear vs. cytoplasmic injection of mRNA, particularly at the highest concentration (10 ng/μL) (*see* Table 15.1).
3. Pups should be born 21 days after transfer. Depending on the delivery method and the ZFN mRNAs concentration used, the birth rate is ranged from 2 to 16%, the lowest levels being observed with cytoplasmic injection of mRNA at the highest concentration (10 ng/μL), however under these conditions we also obtained the highest rate of mutants (*see* Table 15.1).

**Table 15.1**
**ZFN-mediated gene disruption after microinjection[a]**

| Source | Route[b] | Dose (ng/μL) | Rate of surviving embryos (%)[c] | Rate of birth (%) | Rate of mutants (%) |
|---|---|---|---|---|---|
| mRNA | PNI | 10 | 55.9 | 13.5 | 28.6 |
| | | 2 | 61.7 | 14.8 | 19 |
| | | 1.5 | 69 | 20 | 40 |
| | | 0.4 | 64.5 | 16.1 | 5.3 |
| mRNA | ICI | 10 | 72 | 2 | 75 |
| | | 5 | 73.2 | 13.4 | 20 |
| | | 2 | 68 | 12.7 | 11.7 |
| | | 1.6 | 72 | 25 | 14.8 |
| | | 1.5 | 91.2 | 20.5 | 5.9 |
| | | 0.5 | 80 | 13 | 0 |
| Plasmid DNA | PNI | 10 | 80.9 | 11 | 11.1 |
| | | 2 | 77.3 | 17.5 | 9.8 |
| | | 0.4 | 82.8 | 14.7 | 6.5 |

[a]Data is accumulated from injections of ZFNs targeting four distinct genomic loci into embryos from two inbred strains or the outbred Sprague Dawley strain

[b]*PNI* pronuclear injection; *ICI* intracytoplasmic injection

[c]At least 36 embryos were injected

### *3.4. Analysis of Rat DNA for ZFN-Mediated Genome Editing*

#### *3.4.1. Preparation of DNA from Rat Tissue*

1. Add 500 μL tissue digestion buffer to ~50 mg rat tissue (typically tail tissue) in a 1.5 mL tube.
2. Incubate with agitation at 55°C for 4–24 h.
3. When digestion is complete, centrifuge at maximum speed for 1 min to remove the residual hairs.
4. Transfer lysate to a new microcentrifuge tube.
5. Add 500 μL isopropanol.
6. Invert tubes for 20–30 times until nucleic acid precipitates.
7. Centrifuge at the maximum speed for 10 min.
8. Wash nucleic acid pellet with 1 mL 70% ethanol.
9. Allow pellet to dry for 5 min. Do not dry the pellet to completion.
10. Resuspend the pellet in 200 μL TE for >15 min. This is the crude nucleic acid preparation.

#### *3.4.2. Analysis of Rat Tissue for ZFN Modification*

1. Design oligonucleotides that will amplify 200–400 bp the region of interest from rat genome. (*see* Note 18). Always include a sample from the parental rat strain as a control for Surveyor nuclease cleavage.
2. Optimize PCR conditions so that only one specific product is obtained (*see* Note 19).
3. PCR amplify the region of interest from the crude nucleic acid preparation.
4. Transfer 150–300 ng PCR product (5–15 μL) to a fresh tube. Retain excess PCR reaction for potential sequencing (*see* Subheading 15.3.4.3).
5. Denature and re-anneal the PCR product according to the following thermocycler program:

   95°C, 2 min

   95–85°C, –2°C per second

   85–25°C, –0.1°C per second

   4°C, indefinitely
6. With the samples at 4°C, add 0.5 μL of the Surveyor nuclease. Mix well (*see* Note 20).
7. Incubate at 42°C for 20 min.
8. Immediately place the reaction on ice.
9. Add 1/5th volume 6× Surveyor nuclease stop buffer. Mix well (*see* Note 21).
10. Immediately electrophorese the mixture on a 10% polyacrylamide gel at 10–15 V/cm or freeze at –20°C for later analysis (*see* Note 22).

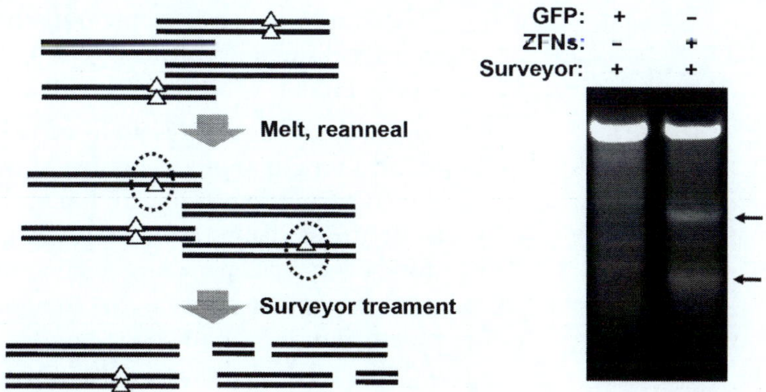


Fig. 15.2. Example Surveyor nuclease assay. (**a**) Surveyor principal. PCR on treated cells generates a mixture of wild-type and mutated PCR products (*triangles*). Melting and reassortment of the PCR strands by annealing creates a mixture of nonheteroduplex and heteroduplex DNA (*circled*) which is a substrate for the Surveyor nuclease. (**b**) Lane 1 – Cells transfected with a GFP plasmid; Lane 2 – cells transfected with plasmids encoding ZFNs. Arrows point to bands of the expected sizes based on the size of the PCR product and the site of ZFN target site cleavage.

11. Visualize gel for evidence of Surveyor nuclease cleavage, which will be evident as new shorter bands not present in the control lane (*see* Fig. 15.2). Another option is to include a lane of untreated PCR product to be certain that the shorter products appear only after the Surveyor nuclease cleavage (*see* Notes 23 and 24).

*3.4.3. Genotyping of ZFN-Modified Rats*

1. Clone the PCR products of Surveyor nuclease-positive rats using the Topo-TA Cloning kit following the manufacturer's instructions or via standard alternate methods.
2. Sequence 8–12 bacterial clones per positive rat (*see* Note 25).
3. Analyze sequencing data for desired alleles.

### 3.5. Obtaining Germline Transmission of ZFN-Modified Chromosomes

1. Select rats for breeding based on the type and frequency of the modified alleles (*see* Notes 26 and 27).
2. Analyze F1 rats as described above.

## 4. Notes

1. Embryo transfer pipettes are prepared according to the detailed procedure in the article by Si-Hoe et al. (11).
2. Holding pipettes are prepared according to the detailed procedure described in this book, in the chapter entitled

"Generation of transgenic rats by ooplasmic injection of sperm cells exposed to exogenous DNA" by M. Hirabayashi and S. Hochi.

3. Microinjection pipettes are prepared by pulling glass capillaries on a pipette puller. The size of the opening will depend on the construct that is injected, e.g., for small DNA constructs, it requires a diameter of less than 1 μm. As the mRNAs solution is very viscous, a larger opening is obtained by breaking the end of the pipette by touching carefully the holding pipette (under the microscope).
4. The injection chamber is prepared as described in this book, in the chapter entitled "Generation of transgenic rats by microinjection of short DNA fragments" by S. Menoret et al.
5. Avoid using waterbaths for incubations. Tissue culture or dry incubators work just fine.
6. PolyA tailing and transcription kits are preferably from the same vendor so that no purifications are necessary between the two steps. If Epicentre's polyA kit is used, make sure that RNase inhibitor is added. It is necessary, not optional as indicated in the user's manual.
7. Our normal yield is about 30 μg/reaction, using Epicentre's transcription and tailing kits and Ambion's purification kit. OD260/280 is always above 2.
8. New running buffer and high voltage (and thus a short running time) are critical for maintaining mRNA integrity during electrophoresis.
9. For injection of rat embryos, either mRNA or plasmid DNA using the CAG promoter (not the CMV promoter) should be used.
10. Under identical transfection conditions, we have constantly observed equal or higher transfection efficiency with mRNA than with plasmid DNA.
11. During nucleofection and electroporation, cells should be washed with saline buffer to reduce residual RNases from serum, and the mRNAs should be added to the cell suspension only immediately before zapping. If DNA is used, the CMV promoter should be used for ZFN expression.
12. The procedure for removal of the oviduct and the position of eggs in the swollen ampulla is well illustrated in the article by Si-Hoe et al. 10.
13. The stock mRNAs solution is diluted in nuclease-free TE 5:0.1. Concentrations of injected mRNA may require optimization for each construct. In our experience, high concentration of nucleic acid resulted in fewer survivors, but a higher yield of mutants (*see* Table 15.1).

14. Injection is made into the male pronucleus which is bigger and more visible than the female pronucleus.
15. The pressure and duration of each injection had to be adjusted. If the pressure is set too high, the embryo will be lysed. The volume injected is dependent of the pressure, the injection time and the size of the hole in the injection pipette.
16. Males are vasectomized by cauterizing the vas deferens at two separate locations. The efficiency of the vasectomy is assessed by mating males with 2–3 females 10.
17. The procedure for transfer of microinjected eggs is well detailed and illustrated in the article by Si-Hoe et al. 10.
18. PCR primers should be designed to amplify a 200–400 bp region overlapping the site of expected ZFN modification. The primers should be designed such that the expected Surveyor nuclease digestion products are not of the same size and that the size of the smallest expected cleavage product is at least 100 bp. As the ZFN modification frequency may be low, it is important to include at least 150 ng of PCR product in the Surveyor nuclease digestion reaction so that the low-abundance digestion products are visible.
19. Non-specific material in the CEL-I reaction will significantly complicate analysis of the digestion products. Low-molecular weight primer-derived bands compete with the desired PCR product for the Surveyor nuclease although small amounts will not negatively affect results. The concentration of the PCR product should be >15 ng/μL.
20. KCl at more than 75 mM, DMSO at >5%, and glycerol at >10% inhibit Surveyor nuclease.
21. The Surveyor nuclease contains 5′ exonuclease activity. Failing to stop the reaction will result in undesirable exonuclease activity and obfuscation of the reaction products by partially degraded DNA.
22. Somewhat better gel resolution can be obtained by adding only 0.25% Orange G to the Surveyor nuclease reaction followed by immediate electrophoresis. Orange G migrates in 10% polyacrylamide gels with an apparent molecular weight of approximately 50 bp.
23. The large majority of ZFN-induced mutations are recognized well by the nuclease. However, the Surveyor nuclease is less sensitive to small mutations than to the larger ones. In particular, single-base changes (e.g., SNPs) often yield digestion products of lower-than-expected intensities. Especially with strongly positive samples such as heterozygous F1 DNAs, 0.5 μL Surveyor nuclease may not be sufficient to completely digest the substrate DNA. We recommend increasing the

amount of Surveyor nuclease to 1.0 µL for these samples. Undigested material typically migrates with a high apparent molecular weight.

24. Analysis of Surveyor nuclease digestion product intensity in F0 animals can be informative, although quantitation of the reaction products is complicated by the reannealing of identical strands (which does not create a Surveyor nuclease substrate). In a simple case, ZFN action on one chromosome at the one-cell stage will result in exactly 50% cleavage of the PCR product by the Surveyor nuclease. Action later in development will result in widely variable signal intensities depending on the time of ZFN action. For example, if every cell in the tissue has two different modified alleles, 100% cleavage will result (0% self-annealing). If 10% of cells have one identical modified allele, 9.5% of the PCR product will be cleaved corresponding to modification of 5% of the tissue's chromosomes (5% self-annealing). Quantitation of Surveyor nuclease digestion gels can be done using NIH Image (NIH), ImageQuant (Molecular Dynamics), or other densitometric software.

25. Samples with relatively low modification rates will require more sequencing to find modified allele(s).

26. The choice of ZFN-modified animals that are used to generate germline animals is not trivial. The genotype of tail tissue may or may not be indicative of the germline genotype in F0 animals. ZFN action early during development is most likely to result in germline transmission of the mutation. Evidence for early ZFN action includes (i) 50% or more digestion of the PCR product by the Surveyor nuclease and (ii) a high frequency of a single modified allele during genotyping.

27. If the F0 germline is uniformly heterozygous for the modification, 50% of the F1 rats should be positive for the mutation. If the F0 has been crossed with a wild-type animal, positive F1s should have a 50% Surveyor nuclease signal (but *see* Note 23).

## References


1. Kim SC, Skowron PM, Szybalski W (1996) Structural requirements for FokI-DNA interaction and oligodeoxyribonucleotide-instructed cleavage. J Mol Biol 258: 638–649
2. Klug A (2005) Towards therapeutic applications of engineered zinc finger proteins. FEBS Lett 579:892–894
3. Pabo CO, Peisach E, Grant RA (2001) Design and selection of novel Cys2His2 zinc finger proteins. Annu Rev Biochem 70: 313–340
4. Porteus MH, Carroll D (2005) Gene targeting using zinc finger nucleases. Nat Biotechnol 23:967–973
5. Urnov FD, Miller JC, Lee YL, Beausejour CM, Rock JM, Augustus S, Jamieson AC, Porteus MH, Gregory PD, Holmes MC (2005) Highly efficient endogenous human gene correction using designed zinc-finger nucleases. Nature 435:646–651
6. Wyman C, Kanaar R (2006) DNA double-strand break repair: all's well that ends well. Annu Rev Genet 40:363–383

7. Lieber MR (2008) The mechanism of human nonhomologous DNA end joining. J Biol Chem 283:1–5
8. Bibikova M, Golic M, Golic KG, Carroll D (2002) Targeted chromosomal cleavage and mutagenesis in Drosophila using zinc-finger nucleases. Genetics 161:1169–1175
9. Carroll D (2008) Progress and prospects: zinc-finger nucleases as gene therapy agents. Gene Ther 15:1463–1468
10 Geurts AM, Cost FC, Miller JC, Freyvert Y, Zeitler B, Choi VM, Jenkins SS, Wood A, Cui X, Meng X, Vincent A, Lam S, DeKelver RC, Michalkiewicz M, Schilling R, Foeckler J, Kalloway S, Weiler H, Menoret S, Anegon I, Davis GD, Zhang L, Rebar EJ, Gregory PD, Urnov FD, Jacob HJ, Buelow R (2009) Knockout rats via embryo microinjection of zinc finger nucleases. Science 325:433
11. Si-Hoe SL, Murphy D (1999) Production of transgenic rodents by the microinjection of cloned DNA into fertilized one-celled eggs. Methods Mol Biol 97:61–100

# Chapter 16

# Application of Microarray-Based Analysis of Gene Expression in the Field of Toxicogenomics

Nan Mei, James C. Fuscoe, Edward K. Lobenhofer, and Lei Guo

## Summary

The field of toxicogenomics, which is becoming an important sub-discipline of toxicology, resulted from the natural convergence of the field of conventional toxicological research and the emergent field of functional genomics. One technology that has played a significant role in the field of toxicogenomics (in addition to many others) is the gene expression microarray. In this chapter, the authors provide an example of the application of gene expression microarrays to the field of toxicogenomics by detailing the strategy that was used for obtaining, analyzing, and interpreting gene expression data generated from RNA isolated from the liver of toxicant-exposed rats.

**Keywords:** Gene expression, Microarray, Mutagens, Toxicogenomics, Pathway analysis, Rat

## 1. Introduction

Since the introduction of the DNA microarray a little over a decade ago, its use in virtually every scientific discipline has become pervasive. DNA microarrays are a research tool that enable the simultaneous evaluation of the relative expression of thousands of genes. Microarray technology has also been proposed as the preferred technology to identify early biomarkers of toxicity and disease (1). When an organism is exposed to environmental stressors (e.g., toxicants), the cells respond by altering the pattern of gene expression. In other words, a subset of genes are "turned on or off" by increasing or decreasing the rate at which the gene is transcribed into mRNA. This, in turn, typically impacts the abundance of the encoded protein, which then results in a variety of cellular changes. With microarray technology, gene expression patterns are measured by extracting RNA from a sample of interest, converting it to a labeled biomolecule, hybridizing

I. Anegon (ed.), *Rat Genomics: Methods and Protocols*, Methods in Molecular Biology, vol. 597
DOI 10.1007/978-1-60327-389-3_16, © Humana Press, a part of Springer Science+Business Media, LLC 2009

it to a microarray containing gene probes, and quantifying the amount of labeled material hybridized to each gene probe using a laser-based scanner. Intensity data, a surrogate for transcript abundance, are then extracted from the hybridized image using specialized software packages. Using statistical and bioinformatics software and databases, the raw data are analyzed and interpreted to derive meaningful biological information, such as the identity of the genes changing in response to exposure and which biological pathways/functions are affected.

The process of determining how the genes interact and influence biological pathways, networks and cellular physiology is facilitated by the field of genomics, which represents the study of the nucleotide sequence, structure, and function of genes within a genome. The application of functional genomics to conventional toxicological research resulted in the emergence of toxicogenomics. Toxicogenomics applies high throughput "omic" tools (such as proteomics and metabolomics, in addition to transcriptomics) to toxicological studies and is becoming an important subdiscipline of toxicology. In many toxicogenomic studies, gene expression profiling using microarrays has been identified as a key component to addressing biological questions and is extensively used (2). The application of microarrays to the field of toxicogenomics is widely varied. For example, the differentially expressed genes that are identified in a study may be used to develop a panel of potential biomarkers that are predictive of certain types of toxicity, to elucidate molecular mechanisms that provide an improved understanding of toxicity, to create gene signatures that could identify classes of chemicals, and to predict the outcome of toxicity prior to the appearance of histological or clinical pathologic changes (3, 4). Transcription is believed to often be an intermediate step in a response to a biological exposure and thus represents a precursor to the ultimate phenotypic outcome. Therefore, the identification of unique gene expression patterns produced by mutagens or carcinogens may facilitate the elucidation of the mechanisms of action leading to disease or tumor induction.

Recently, we have evaluated a number of chemicals and natural agents (comfrey, riddelliine, aristolochic acid, kava, and acrylamide) (5–9) and drugs (pioglitazone, rosiglitazone, troglitazone, fenofibrate and WY-14,643) (10, 11) for both toxic endpoints and gene expression changes using microarray technology. A subset of this data was contributed to the community-wide MicroArray Quality Control (MAQC) project (12), using a set of biologically relevant samples within the context of a standard toxicogenomics study (13). In this chapter, we describe the procedures for the generation of gene expression data using microarrays and also our strategy for obtaining, analyzing and interpreting the resulting data, using the rat liver samples treated with comfrey and riddelliine from the MAQC project as an example.

## 2. Materials

### 2.1. Equipments

1. NanoDrop ND-1000 Spectrophometer (Thermo Fisher Scientific, Waltham, MA), or equivalent device for measuring the absorbance of an RNA sample at 260 and 280 nm.
2. Agilent 2100 Bioanalyzer (Agilent Technologies, Santa Clara, CA).
3. Illumina Hybridization Oven (Illumina, San Diego, CA), or equivalent ovens.
4. Hybex Microarray Incubation System (SciGene Corporation, Sunnyvale, CA).
5. Illumina BeadArray Reader (Illumina).
6. Applied Biosystems 7900HT Fast Real-Time PCR Systems (Applied Biosystems, Foster City, CA).
7. Centrifuge with 96-well plate carriers.
8. Orbital shaker and rocker mixer.
9. Freezers (–20°C and –80°C).

### 2.2. Reagents

1. RNA*later* (Ambion, Austin, TX).
2. RNeasy Mini Kit (Qiagen, Valencia, CA).
3. RNA 6000 Nano LabChip Kit (Agilent).
4. Illumina RNA Amplification Kit (Ambion).
5. MessageAmp II-Biotin *Enhanced* Single Round aRNA Amplification Kit (Ambion).
6. RatRef-12 Expression BeadChip (Illumina).
7. Streptavidin-Cy3 (FluoroLink Cy3; Amersham Biosciences, Piscataway, NJ).
8. TaqMan assays: High-Capacity cDNA Archive Kit; Universal Master Mix (2X) without AmpErase UNG and TaqMan probes (Applied Biosystems).

### 2.3. User Manuals

1. RNeasy Mini Handbook (Qiagen).
2. User's Manual for NanoDrop ND-1000 Spectrophometer (NanoDrop).
3. Reagent Kit Guide for RNA 6000 Nano Assay (Agilent).
4. MessageAmp II-Biotin *Enhanced* Instruction Manual (Ambion).
5. Whole-Genome Gene Expression with IntelliHyb Seal System Manual (Illumina).
6. Gene Expression on Sentrix Arrays Direct Hybridization System Manual for Array Matrix (Illumina).
7. Illumina BeadArray Reader User Guide (Illumina).

8. ArrayTrack User's Manual (http://www.fda.gov/nctr/science/centers/toxicoinformatics/ArrayTrack).
9. Ingenuity Pathways Analysis (http://www.ingenuity.com/products/pathways_analysis.html).
10. TaqMan Gene Expression Assays (http://www3.appliedbiosystems.com/cms/groups/mcb_support/documents/generaldocuments/cms_041280.pdf).

## 3. Methods

In standard toxicology and toxicogenomics studies, an animal model is exposed to a drug or toxicant (chemical, environmental mutagen, etc.) at different doses for different periods of time (Fig. 16.1). Tissues of interest are then harvested at defined time points after the last treatment. The tissues are then flash frozen or treated with an RNA preservative agent and stored in a –80°C freezer. The tissues are stored until total RNA can be extracted using a procedure that has been established for the specific tissue type. An appropriate level of replication within the study design is critical to the proper interpretation of the microarray result. This replication can either be in the form of technical (the same RNA assayed on more than one microarray) (*see* ) or biological (multiple different individuals in the same treatment group, each assayed on a separate microarray) samples. With the commercialization of the manufacturing process, the consistency of the microarrays themselves improved significantly. In addition, as the necessity of absolute consistency in the microarray processing procedures became more appreciated, reproducibility of data became possible. This was clearly demonstrated in the MAQC study, which showed that high quality microarrays coupled with standardized protocols resulted in highly reproducible data (12, 13). Therefore, provided that high quality microarrays are being used and the operators are proficient, toxicogenomic experiments can focus on measuring biological and treatment effects. Typically, in rat toxicogenomics studies, each treatment group has 3–6 rats, each of

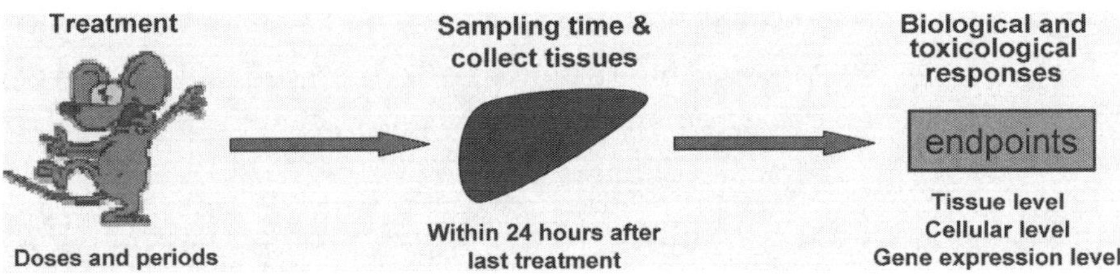


Fig. 16.1. Scheme of the experimental design.

which is profiled on a separate microarray. The following sections detail the methods that were conducted in our laboratories in which gene expression microarrays were used within the context of a rat toxicogenomics study.

### *3.1. Chemicals*

In most standard toxicology studies, the chemical of interest is purchased from a supplier or obtained from another source that either purifies or also studies the chemical. Once obtained, the chemical is dissolved in a solvent at an appropriate concentration, based on chemical features and desired exposure parameters. The exposure doses are selected based on preliminary results, such as those obtained from a chronic tumorigenicity bioassay. In this instance, comfrey roots (*Symphytum officinale*) were purchased from Camas Prairie Products (Trout Lake, WA), ground into powder using a Wiley Mill, and stored at room temperature until use (7). Riddelliine (>97% pure by reversed-phase HPLC analysis) was obtained from the National Toxicology Program (NTP) and dissolved in 0.9% sodium chloride (8).

### *3.2. Animal Treatments*

All animal procedures followed the recommendations of the Institutional Animal Care and Use Committee for the handling, maintenance, treatment, and sacrifice of the rats. Big Blue Fisher 344 transgenic rats (*see* Note 2) were obtained from Taconic Laboratories (Germantown, NY). The rats were received as weanlings, housed 2 or 3 per cage, and fed NIH-31IR chow (Purina Mills, St. Louis, MO). For the comfrey treatment, the treatment schedule was based on the protocol used in a carcinogenesis assay. The base diet was blended with comfrey root powder in a Hobart Mixer to make a diet consisting of 2–8% comfrey root. At 6 weeks of age, one group of rats was maintained on the NIH-311R diet (vehicle control) and other groups were placed on the diet supplemented with 2–8% comfrey root for 12 weeks (7). For the riddelliine treatment, the treatment schedule was based on preliminary results from the NTP 2-year chronic tumorigenicity bioassay. At 6 weeks of age, rats were treated with riddelliine at doses ranging from 0.1 to 1.0 mg/kg body weight (or 0.9% sodium chloride (vehicle control)) by gavage five times a week for 12 weeks (8). Body weight for each individual animal was measured weekly throughout the course of the experiment. Twenty-four hours after the last treatment, 6 rats from each group were sacrificed. Liver and other organ tissues were isolated, frozen quickly in liquid nitrogen, and stored at –80°C (*see* **Note 3**).

### *3.3. RNA Isolation and Quality Control*

The isolation of high quality, intact RNA is critical in obtaining the most sensitivity and reproducibility when performing microarray experiments. RNA is extremely sensitive to destruction by RNases. To avoid RNA degradation during the isolation process, RNase-free reagents, glassware, and plasticware are used.

Total RNA can be isolated from liver (as well as other tissues) using a variety of different approaches; however, we recommend that all samples within a given study are isolated using the same procedure. In this study, RNA from 30 mg of tissue from each of the six biological replicate samples within the control and treated groups was purified using the RNeasy system (Qiagen) according to the manufacturer's protocol. The purity and yield of the extracted RNA is determined spectrophotometrically by measuring the optical density at wavelengths of 260 and 280 nm. The ratio of A260/A280 should be greater than 1.8. The size distribution of the extracted RNA is evaluated using the RNA 6000 LabChip and Agilent 2100 Bioanalyzer (Agilent Technologies). Ideally, only RNA of high quality (e.g., with an RNA Integrity Number (RIN) greater than 8.0) is used for microarray experiments (*see* Note 4).

### 3.4. Microarray Platform

There are numerous high-density DNA microarray platforms commercially available, such as Affymetrix, Agilent, Applied Biosystems, CodeLink, Illumina, and Phalanx. The authors routinely use multiple platforms depending upon their needs for the resulting data as well as the content of the microarray. As detailed in the MAQC manuscript, these samples were assayed using microarrays from Agilent, Affymetrix, Applied Biosystems, and GE Healthcare (13). As the other data has already been described in detail, here, we will use the Illumina platform as a representative to briefly detail the procedures for performing microarray analysis, though readers should be advised to find the most current and complete protocols in the user's manuals associated with the microarray platform that they have selected for use (*see* Note 5). Essentially, the process begins with an isolated RNA sample that is converted into a labeled biomolecule (typically aRNA (also referred to as cRNA) or cDNA). This labeled material is hybridized to the microarray for a defined period of time prior to washing and scanning. The signal intensity associated with each feature (probe) on the microarray is a surrogate for transcript abundance. Intensity data is extracted from the scanned image typically using specialized software that is available for each platform from the manufacturer. The extracted data is then used to gain biological insights into the samples that were profiled.

#### 3.4.1. Sample Labeling and Quality Control of Labeled aRNA

For each sample, 200 ng of total RNA was labeled using the MessageAmp II-biotin enhanced kit (Ambion) according to the manufacturer's instructions. Briefly, double-stranded cDNA was synthesized using T7-oligo (dT) primers. A T7-mediated in vitro transcription (IVT) reaction was then performed in the presence of biotin-nucleotides to form biotin-labeled amplified aRNA. The resulting aRNA was then quantified using a NanoDrop spectrophotometer (Thermo Fisher Scientific) or equivalent.

The fragment size distribution of the aRNA was determined by assaying 200 ng of each sample on the Agilent 2100 Bioanalyzer (Agilent).

*3.4.2. aRNA Hybridization*

The RatRef-12 Expression BeadChip (Illumina) contains 22,523 probes from a total of ~21,900 rat genes and allows parallel processing of 12 arrays. The oligonucleotide probes in the RatRef-12 BeadChip were selected from rat sequences in the National Center for Biotechnology Information collection of reference sequences (NCBI RefSeq). Biotinylated aRNA samples were hybridized to the RatRef-12 Expression BeadChip, according to the manufacturer's instructions. Briefly, 1.5 μg of labeled aRNA from each sample was fragmented into smaller lengths for improved hybridization kinetics via incubation with hybridization buffer (GEX-HYB) for 5 min at 65°C. The fragmented aRNA was then applied to an array and assembled into a BeadChip Hyb Chamber. The BeadChip was hybridized overnight in a 58°C oven, prior to being washed in a SciGene Hybex Microarray Incubation System (SciGene) with a water bath insert. In order to generate a fluorescent signal, the array was exposed to a solution containing 1 μg/ml streptavidin conjugated to the fluorescent dye, Cy3, for 10 min. Once the BeadChips were dry, they were stored in the dark until ready to scan.

*3.4.3. Scanning and Data Outputs*

The BeadChips were scanned using an Illumina BeadArray Reader (Illumina). The microarray images were registered and extracted automatically during the scan process by Illumina's BeadStudio software, using the manufacturer's default settings. The resulting files generated by this software package (e.g., .tif, .csv and .xml) can be exported to most standard databases and gene expression analysis programs. Standard text output contains the following fields: (i) AVG_Signal (representing the average signal intensity associated with each gene), (ii) BEAD_STDEV (representing the standard deviation of the replicate intensity measurements for a given gene), (iii) Avg_NBEADS (representing the number of measurements that were associated with a given gene), and (iv) Detection *P* Value (reflecting the statistical significance of the intensity value associated with a given gene).

***3.5. Microarray Data Analysis***

Many software packages, such as Spotfire DecisionSite for Microarray Analysis (http://spotfire.tibco.com/products/decisionsite_microarray_analysis.cfm), JMP Genomics (http://www.jmp.com/software/genomics/index.shtml), and Rosetta Resolver Gene Expression Data Analysis System (http://www.rosettabio.com/products/resolver), are available commercially for gene expression data management and analysis. Publicly available software tools have also been developed for the management, analysis, visualization, and interpretation of microarray data (http://www.

nslij-genetics.org/microarray/soft.html). In our studies, most of data analysis is performed using ArrayTrack, which was developed by the US FDA's National Center for Toxicological Research (14) and is available to the public (http://www.fda.gov/nctr/science/centers/toxicoinformatics/ArrayTrack).

#### 3.5.1. Data Normalization

The purpose of normalization is to remove the systematic, non-biological variation that can exist within microarray data by adjusting the raw data. The intensities for each probe can be normalized using one of three methods that are recommended by Illumina. These are average normalization, rank invariant normalization, and cubic spline normalization (http://www.illumina.com/downloads/GXBeadStudioNormalization_TechNote.pdf). For this study, the data were normalized with the cubic spline method in BeadStudio software and was then exported to ArrayTrack for further analysis (*see* Note 6).

#### 3.5.2. Correlation, Cluster, and Principal Component Analyses

In order to assess the amount of variability associated with the biological replicate samples within this study, a Pearson's correlation coefficient was calculated for all pair-wise combinations of samples using the $\log_2$ normalized intensity data. Table 16.1 details the resulting values. The values ranged between 0.971 and

**Table 16.1**
**The Pearson correlation coefficients of the $\log_2$ intensity data for all pair-wise sample comparisons**

| | | Comfrey or Riddelliine | | | | | |
|---|---|---|---|---|---|---|---|
| | | #1 | #2 | #3 | #4 | #5 | #6 |
| Comfrey(0.971–0.979) | #1 | 1.000 | 0.978 | 0.978 | 0.971 | 0.976 | 0.975 |
| | #2 | | 1.000 | 0.979 | 0.974 | 0.977 | 0.978 |
| | #3 | | | 1.000 | 0.973 | 0.974 | 0.976 |
| | #4 | | | | 1.000 | 0.975 | 0.973 |
| | #5 | | | | | 1.000 | 0.975 |
| | #6 | | | | | | 1.000 |
| Riddelliine(0.969–0.976) | #1 | 1.000 | 0.969 | 0.972 | 0.970 | 0.973 | 0.973 |
| | #2 | | 1.000 | 0.971 | 0.973 | 0.975 | 0.969 |
| | #3 | | | 1.000 | 0.971 | 0.975 | 0.975 |
| | #4 | | | | 1.000 | 0.976 | 0.972 |
| | #5 | | | | | 1.000 | 0.976 |
| | #6 | | | | | | 1.000 |

0.979 for the biological replicates within the comfrey-treated group and 0.969 and 0.976 for the riddelliine-treated group, suggesting that while there is some individual animal variability within each treatment group, while overall there is a high degree of similarity of the gene expression profiles within each group.

For the purpose of visualizing the different components comprising the variability within this data set, the raw intensity data for all samples were analyzed by Principal Components Analysis (PCA) (*see* Note 7). PCA is a mathematical approach for combining components (transcripts in this instance) in order to reduce the dimensionality of the data such that the greatest amount of variability is represented by the first principal component. The resulting data can then be presented graphically in 2- or 3-dimensional space. As is apparent in Fig. 16.2, the gene expression patterns resulting from comfrey or riddelliine treatments were distinct from those of the control group and there is also a clear separation between comfrey and riddelliine treatments within the first principal component. This indicates that both chemicals impacted the hepatic gene expression profile and that differences existed between the transcriptional response to comfrey exposure as compared to riddelliine exposure. The finding that there was less

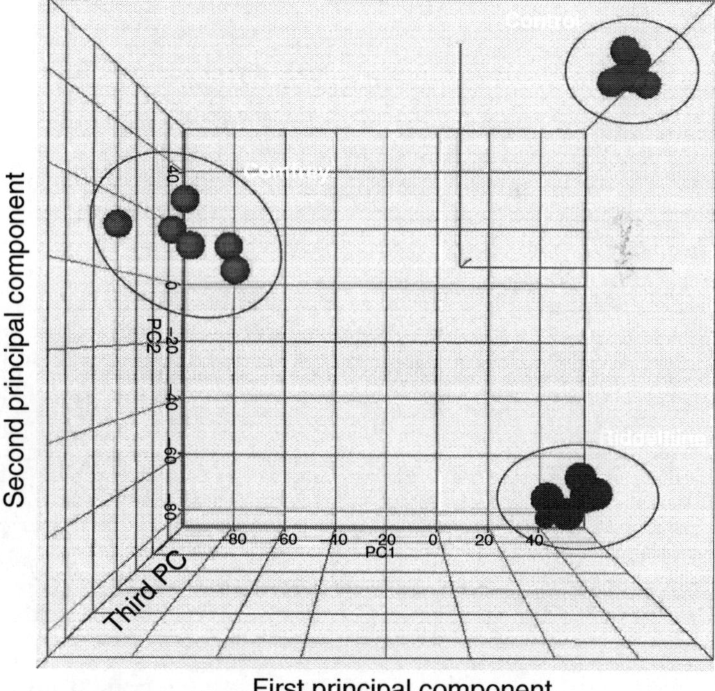


Fig. 16.2. Principal Component Analysis (PCA) of expression profiles for control, 8% comfrey-fed, and 1 mg/kg riddelliine-treated groups. The intensity of the entire gene set was used with no thresholds.

separation between the riddelliine samples and the control samples as compared to the comfrey and control samples was not unexpected as riddelliine is a prototypic tumorigenic pyrrolizidine alkaloid (PA), while comfrey is a mixture of many different substances, including PAs.

*3.5.3. Selection of Differentially Expressed Genes*

Before the selection of differentially expressed genes, the probes with low intensities are usually filtered because the values are near or below the noise level of the assay and therefore represent questionable results. There are a variety of different approaches that can be used to accomplish this. The approach we took was to use the intensity values from the negative control probes on the microarray. We calculated the mean value from the negative controls and then set a criterion of the mean value plus three standard deviations in order to identify the probes that did not have signal values that were significantly above background levels.

One straightforward approach that has been reported for the identification of reproducible gene lists is the use of a fold-change requirement coupled with a nonstringent log ratio *P*-value cutoff (12, 13). Usually, a 1.5- or 2- absolute fold change and a *P*-value less than 0.05 or 0.01 are used to select differentially expressed genes. The impact of different selections for these criteria can be visualized using a volcano plot. Figure 16.3 illustrates the results

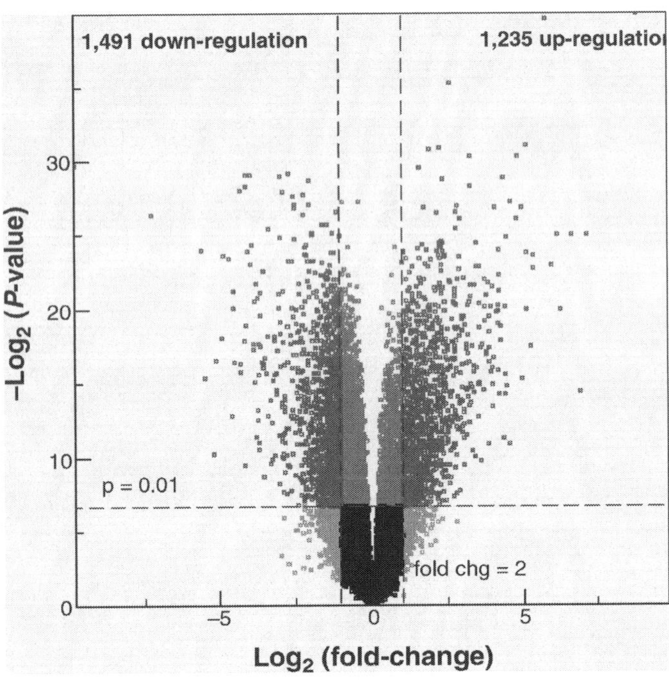


Fig. 16.3. Volcano plot. A gene was identified as significantly changed if the fold change was greater than 2 (*up* or *down*) and the *P*-value was less than 0.01 in comparison to the control group.

from the comfrey study in which we used an absolute value of the fold change greater than 2 and a $P$-value less than 0.01 to select differentially expressed genes. This resulted in the identification of 2,726 genes (1,235 up- and 1,491 down-regulated) that were differentially expressed in response to comfrey as compared to the corresponding control samples.

### 3.6. Validating Microarray Data

Real-time PCR is often used to validate the genes whose expression was found to be altered by the microarray analysis. We performed TaqMan PCR (Applied Biosystems) for a subset of differentially expressed transcripts that resulted from the previous analysis. Triplicate assays for each RNA and gene were performed in which 25 ng total cDNA (input RNA) was used in a 25 μl reaction. We observed good consistency in regard to the directionality of differential expression as well as the relative magnitude of fold-change using this approach and thus are confident that the microarray-based gene expression data is truly representative of the biological sample from which it was derived (9).

### 3.7. Pathway and Functional Analysis

There are many computational gene network prediction tools commercially or publically available, such as Cytoscape, GeneGo MetaCore, GeneSifter, Ingenuity Pathways Analysis, and Kyoto Encyclopedia of Genes and Genomes (KEGG). In our studies, most of data analysis was performed using Ingenuity Pathways Analysis by overlaying a set of differentially expressed genes onto the Ingenuity Knowledge Database, which provides a classification of gene products into molecular functions, biological process, and cellular components. Once these data have been integrated, regulatory networks, functional analysis, and canonical pathways altered in response to the chemical treatments could be explored. For example, we observed 83 and 118 functional processes altered by riddelliine and comfrey treatments, respectively, 46 of which were altered by both chemicals (Fig. 16.4). This suggested that

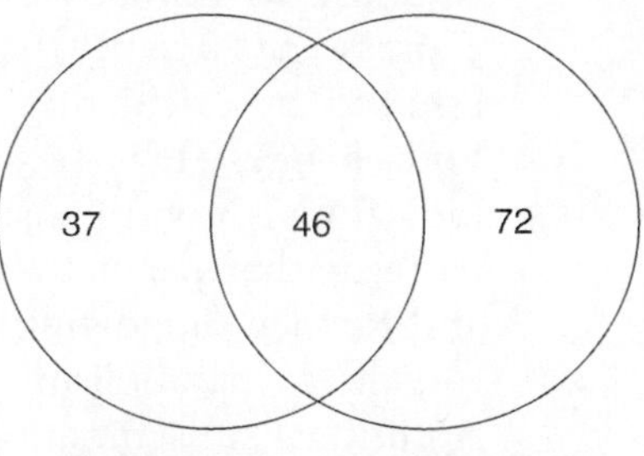


Fig. 16.4. Number of regulated functional processes significantly altered ($P < 0.01$) by comfrey or riddelliine treatments. There were 46 function processes altered by both riddelliine and comfrey identified by Ingenuity Pathways Analysis software.

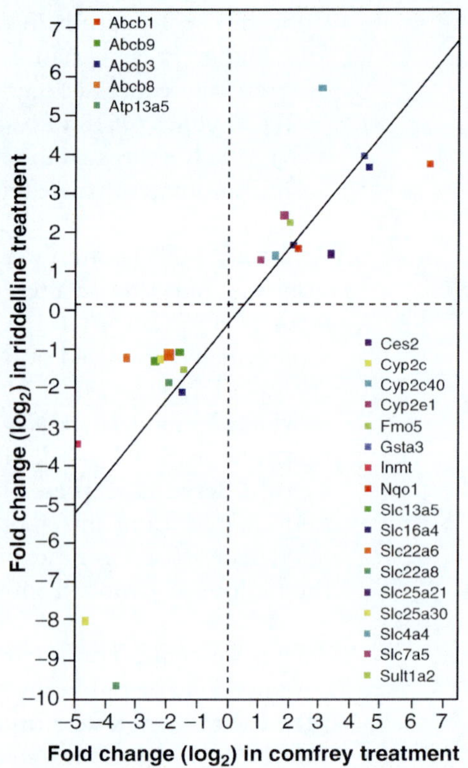


Fig. 16.5. Comparison of fold-change of drug metabolizing genes whose expression was altered by both comfrey and riddelliine treatments.

common mechanism(s) may be responsible for the toxicity associated with exposure to both comfrey and riddelliine (6). Consistent with this hypothesis was the observation that 22 drug metabolizing genes demonstrated differential expression in response to both treatments in the same direction. The gene expression similarity between the two treatments was further assessed by calculating the correlation coefficients of the $\log_2$ fold changes (Fig. 16.5). The correlation coefficient was 0.72, indicating a good consistency in regard to the gene expression patterns that result from exposure to both of these treatments and further supporting the existence of a common mechanism for drug metabolism of both riddelliine and comfrey (6). Additional substantiation for this hypothesis is that both riddelliine and comfrey are pyrrolizidine alkaloids and share some common toxicities (15).

## 4. Notes

1. In the early days of the gene expression microarray technology, technical replicates were extensively used to address variability that arose from (i) the production and manufacturing of the microarrays themselves, which were often fabricated in a researcher's own laboratory, and (ii) the failure to adequately control all of the steps in the complex microarray experimental process.
2. Depending on the study purposes, different rat strains could be used. Big Blue rats are used in this study, because we can perform mutation assay and gene expression on the same tissues. Transgenic gene mutation assays provide a unique opportunity for studying the induction of in vivo mutation, thereby enabling quantitative measurements of mutant frequencies in all tissues/organs as well as the molecular analysis of the induced and spontaneous mutations.
3. An alternative approach that could have been used would have been to immerse the isolated tissues in RNA stabilization solution (e.g., RNA*later* (Applied Biosystems)) and then store them according to the manufacturer's recommendations.
4. Due to the nature of some samples, RIN greater than 8.0 is not always feasible. For example, if exposures result in a high degree of necrosis, the degradation of RNA may be reflective of the cellular death that is occurring within the tissue of interest. Therefore, at times it may be necessary to use samples of lower quality to gain insights into the underlying gene expression patterns. Our experience suggests that lower quality samples are still acceptable for use in a microarray-based gene expression profiling experiment, though the sensitivity of the assay is typically reduced.
5. This approach is the dynamic nature of the field in regard to microarray-based methods. While methods have standardized, they are continually being modified to improve sensitivity and specificity, while also decreasing the input mass requirement.
6. One of our studies performed three technical replicates for a collection of RNA samples (9). In these instances, the intensities for each probe were averaged across the three technical replicates, and the averaged intensities were then normalized as detailed above.
7. Hierarchical Cluster Analysis of the gene expression data is also used for visualizing the data between two or more groups or samples.

## Acknowledgments

The authors thank Drs. Tao Han and Leming Shi from NCTR for their helpful discussions, comments, and criticisms. The views presented in this article do not necessarily reflect those of the U.S. Food and Drug Administration.

### References

1. Casciano DA, Woodcock J (2006) Empowering microarrays in the regulatory setting. Nat Biotechnol 24:1103
2. Lettieri T (2006) Recent applications of DNA microarray technology to toxicology and ecotoxicology. Environ Health Perspect 114:4–9
3. Gatzidou ET, Zira AN, Theocharis SE (2007) Toxicogenomics: a pivotal piece in the puzzle of toxicological research. J Appl Toxicol 27:302–309
4. Mendrick DL (2008) Genomic and genetic biomarkers of toxicity. Toxicology 245: 175–181
5. Chen T, Guo L, Zhang L, Shi L, Fang H, Sun Y, Fuscoe JC, Mei N (2006) Gene expression profiles distinguish the carcinogenic effects of aristolochic acid in target (kidney) and non-target (liver) tissues in rats. BMC Bioinformatics 7(Suppl 2):S20
6. Guo L, Mei N, Dial S, Fuscoe J, Chen T (2007) Comparison of gene expression profiles altered by comfrey and riddelliine in rat liver. BMC Bioinformatics 8(Suppl 7):S22
7. Mei N, Guo L, Zhang L, Shi L, Sun YA, Fung C, Moland CL, Dial SL, Fuscoe JC, Chen T (2006) Analysis of gene expression changes in relation to toxicity and tumorigenesis in the livers of Big Blue transgenic rats fed comfrey (*Symphytum officinale*). BMC Bioinformatics 7(Suppl 2):S16
8. Mei N, Guo L, Liu R, Fuscoe JC, Chen T (2007) Gene expression changes induced by the tumorigenic pyrrolizidine alkaloid riddelliine in liver of Big Blue rats. BMC Bioinformatics 8(Suppl 7):S4
9. Mei N, Guo L, Tseng J, Dial S, Liao W, Manjanatha M (2008) Gene expression changes associated with exnobiotic metabolism pathways in mice exposed to acrylamide. Environ Mol Mutagen 49(9):741–745
10. Guo L, Fang H, Collins J, Fan XH, Dial S, Wong A, Mehta K, Blann E, Shi L, Tong W, Dragan YP (2006) Differential gene expression in mouse primary hepatocytes exposed to the peroxisome proliferator-activated receptor alpha agonists. BMC Bioinformatics 7(Suppl 2): S18
11. Guo L, Zhang L, Sun Y, Muskhelishvili L, Blann E, Dial S, Shi L, Schroth G, Dragan YP (2006) Differences in hepatotoxicity and gene expression profiles by anti-diabetic PPAR gamma agonists on rat primary hepatocytes and human HepG2 cells. Mol Divers 10:349–360
12. Shi L, Reid LH, Jones WD, Shippy R, Warrington JA, Baker SC, Collins PJ, de Longueville F, Kawasaki ES, Lee KY, Luo Y, Sun YA, Willey JC, Setterquist RA, Fischer GM, Tong W, Dragan YP, Dix DJ, Frueh FW, Goodsaid FM, Herman D, Jensen RV, Johnson CD, Lobenhofer EK, Puri RK, Schrf U, Thierry-Mieg J, Wang C, Wilson M, Wolber PK, Zhang L, Amur S, Bao W, Barbacioru CC, Lucas AB, Bertholet V, Boysen C, Bromley B, Brown D, Brunner A, Canales R, Cao XM, Cebula TA, Chen JJ, Cheng J, Chu TM, Chudin E, Corson J, Corton JC, Croner LJ, Davies C, Davison TS, Delenstarr G, Deng X, Dorris D, Eklund AC, Fan XH, Fang H, Fulmer-Smentek S, Fuscoe JC, Gallagher K, Ge W, Guo L, Guo X, Hager J, Haje PK, Han J, Han T, Harbottle HC, Harris SC, Hatchwell E, Hauser CA, Hester S, Hong H, Hurban P, Jackson SA, Ji H, Knight CR, Kuo WP, LeClerc JE, Levy S, Li QZ, Liu C, Liu Y, Lombardi MJ, Ma Y, Magnuson SR, Maqsodi B, McDaniel T, Mei N, Myklebost O, Ning B, Novoradovskaya N, Orr MS, Osborn TW, Papallo A, Patterson TA, Perkins RG, Peters EH, Peterson R, Philips KL, Pine PS, Pusztai L, Qian F, Ren H, Rosen M, Rosenzweig BA, Samaha RR, Schena M, Schroth GP, Shchegrova S, Smith DD, Staedtler F, Su Z, Sun H, Szallasi Z, Tezak Z, Thierry-Mieg D, Thompson KL, Tikhonova I, Turpaz Y, Vallanat B, Van C, Walker SJ, Wang SJ, Wang Y, Wolfinger R, Wong A, Wu J, Xiao C, Xie Q, Xu J, Yang W, Zhong S, Zong Y, Slikker W Jr (2006) The microarray quality control

(MAQC) project shows inter- and intraplatform reproducibility of gene expression measurements. Nat Biotechnol 24:1151–1161

13. Guo L, Lobenhofer EK, Wang C, Shippy R, Harris SC, Zhang L, Mei N, Chen T, Herman D, Goodsaid FM, Hurban P, Phillips KL, Xu J, Deng X, Sun YA, Tong W, Dragan YP, Shi L (2006) Rat toxicogenomic study reveals analytical consistency across microarray platforms. Nat Biotechnol 24:1162–1169

14. Tong W, Cao X, Harris S, Sun H, Fang H, Fuscoe J, Harris A, Hong H, Xie Q, Perkins R, Shi L, Casciano D (2003) ArrayTrack–supporting toxicogenomic research at the U.S. Food and Drug Administration National Center for Toxicological Research. Environ Health Perspect 111:1819–1826

15. Mei N, Guo L, Fu PP, Heflich RH, Chen T (2005) Mutagenicity of comfrey (*Symphytum officinale*) in rat liver. Br J Cancer 92:873–875

# Chapter 17

# Generation of Congenic and Consomic Rat Strains

**Dominique Lagrange and Gilbert J Fournié**

## Abstract

Congenic and consomic rat strains are inbred strains containing in their genome a given genomic region (congenic) or a whole chromosome (consomic) from another strain. They are nowadays invaluable tools for the identification of genes and mechanisms of multifactorial diseases, one of the main goals in biomedicine. They are produced by repeated backcrosses from a donor inbred strain to a recipient inbred strain, and thereafter maintained by conventional brother-x-sister mating. Although their production is lengthy and costly, it only requires a zootechny unit for breeding and tools for genotyping.

**Key words:** Congenic rat, Consomic rat, Genetic dissection, Complex traits, Quantitative trait locus, Positional cloning, Polygenic disease

## 1. Introduction

### *1.1. Definitions, Principles and Aims of Making Congenic and Consomic Rat Strains*

Congenic and consomic rat strains are inbred stains. This means that the individual genomes from all the rats from a given strain are genetically identical and they have on an average less than 0.01 residual heterozygosity. The genome of a congenic rat strain contains a given genomic region from another strain. The genome of a consomic rat strain contains a whole chromosome from another strain (Fig. 17.1).

The making of congenic and consomic rat strains shares common principles and tools. The construction of a congenic strain is a common procedure originating from the pioneer work of Snell on histocompatibility genes in which congenic strains were produced by a series of backcrosses with selection at each generation for rejection of tumor transplants (1). This principle of phenotypic selection is no more used since the advent of molecular genetic markers for genotyping. Thus these strains are produced by repeated backcrosses from a donor inbred strain to a recipient

I. Anegon (ed.), *Rat Genomics: Methods and Protocols*, Methods in Molecular Biology, vol. 597
DOI 10.1007/978-1-60327-389-3_17, © Humana Press, a part of Springer Science+Business Media, LLC 2009

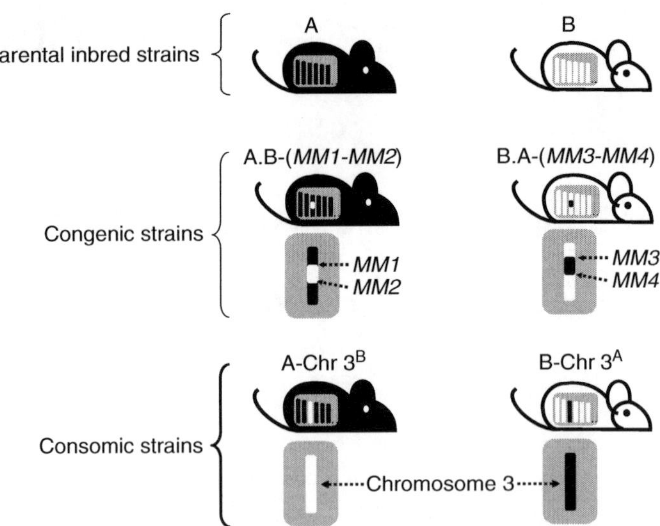


Fig. 17.1. Schematic representation of parental congenic and consomic inbred strains. The *upper part* of the figure shows the two parental inbred strains. The *grey rectangle* symbolizes the genome with the representation of only six chromosomes (each *bar* represents one pair of chromosomes). The chromosomes of the "black" A and of the "white" B strains are represented respectively in *black* and *white*. In the *middle part* are represented one A rat strain (in *black*) congenic for the genomic region of chromosome 3 from the B strain (*white*) spanning from microsatellite marker 1 (MM1) to microsatellite marker 2 (MM2) and one B rat strain congenic for the genomic region of chromosome 3 from the A strain spanning from MM3 to MM4. In the *lower part* of the figure are represented one A rat strain consomic for the chromosome 3 of the B strain and one B rat consomic for the chromosome 3 of the A strain. Nomenclatures for congenic and consomic strains are given in Notes 19 and 27, respectively.

inbred strain, with selection of donor markers for a given chromosomal region (congenic strain) or the whole chromosome (consomic strain).

Congenic and consomic rat strains are tools used to decipher complex genetic traits, which nowadays represent one of the main goals in biomedicine. While they are used for the same purpose, their usefulness and interest in practice are different (2).

### 1.2. The Rat, a Leading Experimental Model for Studies of Human Diseases

The laboratory rat (*Rattus norvegicus*) has been for the last century, and remains, the animal model of choice for physiology, pharmacology, toxicology, behavior and nutrition. Because of its size, ease of manipulation, and breeding characteristics, it is commonly used in models of human diseases for these fields, but also for diabetes, arthritis, infection, allergy, autoimmunity and cancer. All these diseases are multifactorial, i.e., they result from complex interactions between environmental and genetic factors. Due to the genetic heterogeneity of human populations and to the variability of environment, the identification of genes underlying complex traits is difficult in human. By contrast in the rat, as in the mouse, this task is easier because the environment is under control in modern zootechny units and the genetic heterogeneity is abolished due to the use of inbred strains (3). For this purpose, two inbred strains, which differ in the trait to be studied (e.g., the susceptibility to a given disease), are used.

### 1.3. Use of Congenic Rats to Decipher Complex Genetic Traits

The construction of congenic strains usually follows the broad localization of genes controlling the phenotypes of interest within quantitative trait loci (QTLs) of 15–40 centiMorgans (cM). QTLs are identified by linkage analysis in rats issued from crosses between two inbred strains (backcross of F1 rats to parental strain or intercross between F1 rats) (4). In such studies, the simultaneous segregation of several QTLs results in a "phenotypic noise," which limits the precise localization of QTLs.

Congenic strains are constructed to confirm the phenotypic effect of the QTL and to shorten its interval. This is done through mating of the two parental strains followed by successive backcrosses with the host strain. Genotyping of progeny at each generation allows the identification of rats carrying a copy of the desired QTL, which are selected as parents for the next generation. To speed up this process, rats issued from backcrosses at each generation can be genotyped with a set of genetic markers covering the rest of the genome to select, for the next generation, the best rats homozygous for the maximum of host alleles. This strategy of speed congenics allows reducing the number of backcrosses necessary to obtain full congenics (5). During this process, rats characterized by recombination within the region of interest can also be selected to create a panel of congenics for the QTL, which will further allow its genetic dissection and the localization of the gene(s) within a small interval containing a limited number of genes.

Once heterozygous congenics are obtained, they are bred by brother-x-sister mating for the production of a homozygous animal (one-fourth of the progeny), which will be the founder of the congenic strain.

### 1.4. Use of Consomic Rats to Decipher Complex Genetic Traits

Consomic strains, also called chromosome substitution strains, have been proposed as an alternative to identify and localize QTL, particularly in the field of developmental, physiological, metabolic and behavioral processes. Panels of consomic strains allow investigating the effect of any chromosome of one strain as compared to another strain in the same genomic background. Strains are chosen according to marked differences in phenotypes of interest. Consomic panels are constructed by successive backcrosses of a donor strain to a recipient strain as for the construction of congenic strains, but selecting progeny heterozygous for a whole chromosome. A complete consomic rat panel should consist of 22 strains (20 autosomes and sex chromosomes). In addition, strain with the mitochondrial genome of the donor strain, named conplastic strain, can be constructed. Until now several consomic panels have been constructed in the mouse (6–8), while few are available in the rat (2, 9–12).

## 2. Material

### 2.1. Zootechny Unit and Rats

Congenic and consomic strains are created and bred in dedicated zootechny units (*see* Note 1). These units follows the health monitoring recommendations given to the scientific community (13).

1. Room: the design of the breeding facility must provide control of the environmental factors to which the rats are exposed, including temperature, humidity, ventilation, light and noise.
2. Animal housing can be done on conventional racks equipped with cages affording for the presence of three adult rats in accordance with international standards such as the polycarbonate type III H cage (floor area: 806 cm$^2$; height: 18 cm) (Charles River laboratories) or equivalent. All material for housing, including cages, bedding, bottles, disposable filter top, water and diet are sterile (*see* Note 2).
3. Donor and recipient strains: male and female rats, to be ordered at 6–7 weeks of age, and mated at 8–10 weeks of age. Rats from inbred strains, which are thought to be homozygous after 20 generations of brother-x-sister inbreeding are available through several commercial breeders. A single source must be used throughout the entire process of congenic or consomic strain generation. (*see* Note 3).

### 2.2. Genotyping

#### 2.2.1. DNA Extraction

1. Tail biopsy from a 3-week-old rat at weaning: hold rat firmly at base of tail with one hand; with the other cut off 0.5 cm of the tail tip using a single edge razor blade (*see* Note 4).
2. Any validated protocol for DNA extraction such as the protocol of the Jackson laboratory (http://www.jax.org/imr/tail_phenol.html) can be used. It will need:
   - Polypropylene microfuge tube with tight-fitting caps,
   - Digestion buffer: 50 mM Tris–HCl pH 8.0, 100 mM EDTA pH 8.0, 100 mM NaCl, 1% SDS,
   - Proteinase K freshly prepared at 10 mg/ml in sterile, deionised water,
   - Neutralized phenol/chloroform/isoamyl alcohol (25:24:1),
   - 100% and 70% ethanol,
   - 10× TE buffer (100 mM Tris–HCl, 10 mM EDTA, pH 8.0) stored at 4°C, to be diluted 1/10 with sterile, deionised water.

Alternatively one can use a commercial kit such as the High Pure PCR Template Preparation (Roche®), for 100 isolation reactions or the Wizard® SV 96 Genomic DNA Purification System (Promega®) (*see* Note 5).

*2.2.2. PCR Reaction for Agarose Gel Electrophoresis*

1. Template DNA: 10 ng/µl genomic DNA in $H_2O$.
2. 96 well microtiter plates (Abgene®, Epsom, UK).
3. Taq polymerase, buffer for PCR reaction and $MgCl_2$ solution: AmpliTaq Gold® DNA Polymerase, provided with GeneAmp® 10× PCR Buffer II, and 25 mM $MgCl_2$ solution (Applied Biosystems, Foster City, CA). Taq polymerase from other suppliers can be used after validation.
4. Deoxynucleotide (dNTP) mix containing 25 mM of dATP, dCTP, dGTP and dTTP (Invitrogen, Carlsbad, CA, USA).
5. 100 µM forward and reverse primers (stored at −20°C), desalted grade (Eurogentec S.A., Seraing Belgium).
6. Cresol red/sucrose solution: 0.45% cresol red, 65% (w/v) sucrose.

*2.2.3. Agarose Gel Electrophoresis*

1. This procedure assumes the use of a well-adapted horizontal submarine gel system for agarose electrophoresis such as the horizontal midi-gel apparatus with a 14×16 cm gel tray that will accommodate the placement of four combs with 27 teeth (width: 14 cm; thickness: 1 mm) suitable for multipipeting devices (C.B.S. scientific, Del Mar, CA, USA). This plateau will allow the processing of 4×24 DNA samples, standards and DNA size markers. This "one 96-plate – one gel" design is easy to manage and efficient for routine use.
2. Agarose: high resolution MetaPhor® agarose (Cambrex, Rockland, ME, USA) (*see* Note 6).
3. Electrophoresis buffer: TBE 0.5× (45 mM Tris base, 45 mM boric acid, 1 mM EDTA). Prepare 10× stock 108 g Tris base (890 mM), 55 g boric acid (890 mM), 40 ml 0.5 M EDTA, pH 8.0 (20 mM) and $H_2O$ to 1 l. Prepare working solution by dilution of one part with 19 parts of $H_2O$.
4. Ethidium bromide: stock solution at 10 mg/ml (Sigma-Aldrich, St. Louis, MO, USA) is stored at 4°C in the dark. Ethidium bromide is added to the gel (10 µl for 100 ml of gel) and to the electrophoresis buffer (100 µl for 800 ml of buffer) (*see* Note 7).
5. DNA molecular size marker (e.g., 25 bp ; Invitrogen, Carlsbad, CA, USA).
6. Recording of results assumes the use of a suitable apparatus such as the "Molecular Imager Gel Doc XR System" (BioRad Laboratories, Hercules, CA, USA). Any UV trans-illuminator coupled with a Polaroid® camera can also be used.

*2.2.4. PCR Reaction for Capillary Electrophoresis (see Note 8)*

Same material than in Subheading 17.2.2.2, except that one of the two primers (usually the forward primer) is fluorescently labeled (Eurogentec S.A., Seraing, Belgium).

*2.2.5. Capillary Electrophoresis*

1. This procedure assumes the use of capillary electrophoresis system (DNA sequencers) for multicolor fluorescence-based DNA analysis, such as the ABI PRISM® 3100 Genetic Analyzer, which is equipped with 16 capillaries operating in parallel (Applied Biosystems, Foster City, CA).
2. Matrix DS30, to generate the "multicomponent matrix" required when analyzing DNA fragments on the capillary electrophoresis apparatus (Applied Biosystems).
3. Fluorochrome-labeled PCR products.
4. DNA size standard (GeneScan™ -400 ROX™ STANDARD) (Applied Biosystems).
5. Formamide as loading solution.
6. PCR MiroAmp plate (Applied Biosystems).

*2.2.6. SNPs Genotyping by qPCR*

1. This procedure assumes the use of an apparatus suitable for real-time quantitative PCR, such as the ABI prism 7000, and a software suitable for the design of primers such as the PrimerExpress 2.0 (Applied Biosystem) (https://products.appliedbiosystems.com/).
2. Template DNA: 20 ng/μl genomic DNA in $H_2O$.
3. One common primer and two allele specific primers (ASO) at 10 μM for each SNP, to be stored at –20°C (Eurogentec S.A., Seraing Belgium).
4. 96-well plates (Abgene).
5. PCR mix: 20 μl/well containing 12.5 μl of qPCR MasterMix Plus for SYBR® Green I (Eurogentec), 6 μl of water, 0.75 μl of each primer at 10 μM.

## 3. Methods

### *3.1. Genotyping*

*3.1.1. Genotyping Using Microsatellite Markers*

Microsatellite Marker Choice and Validation

1. The microsatellite markers that show polymorphism between the two rat strains are selected using Genome Tools/Genome scanner in the "Rat Genome Data Base" (http://rgd.mcw.edu/GENOMESCANNER/). In the output options, set the "Highlight allele size differences between.." 4 to 40–60 bp to select markers best suited for agarose gel electrophoresis. The two alleles when amplified by PCR will differ in size for at least 4%. Select markers well positioned on the genetic

maps: either "framework markers" (LOD > 3) or "unique placement markers" (LOD > 2). To ensure optimal separation on 4% agarose gel electrophoresis, markers should be selected as far as possible in the range of 100–200 bp in length. A convenient selection should allow covering the region of interest (congenics) or the whole genome (speed congenics and consomics) at intervals of ~10 (range 5–20) centiMorgans (cM) (*see* Note 9).

2. Order the reverse and forward primers to set up detection of alleles using PCR reaction.
3. Extract DNA from the 0.5 cm piece of tail of two parental strains and from F1 hybrids. Follow precisely the protocol which has been set up or the manufacturer's protocol. After extraction, the recovered 100–200 μl solution should contain DNA in a concentration range of ~100–200 ng/μl. It is not routinely necessary to check for the DNA concentration, but it is wise to do so when setting up the technique or when a problem occurs after PCR. For this purpose, dilute the DNA sample 1/50 in $H_2O$ and read optical density at 260 and 280 nm. The ration 260/280 should be 1.8–2.0. Calculate the DNA concentration using a molar extinction coefficient of 20 at 260 nm (light of 260 nm passing 1 cm through a 50 μg/ml DNA solution has an absorbance of 1.0).
4. Using the primers and the DNAs, set up the PCR reaction for each marker using a "touch down" procedure. The standard conditions for the PCR reactions in the 96-well microtiter plates are: buffer: 1×; $MgCl_2$: 2.5 mM; dNTP: 200 μM × 4; AmpliTaq Gold: 0.4U; primers: 1 μM; cresol red 90 μg/ml; sucrose: 13%; in 15 μl final volume containing 5 μl of the DNA solution (50–100 ng). After a denaturation for 5 min at 95°C, 39 cycles are done comprising denaturation 20 s at 95°C, annealing from 64°C to 52°C during 12 cycles (lowering the temperature of annealing of 1°C at each cycle until the temperature for annealing is 52°C) then 27 cycles at 52°C for 30 s, elongation for 30 s at 72°C. The last step of elongation is run on 5 min at 72°C and then the mixture is held at 10°C until study or storage at –20°C. A 96-well plate allows for the simultaneous set-up of 32 markers in the two parental lines and in the F1 hybrid. This method allows for convenient amplifications of 85–90% of the chosen markers. In a few cases the markers are not polymorphic and other markers have to be selected (*see* Note 10). When the amplification is not satisfactory, the amount of Taq polymerase is raised to 0.6U/reaction and two temperatures of annealing (52 and 60°C) are independently checked. This procedure solves the problem in most cases.

Agarose Gel Electrophoresis

1. Dilute 4 g of ResoPhor agarose in 100 ml of 0.5× TBE in a flask. Tare the flask. Place the flask in a boiling water bath under stirring. After 20–25 min, tare the flask and add warm distilled water to obtain the initial weight if needed. Let stand the solution for 15–20 min until the temperature reaches 50–60°C. Eliminate any remaining air bubble by puncture using the tip of a needle or by pushing them to the side using a Pasteur pipette. Add 10 µl of ethidium bromide solution, mix gently and pour the gel. When the gel has solidified (20–25 min in cold room), remove the combs carefully, transfer the gel in the electrophoretic chamber, submerge it in running buffer containing ethidium bromide and load 7 µl of each PCR sample and when needed of marker size, through the buffer. The remaining sample (~8 µl) is stored at 4°C to repeat the analysis if needed. Run the electrophoresis for 2 h at 120 V.
2. Record the result with photography under UV illumination. Correct recording of the genotype is, of course, of major importance. A double lecture is recommended. A practical way of doing so, is to record the result in a file with results from the first reader in normal characters and confirmation from the second reader in bold characters.
3. Set a file with the code of the microsatellite marker, the position of the marker in the genome, the sequences of the primers, the conditions of the PCR reaction (usual or peculiar), the sizes of the microsatellite markers found in data bases for the two strains and the sizes of amplicons as calculated from the DNA size standards, and finally the ease or difficulty of the lecture, which will help to chose if the agarose gel can be re-used or if a new gel should preferably be used.

Capillary Electrophoresis

1. Order the forward primer with a fluorescent label, the reverse unlabelled. Perform the PCR reaction as previously, but lowering the number of cycles to 36. Check the amplicons on agarose gel electrophoresis.
2. Prepare the DNA size marker/running solution by adding 20 µl of the fluorescent DNA size marker to 1 ml of formamide.
3. Add 3.5 µl of the amplicon to 8 µl of the formamide solution containing the DNA size marker in the PCR MicroAmp plate and heat for 3 min at 96°C before running the sample.
4. Run the samples and analyse the result using the "genotyping" program of the system software.

*3.1.2. Genotyping Using SNPs Markers*

1. Design and order the three primers for each SNP to be genotyped.

2. Perform the PCR on the DNA extracted from the rats to be genotyped and as control from the parental strains and F1 hybrid. For each DNA sample, the two combinations of primers are used. Mix in each well 5 µl of DNA solution and 20 µl of the qPCR MasterMix Plus. The following program is used: 50°C 2 min (enzyme activation), 95°C 10 min (denaturation), and then 40 cycles of denaturation 15 s at 95°C followed by annealing/amplification 1 min at 60°C; a final cycle is done at 95°C 15 s, 60°C 20 s and 95°C 15 s.
3. Determine the genotype of each DNA sample according to the kinetics of the PCR reaction given by the apparatus used for real-time quantitative PCR, and record in data file.

### 3.2. Creation of Congenic Strains for a Whole Genomic Region and for Various Part of the Region Using the Speed Congenics Strategy (Fig. 17.2)

A congenic strain is obtained by repeated backcrosses from F1 [donor × recipient] hybrids to a recipient inbred strain. At each backcross, genetic markers are used to select for the next backcross rats heterozygous for the genomic region of interest to be transferred to the receiver inbred strain. After n generations, including the first F1 mating, the genetic background is $100 \times (1-1/2^n)$ % of the recipient strain. When the genetic background is ~100%, at the eighth (99.8%) or the ninth backcross (99.9%), rats heterozygous for the donor region are intercrossed.

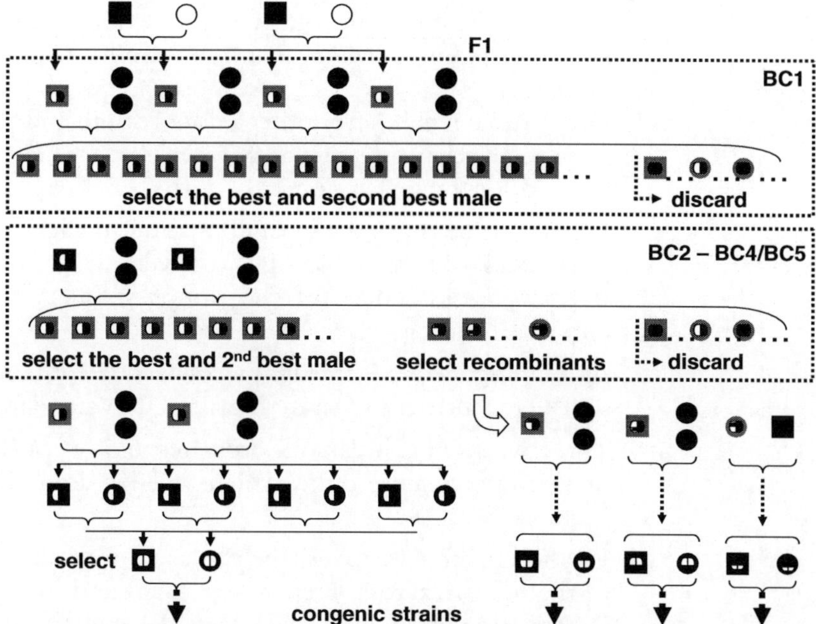


Fig. 17.2. Schematic representation of the breeding procedures for production of congenic strain using the speed congenics technique. *Squares* represent males and *circles* females. *Black* and *white* colors represent, respectively the genomes of the recipient strain and of the donor strain. *BC*, backcross. Selection of recombinants can began at any BC after BC1.

Rats issued from this intercross, homozygous for the region (1/4), are selected for brother-x-sister mating and foundation of the congenic strain. The time needed for each backcross is about 11–13 weeks (3 weeks of gestation; 8–10 weeks to reach the age of mating). Thus the construction of congenic strains in such a way requires at least 2.5 years of breeding and often 3 years or more. To speed up the required time for this construction, the so-called "speed congenics" approach has been proposed (14, 15). This procedure is based on the "marker-assisted" selection, at each generation, of the best breeder for the next generation, i.e., the rat bearing the region to be introgressed and harboring the least amount of donor genome outside this region. Doing so, the number of generations necessary to construct a congenic strain can be reduced to 5–6 and the length of time to 1.5–2 years (*see* Note 11).

During the construction of the congenic strain, rats issued from backcrosses and characterized by recombination within the genomic region of interest are identified by genotyping. These rats can be selected for further backcrossing to develop in parallel a set of congenic strains, each of which being characterized by recombination within the region of interest and bearing different parts of the region. At the end of the process, this set of congenic strains allows the genetic dissection of the region.

Following is the procedure for the development of a set of inbred strains of a given genomic background (the recipient strain), congenic for the whole genomic region and various segments of this genomic region of another strain (the donor strain) (*see* Note 12).

1. Order three females from the A donor strain and three males from the B recipient strain at 6–7 weeks of age. These rats will be used for the generation of F1 rats.
2. Prepare three cages for separate mating in each cage of one 8–10-week-old male of the receiver strain B and one 8–10-week-old female of the donor strain A to produce (A×B) F1 rats (*see* Note 13).
3. At weaning, take 0.5 cm tails pieces of the F1 rats and store at –20°C for further preparation of F1 DNA standard.
4. Three weeks after weaning, order ten 6–7-week-old females of the B recipient strain, which will be used for the first backcross.
5. Five weeks after weaning (the rats are 8 weeks old), proceed to the first backcross. Prepare five cages and mate separately in each cage one 8-week-old (A×B) F1 male to two 8-week-old B females (*see* Note 14).
6. After 2 weeks of mating, pregnancy is detected by looking for abdominal enlargement, and if needed by palpation. One

pregnant female is removed prior to parturition and allowed to deliver in a separate cage since cage size is insufficient to accommodate two lactating dams and their pups.

7. At weaning, when the pups are 3 weeks old determine the sex of rats and sacrifice the female rats (*see* Note 15). Identify each male by a standardized and accurate procedure and cut off 0.5 cm of the tail tip (*see* Note 16).
8. Extract DNAs from the tail.
9. Genotype the males issued from the backcross (8–12 rats/litter: a total of 40–60 male rats are expected) for the markers covering the genomic region of strain A to be introduced into strain B at intervals of 5 cM. Genotyping (~250 reactions for a 20 cM interval) will be completed in ~1 week. Select the males heterozygous for the entire region of the A strain to be introgressed into the B genome. Sacrifice the other males.
10. Genotype the selected heterozygous males (n \simeq 20) for a set of ~100 markers distributed approximately at 20 cM intervals in the genome. Genotyping (~2,000 reactions) will be completed in 2–3 weeks. Select the two males carrying the fewest heterozygous segments of the remaining recipient genome (<25/100), for the next backcross.
11. During the genotyping process, 3 weeks after weaning, order four 6–7-week-old females of the receiver B strain, which will be used in the next backcross.
12. When the selected male rats are at 8–10 weeks of age, proceed to the second backcross. Mate separately in two cages, each selected male with two B females.
13. Repeat the procedures done in the first backcross from steps 6 to 9. The number of expected rats to be genotyped is reduced by half, as the expected number of male rats heterozygous for the entire region of the A donor stain.
14. From these selected male rats, select the best and second best males that carry the fewest heterozygous segments as in step 10. At that stage, genotyping is done on the microsatellite markers that showed heterozygosis in the first backcross (<25/100 markers; usually ~20 markers) and on another set of ~100 markers located between markers used in the first backcross. Thus ~120 markers will be genotyped in ~10 rats (~1,200 reactions that will be completed in 1–2 weeks) and on the whole, the total set of markers will allow a 10 cM interval genome-wide scan. At that stage, it is expected that in the selected rats, more than 90% of the background genome is homozygous for the recipient strain (less than 20 out of the 200 markers are still heterozygous).

15. When the selected male rats are at 8–10 weeks of age, proceed to the third backcross. Mate separately in two cages, each selected male with two B females as in step 12.
16. Repeat the procedures done at step 6.
17. At weaning, when the pups are 3 weeks old do not systematically sacrifice the female rats. At that stage, it could be interesting to select rats showing recombination within the genomic region of interest to further develop in parallel congenic lines characterized by the introgression of various segments of the region, which will allow the genetic dissection of the locus (*see* Note 17). Identify each rat (males and females) and cut off 0.5 cm of the tail tip for DNA extraction.
18. Genotype all the rats (8–12 rats/litter: a total of 32–48 rats is expected) for the markers covering the genomic region of strain A to be introduced into strain B at intervals of 5 cM. Genotyping (~200 reactions for a 20 cM interval) will be completed in 1 week. Sacrifice all the rats (males and females) homozygous for the B genome in the region and the female rats heterozygous for the whole region. Select the male rats heterozygous for the entire region of the A strain, and the male and female rats heterozygous for part of the region (characterized by recombination events within the region).
19. Genotype all the selected rats (20–28) for the markers of the remaining genome, which were found heterozygous at the third backcross (less that 20 out of 200). From the rats heterozygous for the entire region, select the two best males that carry the fewest heterozygous segments of the remaining recipient genome for the next backcross. Rats (male and female) heterozygous for part of the region are also selected, whatever their remaining recipient genome.
20. During the genotyping process, 3 weeks after weaning, order at least four 6–7-week-old females of the receiver B strain, which will be mated with the two best males heterozygous for the whole region. In addition, order two females of the recipient strain for any selected "recombinant" male and one male for any selected "recombinant" female.
21. Five weeks after weaning, proceed to the fourth backcross by mating separately in independent cages, the two best rats heterozygous for the entire region and male rats characterized by recombination within the interval with two females, and each selected "recombinant" female rat with one male.
22. Repeat the procedures done at step 6.
23. At weaning, identify each rat and cut off 0.5 cm of the tail tip for DNA extraction.

24. Genotype all the rats with the markers of the region. The number of rats to be genotyped depends on the number of rats characterized by recombination within the region, which have been selected in addition to the selected two best male rats bearing the entire region (*see* Note 18).
25. Proceed as in steps 18 and 19. Select the two males heterozygous for the entire region that carry the fewest heterozygous segments of the remaining recipient genome (0–4/200 at that stage). Select also the best male heterozygous for each of the selected "recombinant" for further breeding. Keep alive the second best male to be able to rescue the strain in case of problem in breeding the only rat selected for breeding. Select new "recombinant" (males or females) according to their genotypes and to the genotypes of the "recombinants" that have been already selected at the third backcross.
26. At that stage, repeat the procedures of backcrossing and genotyping for all the selected rat strains showing still heterozygosis for some marker in the background recipient genome until all the markers selected for covering the genome at the 10 cM interval show homozygosis for background recipient genome. One or two more backcrosses can be necessary to achieve this goal for all the strains.
27. When homozygosis for the background recipient genome is achieved, rats heterozygous for the whole region or for parts of the region are intercrossed by brother-x-sister (1/1) mating.
28. At weaning, identify each rat and cut off 0.5 cm of the tail tip for DNA extraction.
29. Genotype the rats for the markers of the region of interest to select male and female rats showing homozygosis for the introgressed region.
30. Proceed to the breeding of the created congenic new strains (*see* Note 19) by brother-x-sister mating.

### 3.3. Creation of Consomic Strains

The availability of genetic markers well distributed on the rat genome and covering entirely all chromosomes has allowed the construction of consomic strains. The principle of creating a consomic strain of rat is the same as creating a congenic strain, except an entire chromosome is transferred. The speed procedure of marker-assisted selection can be used. Following is the procedure for the development of a set of inbred strains of a given genomic background (the recipient strain) consomic for chromosomes of another strain (the donor strain).

1. Generate (A×B) F1 rats by mating 8–10-week-old females of the donor strain A and 8–10-week-old male of the receiver B strain and (B×A) F1 rats by mating 8–10-weeks-old females of

the receiver strain B and 8–10-week-old male of the donor A strain (*see* **Note 20**).

2. Three weeks after weaning, order ten 6–7-week-old females of the B recipient strain which will be used for the first backcross (*see* Note 21).
3. Five weeks after weaning (the rats are 8 weeks old), proceed to the first backcross. In three cages, mate separately in each cage one 8-week-old (A × B) F1 male to two 8-week-old B females (*see* Note 22). In two cages, mate separately in each cage one 8-week-old (B × A) F1 male to two 8-week-old B females (*see* Note 23).
4. After 2 weeks of mating, pregnancy is detected by looking for abdominal enlargement, and if needed by palpation. One pregnant female is removed prior to parturition and allowed to deliver in a separate cage since cage size is insufficient to accommodate two lactating dams and their pups.
5. At weaning, when the pups are 3 weeks old, identify each rat (females and males) issued from the [(A × B)F1 × B] BC and from the [(B × A)F1 × B] BC by a standardized and accurate procedure and cut off 0.5 cm of the tail tip.
6. Extract DNAs from the tail of the rats issued from the [(A × B) F1 × B] and [(B × A)F1 × B] BCs beginning with male rats.
7. First, genotype the male rats from the [(A × B)F1 × B] BC (24–36 rats) for ~100 markers covering the whole genome at intervals of 20 cM (*see* Note 24). For each autologous chromosome, select the best two rats heterozygous for the entire chromosome and complete the genotyping for markers allowing the covering of the chromosome at intervals of 10 cM. For each autologous chromosome, select the best rat heterozygous for the entire chromosome, which carries the fewest heterozygous segments of the remaining recipient genome. Keep alive the second best male, in case of problem in breeding the selected male. Genotype the male rats issued from the [(B × A)F1 × B] BC (16–24) for markers of all the autologous chromosomes at intervals of 20 cM. From these rats, select the best rat for the next backcross and keep alive the second best rat. When this first wave of genotyping has been completed, genotype the female rats from the [(A × B)F1 × B] BC (24–36 rats) and from the [(B × A)F1 × B] BC (16–24) at intervals of 10 cM for markers of chromosomes for which no male rat heterozygous for the entire chromosome has been found (*see* Note 25). Select females heterozygous for the entire chromosomes, and then genotype these selected females for the rest of the genome. Select the "best female" for each chromosome for the next

backcross and keep alive the second best female. The females issued from the [(A × B)F1 × B] BC are heterozygous for the entire X chromosome, thus one of these females must be selected for the next backcross and the further creation of the strain consomic for the X chromosome (keep alive the second best female for this strain).

8. Repeat the procedures of backcrossing and genotyping for all the selected rat strains following the same procedures until homozygosis for background recipient genome is obtained for all markers selected for covering the genome at 10 cM intervals. For strains consomic for autologous chromosomes, select as far as possible the best male to be backcrossed to two recipient B females. For the strain consomic for the Y chromosome, backcross one male to two recipient B females. When needed for autologous chromosomes, and for the strain consomic for the X chromosomes, backcross two females to two B males in two separate cages.
9. When homozygosis for the background recipient genome is achieved, rats heterozygous for entire chromosomes are intercrossed by brother-x-sister (1/1) mating (*see* Note 26). Male and female rats showing homozygosis for the introgressed chromosomes are selected and bred by brother-x-sister mating for the generation of the new consomic strains (*see* Note 27).

### 3.4. Construction of Sub-Strains from Congenic Strain or of Congenic Strain from Consomic Strain

When the genetic dissection has to be done or completed for the precise localization of the gene(s), sub-strains with smaller and smaller donor regions are constructed from congenic or consomic strains. This is done using F1 hybrids issued from crossing the congenic or consomic strain to the recipient strain. Some gametes of F1 hybrids will be characterized by recombination events within the introgressed region during meiosis. These F1 rats are backcrossed to the recipient strain or intercrossed to select by genotyping rats from litter mates characterized by recombination events (*see* Note 28). These selected rats are then backcrossed to the recipient strain. Male and female rats issued from this BC, heterozygous for this reduced introgressed region are intercrossed to fix the recombinant chromosome in the homozygous state. Rats issued from this intercross, homozygous for the region, are selected for brother-x-sister mating and foundation of the new congenic sub-strain.

1. Generate F1 rats by mating an 8–10-week-old male of the receiver B strain and 8–10-week-old females of the B.A-MMx-MMy congenic strain or of the B-Chr $n^A$ consomic strain. Rat that will be issued from this backcross generation will become heterozygous on the targeted region (congenic)

or chromosome (consomic) while the remaining genome will be homozygous for the B recipient genome.

2. Five weeks after weaning, proceed to brother-x-sister mating of the F1 rats to produce F2 rats.
3. In cages with trios (one male, two females), pregnancy must be detected after 2 weeks of mating by looking for abdominal enlargement, and if needed by palpation, to remove one pregnant female from the cage prior to parturition.
4. At weaning, when the pups are 3 weeks old, identify each F2 rat (females and males) and cut off 0.5 cm of the tail tip for DNAs extraction.
5. Genotype the rats for markers covering the chromosomal region (congenic) or the chromosome (consomic) to detect rats characterized by chromosomal recombination (*see* **Note 29**).
6. Five weeks after weaning, backcross the selected "recombinant" rats to rats of the B recipient strain.
7. At weaning, when the pups are 3 weeks old, identify each rat (females and males) and cut off 0.5 cm of the tail tip for DNAs extraction.
8. Genotype the rats for the boundary markers that characterize the recombinant chromosomal region to detect rats heterozygous for the same region of the targeted genome.
9. Five weeks after weaning, proceed to brother-x-sister mating of selected rats.
10. At weaning, when the pups are 3 weeks old, identify each rat (females and males) and cut off 0.5 cm of the tail tip for DNAs extraction.
11. Genotype the rats to detect rats homozygous for the donor region and proceed to the breeding of the new congenic strains by brother-x-sister mating.

### 3.5. Quality Controls

To check the constructed congenic and consomic strains for residual contamination of the B receiver genome by sequences of the A donor genome not detected by microsatellite marker genotyping during the strain construction process, additional polymorphic markers leading to the covering of the genome at 5 cM intervals or less can be genotyped. Alternatively, the detailed SNP map launched recently by the STAR Consortium can be used to select an adequate panel of polymorphic A/B SNPs for further genome-wide genotyping control (16).

Moreover, genetic controls and genetic monitoring programs of inbred strains are important to ensure genetic purity of inbreed strains and to track possible accidental breeding errors. For this purpose, DNA of the strains must be regularly archived (*see* Note 30).

## 4. Notes

1. In such units, the environmental factors that could affect animal health and well being, as well as animal physiological and behavioral functions, are under control. The microbial and physical environmental factors, which the rats are exposed to, are standardized and regularly checked. This allows the microbiological quality of animals, particularly a "specific pathogen-free" status. Guidelines for such health monitoring of rodents are given by international quality assurance organisms such as the FELASA working group in Europe (13) (http://www.felasa.eu/recommendations.htm).
2. Care should be given to environmental factors when reproduction rate and efficiency of mating is of great concern. Several environmental factors may affect breeding performance (17). Particularly, living area, i.e., cage size and population density, may influence several areas of well-being, which may in turn impact sexual behaviour and reproduction rate (18). Bedding material must be adapted to breeding. Cottonwood bedding characterized by low dust content is well adapted. Cages and bedding are sterilized by autoclaving before use. 800 ml bottles are autoclaved, filled with sterilized water and equipped with sterilized stainless steel conical caps. Cages must be equipped with disposable filter top that can be sterilized with an adapted system using hydrogen peroxide as sterilizing agent. Diet, sterilized by a 10 kGy irradiation, should provide a high protein content.
3. The couple of strains that will be used for generation of new inbred congenic or consomic strains are chosen according to the phenotypes of interest that will be studied. It is important to know that inbred strains from different sources are not genetically identical, either due to the fixation of spontaneous mutations or from crossbreeding errors during the long-lasting (several decades) inbreeding process. An example of such a possibility has been recently illustrated in a study showing that the difference in susceptibility of two inbred LEW rat strains to autoimmunity was due to genetic contamination after crossbreeding of one of the two studied LEW substrains (19). It is the reason why, as far as possible, a single source must be used throughout the entire process of congenic or consomic strain generation. A list of rat strains holders can be found at the rat genome database ratmap, (http://www.ratmap.org/ratstrains.html).
4. According to most of the institutional animal care and use committees, tail clipping on rats without anesthesia can be

performed once on animals up to the time of weaning (at approximately 21 days of age), removing no more than 0.5 cm of the tail.

5. The Promega method is fast and convenient when a high number of samples must be processed, but it requires specific materials not included in the kit (nuclease free Proteinase K, Vac-Man® 96 Vacuum manifold, Vacuum trap, Vacuum pump and tubing).
6. MetaPhor agarose is best suited when high resolution is needed to separate PCR products differing by ~4% in size. Agaroses from other suppliers, less expensive, can be used, such as Resophor from Eurobio, when the PCR products differs from more that 4% in size. Agarose gels can be reused several times but when high resolution is an absolute requirement, newly made gels have to be used.
7. Ethidium bromine is toxic and a powerful mutagen. Its use necessitates protection procedures when weighing out solid (mask and gloves) and when working with (gloves). Solid waste containers must be used for gels. For contaminated solutions destaining bags such as those provided by Amresco, Solon, OH (code E732-25) are convenient. After ethidium bromide adsorption, the bag can be discarded as solid and the decontaminated buffer can be eliminated as a non-toxic solution.
8. When the size of the PCR products of the genomic DNA from the microsatellite marker to be separated are too close, say less than 4%, agarose gel electrophoresis is not suitable for genotyping. In this case, capillary electrophoresis can be used since it can discriminate two products differing in size by only one base pair. This method can also be used for large-scale genotyping by pooling fluorescently labeled PCR products (20).
9. In all cases, it will be necessary to cover the whole genome at 5–10 cM intervals. This is necessary for the speed congenics procedure, for the generation of consomic rats and for the quality control of any congenic and consomic strains to ensure the homozygosis of the background genome for the receiver strain. Note that the progress in SNPs maps (16) with the advent of large-scale genotyping methods could modify in the future the genotyping procedures for the construction of congenic and consomic rat strains using up-to-date equipped genomic platforms.
10. When some of the chosen markers do not give satisfactory results, go back to step 1 to chose another marker located at the same or near the appropriate position. If the marker gives amplicons differing from less than 4% in length, order

fluorescein-labeled forward marker for the subsequent analysis under capillary electrophoresis.

11. An alternative method to speed up the process of congenic strain construction, named "supersonic congenics," has been proposed (21). This method, presented in this book in the chapter 18, is based on superovulation of prepuberal 3-week-old female to produce fertilized eggs for embryo transfer. Therefore, theoretically, each backcross generation only takes ~6 weeks instead of ~12 weeks.
12. The herein described method can be modified according to the size and practical possibilities of the two teams in charge of breeding procedures and genotyping, respectively. Particularly, the development of the "reciprocal" congenic strain is often needed, which will require the same procedure performed in the opposite way (i.e., with the recipient strain becoming the donor strain and the donor strain becoming the recipient strain) and will double the work to be done.
13. F1 hybrids rats are issued from the first generation of crosses between two inbred strains. They are designated by listing the female strain first and the male strain second with "F1" appended. Note that in the used breeding scheme, male rats from the receiver B strain are used for the generation of (A × B) F1 rats. Thus, the receiver Y chromosome will be fixed in subsequent generations when selected male rats will be backcrossed with female of the recipient B strain.
14. While maximum reproductive performance of male rats is reached at approximately 10 weeks of age, 8-week-old rats are fertile and capable of successful breeding.
15. Determination of sex is easy at 3 weeks of age. It is based on the greater anogenital distance in male compared to female pups. When needed, it can be done earlier in rat pups less than 2 weeks of age.
16. Toe clipping is a cheap and efficient method for identification of small rodents. It is still accepted by some local ethical committees. This method involves removal of the distal phalangeal (coffin) bone of one or more limbs. When needed, it can be performed without anaesthesia on altricial neonates (rats at 7 days of age). Rats older than 7 days must be anesthetized for this procedure to ensure proper technique and to minimize pain and distress. Only one toe per foot may be removed. A pair of sterile sharp scissors can be used for this procedure. It must be disinfected in between uses. Hemostasis can be achieved by gentle pressure using gauze square until bleeding has stopped. The two procedures of rat identification by toe clipping and of tail clipping can be conveniently performed

successively at 3 weeks of age under fluotane anaesthesia. The toe clipping procedure is no more allowed by some ethical local comities. It can be replaced by electronic microchip devices, each ship being encoded with a unique number, implanted subcutaneously and read with a special detector. Other alternative methods include ear notching and punching, and microtattooing.

17. If the genomic region of interest extends on 20 cM, 20% of the spermatozoid bearing the donor genomic A region is expected to be characterized by a recombinant event within the genomic region during meiosis. To increase the further development of congenic strains characterized by various parts of the A region, it could be wise to check males and females for such events of recombination. For 32–48 rats issued from BC2, the identification of several males and females bearing such a recombination within the 20 cM genomic region is expected. If "recombinant" females are selected they will be, of course, backcrossed to males of the recipient B strain.
18. From the two best males, a progeny of 32–48 rats is expected. If four males and four females characterized by recombinations have been selected, the number of rats to be genotyped on the whole will be ~130–190 (32–48 from the two best males; 64–96 from the "recombinant males"; 32–48 from the "recombinant females"). Since at that stage, the percentage of the remaining recipient genome is expected to be ~99% or greater, the number of genotypes to be done will be kept in the order of a few hundred reactions. Of course, the number of "recombinant" rats selected for the creation of strains congenic for part of the region is dependent on the practical possibilities of the team in charge of the genotyping and of the team in charge of breeding in the zootechnic unit.
19. It is necessary to baptize the new strains. For this purpose, the international rules for nomenclature can be followed. Details on these rules and examples for nomenclature of rat (and mouse) strains can be found at http://www.informatics.jax.org/mgihome/nomen/strains.shtml. For congenic strains, the symbol of the recipient strain is separated by a period from the symbol of the donor strain; a hyphen separates the donor strain name from the symbol in italic of the chromosomal region or allele from the donor strain. For example, given the two rat strains A and B, the congenic strain issued from the introgression into the recipient strain B of the genomic region from donor A strain located between the two microsatellite markers MM1 and MM2, will be named B.A-(*MM1-MM2*). However, in the literature, strains are often named in a simpler way, indicating after the recipient strain,

the period and the donor strain, the chromosome of the introgressed region from the donor strain (c for chromosome followed by the number of the chromosome) followed by a hyphen and a letter. Thus one A strain congenic for part of the chromosome 12 (c12) can be named A.Bc12-E (E being in that case any letter from A to Z). A table can be provided showing the correspondence between this simplified denomination and the name given according to the international rules. Alternatively, a figure with a genetic map can show with accuracy the genome of the congenic strain.

20. (A × B) F1 rats will be used for the production of strain B rats consomic for autologous and X chromosomes of strain A, while (B × A) F1 rats will be used for the production of strain B rats consomic for the Y chromosome of strain A.

21. Using recipient B strain as the female parent, ensure the recipient B origin of the mitochondrial genome in any cross. Note that in addition to the construction of strains consomic for the autosomes and sex chromosomes, strain with the mitochondrial genome of the donor A strain in the overall genome background of the recipient B strain can be constructed. Such a strain, named conplastic strain and indicated as B-mt$^{A}$, can be obtained by backcrossing female rats of the mitochondrial A donor strain to males of the recipient B strain (http://jaxmice.jax.org/type/conplastic.html).

22. In this combination, males issued from the backcross will carry the Y and the X chromosomes from the receiver B strain. In females, one of the two X chromosomes will be from the donor A strain.

23. In this combination, males issued from the backcross will carry the Y chromosome from the donor A strain, the X chromosome being from the recipient B strain. In females, the two X chromosomes will be from the recipient B strain.

24. On each chromosome, the most distal markers should be used. However, the possibility cannot be excluded that unwanted segments of recipient origin are retained due to crossovers beyond the most distal markers.

25. Since the percentage of non recombinant chromosomes is expected to be roughly between 12% for the longest chromosome (~150 cM) to 33% for the lowest (~40 cM) (22), it is possible to roughly calculate the expected number of rats (male or female) issued from the backcrosses and bearing non-recombinant chromosome.

26. At that stage, the number of mated couples for a given consomic strain in the next backcross will depend on the size of the chromosome, and accordingly on the number of expected recombination within the given chromosome.

27. Nomenclature: for consomic strains, the symbol of the recipient strain is separated by a hyphen of Chr (for chromosome) followed by the number of the introgressed chromosome with, in superscript and in majuscule, the symbol of the donor strain. For example, given the two rat strains A and B, the consomic strain issued from the introgression into the recipient B strain of the chromosome *n* from the donor A strain will be named B-Chr $n^A$.
28. The F1 × F1 intercross is preferable since meioses can occur in each rat while only one meiose can occur in the backcross.
29. The density of the markers to be genotyped depends on the size of the region. For generation of congenic strains from consomic strains, intervals between markers of 2–5 cM are convenient. For construction of congenic sub-strains from congenic strains, high density maps with markers every ~1 cM are often needed. In this case, genotyping SNPs can be useful. The detailed SNP map launched recently by the STAR Consortium can be used to select the polymorphic SNPs in the chromosomal region of interest (16).
30. Particularly, mix-ups can be a major problem when several congenic strains of the same background are maintained in close proximity. DNA of the strains must be regularly archived to be able, when needed, to control the genomic purity and conformity of any strain. Today, controls of the genotypes can be routinely performed using microsatellite makers. These controls are important when a spontaneous mutation with some phenotypic impact becomes fixed during brother-x-sister inbreeding. Genotyping allows for the discrepancy between a mix-up in the inbreeding process and the occurrence of a spontaneous mutation.

## Acknowledgements

The authors want to thanks the following participants in the works done on rat congenics in our laboratory: Pr Philippe Druet, who has initiated the work on rat genomics; Dr Abdelhadi Saoudi, who has collaborated in most of the studies; Dr Dominique Gauguier, who has helped us to create the first congenic strains; all the students who have participated in the construction of congenic strains and sub-strains, particularly Drs Magali Mas, Jean-François Subra, Pierre Cavaillès, Céline Colacios and Olivier Papapietro and all the persons who are or have been in care of the rats in our Zootechnic unit, particularly Mrs Maryline Calise, Sylvie Appolinaire, Audrey Boyer, Marie-Andrée Daussion,

Corinne Senty, Carine Segui, Magali Toulouse, Aline Tridon, Mr. Patrick Aregui, Thierry Ruiz.

The development of the congenic strains in our laboratory has been supported by grants from Institut National de la Santé et de la Recherche Médicale (INSERM), Genopole® Toulouse Midi-Pyrénées (GenoToul), Agence Nationale de Recherche (ANR), Fondation pour la Recherche Médicale (FRM), Association de Recherche sur la Polyarthrite Rhumatoïde (ARP) and Fondation Arthritis Courtin, Association de Recherche sur la Sclérose en Plaques (ARSEP), Association Française contre les Myopathies (AFM), Etablissement Français des Greffes (EFG), Ligue Contre le Cancer (LCC), Université Paul Sabatier (UPS), Région Midi-Pyrénées, Ministère de l'Aménagement du Territoire et de l'Environnement and Ministère Délégué à la Recherche et aux Nouvelles Technologies.

## References

1. Snell GD (1964) Methods for study of histocompatibility genes and isoantigens. Methods Med Res 10:1–7
2. Cowley AW Jr, Liang M, Roman RJ, Greene AS, Jacob HJ (2004) Consomic rat model systems for physiological genomics. Acta Physiol Scand 181:585–592
3. Lander ES, Schork NJ (1994) Genetic dissection of complex traits. Science 265: 2037–2048
4. Darvasi A (1998) Experimental strategies for the genetic dissection of complex traits in animal models. Nat Genet 18:19–24
5. Wakeland E, Morel L, Achey K, Yui M, Longmate J (1997) Speed congenics: a classic technique in the fast lane (relatively speaking). Immunol Today 18:472–477
6. Gregorova S, Divina P, Storchova R, Trachtulec Z, Fotopulosova V, Svenson KL, Donahue LR, Paigen B, Forejt J (2008) Mouse consomic strains: exploiting genetic divergence between Mus m. musculus and Mus m. domesticus subspecies. Genome Res 18:509–515
7. Singer JB, Hill AE, Burrage LC, Olszens KR, Song J, Justice M, O'Brien WE, Conti DV, Witte JS, Lander ES, Nadeau JH (2004) Genetic dissection of complex traits with chromosome substitution strains of mice. Science 304:445–448
8. Takada T, Mita A, Maeno A, Sakai T, Shitara H, Kikkawa Y, Moriwaki K, Yonekawa H, Shiroishi T (2008) Mouse inter-subspecific consomic strains for genetic dissection of quantitative complex traits. Genome Res 18:500–508
9. Dwinell MR, Forster HV, Petersen J, Rider A, Kunert MP, Cowley AW Jr, Jacob HJ (2005) Genetic determinants on rat chromosome 6 modulate variation in the hypercapnic ventilatory response using consomic strains. J Appl Physiol 98:1630–1638
10. Kunert MP, Drenjancevic-Peric I, Dwinell MR, Lombard JH, Cowley AW Jr, Greene AS, Kwitek AE, Jacob HJ (2006) Consomic strategies to localize genomic regions related to vascular reactivity in the Dahl salt-sensitive rat. Physiol Genomics 26:218–225
11. Kunert MP, Dwinell MR, Drenjancevic-Peric I, Lombard JH (2008) Sex-specific differences in chromosome-dependent regulation of vascular reactivity in female consomic rat strains from a SS×BN cross. Am J Physiol Regul Integr Comp Physiol 295(2):R516–R527
12. Schulz A, Weiss J, Schlesener M, Hansch J, Wehland M, Wendt N, Kossmehl P, Sietmann A, Grimm D, Stoll M, Nyengaard JR, Kreutz R (2007) Development of overt proteinuria in the Munich Wistar Fromter rat is suppressed by replacement of chromosome 6 in a consomic rat strain. J Am Soc Nephrol 18:113–121
13. Nicklas W, Baneux P, Boot R, Decelle T, Deeny AA, Fumanelli M, Illgen-Wilcke B (2002) Recommendations for the health monitoring of rodent and rabbit colonies in breeding and experimental units. Lab Anim 36:20–42
14. Markel P, Shu P, Ebeling C, Carlson GA, Nagle DL, Smutko JS, Moore KJ (1997) Theoretical and empirical issues for marker-assisted

breeding of congenic mouse strains. Nat Genet 17:280–284

15. Weil MM, Brown BW, Serachitopol DM (1997) Genotype selection to rapidly breed congenic strains. Genetics 146:1061–1069
16. Saar K, Beck A, Bihoreau MT, Birney E, Brocklebank D, Chen Y, Cuppen E, Demonchy S, Dopazo J, Flicek P, Foglio M, Fujiyama A, Gut IG, Gauguier D, Guigo R, Guryev V, Heinig M, Hummel O, Jahn N, Klages S, Kren V, Kube M, Kuhl H, Kuramoto T, Kuroki Y, Lechner D, Lee YA, Lopez-Bigas N, Lathrop GM, Mashimo T, Medina I, Mott R, Patone G, Perrier-Cornet JA, Platzer M, Pravenec M, Reinhardt R, Sakaki Y, Schilhabel M, Schulz H, Serikawa T, Shikhagaie M, Tatsumoto S, Taudien S, Toyoda A, Voigt B, Zelenika D, Zimdahl H, Hubner N (2008) SNP and haplotype mapping for genetic analysis in the rat. Nat Genet 40:560–566
17. Lohmiller J, Swing S (2006) Reproduction and breeding. In: Mark AS, Steven HW, Craig LF (eds) The laboratory rat, Elsevier, Amsterdam, pp. 154–156
18. Christian JJ, Lemunyan CD (1958) Adverse effects of crowding on lactation and reproduction of mice and two generations of their progeny. Endocrinology 63:517–529
19. Olofsson P, Johansson A, Wedekind D, Kloting I, Klinga-Levan K, Lu S, Holmdahl R (2004) Inconsistent susceptibility to autoimmunity in inbred LEW rats is due to genetic crossbreeding involving segregation of the arthritis-regulating gene Ncf1. Genomics 83:765–771
20. Hall J, Nanthakumar E (2001) Automated fluorescent genotyping. Curr Protoc Hum Genet Chapter 2: Unit 2.8
21. Behringer R (1998) Supersonic congenics? Nat Genet 18:108
22. Nadeau JH, Singer JB, Matin A, Lander ES (2000) Analysing complex genetic traits with chromosome substitution strains. Nat Genet 24:221–225

# Chapter 18

# Generation of Rat "Supersonic" Congenic/Conplastic Strains Using Superovulation and Embryo Transfer

**Vladimír Landa, Václav Zídek, and Michal Pravenec**

## Abstract

Congenic strains are routinely used for positional mapping of quantitative trait loci; while conplastic strains, derived by substitution of different mitochondrial genomes on the same nuclear genetic background of inbred rodent strains, provide a way to unambiguously isolate effects of the mitochondrial genome on complex traits. Derivation of congenic or conplastic strains using a traditional backcross breeding strategy (10 backcrosses) takes more than 3 years. There are two principal strategies to speed up this process: (1) marker-assisted derivation of "speed" congenic/conplastic strains and (2) derivation of "supersonic" congenic/conplastic strains using in each backcross generation embryos obtained from 4-week-old superovulated females; thus, each backcross generation takes only 7 weeks. Both strategies could also be combined. In the current chapter, a method for derivation of "supersonic" congenic/conplastic rat strains is described.

**Key words:** "Supersonic" congenic/conplastic strains, Superovulation, Embryo transfer, Spontaneously hypertensive rat

## 1. Introduction

The laboratory rat is a widely used animal model of human diseases. In animal models, genetic determinants of complex traits are identified by linkage analyses as quantitative trait loci (QTLs). However, confidence intervals for these putative QTLs are typically very large and do not allow clear identification of candidate genes for follow-up sequencing and functional studies. Derivation of congenic strains and sublines for QTL confirmation and detailed positional mapping is a widely used strategy to limit the number of candidate genes for followup studies. Derivation of congenic strains by a traditional backcross breeding technique requires at least ten generations to achieve 99.8% genetical identity

I. Anegon (ed.), *Rat Genomics: Methods and Protocols*, Methods in Molecular Biology, vol. 597
DOI 10.1007/978-1-60327-389-3_18, © Humana Press, a part of Springer Science+Business Media, LLC 2009

with the background strain (for genes that are not linked to the donor allele) and the expected average length of the differential chromosome segment after 10 backcrosses is about 20 cM (1). Relatively large differential chromosome segments are advantageous at the beginning of positional mapping because of uncertainty in the position of putative QTLs within the large confidence intervals. To narrow these large differential segments, congenic strains are crossed with background strains and F2 animals with recombinant genotypes are selected to derive new congenic sublines with shorter differential chromosome intervals. This strategy might be repeated several times to derive "minimal" congenic strains with differential chromosome segments limited to <2 Mbp (2–4).

Recently, the relationship of mitochondrial DNA (mtDNA) variants to metabolic risk factors for diabetes and other common diseases has begun to attract increasing attention. Substitution of different mitochondrial genomes on the same nuclear genetic background in conplastic strains provides a way to unambiguously isolate effects of the mitochondrial genome on complex traits. Conplastic strains can be developed, similar to congenic strains, by a traditional backcross technique, specifically by backcrossing the mitochondrial genome from the donor strain onto the nuclear genetic background of the recipient strain (1).

Production of both congenic and conplastic strains by a traditional backcross breeding technique is a time-consuming process that usually takes more than 3 years. There are two principal strategies to speed up this process: (1) marker-assisted derivation of "speed" congenics/conplastics (5–7) and (2) the use of superovulation to produce "supersonic" congenic/conplastic strains (8). Both strategies might also be combined. Recently, we derived an SHR/Ola-mt$^{BN/Crl}$ conplastic strain to analyze the effects of different mitochondrial genomes on metabolic and hemodynamic parameters in the spontaneously hypertensive rat (SHR), an animal model of human metabolic syndrome (9) (Fig. 18.1). This conplastic strain was developed by backcrossing the mitochondrial genome of the BN/Crl strain onto the nuclear genetic background of the SHR/Ola strain using a "supersonic" method described by Behringer (8). In each of 10 backcross generations, embryos obtained from 4-week-old superovulated females, mated with SHR males, were transferred to pseudopregnant foster mothers; each backcross generation took only 7 weeks (superovulation in 4 weeks + 3 weeks of pregnancy). Thus, "supersonic" conplastic rats (N10) can be obtained in just about 18 months compared to 40 months when a traditional backcross breeding method is used. In the current chapter, protocols for derivation of rat "supersonic" conplastic strains are described.

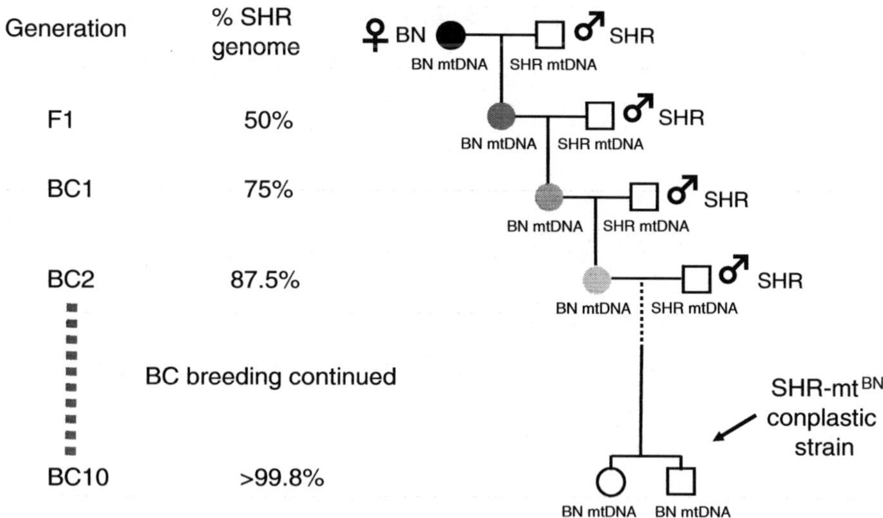


Fig. 18.1. Derivation of a conplastic strain. The contribution of the donor and recipient nuclear genomes is indicated in *black* and *white*, respectively. The relative contributions of donor and recipient alleles at sequential backcross generations are schematically depicted by different shades of *gray*. Since mitochondrial genome is inherited maternally, males of the recipient strain are always used in each backcross generation.

## 2. Materials

### 2.1. Animals

1. Brown Norway (BN/Crl) strain was obtained from Charles River Laboratories, Wilmington.
2. The spontaneously hypertensive rat (SHR/OlaIpcv), originally obtained from OLAC 1976 Ltd. (Bicester, UK), is bred in the Institute of Physiology, Czech Academy of Sciences, Prague since 1980.
3. Outbred Long-Evans males were obtained from the colony bred in the Institute of Physiology, Czech Academy of Sciences, Prague.

### 2.2. Superovulation and Embryo Manipulation

1. Pregnant mare's serum gonadotropin (PMSG) (Intervet).
2. Human chorionic gonadotropin (hCG) (Sigma, St.Louis, MO, USA). One dose of PMSG or hCG was diluted in 0.3 mL of saline.
3. M2 medium (Sigma).
4. Hyaluronidase (0.5 mg/mL) (Sigma).
5. M16 medium (Sigma).
6. 4-well culture dishes (Nunc).

### 2.3. Preparation of Foster-Mothers and Embryo Transfer

1. Narketan (Vetequinol, Nymburk, Czech Republic).
2. Xylapan (Vetequinol, Nymburk, Czech Republic).
3. M16 medium (Sigma).

## 3. Methods

### 3.1. Animals

The mitochondrial genome was transferred from the Brown Norway (BN/Crl) (Charles River Laboratories, Wilmington) strains onto the genetic background of the spontaneously hypertensive rat (SHR/OlaIpcv). The rats were housed in an air-conditioned animal facility and allowed free access to a standard diet and water. All experiments were performed in agreement with the Animal Protection Law of the Czech Republic (311/1997) and were approved by the Ethics Committee of the Institute of Physiology, Czech Academy of Sciences, Prague.

### 3.2. Superovulation and Embryo Manipulation

In each backcross generation, sexually immature female rats (28–35 days old) showing vaginal opening were stimulated for superovulation (10) by an intraperitoneal (IP) injection of 15 IU pregnant mare's serum gonadotropin (PMSG) (Intervet) followed 48 h later by an IP injection of 20 IU human chorionic gonadotropin (hCG) (Sigma, St. Louis, MO, USA) (one dose of PMSG or hCG was diluted in 0.3 mL of saline and injected between 12 p.m. and 1 p.m.). Immediately after the hCG injection, each female was paired with a single SHR male and kept in the separate cage overnight (*see* Note 1). Only the females showing a vaginal plug (Day-1 of pregnancy) were selected (between 8 a.m. and 9 a.m.) and kept in separate cages as the donors of 2-cell stage embryos for further embryo transfer.

The superovulated females with vaginal plugs were killed by cervical dislocation on Day-2 of pregnancy (between 9 a.m. and 10 a.m.), the abdomen of each donor was washed with 70% ethanol, opened with scissors and intact oviducts were excised with the help of fine scissors and forceps, washed in 300–400 μL drop M2 medium (Sigma) and transferred into 100 μL drop of M2 medium, placed on the lid of 60 mm sterile tissue culture petri dish (Fig. 18.2a).

The excised oviducts in the drops of M2 medium were placed under a dissecting microscope (magnification approx. 20–30×), the ampulary portion of the oviduct was fixed with fine forceps and oviducts were flushed with 50–100 μL of M2 medium with the help of sterile 5 mL syringe with 30 G1/2 syringe needles (*see* Note 2).

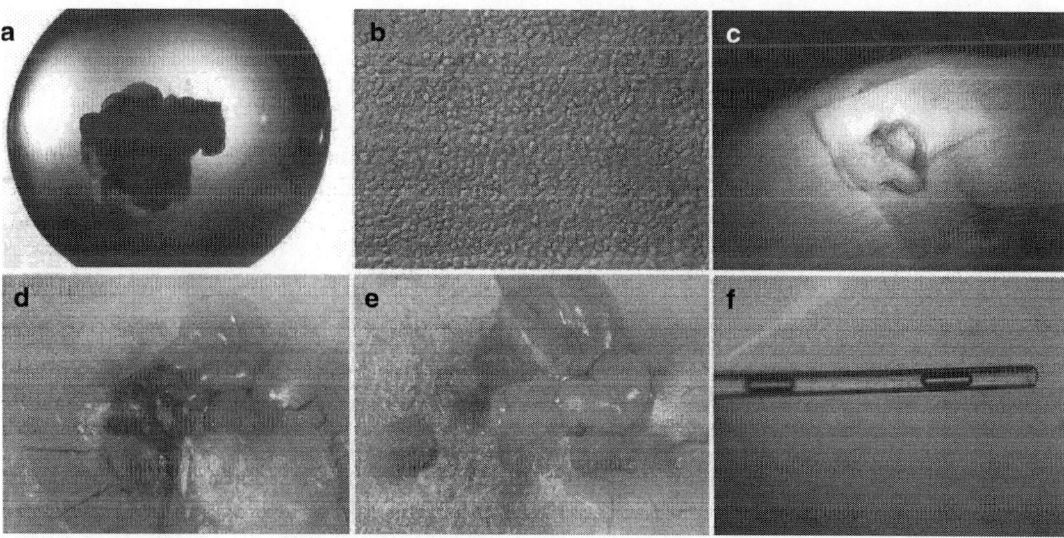


Fig. 18.2. (**a**) The excised oviduct in a drop of the M2 medium ready for recovery of 2-cell stage embryos. (**b**) The flushing from vagina, mass of nucleated epithelial cells typical for proestrus. (**c**) The exposed and fixed fat pad, ovary, oviduct, and proximal part of the uterine horn of the pseudopregnant foster-mother. (**d**) The fixed ovary and oviduct prior to bursa opening. (**e**) The exposed and fixed infundibulum prior to the embryo transfer. (**f**) Two-cell stage embryos in the transfer pipette ready for the transfer into the infundibulum of a foster-mother.

The flushed 2-cell stage embryos and 2-cell stage embryos enclosed in the remnants of cumulus cells were selected and transferred into 100 μL drops of M2 medium supplemented with hyaluronidase (0.5 mg/mL) (Sigma) for 3–5 min. Two-cell stage embryos selected for embryo transfer were freed of attached cumulus cells by gentle pipetting with the hand operated pipette of the diameter slightly bigger than the diameter of embryos. The cumulus cell free 2-cell stage embryos were then washed twice in 100 μL drops of fresh M2 medium. For in vitro culture of 2-cell stage embryos prior to embryo transfer, the selected 2-cell stage embryos were transferred into the 100 μL drop of M16 medium (Sigma) and then placed into 300 μL of M16 medium in 4-well culture dishes (Nunc) and kept in the $CO_2$ incubator (37°C, 5% $CO_2$, 95% air) for 1–2 h prior to embryo transfer into the oviducts of foster-mothers (10).

### 3.3. Preparation of Foster-Mothers and Embryo Transfer

Sexually mature Wistar female rats (90–120 days old) were used as foster-mothers for 2-cell stage embryos in each backcross generation. The suitable foster-mothers were selected in proestrous according to the modified technique described by Marcondes et al. (11). Vaginal secretion was collected (between 1 p.m. and 2 p.m.)

by flushing the vagina with 10 μL of saline (0.9% NaCl) using the pipet tips (yellow 1–200 μL), flushings were expelled onto the marked microscope slide and inspected under the microscope (magnification 440×, Olympus IX70 equipped with DIC optics). The rats showing strong sign of proestrous (mass of nucleated epithelial cells, Fig. 18.2b) were selected and placed (between 2 p.m. and 3 p.m.) into the cage with two vasectomized Long-Evans males overnight (*see* Note 3). Only the rats showing vaginal plugs (next morning between 9 a.m. and 10 a.m.) were used as the foster-mothers.

Sterile male Long-Evans rats (90–120 days old) used for mating with females selected in proestrous were prepared by vasectomy 2 weeks in advance. Selected Long-Evans males of proven fertility were anesthetized by IP injection of Narketan (150 mg/kg) (Vetequinol, Nymburk, Czech Republic) and Xylapan (2 mg/kg) (Vetequinol), the abdomen was shaved, sterilized with 70% ethanol and the body cavity was opened (2 cm cut in the skin and body wall, starting from the base of the penis). The vas deferens was pulled out with fine blunt forceps, 2–3 mm were cutted off, cauterized with red-hot tips of a second pair of fine forceps and placed back inside the body cavity. The second vas deferens was cut off in the same way. The abdominal wall and skin were sewed shut with 3–4 stitches each. The sterility of vasectomized males was proven 10–12 days later.

The selected pseudopregnant foster-mothers were used for the transfer of 2-cell stage embryos into the oviduct on the same day when vaginal plugs were detected (D1 of pseudopregnancy, 10 a.m–11 a.m.). The recipient female was anesthetized by IP injection of Narketan (150 mg/kg) (Vetequinol) and Xylapan (2 mg/kg) (Vetequinol), the abdomen shaved (left flank, 4 × 2 cm) and sterilized with 70% ethanol and the abdominal cavity was opened through the cut made in skin and body wall (1.5–2 cm). The ovary with the oviduct and the proximal part of the uterine horn were exposed and fixed on a strip (3 × 2 cm) cut from non-woven wound dressing (Fig. 18.2c), soaked in saline (0.9% NaCl) so that the infundibulum was clearly visible (Fig. 18.2d). With the help of two pairs of fine forceps, a hole was made in the bursa over the oviduct opening, the infundibulum was gently squeezed by forceps and fixed on a fine paper strip cut off from a paper towel so that the oviduct opening was steady and visible for easy insertion of the transfer pipette (Fig. 18.2e). Two-cell stage embryos selected for embryo transfer were drawn up under the dissecting microscope (magnification approx. 20–30×) into the transfer pipette in a minimal volume of M16 medium and between the small air bubbles (Fig. 18.2f). The transfer pipette (*see* Note 4) was inserted into the infundibulum and the content was blown inside the oviduct. Ten to fourteen 2-cell embryos were transferred into a single oviduct. The fat pad, uterus, oviduct, and ovary were placed back into the abdominal cavity, and the body wall and skin were sewed (2–3 stitches each of them). The foster-mothers were than placed in a cage and kept in the warm place for 1–2 h.

### 3.4. Genotyping

Genotype analysis of polymorphic markers (e.g., microsatellites) distributed at about 10 cM intervals across the genome that distinguish the donor and recipient strains can be used to confirm the correct nuclear genetic status of the conplastic strain (9). More extensive genotyping can also be performed using array technologies to screen thousands of single nucleotide polymorphisms (SNPs) across the nuclear genome. Variants in mtDNA that distinguish between the donor and recipient strains can be used to confirm satisfactory transfer of the correct mitochondrial genome.

## 4. Notes

1. *Superovulation:* Female rats of different genetic origin (e.g., different inbred strains, outbred stocks) might respond differently to the PMSG/hCG dosage. The optimal number of recovered embryos should be 30–50 2-cell stage embryos per superovulated female. In the case of a lower number of recovered 2-cell stage embryos (10–20 embryos), the dose of the PMSG should be higher (20 IU), while in the case of a massive superovulatory response (60–120 ovulated oocytes) characterized by low fertilization rate (lower than 10% of 2-cell stage embryos), the dose of the hCG should be decreased to 15 IU or 10 IU per injected female. Only healthy females should be used as embryo donors. The intraperitoneal injection should be done using a tuberculine needle and the needle should be passed through the abdominal wall in the middle of the fourth quadrant of the abdomen.
2. *Flushing of the oviduct:* To ensure that the excised oviduct is completly flushed, it is recommended to aspirate a minimal amount of the air into the 30 G1/2 needle before insertion of the needle into the infundibulum. The passage of the air bubble along the whole oviduct can be easilly controlled and the rupture of the oviduct is clearly visible. In the case of a rupture in the oviduct wall, the flushing should be done also from the place of the rupture or from the opposite side (utero-tubal junction). The manipulation of the 2-cell stage embryos in the M2 medium (flushing the oviduct, hyaluronidase treatment, washing, selection of intact 2-cell stage embryos, and transfer into the M16 medium) should be as short as possible

(10 min). Only morphologically normal 2-cell stage embryos (well-formed blastomeres of equal size, intact second polar body, without cellular remnants or several spermatozoa in the periviteline space) should be selected for further use (in vitro culture, embryo transfer).

3. *Selection of suitable foster-mothers:* Only healthy females showing compact mass of nucleated epithelial cells typical for proestrus in vaginal flushings should be selected as the embryo recipients. Rats showing any signs of dioestrus (mainly polymorphonuclear leucocytes, a few nucleated epithelial cells) or oestrus (large non-nucleated cornified epithelial cells) are not suitable as recipients. The following three conditions should be fulfilled before the embryo transfer: (1) the presence of "fresh" (corpus hemorrhagicum) ovulation points on the ovary, (2) the presence of cumulus-oocyte complexes in the oviduct (might be clearly visible through the oviduct wall), and (3) increased number of blood capillaries in the ovarian bursa.
4. *Embryo transfer:* An ideal transfer pipette (hand pulled over the flame from glass tubes of 1 mm diameter) should have a tip 1–1.5 cm long with the opening slightly wider than the diameter of the 2-cell stage embryos, and should be polished by flame to avoid sharp edges and enable smooth insertion into the infundibulum. During the embryo transfer, the pipette should be inserted close to the first bend of the oviduct, gently fixed with forceps and the medium with embryos expelled into the oviduct while the pipette is slowly pulled back. The second air bubble should stop just past the first oviductal bend to prevent reverse leakage of the medium with embryos. The 2-cell stage embryos in the M16 medium in the 4-well culture dishes (Nunc) should not be kept outside the $CO_2$ incubator for longer than 10 min to avoid extreme pH changes.

## Acknowledments

This work was supported by grants 1P05ME791 and 1 M6837805002 from the Ministry of Education of the Czech Republic, grants 301/06/0028 and 301/08/0166 from the Grant Agency of the Czech Republic, grant IAA500110604 from the Grant Agency of the Academy of Sciences of the Czech Republic, and by the European Commission within the Sixth Framework Programme through the Integrated Project EURATools (contract no. LSHG-CT-2005-019015). M.P. is an international research scholar of the Howard Hughes Medical Institute.

## References

1. Silver LM (1995) Coisogenics, congenics, and other specialized strains. In: Silver LM (ed) Mouse genetics: concepts and applications. New York, Oxford University Press, pp 32–57
2. Saad Y, Garrett MR, Manickavasagam E, Yerga-Woolwine S, Farms P, Radecki T et al (2007) Fine-mapping and comprehensive transcript analysis reveals nonsynonymous variants within a novel 1.17 Mb blood pressure QTL region on rat chromosome 10. Genomics 89:343–353
3. Lee SJ, Liu J, Westcott AM, Vieth JA, DeRaedt SJ, Yang S et al (2006) Substitution mapping in dahl rats identifies two distinct blood pressure quantitative trait loci within 1.12- and 1.25-mb intervals on chromosome 3. Genetics 174:2203–2213
4. Seda O, Liska F, Sedová L, Kazdová L, Krenová D, Kren V (2005) A 14-gene region of rat chromosome 8 in SHR-derived polydactylous congenic substrain affects muscle-specific insulin resistance, dyslipidaemia and visceral adiposity. Folia Biol (Praha) 51:53–61
5. Markel P, Shu P, Ebeling C, Carlson GA, Nagle DL, Smutko JS et al (1997) Theoretical and empirical issues for markerassisted breeding of congenic mouse strains. Nat Genet 17:280–284
6. Wakeland E, Morel L, Achey K, Yui M, Longmate J (1997) Speed congenics: a classic technique in the fast lane (relatively speaking). Immunol Today 18:472–477
7. Fournié, G.J. (2009) Generation of consomic and congenic rat strains (including speed congenics) (technical). Chapter 15, Current Issue
8. Behringer R (1998) Supersonic congenics? Nat Genet 18:108
9. Pravenec M, Hyakukoku M, Houstek J, Zidek V, Landa V, Mlejnek P et al (2007) Direct linkage of mitochondrial genome variation to risk factors for type 2 diabetes in conplastic strains. Genome Res 17:1319–1326
10. Popova M, Krivokharchenko A, Ganten D, Bader M (2002) Comparision between PMSG- and FSH-induced superovulation for the generation of transgenic rats. Mol Reprod Dev 63:177–182
11. Marcondes FK, Bianchi FJ, Tanno AP (2002) Determination of the estrous cycle phases of rats: Some helpful consideration. Braz J Biol 62:609–614

# Chapter 19

# Analysis by Quantitative PCR of Zygosity in Genetically Modified Organisms

**Laurent Tesson, Séverine Rémy, Séverine Ménoret, Claire Usal, and Ignacio Anegon**

## Abstract

It is extremely useful to define a rapid and accurate method for identifying homozygous and heterozygous transgenic animals prior to setting up breeding programs for transgenic colonies and in experiments in which gene dosage effects could have a functional impact. Southern-blotting is a means of identifying zygosity, but such a method is time consuming and produces a high level of ambiguous results. Some years ago, we described the rapid, precise, non-ambiguous, and high-throughput identification of zygosity in transgenic animals by real-time PCR. This technique allows us to make a clear-cut identification of transgenic rats, transgenic mice, and double-transgenic pigs. Since 2002, however, several authors have made improvements to this method. The following paper describes the ease with which zygosity is determined using real-time PCR.

**Key words:** Rats, Real-time PCR, Transgenic, Zygosity

## 1. Introduction

Zygosity information is always required for effective breeding and colony maintenance and the heterozygous or homozygous status for the transgene has been shown to correlate with gene expression levels. Genotyping by real-time PCR is now widely recognized as a more effective method than Southern-blot hybridization or mating. Since 2002, the use of TaqMan probes has been reported by both ourselves (1) and several other groups (2–4). From 2002 onwards, the Sybr Green real-time PCR reaction has also been applied to analyze zygosity not only at lower cost but with greater simplicity (5, 6).

To determine zygosity, it is necessary to follow certain simple rules such as preparing high-quality DNA, choosing primers to attain a PCR efficiency close to 95–100% and by using simple

I. Anegon (ed.), *Rat Genomics: Methods and Protocols*, Methods in Molecular Biology, vol. 597
DOI 10.1007/978-1-60327-389-3_19, © Humana Press, a part of Springer Science+Business Media, LLC 2010

mathematical calculations. The comparative cycle threshold method (ΔΔCt method) can discriminate animals differing only by one factor out of two in the amount of target sequences. Using a generic dye such as SybrGreen, different PCR amplicons and/or non specific amplification products such as primer dimers can accurately be distinguished by the generation of DNA melting curves and first derivative melting peaks (7).

This method is applicable to transgenic animals generated by both microinjection of naked DNA or lentiviral transgenesis (after segregation of the many insertion sites).

In the case of lentiviral transgenesis, zygosity status can also be achieved by non-quantitative PCR after characterization of the insertion sites (8) (see the chapter on the characterization of transgene insertion sites by E. C. Bryda et al. in this book). In the case of lentiviral transgenesis, PCR is performed with three specific primers: one forward primer on the 5′ side of the genomic sequence, one reverse primer on the 3′ side of the genomic sequence for wild type allele amplification, and one reverse primer on the 5′ end of the transgene (8). Homozygous animals display a specific amplification product (genomic–transgene primers), heterozygous animals display two amplification products (genomic–genomic and genomic–transgene primers) and wild type animals display a specific amplification product (genomic–genomic primers).

Quantitative PCR can also be used for determining transgene copy number in transgenic animals (9).

## 2. Materials

## 3. Genomic DNA Preparation

1. Lysis solution (0.2 M NaCl, 0.1 M Tris pH 8.3, 5 mM EDTA, 100 μg/ml PK, and 0.2% SDS).
2. Phenol Tris buffered at pH 7.9.
3. Chloroform.
4. 100 and 70% Ethyl alcohol.
5. 10–1 mM Tris–EDTA.
6. Nanodrop or another spectrophotometer.

## 4. Primer Design

1. Electronic version of the vector.
2. Program for primer design (any program can be used: Oligo, Primer Express, etc.).

## 5. Real-Time PCR: Instruments and Reagents

1. MicroAmp Optical 96-well reaction plate with barcode.
2. Optical caps or Optical Adhesive film.
3. Pipette tips with aerosol filter (nuclease free).
4. Power SybrGreen PCR Master mix (Applied Biosystems).
5. TaqMan 2× Universal PCR Master Mix, No AmpErase UNG.
6. ABI PRISM 7700 Sequence Detection System.
7. SDS 1.9.1 program with dissociation curve.
8. Any real-time instrument can be used.

## 6. Methods

## 7. DNA Preparation

1. Cut a small piece (1–2.5 mm) from the tip of the tail of 10–15-day-old rats with a pair of sharp scissors. The identification of each rat can be carried out using any type of labeling.
2. Place the cut tip into a 1.5 ml microtube (Eppendorf type).
3. Add 450 μl of lysis solution with proteinase K (*see* **Note 1**), mix thoroughly by vortexing and incubate overnight at 56°C.
4. Centrifuge tubes for 30 s at 12,000 rpm.
5. Take 250 μl of the above into a new tube with 250 μl phenol-chloroform V/V, shake for 1 min and centrifuge for 5 min at 12,000 rpm.
6. Pour the upper phase (200 μl) into a new tube and add 500 μl of 100% ethanol (stored at –20°C) to precipitate the DNA. Mix gently and the DNA pellet will appear.
7. Centrifuge for 5 min at 12,000 rpm. Remove the supernatant (n.b. the pellet must remain at the bottom of the tube). Add 250 μl of 70% ethanol (stored at –20°C) and shake the tube for a few seconds.
8. Centrifuge for 5 min at 12,000 rpm. Remove the supernatant (make sure the pellet remains at the bottom of the tube). Air-dry the pellet and re-dissolve the DNA with 100 μl of TE 10/1 (*see* **Note 2**).

9. Measure the DNA concentration with the Nanodrop instrument (1 µl is sufficient) or with any spectrophotometer.

## 8. Primer Design

1. The primers are carefully chosen using any specialized software based on their Tm as calculated by the nearest neighbor method (as close as possible to 66°C) with less than 2°C difference between them and all primer duplexes kept to a minimum (less than four nucleotides), with no hairpin and no G, C, nor GC extending more than four nucleotides. Make sure that the primers bind to a "balanced" region (i.e., nearly equal amounts of all four bases), that they contain neither monotonous nor repetitive sequences, and that they are not self complementary. The length of the product should be between 100 and 400 base pairs. Finally, submit primer sequences to BLASTn analysis (NCBI) to confirm their specificity.
2. If the transgene is a cDNA and the species origin is the same as the transgenic animal (i.e., a rat transgene in a transgenic rat) and in order to amplify the transgene cDNA and avoid amplification of the genomic counterpart, primers should be chosen on two exons separated by an intron. For the reference gene, choose one that is present at one copy per chromosome. In order to simplify the procedure, care should be taken that this does not take place on an X chromosome, i.e., HPRT on rat which necessarily means correcting the calculation between males (one copy) and females (two copies) (1).
3. Primers need to be validated under the same PCR conditions used for real-time PCR to obtain a single product of the expected size (*see* **Note 3**). The efficiency of the PCR is determined as described below.

## 9. Adjusting Parameters

1. The efficiency of PCR amplifications has to be greater than 95% as assessed by the slope of the curve Ct = *f*(–log[DNA]). Thus, serial two-fold dilutions of microinjection fragment diluted in wild type genomic DNA for the transgene or genomic DNA for the reference gene (100–6.25 ng), are tested in the standard PCR conditions. Theoretically, a PCR with 100% efficiency should exhibit a slope of –3.32 (*see* **Note 4**). To use the $2^{-\Delta\Delta Ct}$ calculation, the slope of the curve

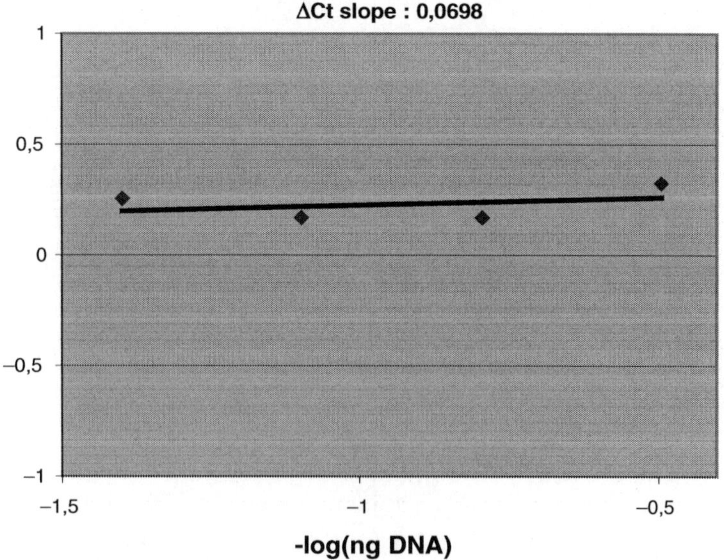


Fig. 19.1. The difference in Ct values (ΔCt) was plotted against the amount of log template (DNA). The difference in PCR efficiencies was determined by calculating the gradient of the graph.

$\Delta Ct(Ct_{gene\ target} - Ct_{reference\ gene})$ vs. (–log[DNA]) has to be lower than 0.1 (Fig. 19.1), meaning the Cts measured are independent of target DNA (*see* **Note 5**).

## 10. Real-Time PCR Procedure

1. All samples must be measured in duplicate. Change pipette tips for each well.
2. In an Eppendorf microtube, prepare a pre-mix for the number of samples which are measured in duplicate. For a complete microplate (96 wells), prepare a pre-mix for 100 wells.

   2× Power SybrGreen PCR Master mix: 12.5 µl
   forward primer (10 µM): 0.75 µl
   reverse primer (10 µM): 0.75 µl
   $H_2O$: 6 µl
3. In each well (two wells per sample) add 5 µl of diluted sample (2.5 ng/µl).
4. Add 20 µl of pre-mix to each well.
5. Cover the microplate with the optical caps. *Important*: Do not touch the surface of the caps. Gloves must be worn for this operation.

6. Shake the plate and centrifuge it at 2,000 rpm for 2 min.
7. Program a run on the ABI Prism 7700 SDS as follows:

Open the program file SDS 1.9.1.
On the 7700 Single Reporter Standard Plate:
- Select the dye layer as Sybr.
- In Sample Type, go to Sample Type Set Up and select Sybr for Unknown (UNKN) and No Template Control (NTC).
- The reference (ROX) should be indicated and the quencher is not necessary.

Go to thermal cycler conditions:
Stage 1: 2′ at 50°C
Stage 2: 5′ at 95°C
Stage 3 (40 cycles): 15″ at 95°C
1′ at 60°C
Optional: a supplementary step (15″) can be added to collect data at "Tm-3°C" (Fig. 19.2) (*see* **Note 3**).
Stage 4: melting curve by heating the PCR product from 60 to 95°C in 20′.
Sample volume: 25 μl
Collect data at 60°C, "Tm-3°C" and during the last heating phase.
Save and name your experiment.

8. Place the plate in the machine.

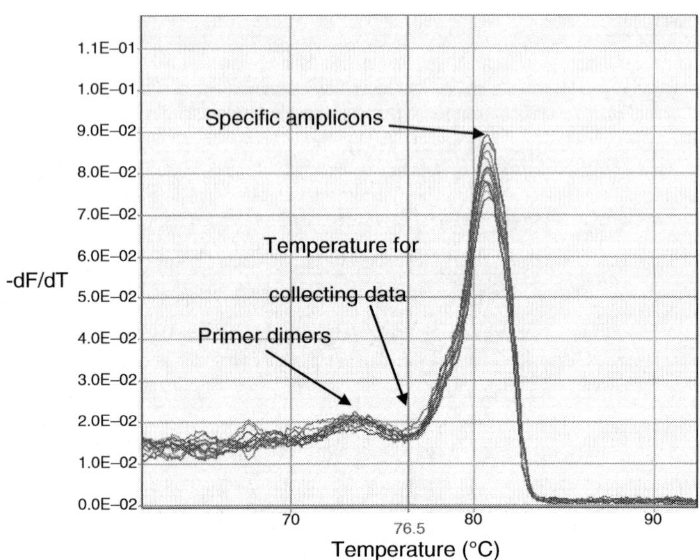


Fig. 19.2. –d*F*/d*T* vs. *T*. Specific PCR products give high and sharp melting peaks, whereas primer-dimers give a small melting peak at a lower temperature. Using the data at 80°C should provide the true result without any contamination by nonspecific fluorescence.

9. Click on Show Analysis, run the plate and wait for the end of the cycles.
10. Save the results and analyze.
    10.1. In the option menu, select the stage and the step you wish to analyze (Stage 3 Step 2 for 60°C or 3 if "Tm-3°C").
    10.2. Analyze and place the threshold in the middle of the linear slope of the curve (often 0.1).
    10.3. Export data to an Excel file to make calculations.
    10.4. Export the multicomponent and open it by "dissociation curve" software to obtain the melting curve analysis of PCR products. The primer-dimers accumulation is greater when the target is in low abundance or absent as in the NTC (no template control). "Tm-3°C" provides the correct result in quantitative assays.

## 11. Calculation

In order to make the $2^{-\Delta\Delta Ct}$ calculation, open the result file. Calculate the mean of each duplicate.

Calculate $\Delta Ct = Ct_{\text{gene target}} - Ct_{\text{reference gene}}$

Calculate $\Delta\Delta Ct = (Ct_{\text{gene target}} - Ct_{\text{reference gene}})_{\text{unknown}} - (Ct_{\text{gene target}} - Ct_{\text{reference gene}})_{\text{known}}$

"Known" refers to an animal with a heterozygous status. The $F_1$ transgenic animal is the right choice and it is better than using the founder animal because of mosaicism.

Calculate the $2^{-\Delta\Delta Ct}$ (Table 19.1). The known animal should give a result of 1 and the heterozygous animal should produce the same result (often between 0.8 and 1.3). A homozygous animal should give a result close to 2 (1.8–2.3) (*see* **Note 6**).

## 12. Notes

1. For each experiment, prepare the required volume of lysis solution. PK solution stock (10 mg/ml) must be stored at –20°C.
2. Any comparable method for high-quality DNA isolation or crude DNA solution can be used (6).
3. At the end of the real-time PCR, an agarose gel electrophoresis can be run and/or an additional step can be added to the analysis of the results, which leads to a melting curve (Fig. 19.2).

**Table 19.1**
**Calculation to determine the zygosity status of transgenic rats**

| Rat no. | Ct gene target | Ct reference gene | ΔCt | ΔΔCt | $2^{-\Delta\Delta Ct}$ | Status |
|---|---|---|---|---|---|---|
| 49 | 22,13 | 25,2 | −3,07 | −1,115 | 2,17 | Homozygous |
| 50 | 23,1 | 25,055 | −1,955 | 0 | 1,00 | Heterozygous |
| 52 | 23,475 | 25,6 | −2,125 | −0,17 | 1,13 | Heterozygous |
| 211 | 23,42 | 25,49 | −2,07 | −0,115 | 1,08 | Heterozygous |
| 214 | 24,45 | 26,53 | −2,08 | −0,125 | 1,09 | Heterozygous |
| 218 | 22,94 | 25,975 | −3,035 | −1,08 | 2,11 | Homozygous |
| 220 | 24,095 | 26,09 | −1,995 | −0,04 | 1,03 | Heterozygous |
| 223 | 23,59 | 25,94 | −2,35 | −0,395 | 1,31 | Heterozygous |
| 226 | 25,105 | 27,285 | −2,18 | −0,225 | 1,17 | Heterozygous |
| 272 | 23,6 | 26,73 | −3,13 | −1,175 | 2,26 | Homozygous |
| 278 | 22,635 | 25,81 | −3,175 | −1,22 | 2,33 | Homozygous |
| 282 | 22,145 | 25,18 | −3,035 | −1,08 | 2,11 | Homozygous |
| 283 | 23,29 | 25,525 | −2,235 | −0,28 | 1,21 | Heterozygous |

The PCR product is heated from 60 to 95°C and the fluorescence data produces a curve $-dF/dT$ vs. $T$ (negative first derivative of fluorescence intensity with respect to temperature $T$), which shows the exact Tm of this PCR product. This could be helpful if there are primer-dimers during the amplification step, which result in non specific fluorescence. However, it is easy to circumvent this by adding a "Tm-3°C" step (15 s) during the amplification step and by using this data for analysis. The optimal temperature to collect data, the so-called "Tm-3°C" (generally, 3°C above the Tm product) is the temperature where primer-dimers are denatured (no fluorescence) and specific products are double stranded (fluorescent). The melting curve analysis shows that all reactions are free of primer-dimers or other non-specific products.

4. If the PCR efficiency is not sufficient, 1.75 µl per well of BSA (bovine serum albumin) at 10 mg/ml can be added to the pre-mix. This treatment should overcome the effect of any potential inhibitors.
5. Prior et al. (10) demonstrated that "neither an efficiency correction nor closely matched efficiencies for the two reactions are required for accurate determination of zygosity."

In our experience, the efficiencies of each PCR (target gene and reference gene) should be as similar as possible and not less than 90% (slope –3,58). A difference as much as 0.15–0.2 between both slopes is considered to be acceptable (6).

6. Haurogne et al. (5) suggested avoiding calculations. They measured the DNA concentration obtained from the tail biopsies, diluted to 50 ng/μl and measured again. Two dilutions are made from these solutions and submitted to the real-time PCR procedure. The reference results between all the samples are really close and the results of the transgene differ by 1 Ct between heterozygous and homozygous.

## References


1. Tesson L, Heslan JM, Menoret S, Anegon I (2002) Rapid and accurate determination of zygosity in transgenic animals by real-time quantitative PCR. Transgenic Res 11(1):43–48
2. Bubner B, Gase K, Baldwin IT (2004) Two-fold differences are the detection limit for determining transgene copy numbers in plants by real-time PCR. BMC Biotechnol 4:14
3. Mitrecic D, Huzak M, Curlin M, Gajovic S (2005) An improved method for determination of gene copy numbers in transgenic mice by serial dilution curves obtained by real-time quantitative PCR assay. J Biochem Biophys Methods 64(2):83–98
4. Ji W, Zhou W, Abruzzese R, Guo W, Blake A, Davis S, Davis S, Polejaeva I (2005) A method for determining zygosity of transgenic zebrafish by TaqMan real-time PCR. Anal Biochem 344(2):240–246
5. Haurogné K, Bach JM, Lieubeau B (2007) Easy and rapid method of zygosity determination in transgenic mice by SYBR Green real-time quantitative PCR with a simple data analysis. Transgenic Res 16(1):127–131
6. Sakurai T, Kamiyoshi A, Watanabe S, Sato M, Shindo T (2008) Rapid zygosity determination in mice by SYBR Green real-time genomic PCR of a crude DNA solution. Transgenic Res 17(1):149–155
7. Ririe KM, Rasmussen RP, Wittwer CT (1997) Product differentiation by analysis of DNA melting curves during the polymerase chain reaction. Anal Biochem 245:154–160
8. Pillai MM, Venkataraman GM, Kosak S, Torok-Storb B (2008) Integration site analysis in transgenic mice by thermal asymmetric interlaced (TAIL)-PCR: segregating multiple-integrant founder lines and determining zygosity. Transgenic Res 17(4):749–754
9. Ballester M, Castello A, Ibanez E, Sanchez A, Folch J (2004) Real-time quantitative PCR-based system for determining transgene copy number in transgenic animals. Biotechniques 37:610–613
10. Prior FA, Tackaberry ES, Aubin RA, Casley WL (2006) Accurate determination of zygosity in transgenic rice by real-time PCR does not require standard curves or efficiency correction. Transgenic Res 15(2):261–265

# Chapter 20

# A Restriction Enzyme-PCR-Based Technique to Determine Transgene Insertion Sites

**Elizabeth C. Bryda and Beth A. Bauer**

## Abstract

Currently, most genetically engineered rat strains are created by methods that involve random integration of transgenes into the genome. The ability to identify the chromosomal location of the transgene insertion site enables the development of efficient genotyping assays, allows segregation of multiple transgene integration sites to be followed while breeding, and facilitates characterization of possible positional effects on phenotype. Here we describe a method for determining the chromosomal location of transgene insertion that combines restriction endonuclease enzyme digest with subsequent rounds of PCR amplification to produce amplicons representing the chromosomal regions flanking the integrated transgene. This method provides a reliable means for determining the exact location of insertion of transgenes within the genome.

**Keywords:** Transgene, Integration site, Chromosome walking, PCR, Restriction endonuclease digestion, GFP transgene

## 1. Introduction

Currently, the primary means for genetically engineering rat models to study disease or gene function involves the use of pronuclear injection or retroviral delivery systems (1). In transgenic animals created by embryo microinjection, the site of integration of the transgene within the genome is a random event (2). Determining transgene integration sites is challenging yet without this information, genotyping of the rat strain is limited to identifying whether animals carry the transgene or not. The ability to determine if transgene-positive animals are homozygous or heterozygous for the transgene is often not reliably possible. Additionally, when the rat strain has been created using lentiviral approaches, the ability to efficiently identify and follow segregation of the multiple insertion sites that commonly occur is difficult.

I. Anegon (ed.), *Rat Genomics: Methods and Protocols*, Methods in Molecular Biology, vol. 597
DOI 10.1007/978-1-60327-389-3_20, © Humana Press, a part of Springer Science+Business Media, LLC 2010

A number of PCR-based methods, often referred to as "chromosome walking" techniques have been developed to isolate DNA fragments adjacent to known sequences, including inverse PCR (3–5), ligation mediated PCR (LM-PCR) (6), randomly primed PCR (RP-PCR) (7, 8), and T-linker PCR (9). The method described here incorporates several elements of these techniques in a unique way that allows the capture of DNA fragments containing the chromosomal region flanking the transgene. This method enables quick and inexpensive determination of transgene insertion sites and has been successfully used to identify the integration sites of genetically engineered rodents created by both pronuclear injection and lentiviral delivery (10).

To demonstrate how this method can be used, we provide an example of our determination of the integration site of a green fluorescent protein (GFP) transgene in a rodent strain. The key elements of the method are appropriate primer design, the use of a panel of restriction enzymes to maximize the ability to create small fragments containing the junction between the transgene and the flanking chromosomal sequence, and multiple rounds of PCR to enrich for amplicons corresponding to the insertion site region. The method involves common equipment and standard techniques that can be performed readily in any lab with minimal molecular biology expertise.

## 2. Materials

### 2.1. Primers and Linkers

1. Primer D: 5′ GCAAACGATAAATGCGAGGACGGT 3′ (*see* Note 1).
2. Primer G: 5′ ATGCGAGGACGGTACACGGCGACC 3′.
3. Y-Linker A: 5′ GTGCAGCCTTGGGTCGCCGTGT /3InvdT (*see* Note 2).
4. Y-Linker E: 5′ GCAAACGATAAATGCGAGGACGGTACA CGGCGACCCAAGGCTGCACT 3′.
5. The primers used for the example case (GFP transgene) are listed in Fig. 20.1.

### 2.2. DNA Extraction and Restriction Endonuclease Digestion

1. DNeasy Blood & Tissue kit (Qiagen, Valencia, CA).
2. All restriction enzymes are available from New England BioLabs (Ipswich, MA). These enzymes come with the specific 10× reaction buffer for the enzyme as well as 100× BSA if it is required for the reaction.
3. Nuclease-free water (Hyclone, Logan, UT).

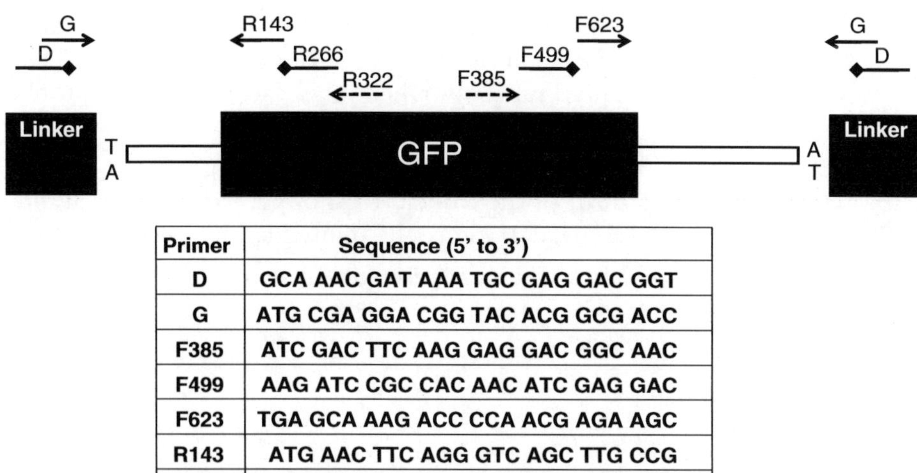


| Primer | Sequence (5' to 3') |
|---|---|
| D | GCA AAC GAT AAA TGC GAG GAC GGT |
| G | ATG CGA GGA CGG TAC ACG GCG ACC |
| F385 | ATC GAC TTC AAG GAG GAC GGC AAC |
| F499 | AAG ATC CGC CAC AAC ATC GAG GAC |
| F623 | TGA GCA AAG ACC CCA ACG AGA AGC |
| R143 | ATG AAC TTC AGG GTC AGC TTG CCG |
| R266 | ATG GCG GAC TTG AAG AAG TCG TGC |
| R322 | TGT AGT TGC CGT CGT CCT TGA AGA AG |

Fig. 20.1. Primers used for determining insertion site sequence flanking a GFP transgene. ■■ represent known sequences of the GFP transgene and linker sequence. The open boxes flanking the GFP gene represent unknown sequences. Gene-specific primer 1 is represented by a ←--- . The base pairing between the A overhang produced during the first PCR amplification (Subheading 20.3.4) and the T overhang on the linker sequence is shown. Gene-specific primer 2 and Y-linker primer D are used in the second PCR reaction (Subheading 20.3.5) and are indicated by —◆. Gene-specific primer 3 and Y-linkerV primer G used in the third PCR reaction (Subheading 20.3.6) are indicated by ←—. The nucleotide sequences (5′–3′) for all primers are listed. The R series primers are used to capture unknown sequences upstream of the 5′ end of the GFP transgene, whereas the F series primers are used to capture unknown sequences downstream of the 3′ end of the GFP transgene.

### 2.3. PCR and Y-Linker Ligation

1. FastStart *Taq* (5 U/μl) and 10× FastStart *Taq* with $MgCl_2$ Buffer (Roche, Indianapolis, IN).
2. 10 mM dNTP mix (Promega, Madison, WI).
3. T4 DNA Ligase (3 U/μl) (Fisher Scientific, Pittsburgh, PA). This enzyme is supplied with 10× T4 DNA ligase buffer.

### 2.4. Recovery of Amplification Products

1. 5× Tris Borate (TBE), for 1 L, add 54 g Tris Base, 27.5 g boric acid, and 20 ml 0.5 M EDTA, pH 8.0. All reagents are available from Fisher Scientific (Pittsburgh, PA). Store in a glass bottle at room temperature. Discard if a precipitate forms. Dilute to 1× before use.
2. 3% ReadyAgarose gels (Bio-Rad, Hercules, CA). These pre-cast agarose gels contain ethidium bromide, which is a mutagen and should be handled appropriately (wear gloves). Electrophoresis is done in a Bio-Rad Sub-Cell GT (*see* Note 3).
3. Gel loading buffer, prepare a bromophenol blue stock solution by mixing 25 mg of bromophenol blue (Sigma, St. Louis, MO)

in 10 ml of 1× TBE. Store at 4°C. To make gel loading buffer, combine 7 ml of 100% glycerol (Fisher Scientific, Pittsburgh, PA) and 3 ml bromophenol blue stock solution. Store at 4°C.

4. Molecular size standard: 1 Kb Plus DNA Ladder (1 μg/μl) (Invitrogen, Carlsbad, CA). A working stock is made by mixing 20 μl of the ladder, 40 μl of gel loading buffer and 340 μl of TE, pH 8.0. Store the diluted ladder at 4°C.
5. Qiaquick Gel Extraction kit (Qiagen, Valencia, CA).

## 3. Methods

To help facilitate understanding of the technique, the locations and sequences for primers designed to a GFP transgene are provided (Fig. 20.1), as well as several examples of outcomes experienced along the way (Figs. 20.2 and 20.3). The GFP primers shown here can be used for any project where the GFP gene sequence is found at the 5′ or 3′ end of the transgene. The technique itself can be adapted for use in determining the integration site of any transgene as long as some information about the DNA sequence of the transgene is known.

### 3.1. Primer Design

1. Determine the nucleotide sequence of the transgene. Some information about the sequence of the transgene must be available in order to design PCR primers.
2. Primers for identification of flanking sequence upstream of the 5′ end of the transgene. Design three transgene-specific PCR primers to the 5′ end of transgene sequence such that each is immediately adjacent to the previous primer but not overlapping. Gene-specific primer 1 is farthest from the junction between the known transgene sequence and the unknown integration site, whereas Gene-specific primer 3 is closest (Fig. 20.1, R series primers). The 5′ to 3′ transgene sequence is used as the template to design these primers and thus the sequences of the primers themselves will be the reverse complement of the 5′ to 3′ transgene sequence. Gene-specific primer 1 is typically designed no more than 400 bp from the 5′ end of the transgene sequence.
3. Primers for identification of flanking sequence downstream of the 3′ end of the transgene. Design three transgene-specific PCR primers to the 3′ end of transgene sequence such that each is immediately adjacent to the previous primer but not overlapping. Gene-specific primer 1 is farthest from the junction between the known transgene sequence and the unknown integration site, whereas Gene-specific primer 3 is closest (Fig. 20.1, F series primers). The sequences of the

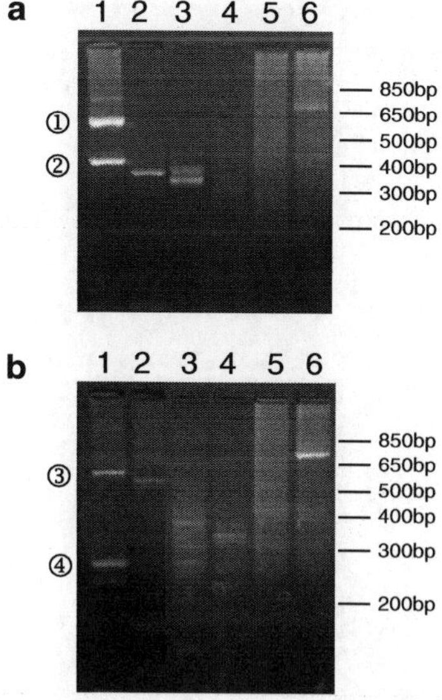


c

TGAGCAAAGACCCCAACGAGAAGCGCGATCACATGGTCCTGCTGGAGT
TCGTGACCGCCGCCGGGATCACTCTCGGCATGGACGAGCTGTACAAGT
AAAGCGGCCCTAGAGCTCGCTGATCAGCCTCGACTGTGCCTTCTAGTTG
CCAGCCATCTGTTGTTTGCCCCTCCCCCGTGCCTT④CCTTGACCCTGGA
AGGTGCCACTCCCACTGTACTTTCCTAATAAAATGAGGAAATTGCATCG
CATTGTCTGAGTAGGTGTCATTCTATTCTGGGGGGTGGGGTGGGGCAG
GACAGCAAGGGGGAGGATTGGGAAGACAAT②AGCAGGCATGCTGGG
GATGCGGTGGGCTCTATGGCTTCTGAGGCGGAAAGAACCAGCTAGGG
CTCGACGACTAGTGCGGTAAATA①GCAATAAATTGGCCTTTTTTATCG
GCAAGCTCTTTTAGGTTTTTCGCATGTATTGCGATATGCATAAACCAGC
CATTGAGTGCAGCCTTGGGTCGCCGTGTACCGTCCTCGCATA ③

Fig. 20.2. Typical results obtained using GFP primers. (**a**) 2 h restriction enzyme digestion followed by single round PCR amplification, linker ligation, and subsequent second- and third- round PCR reactions. Odd numbered lanes are from reactions utilizing restriction enzyme *Pst*I. Even numbered lanes are from reactions utilizing restriction enzyme *EcoR*I. Results are from three different GFP transgenic rodent lines (lanes 1–2 = unknown site transgene #1, 3–4 = unknown site transgene #2, 5–6 = unknown site transgene #3). Size in base pairs (bp) based on a molecular size standard (not shown) is indicated on the right hand side of each panel. (**b**) Same procedure and samples as (**a**) except an overnight restriction enzyme digestion was performed. Note that different-sized amplicons were recovered (compare **a** and **b**) depending on the restriction endonuclease used and the length of enzyme digestion incubation. (**c**) Sequencing results obtained from analysis of bands ① and ② from Panel (**a**) and bands ③ and ④ from Panel (**b**). The same flanking sequence was obtained from all 4 amplified bands and only the length of the sequence varied (numbers in circles indicate end of sequence obtained from each product). The forward primer used for sequencing is indicated by the arrow. The end of the GFP transgene is underlined and the rectangle is the linker sequence found at the end of each product and is included for illustration purposes on only the longest product, but was found at the end of all four products.

primers themselves will be identical to the 5′ to 3′ sequence of the transgene. Gene-specific primer 1 is typically designed no more than 400 bp from the 3′ end of the transgene sequence.

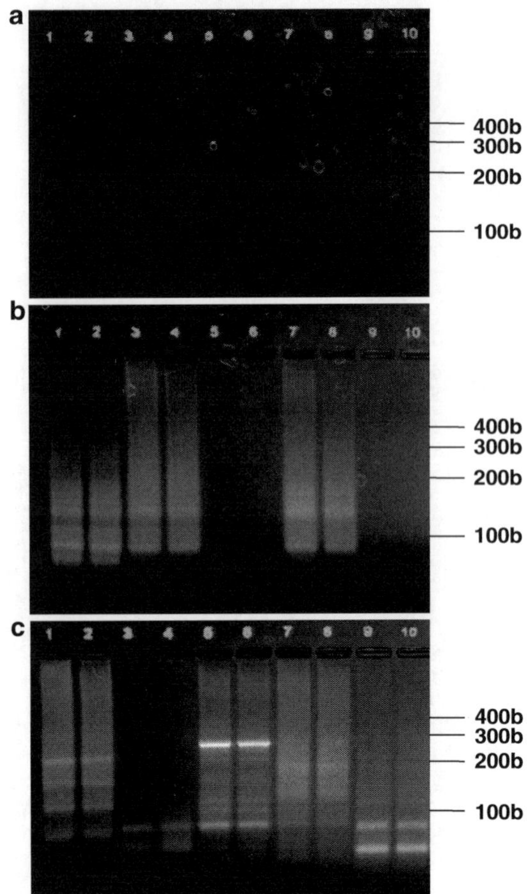


Fig. 20.3 Troubleshooting. In this series of experiments, the objective was to determine the integration site of a GFP transgene in a rodent transgenic line. In all three panels, each reaction was run in duplicate and five different restriction enzymes were used (lane 1–2:*Bgl*III, 3–4:*EcoR*I, 5–6:*Hha*I, 7–8:*Hind*III, 9–10:*Msp*I). The reactions in each panel were identical except for the Gene-specific primer 1 that was used. Each panel shows the results of gel electrophoresis to separate and recover amplicons (Subheading 20.3.7). Size in base pairs (bp) based on a molecular size standard (not shown) is indicated on the right hand side of each panel. (**a**) No amplification (blank gel) is the typical result obtained if Gene-specific primer 1 fails to provide adequate amplification in subheading 20.3.3. Other causes can include using a polymerase enzyme that does not add an "A" overhang or failure of the linker ligation (Subheading 20.3.4). (**b**) The only difference between the reactions shown here and Panel a was that a new Gene-specific primer 1 was used. Faint bands or smeary patterns are the typical results obtained if Gene-specific primers are not specific for the transgene and amplify areas of the genome in addition to the transgene. (**c**) The only difference between the reactions shown here and Panels (**a**) and (**b**) was that a new Gene-specific primer 1 was used. Redesign of Gene-specific primer 1 resulted in distinct amplicons for numerous reactions. *Hha*I digestion reactions yielded PCR products for which flanking sequence was obtained (Lanes 5 and 6). *Bgl*III (Lanes 1 and 2) and *Msp*I (Lanes 9 and 10) reactions also yielded distinct amplicons, which were not analyzed further in this experiment, but could have been sequenced to confirm the results obtained with the amplicons recovered from the *Hha*I reactions.

4. All gene-specific primers are designed to have an optimum $T_M$ of 63°C, optimum primer length of 24 bp, and an optimum %GC content of 50. Gene-specific primers 2 and 3 should be designed to minimize any complementarity with Linker primers D and G, respectively (*see* Note 4).

5. Once primers are designed, Basic Alignment Search Tool (BLAST)(11) or BLAST-like Alignment Tool (BLAT)(12) are used to confirm that primers are specific for the transgene sequence and do not have significant homology to random sequences in the rat genome.

### 3.2. DNA Extraction and Restriction Endonuclease Digest

1. Take 2 mm tail biopsies from 2 week old rats or ear punch biopsies from adult rats (*see* Note 5).
2. Prepare genomic DNA using Qiagen DNeasy Blood & Tissue Kit following the manufacturer's instructions for preparing genomic DNA. Elute into 200 μl AE solution, which is supplied with the kit (*see* Note 6).
3. Adjust concentration of genomic DNA to 50 ng/μl.
4. Digest 1.0 μg of genomic DNA with each of a panel of eight different restriction endonucleases: *Bgl*II, *Eco*RI, *Hha*I, *Hind*III, *Msp*I, *Pst*I, *Sac*I or *Xba*I (*see* Note 7) in a total reaction volume of 30 μl by combining the following in a 0.2 ml microfuge tube:

For *Hha*I, *Pst*I, *Sac*I, and *Xba*I:

3.0 μl 10× BSA

3.0 μl 10× enzyme-specific buffer

1.0 μl restriction enzyme (20 U/μl)

20.0 μl 50 ng/μl genomic DNA

3.0 μl nuclease-free water

For *Bgl*II, *Eco*RI, *Hind*III, and *Msp*I:

- 3.0 μl 10× enzyme-specific buffer
- 1.0 μl restriction enzyme (20 U/μl)
- 20.0 μl 50 ng/μl genomic DNA
- 6.0 μl nuclease-free water (*see* Note 8)

5. Incubate at 37°C for 2 h to overnight (*see* Note 9)

### 3.3. First Round PCR

1. In a total reaction volume of 25 μl, combine the following in a 0.2 ml microfuge tube (*see* Note 10):

   2.50 μl 10× FastStart *Taq* with 20 mM $MgCl_2$ Buffer

   4.00 μl 1.25 μM each dNTP

   1.60 μl 25 μM Gene-specific primer 1

   0.25 μl 5 U/μl FastStart *Taq*

   13.45 μl nuclease-free water

   3.2 μl digested DNA from subheading 20.3.2
2. Perform a single round of PCR in a thermal cycler as follows:

   94°C for 10 min

   60°C for 1 min

72°C for 10 min

4°C hold

### 3.4. Y-linker Ligation

1. Prepare Y linker by combining 250 μl of 8 μM Y-linker A and 250 μl 8 μM Y-linker E in a 1 ml microfuge tube and incubate at 95 °C for 5 min and allow the reaction to cool to room temperature on the benchtop. Use immediately once cooled or store Y linker at –20°C.
2. In a 0.2 ml microfuge tube, combine the following:
   7.0 μl of PCR product from subheading 20.3.3
   1.0 μl 4 μM Y-linker
   1.0 μl 400 U/μl T4 DNA Ligase
   1.0 μl 10× T4 DNA Ligase Buffer
3. Incubate at 16°C for a minimum of 16 h or longer (*see* Note 11).

### 3.5. Second Round PCR

1. In a total reaction volume of 25 μl, combine the following in a 0.2 ml microfuge tube:
   2.50 μl 10× FastStart *Taq* with 20 mM $MgCl_2$ Buffer
   4.00 μl 1.25 mM dNTP
   0.40 μl 25 μM Y-linker primer D
   0.40 μl 25 μM Gene-specific primer 2
   0.25 μl 5U/μl FastStart *Taq*
   16.45 μl nuclease-free water
   1.00 μl reaction mix from subheading 20.3.4
2. Perform the PCR in a thermal cycler as follows:
   1 cycle: 94°C for 5 min
   20 cycles:
   94°C for 30 s
   60°C for 30 s
   72°C for 1 min 30 s
   1 cycle: 72°C for 10 min

### 3.6. Third Round PCR

1. In a total reaction volume of 25 μl, combine the following in a 0.2 ml microfuge tube (*see* Note 12):
   2.50 μl 10× FastStart *Taq* with 20 mM $MgCl_2$ Buffer
   4.00 μl 1.25 mM dNTP
   0.40 μl 25 μM Y-linker primer G
   0.40 μl 25 μM Gene-specific primer 3
   0.25 μl 5U/μl FastStart *Taq*
   16.45 μl nuclease-free water
   1.00 μl reaction mix from subheading 20.3.5

2. Perform the PCR in a thermal cycler as follows:
   1 cycle: 94°C for 5 min

   20 cycles:

   94°C for 30 s

   60°C for 30 s

   72°C for 1 min 30 s

   1 cycle: 72°C for 10 min

### *3.7. Recovery of Amplication Products*

1. Add 2.5 μl of gel loading buffer to the entire reaction from subheading 20.3.6.
2. Load 20 μl of sample mixed with loading buffer onto a 3% precast agarose gel (*see* Note 13).
3. Load 15 μl of 1 Kb Plus DNA ladder working stock in one lane of the gel.
4. Perform gel electrophoresis in 1× TBE buffer at 100 V or less for a minimum of 30–45 min (*see* Note 14).
5. Excise individual discrete amplification products from the gel and gel-purify using QIAquick Gel Extraction kit. Elute the recovered DNA into 10 μl EB Buffer, which is supplied with the kit (*see* Note 15).

### *3.8. Nucleotide Sequence Analysis of Recovered PCR products*

1. In 0.5 ml microfuge tube, combine:
   4.4 μl DNA, gel-purified product from subheading 20.3.7

   1.6 μl 25 μM Gene-specific primer 3

   6.0 μl water
2. Perform nucleotide sequence analysis (*see* Note 16).
3. Analyze nucleotide sequences. The initial bases identified should correspond to sequence immediately adjacent to the gene-specific primer used for sequence analysis and should match sequence from the transgene construct.
4. Use the BLAT (BLAST-like Alignment Tool)(12) at Ensembl (http://www.ensembl.org) to compare the nucleotide sequence of the recovered amplicon to the rat genome sequence. The nucleotide sequence immediately adjacent to sequence identifiable as part of the transgene should represent the chromosomal sequence of the insertion site and will show high homology to rat genomic DNA. As little as 20–40 bp of flanking sequence (i.e., sequence immediately adjacent to the transgene sequence) allows identification of the chromosomal region into which the transgene has been inserted.
5. If amplicons have been recovered and sequenced for both the 5′ and 3′ flanking regions adjacent to the transgene, they should match the same region of the same chromosome

and will define the precise boundaries of the transgene insertion site.

6. The last region of the amplicon nucleotide sequence should match the sequence of the Y-linker Primer G (*see* Note 17).

### 3.9 Troubleshooting

1. No amplification; gel is "blank" following gel electrophoresis to recover amplicons (Fig. 20.3, a). Assuming that all steps of the procedure have been followed, the primary reason for failure of the technique seems to be failure of Gene-specific primer 1 to amplify well in the initial round of PCR amplification. Redesigning the primer typically solves the problem. Additional issues to address include checking for restriction enzyme recognition sites in the region between Gene-specific primer 1 and the transgene/insertion site junction; failure to use a *Taq* polymerase that adds A overhangs; failure to prepare the linker properly or failure of the linker ligation step itself.
2. No distinct amplicons recovered, smeary bands on gel (Fig. 20.3, b). This result is seen when the Gene-specific primers are not specific for the transgene and amplify other regions of the genome. Redesign of the primers, particularly Gene-specific primer 1 typically solves this problem. A different choice of restriction enzymes may also help.

## 4. Notes

1. All oligonucleotides (primers and linkers) should be desalted.
2. Y-Linker A has an inverted base (3InvdT) incorporated at the 3′ end of the oligonucleotide. Inverted bases involve the modification of the natural 3'-5' linkage to a 3'-3' linkage. This prevents the oligo from being extended beyond this base. The modified base is included to ensure that the complete annealed Y-linker will have a defined region that is double-stranded and a defined region that is single-stranded (10).
3. Precast gels are used for convenience but hand-poured 3% agarose gels work equally well. Likewise, we use the recommended Bio-Rad electrophoresis system because of its compatibility with the precast gels, but any standard electrophoresis system is suitable.
4. Any primer design program may be used to facilitate good primer design. We routinely use Primer Express (Applied Biosystems, University Park, IL), Primer 3 at http://frodo.wi.mit.edu (13) and PrimerQuest (Integrated DNA Technologies, Coralville, IA) (13). In our experience, Gene-specific primer 1 is critical for the success of the protocol.

On occasions where no amplicons have been recovered, redesign of Primer 1 alone, usually by relocating the primer to a slightly different region of the sequence, was sufficient to overcome the initial failure (Fig. 20.3). We speculate that this is due to the importance of having successful amplification in the initial single round of PCR (Subheading 20.3.3).

5. While tail snips and ear punches are used as tissue sources from live animals, any tissue from which genomic DNA can be extracted may be used.
6. When using the DNeasy kit, the protocol "Purification of Total DNA from Animal Tissues" is followed. However, any method or kit that allows extraction of high-quality genomic DNA may be used.
7. Whenever possible, first analyze the 5′ and 3′ ends of transgene construct sequences for the presence of restriction endonuclease recognition sites. Avoid using a restriction enzyme that has recognition sites within the 5′ or 3′ ends of the transgene sequence. In our experience, at least one of the eight enzymes listed will give good results. However, it is possible to explore the use of other restriction endonucleases if necessary.
8. The amount of water added to the reaction is based on using 1 μl of an enzyme at a concentration of 20 U/μl. If the enzyme concentration is lesser or greater, add the amount of enzyme that represents 20 U and adjust the amount of water added accordingly.
9. There does not seem to be an appreciable difference in the results obtained from short incubations (2 h) vs. overnight incubation although in some cases, the sizes of the most prominent amplicons recovered at the end of the procedure differ (Fig. 20.2).
10. It is essential that the *Taq* polymerase used for this step of the procedure is one that can perform non-templated 3′ addition of nucleotides. This addition primarily involves addition of adenosine (A) overhangs (14). The ability in subheading 20.3.4 of the linker to base pair with the PCR-generated amplicons recovered in this step relies on the presence of 3′ A overhangs on the amplicons. Use of a *Taq* polymerase lacking the ability to perform non-templated addition of A overhangs will result in failure of the entire method and is the most common mistake noted when troubleshooting lack of success by new users of this protocol.
11. The linker is created by allowing base pairing of two complementary oligonucleotides (Y-linker A and Y-linker E) such that the resulting double-stranded Y linker contains a 5′ T overhang that is then able to base pair with the terminal 3′ A

overhang on the amplicon generated after the single round of PCR amplification (Fig. 20.1). Ligation of the linker can occur on either the 3′ or 5′ end of the amplicon or both. Because the linker sequences now define the unknown regions at the ends of the amplicon, linker-specific primers (Primers D and G) can be used in conjunction with gene-specific primers to amplify PCR products containing the regions flanking the transgene (Fig. 20.1).

12. The second and third rounds of PCR represent two rounds of nested PCR, which are designed to increase the specificity of the major amplicons being produced as well as generate sufficient amounts of these amplicons for efficient recovery for subsequent sequence analysis.
13. The major amplicons that tend to be recovered are typically in the 200–800 bp range, therefore, the use of 3% agarose gels allows better separation of these products. However, if larger products are expected, the use of gels containing less agarose, such as a 1% gel, is advised.
14. The electrophoresis conditions can vary depending on the size of the gel and electrophoresis equipment used. The goal is to use a voltage and running time that allows sufficient separation of the resulting amplicons to facilitate clean excision of the bands from the gel. This can be accomplished by using 5.5–6.5 V/cm during gel electrophoresis.
15. It is prudent to excise all intensely staining discreet bands for analysis since recovery of multiple amplicons containing nucleotide sequence corresponding to the same chromosomal region lends greater confidence that the bona fide integration site has been identified (see examples, bands 1–4 in Fig. 20.2a & 20.2b). Any commercially available kit for purifying DNA from agarose gels may be used. It is important to resuspend the recovered DNA in a small enough volume of water or an appropriate buffer that is compatible with the requirements for the nucleotide sequencing method chosen for subsequent analysis.
16. Our nucleotide sequence analysis is performed at the University of Missouri DNA Core and the amounts of reagents listed reflect the Core's requirements for sample submission based on the instrumentation they use for the analysis. Any method for direct sequencing of PCR products may be used. The critical feature is the use of Gene-specific primer 3 as the sequencing primer.
17. An example of recovered sequences is shown in Fig. 20.2c. Four different-sized amplicons were recovered for one GFP transgenic rodent strain (labeled 1–4, Figs. 20.2a & 20.2b). In Fig. 20.2c, the arrow indicates the sequence corresponding to Gene-specific primer 3 (F623 from Fig. 20.1).

The underlined sequence corresponds to the 3′ end of the GFP transgene sequence. The sequence after that corresponds to genomic sequence found immediately flanking the transgene. The sequences recovered from each of the 4 amplicons are indicated and all 4 contained overlapping sequence. Amplicon 3, which was the largest, contained the longest sequence. It should be noted that the sequences obtained for all amplicons had linker sequence (boxed region), but this is only shown for amplicon 3. BLAT analysis of the flanking sequence was used to successfully identify the insertion site of the GFP transgene in this rodent transgenic strain.

## References


1. Tesson L, Cozzi J, Menoret S, Remy S, Usal C, Fraichard A, Anegon I (2005) Transgenic modifications of the rat genome. Transgenic Res 14:531–46
2. Dellaire G, Chartrand P (1998) Direct evidence that transgene integration is random in murine cells, implying that naturally occurring double-strand breaks may be distributed similarly within the genome. Radiat Res 149:325–9
3. Liang Z, Breman AM, Grimes BR, Rosen ED (2008) Identifying and genotyping transgene integration loci. Transgenic Res 17:979–83
4. Ochman H, Gerber AS, Hartl DL (1988) Genetic applications of an inverse polymerase chain reaction. Genetics 120:621–3
5. Rosenthal A (1992) PCR amplification techniques for chromosome walking. Trends Biotechnol 10:44–8
6. Rosenthal A, Jones DS (1990) Genomic walking and sequencing by oligo-cassette mediated polymerase chain reaction. Nucleic Acids Res 18:3095–6
7. Shyamala V, Ames GF (1989) Genome walking by single-specific-primer polymerase chain reaction: SSP-PCR. Gene 84:1–8
8. Parker JD, Rabinovitch PS, Burmer GC (1991) Targeted gene walking polymerase chain reaction. Nucleic Acids Res 19:3055–60
9. Yuanxin Y, Chengcai A, Li L, Jiayu G, Guihong T, Zhangliang C (2003) T-linker-specific ligation PCR (T-linker PCR): an advanced PCR technique for chromosome walking or for isolation of tagged DNA ends. Nucleic Acids Res 31:e68
10. Bryda EC, Pearson M, Agca Y, Bauer BA (2006) Method for detection and identification of multiple chromosomal integration sites in transgenic animals created with lentivirus. Biotechniques 41:715–9
11. Altschul SF, Gish W, Miller W, Myers EW, Lipman DJ (1990) Basic local alignment search tool. J Mol Biol 215:403–10
12. Kent WJ (2002) BLAT–the BLAST-like alignment tool. Genome Res 12:656–64
13. Rozen S, Skaletsky H (2000) Primer3 on the WWW for general users and for biologist programmers. Methods Mol Biol 132:365–86
14. Brownstein MJ, Carpten JD, Smith JR (1996) Modulation of non-templated nucleotide addition by Taq DNA polymerase: primer modifications that facilitate genotyping. Biotechniques 20(1004–6):1008–10

# Chapter 21

# Cryopreservation and Orthotopic Transplantation of Rat Ovaries


## Martina Dorsch and Dirk Wedekind



## Abstract

The number of rat strains increased considerably in the last decade and will increase continuously during the next years. This requires enough space for maintaining vital strains and techniques for cryobanking, which can be applied not only in specialised rat ressource centres but also in regular animal houses. Here we describe an easy and fast method for the cryopreservation and transplantation of frozen-thawed ovaries of the rat.

With dimethyl sulfoxide as cryoprotectant rat ovaries can be stored at −196°C for unlimited time. For revitalisation thawed ovaries have to be orthotopically transplanted into appropiate ovarectomised recipients. Reestablishment of the reproductive cycle in the recipients can be confirmed by vaginal cytology shortly after transplantation. The recipients are able to produce 2–3 litters after mating with males of an appropriate strain.

Cyropreservation of ovaries thus can be considered a reliable method to preserve scientifically and economically important stocks and strains of rats that are currently not required.



**Key words:** Rat, Ovary, Orthotopic transplantation, Cryopreservation


## 1. Introduction

The diversity of rat models is composed of a huge amount of genetically different inbred strains that have been derived from some outbred stocks and wild rats (1, 2) and still is extended by strains derived from colonies with spontaneous mutations and those established by special breeding programmes such as coisogenic, congenic, consomic, and recombinant strains (3–9). Furthermore, genetically modified rats become more and more significant for biomedical research (10). Maintenance of the increasing number of rat strains exceeds holding capacities and costs.


I. Anegon (ed.), *Rat Genomics: Methods and Protocols,* Methods in Molecular Biology, vol. 597
DOI 10.1007/978-1-60327-389-3_21, © Humana Press, a part of Springer Science+Business Media, LLC 2010

This led to the establishment of rat resource centres in the USA (11), Japan (12), Germany (13), and the Czech Republic (14).

Nonviable storage (cryobanking) of strains can reduce the demand for space and the actual shelf costs. Cryopreservation of preimplantation embryos is considered the method of choice for cryobanking of rats. However, this method is expensive and time consuming. Cryopreservation of rat spermatozoa (15, 16) might be one alternative.

The cryopreservation of whole ovaries is an additional tool for preserving the genetic pool of valuable strains (17) and for the rescue of strains with impeded reproductive performance (18). After thawing, ovaries can be transplanted into syngenic recipients or, if not available into athymic rats (unpublished data).

Histological preparations made directly after thawing confirmed the morphological integrity of the organ is not affected by the procedure of freezing and thawing. Post transplantation frozen-thawed ovaries never show signs of rejection or necrosis.

The production of live offspring by transferring frozen-thawed parts of the reproductive tract of female LEW rats using vascular anastomosis has also been reported by Wang et al. (19, 20) and Yin et al. (21).

The cryopreservation of ovaries in mice is currently being performed as a service at the Jackson Laboratory (http://www.jax.org/cryo/ovary.html) following a protocol first described by Sztein et al. (22, 23).

Cryobanking of rat ovaries as described here could be provided by rat resource centres as service and can also be established easily at regular animal houses.

## 2. Materials

### 2.1. Animals

1. *Ovary donors*: Female rats of the strain that should be stored. Preferably this should be nulliparous and young females (*see* Note 1).
2. *Ovary recipients for revitalisation:* female rats at the age of 4 weeks. The choice of the recipient strain depends on the strain of the ovary donor. Best solution is to use a syngenic strain. If this is not possible an immunincompetent athymic strain has to be used (*see* Note 2).

### 2.2. Buffer and Media for Cryopreservation and Thawing

1. Phosphate buffered medium (PB1) is mixed according the protocol described by Whittingham (24): 100.93 mM NaCl, 2.73 mM KCl, 0.33 Na-pyruvate, 1.42 mM $KH_2PO_4$, 0.48 mM $MgCl_2 \times 6H_2O$, 8.05 mM $Na_2HPO_4 \times 12\ H_2O$, 0.94 mM

$CaCl_2 \times 2H_2O$, 1 mg/ml Penicillin G, 0.05 mg/ml Strepto-mycin, 1 mg/ml glucose, 3 mg/ml BSA. As ph-indicator add some phenol red. Mix the substances in the indicated order and filtrate it through a 0.45 μm disposable filter unit. PB1 can be used up to 3 weeks if stored at 4°C (*see* Note 3).

2. Cryoprotectant solution (CPS), 1.5 M dimethylsulfoxide (DMSO, Sigma Germany) in PB1, supplemented with 10% foetal calf serum (Sigma, Taufkirchen, Germany). Prepare shortly before use.
3. Liquid nitrogen.

### 2.3. Equipment for Freezing, Storage and Thawing

1. Petri dish (ø 30 mm).
2. 1.5 ml cryotubes.
3. Adequate labelling system for the permanent labelling for cryotubes.
4. Pasteur pipettes.
5. Programmable cooler (Haake C50P, Thermo Electron Corporation, Germany) with alcohol bath.
6. Liquid nitrogen storage tank with storage system.
7. Dewar-jar.

### 2.4. Equipment and Material for Ovary Preparation and Transplantation

1. Electric shaver.
2. Stereomicroscope.
3. Microscope.
4. Warming plate or heating pad (25–30°C).
5. Anaesthesia for rats that will produce narcosis for about 30 min (*see* Note 4).
6. Eye ointment (e.g., Dexapanthenol: Bepanthen®, Hoffmann-LaRoche Leverkusen, Germany).
7. Analgesia: we recommend metamizol (Novalgin®, Hoechst Vetrinär, Germany).
8. Epinephrine (Suprarenin®, Aventis-Pharma).
9. Basic equipment of surgical instruments.
10. Absorbable suture material (Vicryl®-rapid, 4-0, Ethicon, Norderstedt, Germany).
11. Michel clips (7.5×1.75 mm, Medicon EG, Tuttlingen, Germany).
12. Disinfectant (Braunol® 2000, Braun Melsungen AG, Melsungen, Germany).
13. Saline (0.9 NaCl).
14. Sterile loop or plastic scoop to take vaginal smears.
15. Microscope slides.

## 3. Methods

For all the methods described below, it is important to use sterile material. Preferably, all procedures should be performed under a clean workbench. This regime is recommended to prevent germ transmission.

### 3.1. Preparation of Donor Ovaries

Donor females are sacrificed after $CO_2$ anaesthesia by cervical dislocation.

After disinfection of the ventral body side, the cavity is opened and the ovaries are removed aseptically and placed into a Petri dish containing two drops of PB1.

Fat pad and periovarial sac (bursa) have to be removed carefully as far as possible under a stereomicroscope using fine forceps and scissors.

We suggest not to dissipate the ovaries into halves, because too much follicles may be hurt and loose their functionality (*see* Note 5).

### 3.2. Cryopreservation and Thawing of Ovaries

#### 3.2.1. Cryopreservation

The procedure for cryopreservation is adapted from Sztein et al. (22) with some modifications.

First thing to do is the permanent labelling of the cryotubes and sterilisation.

After preparation, the ovaries are transferred into 1.5 ml cryotubes containing 500 µl cryoprotectant solution (CPS) at room temperature (20 ± 2°C). Use one vial for each ovary.

The ovaries are left for 10 min at room temperature and then 45 min on ice (0–4°C).

Transfer the cryotubes to a pre-cooled alcohol bath (–6°C) of a programmable cooler.

Induce seeding mechanically after 10 min by touching the surface of CPS with the tip of a Pasteur pipette pre-cooled at –196°C in liquid nitrogen.

Reduce the temperature to –60°C with a cooling rate of 0.4°C/min.

After 10 min at –60°C transfer the cryotubes into liquid nitrogen for storage.

#### 3.2.2. Thawing

Remove the cryotube from the liquid nitrogen to room temperature and wait until ice has melted.

The CPS is then removed carefully with a micropipette and replaced by 500 µl PB1.

After 10 min the ovaries can be transferred to another Petri dish containing 2 ml PB1.

After another 10 min the ovaries are ready for transplantation.

### 3.3. Ovary Transplantation

The technique is adapted from Jones and Krohn (25) who performed orthotopic transplantation of the ovary in the mouse.

During surgery place the female on a heating pad.

Anaesthetise the recipient female as described.

Apply eye ointment for corneal protection as soon as the toe pinch reflex has disappeared.

The anaesthetised animal is laid on its ventral surface with the tail looking to the investigator.

Shave and disinfect the surgical field. A single midline dorsal incision of the skin, across the lumbar area is made (Fig. 21.1). The scissors are inserted subcutaneously through the incision and pushed down to either side of the animal and opened to blunt dissect the connective tissue to allow access to the muscle.

The peritoneal cavity is opened on the right body side by a 5 mm muscle incision a little below the kidney. The ovary, surrounded by fat, should be found easily.

Pull out the ovary through the incision by grasping the periovarian fat with forceps. The ovary is hold in place using fine artery-clamps (Bulldog-Clamp).

Make a ligature between the fallopian tube and right uterine horn with absorbable suture material and excise the ovary with a single cut (Fig. 21.2). Bleeding is usually slight and will soon stop.

The right uterus-horn is returned into the cavity and the muscle incision is closed with one or two stitches using absorbable material.

To remove the left ovary, open the cavity on the left body side as described above.

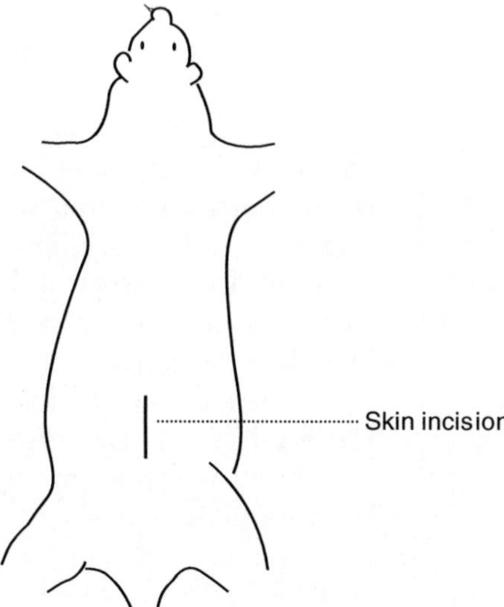


Fig. 21.1. Dorsal skin incision across the lumbar area gives access to both ovaries.

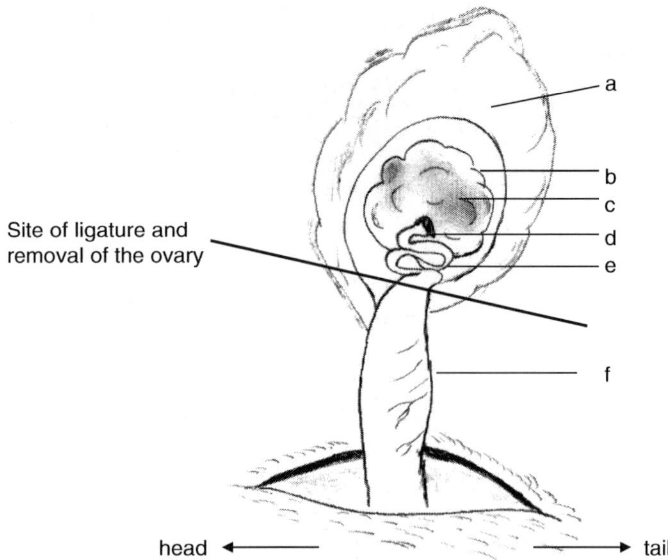


Fig. 21.2. Right ovary. Indicated is the site of ligature and cut to remove the ovary that should not be replaced by a grafted ovary. (**a**) fat, (**b**) periovarian sac (bursa), (**c**) ovary, (**d**) infundibulum, (**e**) oviduct, (**f**) uterus.

Pull out the left ovary carefully by grasping the periovarian fat and hold it in position using fine artery-clamps.

Apply some drops of Epinephrine onto the bursa to minimise the risk of bleeding.

A small incision is made into the bursa with spring type micro-scissors (Fig. 21.3) and the ovary is then removed by a cut on its hilum. Carefully avoid small blood vessels, because agglutinated blood will block the infundibular opening and the female will never become pregnant despite displaying normal estrus cycles.

The removed ovary is replaced with the donor ovary once haemostasis has been obtained by the use of swabs. The incision of the bursa is sutured with one reef knot using fine absorbable material.

The donor-ovary and the uterus-horn are returned into the cavity and the muscle incision is closed by continuous suture with absorbable material.

The skin can be closed with Michel clips which have to be removed 10–12 days after surgery.

In order to prevent hypothermia, the cages holding anaesthetised animals have to be kept warm until total recovery from anaesthesia.

For analgesia we apply 0.2–0.4 ml metamizol s.c per animal if signs of discomfort are visible after recovery from anaesthesia.

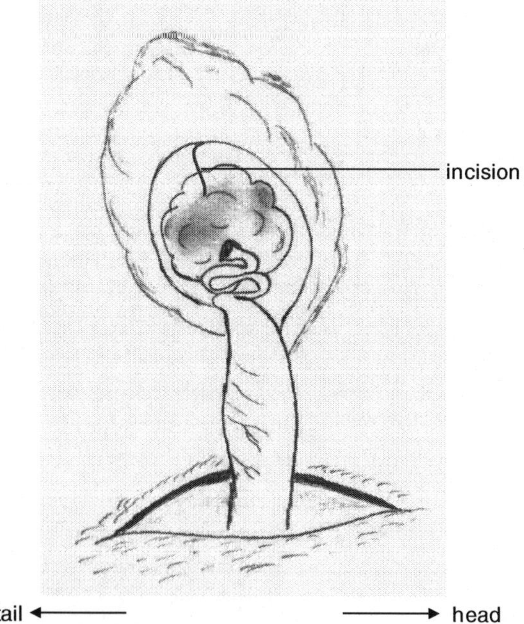


Fig. 21.3. Left ovary. Indicated is the site for the inc ision into the bursa to remove the ovary and replace it by the grafted ovary.

### 3.4. Vaginal Cytology

Two weeks after surgery vaginal smears can be taken to determine the restoration of the reproductive cycle.

For this, loose cellular material is removed from the vagina by gently scraping the dorsal vaginal wall with a sterile loop or scoop, dispersed in a drop of saline on a slide, air dried, stained, and examined microscopically (*see* Note 6).

### 3.5. Mating Regimen

Females that display a regular reproductive cycle of 4–5 days can be mated.

Females should become pregnant within 3 months after transplantation (*see* Note 7).

## 4. Notes

1. Young, nulliparous females will produce the best reproductive performance, but we also produced offspring with ovaries from old breeders. However, such animals can only be the last chance to save a special strain.
2. In case no syngenic recipient is available an immunodeficient recipient has to be used. For this purpose athymic rats are the best choice. However, we have to mention that the reproductive performance of such strains is superior.

3. Unless stated otherwise, all buffer and media are prepared by using aqua irrigation solution (Delta Select GmbH, Germany). Also sterile water with an electric resistivity of 0.055 mS (milli Siemens) can be used.
4. Anaesthesia for rats: 950 μl Ketamine 10% (WDT, Garbsen, Germany) and 50 μl Xylazin (Rompun 2%, Bayer, Leverkusen, Germany) are mixed shortly before use. Recipient females receive an injection (i.p.) of 0.1 ml/100 g body weight of this mixture. This achieves sufficient anaesthetic and analgesic effect by acting on *N*-methyl-d-aspartate- and adrenergic alpha2-receptors respectively (26, 27).
5. In case older donors are used, it could be necessary to dissect the ovaries into halves that they fit into the bursa of a young recipient.
6. Determination of oestrus cycle with vaginal smears: Loose cellular material can be removed from the vagina using an inoculating loop or a scoop, dispersed in a drop of saline on a slide and air dried. The cells are fixed to the slide by incubation for 2 min in methanol. The cells are stained for 2 min with Löffler's methylene blue and washed twice in $H_2O$. After drying you can examine the slides immediately, embedding is not necessary.
   i. In *diestrus* (duration 2–3 days) you will find leucozytes, some round or oval epithelial cells, and mucus as a typical sign.
   ii. In *proestrus* (duration 0.5 days) there are numerous club like or spindle shaped cells often with vacuoles, nucleated cells, and a few leucocytes.
   iii. In females in *estrus* (duration 0.5 days) exclusively cornified epithelial cells are abound.
   iv. In *metestrus* (duration 1 day) you find leucocytes, some cornyfied epithelial cells, and cells containing a nucleus.
7. Be careful not to induce pseudoprgnancy by introducing the loop or scoop to deep into the vagina and stimulate the cervix. The female will not become pregnant.
8. If a female does not become pregnant during 3–4 months after transplantation, the transplantation has failed. This can have several reasons:
   i. Female does not perform reproductive cycles due to lost functionality of the donor ovary.
   ii. Female shows reproductive cycle, but infundibulate opening is blocked by agglutinated blood.
   iii. Female shows reproductive cycle, but donor ovary dropped out of the bursa.
   iv. The only way to reduce failures is practise. Only skilled personal will produce reproducible results.

## Acknowledgments

The authors like to thank Mrs. R. Eisenblaetter and C. Elfers for technical assistance.

### References

1. Canzian F (1997) Phylogenetics of the laboratory rat Rattus norvegicus. Genome Res 7:262–267
2. Hedrich HJ (ed) (1990) Genetic monitoring of inbred strains of rats. Gustav Fischer Verlag, Stuttgart, New York
3. Chwalisz WT, Koelsch BU, Kindler-Röhrborn A, Hedrich HJ, Wedekind D (2003) The circling behavior of the deafblind LEW-ci2 rat is linked to a segment of RNO10 containing Myo15 and Kcnj12. Mamm Genome 14:620–627
4. Cowley AW Jr, Roman RJ, Jacob HJ (2004) Application of chromosomal substitution techniques in gene-function discovery. J Physiol 554:46–55
5. Kwitek-Black AE, Jacob HJ (2001) The use of designer rats in the genetic dissection of hypertension. Curr Hypertens Rep 3:12–18
6. Länger B, Dorsch M, Gärtner K, Wedekind D, Kamino K, Hedrich HJ (2004) WKY/Ztm-ter: a new rat inbred strain on the WKY/Ztm genetic background with congenital teratomas. Lab Anim 38:425–431
7. Lenzen S, Tiedge M, Elsner M, Lortz S, Weiss H, Jörns A, Klöppel G, Wedekind D, Prokop CM, Hedrich HJ (2001) The LEW.1AR1/Ztm-iddm rat: a new model of spontaneous insulin-dependent diabetes mellitus. Diabetologia 44:1189–1196
8. Lessenich A, Lindemann S, Richter A, Hedrich HJ, Wedekind D, Kaiser A, Löscher W (2001) A novel black-hooded mutant rat (ci3) with spontaneous circling behavior but normal auditory and vestibular functions. Neuroscience 107:615–628
9. Pravenec M, Klír P, Kren V, Zicha J, Kunes J (1989) An analysis of spontaneous hypertension in spontaneously hypertensive rats by means of new recombinant inbred strains. J Hypertens 7:217–221
10. Tesson L, Cozzi J, Ménoret S, Rémy S, Usal C, Fraichard A, Anegon I (2005) Transgenic modifications of the rat genome. Transgenic Res 14:531–546
11. http://www.nrrrc.missouri.edu/
12. http://www.anim.med.kyoto-u.ac.jp/NBR/
13. http://www.mh-hannover.de/2652.html
14. http://www.euratools.eu
15. Nakatsukasa E, Inomata T, Ikeda T, Shino M, Kashiwazaki N (2001) Generation of live rat offspring by intrauterine insemination with epididymal spermatozoa cryopreserved at −196 degrees C. Reproduction 122: 463–467
16. Nakatsukasa E, Kashiwazaki N, Takizawa A, Shino M, Kitada K, Serikawa T, Hakamata Y, Kobayashi E, Takahashi R, Ueda M, Nakashima T, Nakagata N (2003) Cryopreservation of spermatozoa from closed colonies, and inbred, spontaneous mutant, and transgenic strains of rats. Comp Med 53:639–641
17. Dorsch M, Wedekind D, Kamino K, Hedrich HJ (2007) Cryopreservation and orthotopic transplantation of rat ovaries as a means of gamete banking. Lab Anim 41:247–254
18. Dorsch M, Wedekind D, Kamino K, Hedrich HJ (2004) Orthotopic transplantation of rat ovaries as a tool for strain rescue. Lab Anim 38:307–312
19. Wang X, Bilolo KK, Qi S, Xu D, Jiang W, Vu MD, Chen H (2002) Restoration of fertility in oophorectomized rats after tubo-ovarian transplantation. Microsurgery 22:30–33
20. Wang X, Chen H, Yin H, Kim SS, Lin Tan S, Gosden RG (2002) Fertility after intact ovary transplantation. Nature 415:385
21. Yin H, Wang X, Kim SS, Chen H, Tan SL, Gosden RG (2003) Transplantation of intact rat gonads using vascular anastomosis: effects of cryopreservation, ischaemia and genotype. Human Reprod 18:1165–1172
22. Sztein J, Sweet H, Farley J, Mobraaten L (1998) Cryopreservation and orthotopic transplantation of mouse ovaries: new approach in gamete banking. Biol Reprod 58: 1071–1074
23. Sztein J, McGregor TE, Bedigian HJ, Mobraaten L (1999) Transgenic mouse strain rescue by frozen ovaries. Lab Anim Sci 49:99–100
24. Whittingham DG (1974) Embryo banks in the future of developmental genetics. Genetics 78:395–402

25. Jones EC, Krohn PL (1960) Orthotopic ovarian transplantation in mice. J Endocrinol 2:135–146
26. Maurset A, Skoglund LA, Hustveit O, Oye I (1989) Comparison of ketamine and pethidine in experimental and postoperative pain. Pain 36:37–41
27. Maze M, Tranquilli W (1991) Alpha-2 adrenoceptor agonists: defining the role in clinical anesthesia. Anesthesiology 74:581–605

# Chapter 22

# Techniques for In Vitro and In Vivo Fertilization in the Rat

**Naomi Kashiwazaki, Yasunari Seita, Akiko Takizawa, Naoki Maedomari, Junya Ito, and Tadao Serikawa**

## Abstract

Although in vitro and in vivo fertilization are powerful tools for restoring conserved sperm as well as stocked males in the rat, the techniques have progressively gained importance. However, the techniques are not used extensively for efficient production of rat offspring, because the techniques require a great deal of skill. This chapter describes the protocols for in vitro and in vivo fertilization in the rat. Namely, sperm collection, sperm cryopreservation, pre-incubation of sperm, and insemination (co-culture with sperm and oocytes) for in vitro fertilization and intrauterine insemination for in vivo fertilization with fresh or frozen/thawed spermatozoa are provided.

**Key words:** Intrauterine insemination, In vitro fertilization, Rat sperm, Sperm cryopreservation

## 1. Introduction

The laboratory rat has been used extensively in many fields of medical and biological research because of body size, ease of manipulations and breeding characteristics. In addition, there is a lot of unique and invaluable rat models that resemble human diseases such as cardiovascular disease, diabetes, arthritis and many behavioral disorders (1). Recently, the rat genome sequence of the Brown Norway that is a well characterized inbred strain and the founder strain for several important genetic panels was released (2). Although gene targeting in the rat is still limited, recent advances in molecular and reproductive technologies (3) have enabled generating significant transgenic rats by DNA microinjection (4), sperm mediated gene transfer (5), lentivirus method (6) and *N*-ethyl-*N*-nitrosourea (ENU) mutagenesis (7). To collect, preserve and supply unique and invaluable rats for medical

I. Anegon (ed.), *Rat Genomics: Methods and Protocols,* Methods in Molecular Biology, vol. 597
DOI 10.1007/978-1-60327-389-3_22, © Humana Press, a part of Springer Science+Business Media, LLC 2010

and biological research at the international and local research communities, the national rat genetic resources have been established (3).

Techniques for in vitro and in vivo fertilization in the rat provide the genetic resources and its users with efficiency of restoring conserved sperm as well as stocked males, and also enable generation of fertilized ova and embryos for various purposes such as ENU-mutagenesis, DNA-microinjection and cryopreservation. The techniques in the rat therefore have progressively gained importance.

This chapter provides detailed description of the protocols for in vitro and in vivo fertilization in the rat. Namely, sperm collection, sperm cryopreservation, pre-incubation of sperm, and insemination (co-culture with sperm and oocytes) for in vitro fertilization (IVF), and intrauterine insemination for in vivo fertilization with fresh or frozen/thawed spermatozoa are provided.

## 2. Materials

### *2.1. Collection of Epididymal Sperm*

1. Modified R1ECM for fertilziation (F-mR1ECM) (*see* Table 22.1: 110 mM NaCl, 3.2 mM KCl, 0.5 mM $MgCl_2$, 2.0 mM $CaCl_2$, 25.0 mM $NaHCO_3$, 7.5 mM D-glucose, 0.5 mM sodium pyruvate, 10.0 mM sodium lactate, 0.1 mM glutamine, 2% (v/v) minimal essential medium (MEM) essential amino acid solution (50×; Gibco BRL, Grand Island, NY), 1% (v/v) MEM nonessential amino acid solution (100×; Gibco BRL) and 4 mg/mL of fatty acid free bovine serum albumin (BSA)) (8, 9).
2. For dissection of cauda epididymis: fine scissors, fine forceps, watchmaker's forceps, filter papers (about 5 cm×5 cm), a 26-gauge needle, a 35-mm plastic dish (Falcon 35-1008, Becton Dickinson Labware, Franklin Lakes, NJ) supplemented with 500 μL F-mR1ECM.
3. Fertile male rats (We use over 15 weeks of aged rats, although around 7–8 weeks aged rats are used as sperm donor in general).
4. For sperm quality evaluation: examination chamber on a warm-plate at 37°C and a hemocytometer.

### *2.2. Preparation of Recipient Females, and Vasectmized Males*

1. Light cycle-controlled rat room.
2. Recipient females (10–14 weeks age; we use Crlj: Wistar rats) for embryo transfer and intrauterine insemination.
3. Vasectomized males (15–30 weeks age, Wistar (any strain)).

**Table 22.1**
**Modified R1ECM media for rat sperm, oocytes and in vitro fertilization (IVF)[a]**

| Ingredients | Fw. | Product No.[b] | F-mR1ECM | C-mR1ECM |
|---|---|---|---|---|
| NaCl | 58.44 | | 110 mM | 76.7 mM |
| KCl | 74.55 | | 3.2 mM | 3.2 mM |
| $CaCl_2$ | | | 2.0 mM | 2.0 mM |
| $MgCl_2$ | 95.21 | | 0.5 mM | 0.5 mM |
| $NaHCO_3$ | 84.1 | | 25.0 mM | 25.0 mM |
| Sodium lactate | 72.06 | | 10.0 mM | 10.0 mM |
| Sodium pyruvate | 110.04 | | 0.5 mM | 0.5 mM |
| Glucose | 180.16 | | 7.5 mM | 7.5 mM |
| Glutamine | 146.1 | | 0.1 mM | 0.1 mM |
| Polyvinyl alcohol | Av Mol wt 30,000–70,000 | P8136 | – | 1.0 mg/mL |
| BSA[c] | | A6003-10G | 4.0 mg/mL | – |
| EAA[d] (Invitrogen) | | 11130-051 | 2% (v/v) | 2% (v/v) |
| NEAA[e] (Invitrogen) | | 11140-050 | 1% (v/v) | 1% (v/v) |
| Osmolarity | | | 310 mOsM | 246 mOsM |

[a]According to the methods described in **refs.** 8 and 9
[b]The product No of the ingredients except BSA and amino acid solutions is on Sigma Life Science (St. Louis, MO)
[c]BSA; albumin, from bovine serum, essentially fatty acid free, Sigma-Aldrich (St. Louis, MO)
[d]Minimal essential medium (MEM) amino acid solution (50×; Gibco BRL, Grand Island, NY)
[e]MEM nonessential amino acid solution (100×; Gibco BRL)

### 2.3. In Vitro Fertilization (IVF)

#### 2.3.1. Sperm Pre-incubation

1. F-mR1ECM (*see* Table 22.1). When using cryopreserved sperm, F-mR1ECM including 200 μM of 3-isobutyl-1-methyl-xanthin (IBMX) is superior (see Note **1**) (10).
2. Paraffin or mineral oil (*see* Note 2).
3. Oil-covered droplets of 100 μL F-mR1ECM in 35-mm dishes (Falcon 35-1008).
4. Incubator (37°C, 5% $CO_2$ in air).

#### 2.3.2. Superovulation

1. Immature female rats (4–5 weeks age; we use Crlj: Wistar rats).
2. Equine chorionic gonadotropin (eCG).
3. Human chorionic gonadotropin (hCG).
4. Physiological saline.
5. For injection: a 1-mL syringe and a 26-gauge needle.

*2.3.3. Collection of Cumulus Oocyte Complexes (COCs) and Insemination (Co-culture with Sperm and COCs)*

1. F-mR1ECM (*see* Table 22.1).
2. Paraffin or mineral oil (*see* Note 2).
3. For dissection of oviductal ampulla: fine scissors, fine forceps, and preparation needles.
4. For collection: sperm suspended 100 μL droplets of F-m-R1ECM in a 35-mm plastic dish (Falcon 35-1008) (*see* Subheading 22.2.3.1).
5. Incubator (37°C, 5% $CO_2$ in air).

*2.3.4. Culture of Putative Zygotes*

1. Modified R1ECM for culture (C-mR1ECM) (*see* Table 22.1: 76.7 mM NaCl, 3.2 mM KCl, 0.5 mM $MgCl_2$, 2.0 mM $CaCl_2$, 25.0 mM $NaHCO_3$, 7.5 mM D-glucose, 0.5 mM sodium pyruvate, 10.0 mM sodium lactate, 0.1 mM glutamine, 2% (v/v) minimal essential medium (MEM) essential amino acid solution (50×; Gibco BRL), 1% (v/v) MEM non-essential amino acid solution (100×; Gibco BRL) and 1.0 mg/mL of polyvinyl alcohol).
2. Paraffin or mineral oil (*see* Note 2).
3. Oil-coverd droplets of 100 μL C-mR1ECM supplemented with 0.1% (w/v) hyaluronidase in a 35-mm plastic dish (Falcon 35-1008).
4. Incubator (37°C, 5% $CO_2$ in air).

*2.3.5. Embryo Transfer of Zygotes (IVF embryos)*

1. The zygotes (IVF embryos) in droplets of C-mR1ECM) (*see* Table 22.1).
2. Stereo microscopy with transmitted light base.
3. Anesthetic.
4. Seventy % Ethanol in squeeze bottle.
5. Surgical apparatus; scissors, forceps.
6. Sterile gauze.
7. Sterile Pasteur pipettes with cotton plug for embryo transfer capillary.
8. Mouthpiece.
9. Suture thread (silk, No. 2).
10. Clips for suture (Michel suture clips; 7.5 mm × 1.5 mm, Harvard Apparatus, MA, US).

### 2.4. In Vivo Fertilization: Intrauterine Insemination

*2.4.1. Sperm Freezing Medium*

1. Components; Lactose, trishydroxymethylaminomethane (Tris), penicillin G, streptomycin sulfate, Orvus ES Paste (OEP; Equex STM™, Nova Chemical Sales Inc., Scituate, MA; *see* Note 3), chicken eggs, sterile distilled water or extra-pure water, sterile filter paper (about 15 cm × 15 cm).
2. Preparation of primary sperm freezing medium (SFM-1):

2.1. Lactose (10.4 g) is dissolved with water, and the volume is adjusted to 100 mL.

2.2. One chicken egg is washed well, and disinfected with 70% alcohol. The egg-shell is broken carefully and the egg-yolk with small amount of egg-white is placed on the sterile filter paper. Then the egg-white is removed by rolling the egg-yolk. The surface membrane of the egg-yolk is ruptured, and the inner yolk is aspirated into a sterile syringe. Thirty mL of the inner yolk is added to the lactose solution (*see* item 2.1).

2.3. Penicillin G (1,000 unit/mL) and streptomycin sulfate (1 mg/mL) are added to the egg-yolk and lactose solution (*see* item 2.2).

2.4. The pH of the egg-yolk and lactose solution is measured using pH meter or BTB test paper. In general, the pH of the solution shows around 7.8.

2.5. The pH is adjusted to 7.4 by dropping of 10% Tris (v/v) solution.

2.6. The solution adjusted pH 7.4 is aliquoted to sterile 50-mL centrifuge tubes, and centrifuged at 1,250–1,750×$g$ (around 3,000 rpm) for 15 min. The upper layer of the solution is recovered into sterile 15-mL tubes. After centrifugation, when a thin membrane is formed in the top of the solution, the membrane is removed using a sterile needle.

2.7. The recovered solution is used as SFM-1, and stored at below –20°C.

3. Preparation of secondary sperm freezing medium (SFM-2):

3.1. Using 1-mL syringe with needle, OEP (*see* **Note 3**) is added to SFM-1 at 1.4% (v/v), and the solution is mixed well using a sterile pipette.

3.2. The solution included OEP is aliquoted into sterile 15-mL tubes as SFM-2, and stored at below –20°C.

*2.4.2. Freezing of Epididymal Sperm*

1 Sperm freezing medium 1 and 2 (SFM-1 and SFM-2) (*see* Subheading 22.2.4.1) (11).

2 For dissection of cauda epididymis: fine scissors, fine forceps, watchmaker's forceps, filter papers (about 5 cm×5 cm), a 26-gauge needle, a 35-mm plastic dish (Falcon 35-1008) supplemented with 1.5 mL of SFM-1.

3 For sperm quality evaluation: examination chamber on a warm-plate at 37°C and a hemocytometer (*see* Subheading 22.2.1).

4 Incubator (15°C in air).

5 Refrigerator (5°C in air).

6 0.25-mL palastic straws for bull semen.

7 Programmable freezer.

8 Liquid nitrogen ($LN_2$) tank.

*2.4.3. Intrauterine Insemination*

1. Cryopreserved sperm (*see* Subheading 22.2.4.2).
2. Water bath (37°C).
3. Straw cutter (Fujihira Industry, Japan; Cryo Bio System/ IMV, France).
4. Anesthetic.
5. Surgical apparatus; scissors, forceps.
6. Sterile 1.5-mL tubes.
7. Seventy % Ethanol in squeeze bottle.
8. Sterile gauze.
9. Sterile 21-gauge needles.
10. Sterile Pasteur pipettes with cotton plug for insemination.
11. Mouthpiece.
12. Suture thread (silk, No 2).
13. Clips for suture (Michel suture clips; 7.5 mm×1.5 mm, Harvard Apparatus).

## 3. Methods

### *3.1. Collection of Epididymal Sperm*

1. Males as sperm donors are anesthetized appropriately, and sacrificed by cervical dislocation.
2. After disinfection, the abdomen is opened using scissors and forceps, then the bilateral testes and epididymides are excised.
3. The caudal epididymides are placed on sterile filter paper (about 5 cm×5 cm), and its fat adhered to the caudal region are removed thoroughly and carefully.
4. Cauda epididymides are excised from mature males under room temperature (22–24°C), and are placed on sterile filter paper (about 5 cm×5 cm), and its fat adhered the caudal region are removed thoroughly and carefully.
5. Both the caudal epididymides are sticked by a 26-gauge needle, a few droplets of sperm mass is collected and introduced into a 100 μL-drop of F-mR1ECM (Table 22.1).
6. After a few minutes, sperm motility and concentration of the drop of F-mR1ECM are microscopically assessed.

7. Sperm suspension was transferred into another 200 μL-drop covered with oil for sperm pre-incubation to adjust to $5 \times 10^5$ sperm/mL.

### 3.2. Preparation of Recipient Females and Vasectmized Males

1. A mature female confirmed to be proestrus by smear test (nucleated epithelia cells are noted) during the light-on period (in the morning) is selected as a candidate of recipient female (for embryo transfer of IVF embryos or intrauterine insemination).
2. The female is caged together with a vasectomized male that has copulation ability 2 h before the dark period (2 h before light-off) to induce pseudo-pregnancy.
3. On Next morning (for embryo transfer of IVF embryos) or at 2–5 h after light-off (for intrauterine insemination), vaginal plug(s) is (are) checked to confirm mating with the vasectomized male.

### 3.3. IVF

#### 3.3.1. Sperm Pre-incubation

1. Spermatozoa are collected from cauda epididymides and transfered into another 200 mL-drop covered with oil for sperm pre-incubation to adjust to $5 \times 10^5$ sperm/mL (*see* Subheading 22.2.1).
2. The sperm suspension in a plastic dish is cultured for 5 h at 37°C under 5% $CO_2$ in air to induce sperm capacitation in vitro (*see* Note 1).

#### 3.3.2. Superovulation

Animals are hosed with free access of water and standard chow under a 12-h day/night cycle, most often comprising a 6 a.m. to 6 p.m. light period. We have good experience using 4–5 week-age females of Crlj: Wistar and Slc: Sprague-Dawley for superovulation (12).

1. Preparations of hormones: lyophilized eCG and hCG were diluted to a final concentration of 166.7 inter unit (IU)/mL by physiological saline.
2. Three days before IVF, eCG of 300 IU/kg (body weight: BW) is intraperitoneally injected into the females.
3. After 48 h of the eCG injection (300 IU/kg (BW)), hCG is injected in the same way.

#### 3.3.3. Collection of Cumulus Oocyte Complexes (COCs) and Insemination (Co-culture with Sperm and COCs)

1. Twelve to fourteen h after the hCG injection, the oviducts of the donors are excised, and placed on sterile filter paper (about 5 cm × 5 cm).
2. The fat adhered to the oviducts are removed thoroughly and carefully. The oviducts are placed into oil-covered F-mR1ECM drop (*see* Table 22.1, Note 4) in the dish for sperm pre-incubation, and moved to near the drop.

3. The ampullae are broken using a 26-gauge needle and fine forceps under stereoscopic microscope. When the ampullae are broken, cumulus-oocyte-complexes (COCs) are released.
4. The COCs are introduced into the drop containing sperm, by a 26-gauge needle, and then co-cultured for 10 h at 37°C under 5% $CO_2$ in air.

#### 3.3.4. Culture of Putative Zygotes

1. After co-culture for 10 h, putative zygotes are transferred to C-mR1ECM (Table 22.1) supplemented with 0.1% hyaluronidase for 1 min.
2. The zygotes are washed three times in C-mR1ECM. The putative zygotes are transferred into oil-covered 100 μL droplets of C-mR1ECM in a 35-mm plastic dish (Falcon 35-1008), and cultured at 37°C under 5% $CO_2$ in air.

#### 3.3.5. Embryo Transfer of Zygotes (IVF Embryos)

Only the general techniques for embryo transfer are described in this chapter. The putative zygotes (IVF embryos) are transferred into both oviducts of the recipient on 0.5-day post coitum. The recipient is naturally mated with a vasectomized male to induce pseudo-pregnancy.

1. Anesthetize the recipient female for embryo transfer.
2. The zygotes (IVF embryos) are divided to two groups in droplets of C-mR1ECM.
3. The female is retained in the prone position, and the coat is removed using hair-clipping scissors in the bilateral lateral abdominal regions (about 3 cm × 3 cm) corresponding to the positions of the ovaries located caudal to the last costales.
4. The clipped surgical areas are disinfected with 70% ethanol.
5. A small incision (about 1 cm) is cut in the skin of the surgical area.
6. The ovary and oviduct are carefully pulled out by pinching the ovarian fat tissue with forceps, and retained on sterile gauze.
7. Pinching the ovarian bursa with fine forceps, the membrane is insiced along equatorial plane of the bursa (*see* Note 5).
8. The capillary already containing the zygotes (IVF embryos) is carefully introduced into infundibulum (the top end of oviducts).
9. The zygotes (IVF embryos) are expelled from the capillary in the infundibulum.
10. The ovary and oviduct are carefully returned to abdomen.
11. The muscle tissue is sutured corresponding to the size of the incision, and then the skin closed with auto clips or suture. The remaining half zygotes (IVF embryos) are transferred into the opposite oviduct in the same way.

### 3.4. In Vivo Fertilization: Intrauterine Insemination

#### 3.4.1. Freezing of Epididymal Sperm

1. Before sperm collection, SFM-1 is warmed at room temperature (22–24°C), and SFM-2 is also kept at 5°C.
2. Cauda epididymides are collected (*see* Subheading 22.2.1).
3. Both the caudal epididymides are put into 1.5 mL of SFM-1 (*see* Subheading 22.2.2) in a 35-mm plastic dish (Falcon 35-1008), then cut about 15 times in each using minisize scissors, and kept for 5 min (sampling time) under room temperature.
4. After the cutting of caudal epididymides, sperm motility and concentration are micrscopically checked, and recorded.
5. After the sampling time, the dish contained sperm samples are transferred into a 15°C incubator for 30 min.
6. The dish is moved into a refrigerator or incubator at 5°C, and kept for 30 min.
7. After cooling to 5°C, 2.0 mL of the cooled SFM-2 is gently dropped into the sperm sample in SFM-1 using a micropipette.
8. The sperm suspension is loaded into 0.25-mL plastic straws, and exposed to vapor of $LN_2$ around −150°C (at about 4 cm above the level of $LN_2$) for 15 min (pre-freezing).
9. The straws contained sperm are then plunged into $LN_2$, and stored at −196°C.

#### 3.4.2. Intrauterine Insemination

1. For thawing cryopreserved sperm, the straw is immediately immersed in 37°C water for 15 s.
2. The surface of thawed straw is completely wiped off with tissue paper.
3. The straw is cut using straw cutter for bovine artificial insemination, the sperm suspension is transferred into 1.0 mL of F-mR1ECM (Table 22.1) in a 1.5-mL tube kept at 37°C.
4. Sperm sample in the tube is kept at 37°C, and motility of the sample is checked as mentioned above (*see* Subheading 22.2.1 and Note 6).
5. After confirming vaginal plugs (mating), the female is appropriately anesthetized.
6. The female is retained in the prone position, and the coat is removed using hair-clipping scissors in the bilateral lateral abdominal regions (about 3 cm × 3 cm) corresponding to the positions of the ovaries located caudal to the last costales.
7. The clipped surgical areas are disinfected with 70% ethanol.
8. A small incision (about 1 cm) is cut in the skin of the surgical area.
9. The ovary and upper uterine horn are carefully pulled out by pinching the ovarian fat tissue with forceps, and retained on sterile gauze.

10. In the bilateral uterine horns, a 21-gauge needle is inserted into the uterine wall at a site 1.0–1.5 cm lower (vaginal side) from the oviduct-uterine junction to make a small hole penetrating into uterine lumen.
11. On bilateral sides, about 50 μL of sperm suspension which is contained $1–5 \times 10^6$ sperm and prepared beforehand is aspirated into a injection pipette made from a Pasteur pipette, using a mouth-piece (the degree of inhalation that suck 50 μL is to be investigated beforehand when washing the pipette with the sperm suspension).
12. The pipette tip is precisely inserted into the uterine lumen via the small hole, and the sperm suspension is gently injected into the uterine lumen.
13. After injection, the muscle tissue is sutured corresponding to the size of the incision, and then the skin closed with auto clips or suture. This insemination is performed on the opposite uterine horn in the same way.

## 4. Notes

1. When using cryopreserved (frozen/thawed) sperm, F-mR1ECM including 200 μM of 3-isobutyl-1-methylxanthin (IBMX) is suitable to induce in vitro capacitation of cryopreserved sperm. Recently we found that treatment with IBMX dramatically increased cAMP and tyrosine phosphorylation levels in frozen/thawed rat sperm. When the IBMX-treated frozen/thawed sperm was used for IVF, the proportions of pronuclear and blastocyst formation were significantly higher than those of frozen/thawed sperm treated without IBMX. The IVF embryos can develop to term at a high success rate equivalent to the rate obtained with IVF using fresh rat sperm (10).
2. We recommend high quality of paraffin or mineral oil (mouse embryo tested) for covering micro-drops of medium, because of toxic contamination or deterioration of oil quality. If you use standard quality oil, before use you had better check the toxicity of oil using mouse embryo culture. When using high quality oil, mouse embryos show more than 70% as blastocyst formation rate (13).
3. Before adding OEP into SFM-2, when the OEP does not look transparent, the container of OEP is kept in about 50°C for 30 min to remove air bubbles. After this treatment, the OEP shows transparent amber.

4. There is a genetic difference between Wistar and Sprague-Dawley rats in fertilization rates of IVF. In Sprague-Dawley rats, the F-mR1ECM containing 140 mM NaCl (360 mOsm) is superior to that containing 110 mM NaCl (310 mOsm) in fertilization rates during insemination (co-cultuture with sperm and oocytes) (14).
5. When making an incision of the bursa, Noyes scissors is preferable. It is key for embryo transfer into oviducts to diminish bleeding in the rat.
6. Even if sperm motility of inseminated sample showed less than 5%, there is a possibility that the recipient become pregnant through intrauterine insemination on the basis of our previous data. In the study (15), we obtained nine pups derived from sperm sample whose motility showed 2%, from three pregnancies of five inseminated recipients.

## Acknowledgments

We also thank all members in Laboratory of Animal Reproduction, Azabu University. This work was supported in part by the Promotion and Mutual Aid Corporation for Private School of Japan and a Matching Fund Subsidy for Private Universities to N. K.

### References

1. Moreno C, Jacob HJ (2006) Interating biology with rat genomic tools. In: Suckow MA, Weisbroth SH, Franklin CL (eds) The laboratory rat. Elsevier Academic Press, Burlington, MA, pp 679–692
2. Gibbs RA, Weinstock GM, Metzker ML, Muzny DM, Sodergren EJ, Scherer S, Rat Genome Sequencing Project Consortium et al (2004) Genome sequence of the Brown Norway rat yields insights into mammalian evolution. Nature 428:475–476
3. Aitman TJ, Critser JK, Cuppen E, Dominiczak A, Fernandez-Suarez XM, Flint J et al (2008) Progress and prospects in rat genetics: a community view. Nat Genet 40:516–522
4. Hirabayashi M (2008) Technical development for production of gene-modified laboratory rats. J Reprod Dev 54:95–99
5. Kato M, Ishikawa A, Kaneko R, Yagi T, Hochi S, Hirabayashi M (2004) Production of transgenic rats by ooplasmic injection of spermatogenic cells exposed to exogenous DNA: a preliminary study. Mol Reprod Dev 69:153–158
6. Dann CT, Garbers DL (2008) Production of knockdown rats by lentivial transduction of embryos with short hairpin RNA transgenes. Methods Mol Biol 450:193–209
7. Mashimo T, Yanagihara K, Tokuda S, Voigt B, Takizawa A, Nakajima R et al (2008) An ENU-induced mutant archive for gene targeting in rats. Nat Genet 40:514–515
8. Miyoshi K, Kono T, Niwa K (1997) Stage-dependent development of rat 1-cell embryos in a chemically defined medium after fertilization in vivo and in vitro. Biol Reprod 56:180–185
9. Oh SH, Miyoshi K, Funahashi H (1998) Rat oocytes fertilized in modified rat 1-cell embryo culture medium containing a high sodium chloride concentration and bovine serum albumin maintain developmental ability to the blastocyst stage. Biol Reprod 59: 884–889
10. Seita Y, Sugio S, Ito J, Kashiwazaki N (2009) Generation of live rats produced by in vitro fertilization using cryopreserved spermatozoa. Biol Reprod 80:503–510

11. Nakatsukasa E, Inomata T, Ikeda T, Shino M, Kashiwazaki N (2001) Generation of live rat offspring by intrauterine insemination with epididymal spermatozoa cryopreserved at −196°C. Reproduction 122:463–467
12. Sano D, Yamamoto Y, Samejima T, Seita Y, Inomata T, Ito J et al (2009) A combined treatment with ethanol and 6-dimethylaminopurine is effective for the activation and further embryonic development of oocytes from Sprague-Dawley and Wistar rats. Zygote 17:29–36
13. Otsuki J, Nagai Y, Chiba K (2007) Peroxidation of mineral oil used in droplet culture is detrimental to fertilization and embryo development. Fertil Steril 88:741–743
14. Jiang J-Y, Tsang BK (2004) Optimal conditions for successful in vitro fertilization and subsequent embryonic development in Sprague-Dawley rats. Biol Reprod 71: 1974–1979
15. Nakatsukasa E, Kashiwazaki N, Takizawa A, Shino M, Kitada K, Serikawa T et al (2003) Cryopreservation of spermatozoa from closed colonies, and inbred, spontaneous mutant, and transgenic strains of rats. Comp Med 53:639–641

# Chapter 23

# Rat Strain Repositories

## Birger Voigt

## Abstract

More than 500 inbred rat strains have been developed during the past 100 years for a wide range of biomedical applications. In addition to these traditionally bred strains, many induced mutants and several thousand mutagenized sperm samples have recently been generated. At present this huge number of strains is mainly managed by two rat resource centers, the National Bio Resource Project for the Rat in Japan (NBRP-Rat) and the US based Rat Resource and Research Center (RRRC). These resource centers not only collect, maintain and distribute rat strains as animals or cryopreserved embryos and spermatozoa, but also perform additional tasks such as phenotypic and genetic characterization as well as microbiological cleaning. Furthermore, they support researchers through informative databases in the selection of rat strains for specific research purposes. These global rat resource centers are essential for successful and sustainable research using the rat as a model species.

**Key words:** Rat, *Rattus norvegicus*, Repository, ENU, Phenome, Genotyping, Resource center

## 1. Introduction

The breeding and commercial supply of rat strains began long before the development of the first rat inbred strain for scientific purposes by King in 1909 (1). The first known manual on the breeding and maintenance of fancy rats was published in Japan in 1787 (Fig. 23.1). There was no application of rats in science at that time but various colored animals could be obtained from pet shops, which were popular in urban communities in Japan during the Edo period.

The history of the laboratory rat began approximately 150 years ago. Rat strains have been maintained for a wide range of research applications for about 100 years on the basis of individual efforts by many scientists across several continents. Such effort is intrinsically inefficient and susceptible to unexpected changes in funding and local interests. The NIH Rat Model

I. Anegon (ed.), *Rat Genomics: Methods and Protocols,* Methods in Molecular Biology, vol. 597
DOI 10.1007/978-1-60327-389-3_23, © Humana Press, a part of Springer Science+Business Media, LLC 2010

Fig. 23.1. Illustrations from *Chingansodategusa*, a Japanese guidebook on the breeding of fancy rats and mice for color *circa* 1787.

Repository Workshop was held in August 1998, and 58 scientists from the USA and abroad discussed the needs, use, opportunities, and parameters for the optimal importation, standardization, maintenance, and distribution of genetically defined rat strains. This group of internationally recognized scientists strongly encouraged the NIH to fund a National Rat Genetics Resource Center. As a result, the Rat Resource and Research Center (RRRC) was established in 2001. A similar approach on a larger scale was undertaken in Japan in 2002 with the National Bio Resource Project for the Rat, (NBRP-Rat), to address the need for collecting, preserving, and supplying unique rat strains that resemble human diseases or are of other value for biomedical research and which have mainly been developed in Japan (2).

Besides the two major rat repositories, RRRC (http://www.nrrrc.missouri.edu/) and NBRP-Rat (http://www.anim.med.kyoto-u.ac.jp/NBR/), two smaller repositories were established in Germany (http://www.mh-hannover.de/2652.html) and the Czech Republic (http://www.euratools.eu). The latter two do not presently supply rat strains or are not yet operational.

The general workflow of the repositories consists of the collection of rat strains, the maintenance of live stock under specific pathogen-free conditions, the cryopreservation of embryos and sperm, the preparation and maintenance of a publicly accessible database on the deposited rat strains, and the global supply of these strains. NBRP-Rat and RRRC also perform additional tasks to promote the rat as an animal model for physiology, pharmacology, toxicology, nutrition, behavior, and other life sciences.

## 2. Repositories

### *2.1. The National Bio Resource Project for the Rat in Japan (NBRP-Rat)*

The National Bio Resource Project for the Rat in Japan was established in 2002 as part of a national effort to preserve strategically important biogenetic resources. NBRP not only comprises the species "rat" but also other animals, plants, and microorganisms

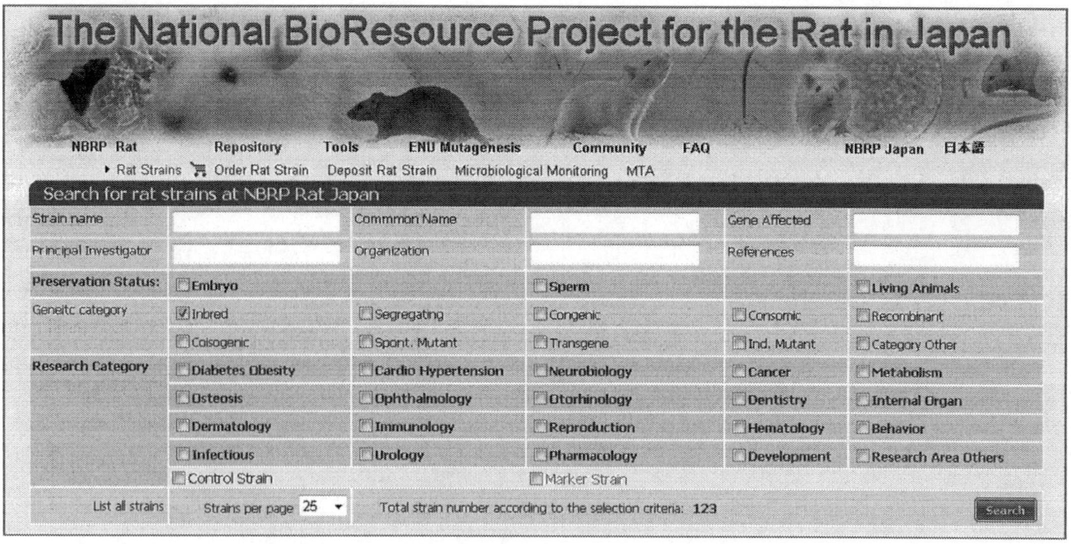

Fig. 23.2. Search interface of the NBRP-Rat Strain Database. Several categories like "General strain information," "Preservation status," "Genetic category," or "Research Category" can be combined. The search fields and checkboxes in the database queries are "AND" -linked, which supports the selection of specific rat strains.

(http://www.nbrp.jp). By October 2008, 562 rat strains including 123 inbred strains, 83 mutant strains, 195 congenic strains, 4 consomic strains, 40 recombinant inbred strains and 16 ENU-mutageneized strains, were deposited at the NBRP-Rat (Fig. 23.2).

In contrast to other repositories, NBRP-Rat is also characterizing its deposited strains. Original data that were generated through the Japanese Rat Phenome Project, which was launched shortly after the establishment of the repository, is one of the major efforts of this resource center (3). Six male and female rats are phenotypically characterized with respect to 109 parameters (Table 23.1).

This systematic characterization is the first large-scale strain survey of many phenotypes of biological importance in the rat. A major feature of this phenome project is the development of phenotypic "Strain Ranking," which allows for visual data scoring, shows the biological range of various phenotypic parameters, and reveals normal and abnormal values for various rat strains (Fig. 23.3). Strain ranking at NBRP-Rat affords the unique opportunity to easily and simultaneously compare phenotypic values for multiple rat strains. Many strains have been developed for disease-based phenotypes and used as models of complex human diseases, such as hypertension, diabetes, and neurological disorders. The strain ranking of this phenome project could replicate well-known characteristics such as high blood pressure, high glucose levels, and behavioral anomalies, implying the reliability and reproducibility of the measurements. In addition, it could reveal unique and unknown characteristics in even well-characterized strains.

**Table 23.1**
**Phenotypic measurements of the NBRP-Rat phenome project**

| Category | Characterization | Measurements |
|---|---|---|
| Functional Observational Battery (FOB) | Home cage observations (6) | Body position, Respiration, Clonic and tonic involuntary movement, Vocalization, Palpebral closure |
| | Hand-held observations (8) | Reactivity, Handling, Palpebral closure, Lacrimation, Salivation, Piloerection, Skin color, Others |
| | Open field activity (10) | Rearing, Clonic and tonic involuntary movement, Gait, Movements, Arousal, Occurrence of stereotype, Abnormal behavior, Defecations, Urinations |
| | Stimulus response (8) | Approach response, Touch, Eyelid reflex, Pinna, Sound, Tail pinch, Pupillary, Righting |
| | Nervous and muscle observations (5) | Abdominal and limb tone, Forelimb and hindlimb grip strength (kgf), Landing foot splay (mm) |
| Behavioral studies | Locomotor activity (4) | 0–10, 10–20, 20–30 min, and total (0–30 min) |
| | Passive avoidance (2) | Training (s), Retention (s) |
| Blood Pressure | Blood pressure (2) | Systolic Blood Pressure (mmHg), Heart Rate (1/min), |
| | Body temperature (1) | Body Temperature (°C) |
| Biochemical Blood Tests | Biochemistry (16) | GOT, GPT, ALP, TP, ALB, A/G, GLU, T-CHO, HDL, LDL, TG, T-BIL, BUN, CRE, IP, Ca |
| | Plasma electrolytes (3) | $Na^+$, $K^+$, $Cl^-$ (mEq/l) |
| Hematology | Blood counts (8) | RBC, Hb, Ht, MCV, MCH, MCHC, WBC, Platelet, |
| | White blood cells (7) | Bas. Eos. St. Seg. Lym. Mon. Other (%) |
| | Bleeding value (2) | PT (s), APTT (s) |
| Urine parameters | Urine (2) | Volume (ml) |
| | Urinary electrolytes (6) | $Na^+$, $K^+$, $Cl^-$ (mEq/l) |
| Anatomy | Body weight (3) | 5, 6, 10 weeks (g) |
| | Organ weights (16) | Brain, Heart, Lung, Liver, Kidneys, Adrenals, Spleen, Testes (g) |

The numbers for each measurement taken are shown in parenthesis

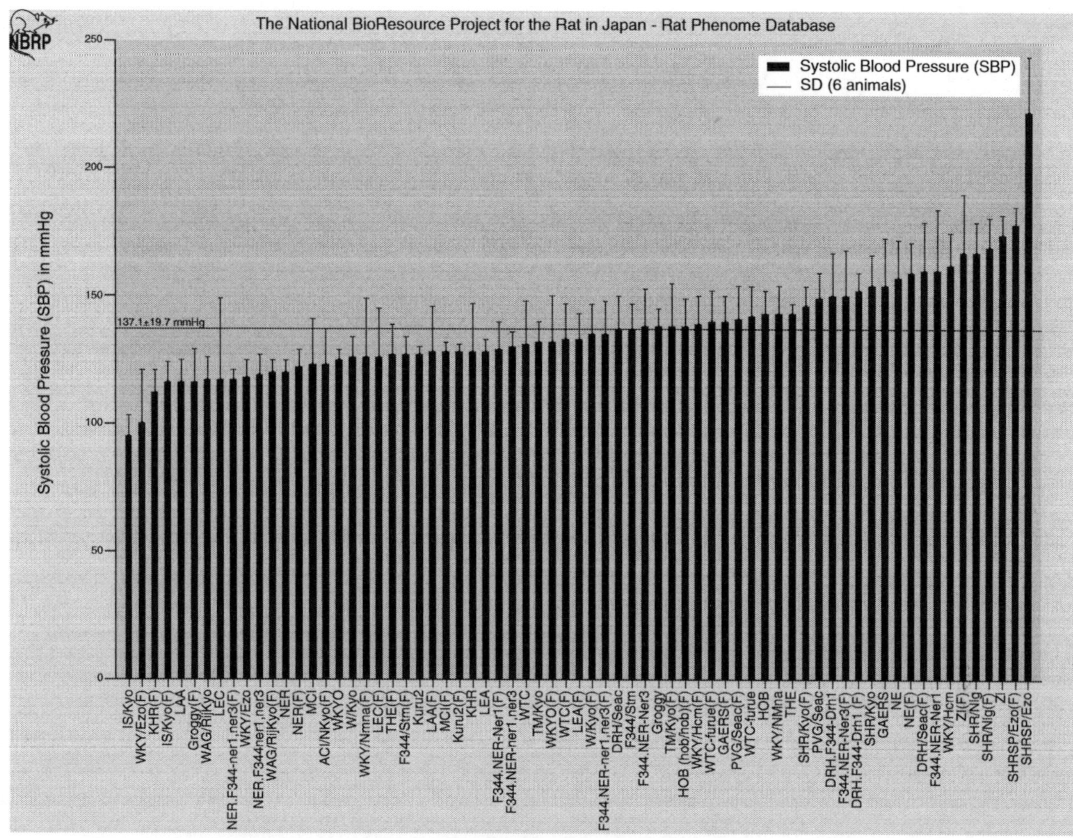


Fig. 23.3. NBRP-Rat Phenome Project. Strain ranking for Systolic Blood Pressure for selected male and female pairs of rat strains. Data on more than 160 strains are available in this phenome database.

In addition to phenotyping, NBRP-Rat is also genotyping selected rat strains and has developed a pedigree that integrates all major rat strains as well as several mutants and even wild rats from different continents (http://www.anim.med.kyoto-u.ac.jp/NBR/phylo.aspx). The genotyping comprises 357 microsatellite markers and includes animals from some 180 different origins (4). Furthermore, a charting tool (5) utilizes these data to draw strain specific phylogenetic charts that refer to a freely selectable rat strain (Fig. 23.4)

Several inbred strains, including the set of 34 LEXF/FXLE recombinant inbred (RI) strains, have been typed with more than 20,000 SNPs in collaboration with the STAR Consortium (6). This is especially beneficial for experiments involving the above mentioned RI strains, since it enormously simplifies their application in the dissection of complex traits and reduces the experimental effort that is required for such studies because of the phenotyping of selected RI strains. NBRP-Rat also provides another tool in this context: the BAC libraries and its sequence data of the RI founder strains, LE/Stm and F344/Stm. The genome browser and ordering information for these BAC clones are publicly

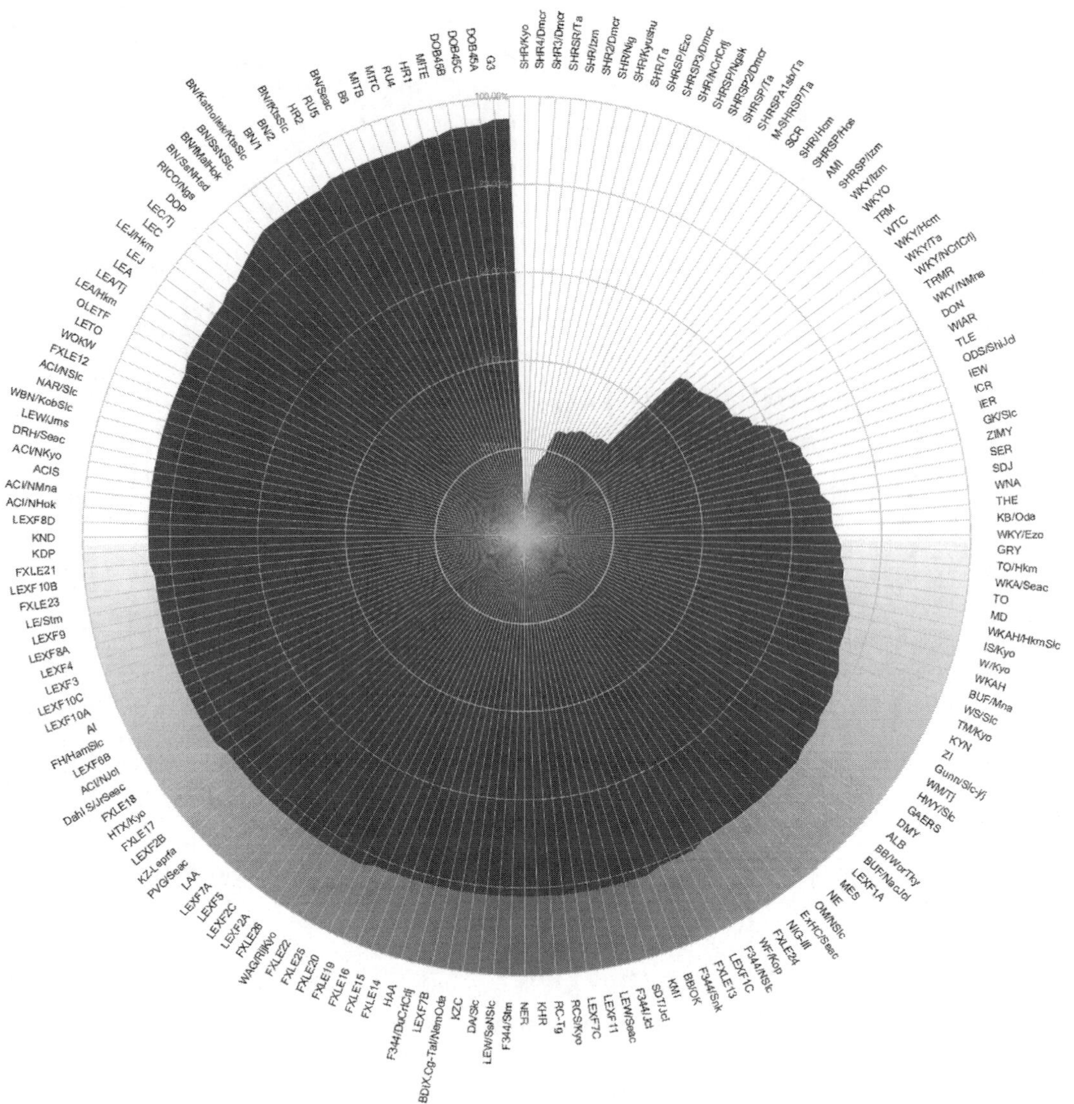


Fig. 23.4. Pedigree charting tool. The genetic differences, based on 357 microsatellite markers, are shown in each chart. The interface allows individual selection of reference strains and displays the "genetic distance" to the selected strain as a percentage (http://www.anim.med.kyoto-u.ac.jp/nbr/pedigree/).

accessible from the NBRP website at http://www.anim.med.kyoto-u.ac.jp/NBR/tools.aspx.

NBRP-Rat has also published a DVD on cryopreservation and assisted reproduction techniques, which is freely available for those in the research community who are interested. Since the reliability of these techniques has been improved using the original research of this repository and collaborative partner institutes, it was possible to integrate the Kyoto University Rat Mutant Archive (KURMA) into NBRP-Rat (7). KURMA consists of a set

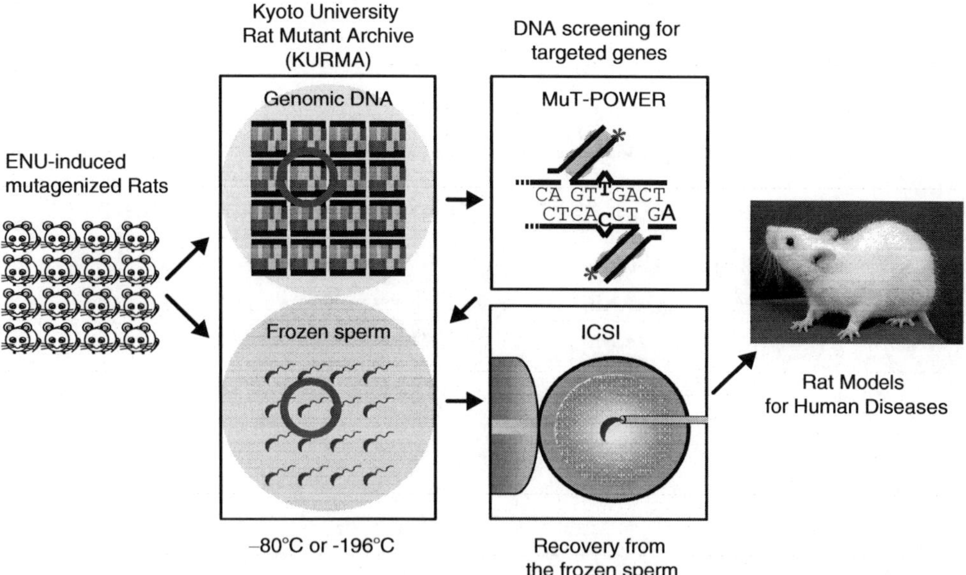


Fig. 23.5. Diagram of the generation of gene-targeted rat models of human diseases. ENU is injected intraperitoneally into 9 and 10 weeks old male F344/NSlc rats. They are then mated 10 weeks after injection. DNA and sperm of their offspring (G1) are stored; the DNA can be screened through a newly developed method (MuT-POWER) and affected sperm revitalized by Intracytoplasmic Sperm Injection (ICSI) to derive viable offspring from the G1 sperm.

of sperm and DNA samples taken from some 10,000 ENU mutagenized F344 G1 animals (Fig. 23.5).

Figure 23.5: Diagram of the generation of gene-targeted rat models of human diseases. ENU is injected intraperitoneally into 9 and 10 weeks old male F344/NSlc rats. They are then mated 10 weeks after injection. DNA and sperm of their offspring (G1) are stored; the DNA can be screened through a newly developed method (MuT-POWER) and affected sperm revitalized by Intracytoplasmic Sperm Injection (ICSI) to derive viable offspring from the G1 sperm.

### 2.2. Rat Resource and Research Center

The RRRC began the distribution of rat models in 2002 and continues to supply an increasing number of animals each year. Currently, 244 rat strains are available from this resource (Fig. 23.6).

Most recently, the RRRC has initiated the importation of transposon induced knock-out rat models. As with the NBRP-Rat repository, the goals of the RRRC are to actively recruit valuable rat models into the Center, increase the efficiency for archiving and reanimating rat lines, and continue fostering the international network of rat repositories. The RRRC is also committed to continue the distribution of high quality rats with confirmed genotypes and health status, make rat tissues and reagents available, and aggressively disseminate information about the rat models and services available to the worldwide biomedical community. In addition to its repository function, the RRRC

Search [ ] GO

|| Home || About RRRC || Contact Us ||

**Search the RRRC Rat Database for available strains**

Search | View all GFP rats | View All Records | Clear Form

| Select and enter your search criteria below to retrieve a list of available rat strains deposited at the RRRC. | | | | |
|---|---|---|---|---|
| **Strain Name** (contains) | | | **Donating Investigator** | |
| **Common Name** (contains) | | | **Donating Institution** | |
| **Gene** (contains) | | | **RRRC #** | |
| **Disposition Status** | Embryo | Sperm | Live Animals | Other |
| **Genetic Category** | Inbred | Segregating | Congenic | Consomic |
| | Recombinant | Coisogenic | Transgene | Spontaneous Mutant |
| | Induced Mutant | Outbred | Transposon Mutation | Other |
| **Research Uses** | Diabetes/Obesity | Cardio/Hypertension | Neurobiology | Onocology |
| | Metabolism | Osteosis | Ophthalmology | Otorhinology |
| | Dentistry | Internal Organ | Dermatology | Immunology |
| | Reproduction | Hematology | Behavior | Infectious |
| | Urology | Pharmacology | Development | Reporter Gene |
| | Transplantation | Alcoholism | Others | |
| **Records per page** | 10 | | | |

Search | View all GFP rats | View All Records | Clear Form

Fig. 23.6. Search interface of the RRRC available at http://www.nrrrc.missouri.edu/SearchStrainsParms.asp. Both NBRP-Rat and RRRC provide very similar search forms to simplify their use for the end-users. The search categories are widely homologized between both repositories.

performs research to improve the operation of the Center in areas including the following cryopreservation of rat embryos and spermatozoa, rat assisted reproductive technologies such as intra-cytoplasmic sperm injection, super-ovulation and embryo culture, infectious pathogen screening, and genotyping (8, 9). The RRRC also provides training and educational activities related to the use and development of rat models.

## 3. Remarks

Rat resource centers' major aim is the supply of bio materials to meet the demands of the research community. Progressing technologies require existing repositories to preserve and maintain not only animals or materials of traditionally derived rat strains but also genetically engineered rats. ENU or transposon mediated mutagenized archives with thousands of individual samples need to be stored and made available in the interest of biomedical research. Such repositories would ideally be managed across core centers in the USA, EU, and Japan to share resources and to standardize cryopreservation technologies.

## References


1. Lindsey JR (1979) Historical foundations. In: Baker HJ, Lindsey JR, Weisbroth SH (eds) The laboratory rat. Academic, New York, pp 1–36
2. Serikawa T (2004) Colourful history of Japan's rat resources. Nature 429:15
3. Mashimo T, Voigt B, Kuramoto T, Serikawa T (2005) Rat Phenome Project: the untapped potential of existing rat strains. J Appl Physiol 98:371–379
4. Mashimo T, Voigt B, Tsurumi T, Naoi K, Nakanishi S, Yamasaki K, Kuramoto T, Serikawa T (2006) A set of highly informative rat simple sequence length polymorphism (SSLP) markers and genetically defined rat strains. BMC Genet 7:19
5. Choose Your Rat (2005) Science 309:361
6. STAR Consortium, Saar K, Beck A, Bihoreau MT, Birney E, Brocklebank D, Chen Y, Cuppen E, Demonchy S, Dopazo J, Flicek P, Foglio M, Fujiyama A, Gut IG, Gauguier D, Guigo R, Guryev V, Heinig M, Hummel O, Jahn N, Klages S, Kren V, Kube M, Kuhl H, Kuramoto T, Kuroki Y, Lechner D, Lee YA, Lopez-Bigas N, Lathrop GM, Mashimo T, Medina I, Mott R, Patone G, Perrier-Cornet JA, Platzer M, Pravenec M, Reinhardt R, Sakaki Y, Schilhabel M, Schulz H, Serikawa T, Shikhagaie M, Tatsumoto S, Taudien S, Toyoda A, Voigt B, Zelenika D, Zimdahl H, Hubner N (2008) Nat Genet 40(5):560–566
7. Mashimo T, Yanagihara K, Tokuda S, Voigt B, Takizawa A, Nakajima R, Kato M, Hirabayashi M, Kuramoto T, Serikawa T (2008) Nat Genet 40(5):514–515
8. Woods EJ, Benson J, Agca Y, Critser JK (2004) Fundamental cryobiology of reproductive cells and tissues. Cryobiology 48:146–156
9. Agca Y, Critser JK (2005) Assisted Reproductive Technologies and Genetic Engineering in Rats. Suckow MA, Weisbroth SH, Franklin CL, second edition. The Laboratory Rat, Academic Press, pp 165–190

# Chapter 24

# Neurobehavioral Tests in Rat Models of Degenerative Brain Diseases

**Yvonne K. Urbach, Felix J. Bode, Huu Phuc Nguyen, Olaf Riess, and Stephan von Hörsten**

## Abstract

Each translational approach in medical research forces the establishment of neurobehavioral screening systems, dedicated to fill the gap between postgenomic generation of state-of-the-art animal models (i.e. transgenic rats) on the one hand and their added value for really predictive experimental preclinical therapy on the other. Owing to these developments in the field, neuroscientists are frequently challenged by the task of detecting discrete behavioral differences in rats. Systematic, comprehensive phenotyping covers these needs and represents a central part of the process. In this chapter, we provide an overview on theoretical issues related to comprehensive neurobehavioral phenotyping of rats and propose specific classical procedures, protocols (similar to the SHIRPA approach in mice), as well as techniques for repeated, intraindividual phenotyping. Neurological testing of rats, motorfunctional screening using the accelerod approach, emotional screening using the social interaction test of anxiety, and testing of sensorimotoric gating functions by prepulse inhibition of the startle response are provided in more detail. This description is completed by an outlook on most recent developments in the field dealing with automated, intra-home-cage technologies, allowing continuous screening in rats in various behavioral and physiological dimensions on an ethological basis.

**Key words:** Transgenic rat, Neurodegenerative diseases, Standardization, Phenotyping, Automated home-cage technology

## 1. Neurological Diseases in Rat Models

Aging-related neurological and neurodegenerative diseases are the culmination of many different genetic and environmental influences, and their modeling and treatment represent one of the most important challenges for mankind today. Neurodegenerative diseases broadly include many different disorders, such as Alzheimer's, Parkinson's, amyotrophic lateral sclerosis, multiple

I. Anegon (ed.), *Rat Genomics: Methods and Protocols,* Methods in Molecular Biology, vol. 597
DOI 10.1007/978-1-60327-389-3_24, © Humana Press, a part of Springer Science+Business Media, LLC 2010

sclerosis, Huntington's, Ataxia's, dementias with Lewy bodies, traumatic brain injury, epilepsy, and stroke.

Avenues for a better understanding of their pathogenesis, improved therapy, identification of biomarkers, and diagnosis are found in the development of highly predictive animal models, their comprehensive phenotyping/characterization, and their use in systematic screens of experimental therapeutics. This translational approach urgently forces the establishment of neurobehavioral screening systems, dedicated to fill the gap between postgenomic generation of state-of-the-art animal models (i.e. transgenic rats) on the one hand and their added value for really predictive experimental preclinical therapy on the other. *A priori* the rat provides several advantages as a model for human diseases over mice, as rats are bigger in size – facilitating imaging, surgery and tissue collection–, live longer, and exhibit a more complex behavioral phenotype (1–3). The fundamental organization of a rat brain is more similar to the human brain (4, 5) compared with mice, and several other differences such as metabolism and aging favor the species *Rattus norvegicus* with regard to modeling of human neurodegenerative disorders.

In contrast to lesioning or neurotoxic compounds, genetic models of human diseases provide the advantage of mirroring adequate pathomechanisms and showing progression. Consequently, transgenic rats have been developed, with the first examples being found in the rat expressing the mouse Ren-2 renin gene and developing severe hypertension (6) as a model for hypertension, the transgenic rat expressing HLA-B27 and human b2M (7) as a model for spondyloarthropathies, and the Huntington's disease rat representing the first transgenic rat model of a human neurodegenerative disorder (8).

Alzheimer's disease (AD) is characterized by loss of neurons and synapses in the cerebral cortex and certain subcortical regions. This loss results in gross atrophy of the affected regions, including degeneration in the temporal lobe and parietal lobe, and parts of the frontal cortex and cingulate gyrus. With regard to the other neurodegenerative disorders, a recently published transgenic rat model for Alzheimer's disease (9) is overexpressing transgenes with three mutations found in familial AD. Two mutations are detected in the amyloid precursor protein (APP) and one mutation in the presenilin-1 (PS1). This rat develops Aβ deposits in both diffuse and compact forms, as well as impaired spatial learning and memory, and therefore recapitulates AD-like amyloid pathology and cognitive impairment.

A loss of dopaminergic neurons in the substantia nigra (SN) and the existence of Lewy bodies are key features of Parkinson's disease (PD). "Classical" transgenic rat models (as being generated by pronucleus microinjections) for PD have not been published so fare though several groups are presently characterizing such tgPD rat models. In contrast to the other "classical" transgenic rat models, most transgenic rat models for PD are induced by topical viral vector-based gene-transfer into the SN and/or striatum (5, 10). Chemically induced rat models produced by dopamine depletion after the administration of 6-hydroxydopamine into the nucleus accumbens septi, the caudate nucleus (11–13), the medial forebrain (14–16), or the substantia nigra itself (17) are used to study the disease. Another rat model develops the disease because of administration of the Japanese encephalitis virus (18). A retinoid pesticide called Rotenone has recently been found to reproduce the nerve cell loss in the substantia nigra and the occurrence of Lewy bodies in the rat (19).

Amyotrophic lateral sclerosis (ALS) is a fatal neurodegenerative disease caused by selective motor neuron death (20, 21), the nerve cells in the central nervous system that control voluntary muscle movement. A widely used transgenic rat model for ALS is the cytosolic copper-zinc superoxide dismutase (SOD1) transgenic rat. Two mutations in the gene encoding for SOD1, G93A and H46R, have been associated with familial ALS (20, 21).

Multiple sclerosis (MS), also known as disseminated sclerosis or encephalomyelitis disseminata, is an autoimmune condition in which the immune system attacks the central nervous system, leading to demyelination. The experimental autoimmune encephalomyelitis (EAE) rat is a model for Multiple Sclerosis with two inbred strains being frequently used, Lewis (LEW) and Dark Agouti (DA).

Huntington's disease (HD) is a neurological disorder characterized by uncoordinated body movements and decline in some mental abilities, which affects several aspects of behavior. The cause of this disease is a CAG repeat expansion in the gene coding for the huntingtin protein (22). For Huntington's disease, neurotoxic animal models have focused on the striatal pathology and do not show disease progression, but in humans, the range of symptoms is much broader. A transgenic rat model of HD carries a truncated huntingtin cDNA fragment with 51 CAG repeats, under the control of the native rat huntingtin promoter (8).

Ataxias are characterized by lack of coordination of muscle movement and behavioral abnormalities as well as psychological symptoms. Ataxias are clinical manifestations implying parts of the nervous system that coordinate movement, such as the cerebellum. There are several possible causes for these patterns of neurological dysfunction (23). Some hereditary ataxias are autosomal dominantly inherited neurodegenerative diseases resulting from CAG/polyglutamine repeat expansions in different genes/proteins. For hereditary ataxias, two rat models, one transgenic rat model of spinocerebellar ataxia type 17 (SCA17) and one virus-based rat model of spinocerebellar ataxia type 3 (SCA3), exist. The transgenic rat SCA17 rat model carries a full-length human cDNA construct encoding for the mutant TATA-box binding protein (Riess et al., unpublished). The other one is the rat model for Machado–Joseph disease (MJD) or spinocerebellar ataxia type 3 (SCA3) (24). To study SCA3, ataxin-3 was overexpressed in the rat brain by using lentiviral vectors (LV). LV encoding human mutant ataxin-3 (Atx3-72Q) were stereotaxically injected into the substantia nigra, cortex, and striatum.

Epilepsy is a common chronic neurological disorder that is characterized by recurrent unprovoked seizures. These seizures result from excessive or synchronous neuronal activity in the brain. For epilepsy, several mutant rat models exist, for example, the genetically epilepsy-prone rat (25), the WAG/Rij rat created by Bachararch and colleagues at Glaxo in 1924, and the genetic absence epilepsy rat from Strasbourg (26). Furthermore, ***kindling models*** are available (27).

However, so far no transgenic rat models with an epilepsy-like symptomatology were published.

Stroke is the rapidly developing loss of brain functions, due to a disturbance in the blood vessels supplying blood to the brain. There are several models of stroke, which can be divided into two general categories: focal ischemia and global ischemia. These ischemic models have been used extensively to study the efficacy of neuroprotective agents (28).

In order to gain full advantage of this recently intensified generation of several transgenic rat models for human neurodegenerative disorders, sufficient and standardized (neurobehavioral) phenotyping procedures are essential.

## 2. The Problem of Standardized Behavioral Phenotyping

Nowadays, advances in transgenic technology more readily allow for linking the gap from genes to behavior. The human, rat, and mouse genome projects, coupled with an impressive array of molecular genetic technologies to manipulate expression of individual genes within the brain, have provided unprecedented opportunities for investigating the influence of genes on behavior. Consequently, the field of behavioral neurogenetics rapidly evolves and bridges a "post-genomic-gap" between genomics and phenomics (29). Owing to these developments in the field, neuroscientists are frequently challenged by the task of detecting discrete behavioral differences in mice and rats.

While there are considerable efforts to promote concepts and techniques for behavioral analysis of mice (30–46) and genetic analysis of complex traits (47, 48), there have also been major drawbacks, (46, 49–54) along with an intense and on-going conceptual discussion on the throughput of such screening systems (55–57). However, such efforts and discussions have not been undertaken for the rat as an experimental animal model.

This forbearance is not comprehensible to all intents and purposes because under many circumstances, the rat as a model for human diseases provides considerable advantage over mice and also most of the behavioral tests have initially been developed in rat. One possible explanation for the intensive use of mice is the much earlier and straightforward development of transgenic technologies in this species. However, only after more than 10 years the first transgenic rats were generated (6, 7, 58). Today, these techniques become more and more accessible for rats, being applied to CNS diseases (8, 59–62), and soon will provide a most interesting perspective for a parallel development of transgenic rat models along with their mouse counterpart. Thus, neurobehavioral assessment and phenotyping of rodents and in particular of rats need further development on a practical and theoretical basis (Fig. 24.1).

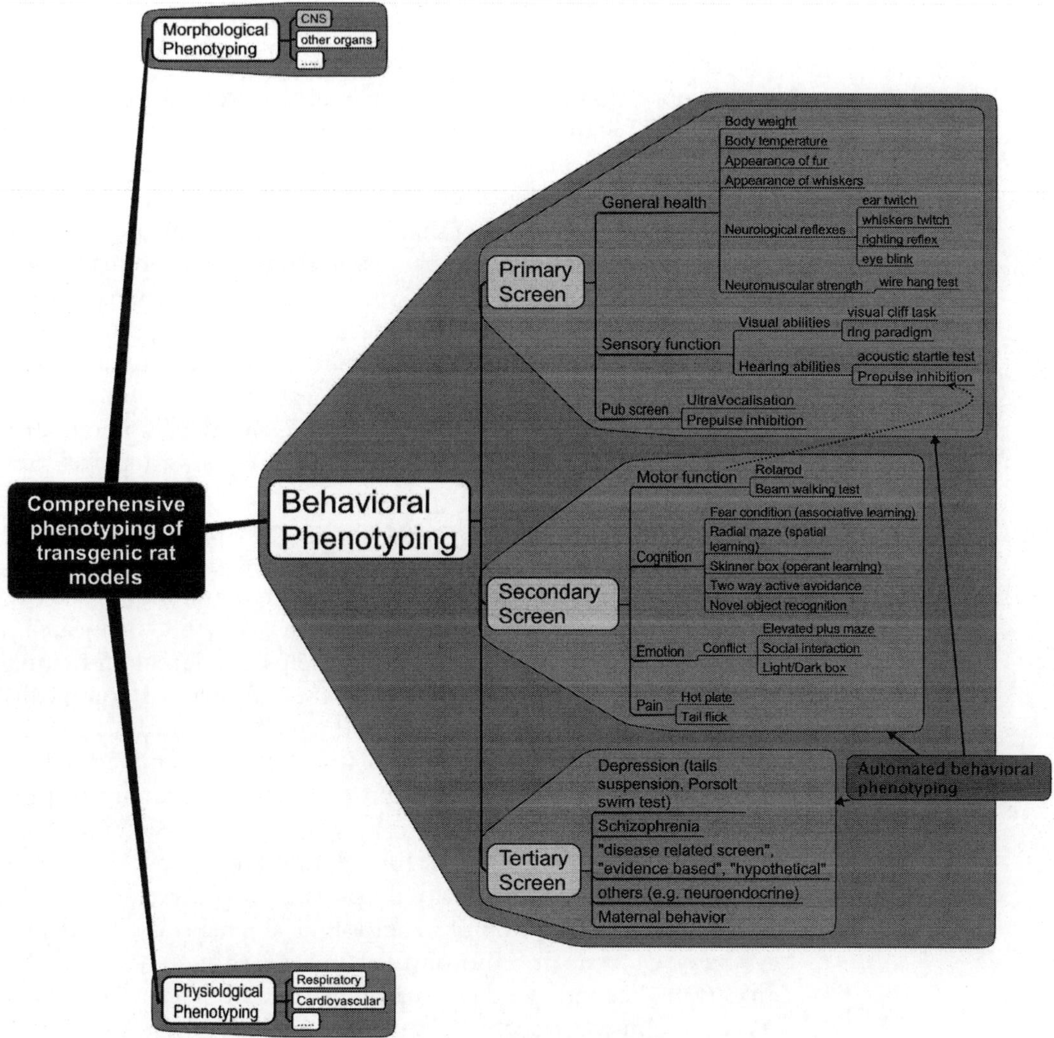


Fig. 24.1. Comprehensive phenotyping approach for transgenic rat models

As already can be learned from the field of mouse and rat phenotyping, all further efforts on behavioral assessment of laboratory animals should be done under certain directives in order to circumvent limitations in validity, reliability, and replicability across laboratories (49, 50), and throughput (55–57).

However, while the characterization of behavioral differences in rats has a long history, in contrast to mice neither a guideline for the systematic evaluation of an unknown phenotype in rats is available at present nor any attempt to increase the throughput of such phenotyping has been undertaken. Obviously, therefore, also no concepts and efforts on rat model specific data mining are available. These three interdependent needs (a) *systematic comprehensiveness of phenotyping*, (b) *high throughput*, and (c) *data mining/analysis* must urgently be fulfilled in the view of the upcoming transgene techniques for rat models, which are presently available only for mice.

*Ad (a) Systematic and comprehensive phenotyping as a secondary screening*:

The importance of a rational strategy in the design, procedure, and composition of different behavioral tasks represents to a large extent the "State-of-the-art" in classical phenotyping of mice nowadays (30, 34, 35, 44, 45, 50). The basic necessity for "systematic comprehensiveness" in the process of phenotyping derives from the simple fact that it is most of the time not possible to foresee what phenotypic changes a certain mutation may cause. The disruption of a single gene may lead to a cascade of events, e.g. there may be compensatory effects of other genes, the presence or absence of which may vary depending on the genetic background used.

Thus, in regard to the necessity for the development of a guideline for a comprehensive behavioral assessment of rats, several things must be learned and adopted from the field of mouse phenotyping. As has been suggested for phenotyping of mice, also in rats, a rational strategy in the design, procedure, and composition of different behavioral tasks minimizes the risk of false positive and false negative results (45, 63, 64). Such a multi-tiered strategy (33, 63) also helps to detect individual differences, which result from an increased level of complexity immanent within a living animal and which would remain undetected if only a hypothesis-driven test-battery is applied. These basic considerations in planning experiments, screening of general health and neurophysiologic functions as well as composition of groups of behavioral tasks suitable for a reliable detection of differences in various behavioral domains (35) also apply to rats (under the consideration of the ethological perspective) and span an array of tests for motor function, learning and memory, anxiety, nociception, and psychiatric-like conditions among others. As has recently been successfully documented (8, 38), it is very likely that the transfer of these theoretical and practical considerations will readily increase validity and reliability of any conclusion drawn from rat phenotyping work, as it has become increasingly clear that accurate interpretation of individual behavioral tests can only be achieved in the context of a thorough set of behavioral assessments, which are not used in the vast majority of publications in the field.

However, while this "classical" approach of phenotyping of rats (and mice) per se may help dissecting genes and behavior, these tasks remain time consuming, cost intensive, and laborious. Furthermore, scientists are challenged by a large amount of different techniques and may experience problems in developing a meaningful and rational strategy in planning their experimental design. Thus, a high-throughput screening is not possible, and also, inter-laboratory comparability is not warranted (34, 54, 56). Last but not least, even due to these latter constraints, data mining on an inter-laboratory level is impossible. Finally, prior behavioral testing influences test performance, as has been demonstrated in the Morris water maze or elevated plus maze performance of naïve mice compared to mice with prior behavioral testing (65, 66).

For a classical phenotyping of mice and rats, at least ten major guidelines result from general considerations, examples in the literature, and our own experience:

1. Every firm conclusion based on certain differences in a specific behavioral performance must be substantiated by the proof that not only is the experimental animal healthy, but also is equipped with the corresponding sensory and sensorimotoric abilities (35–37, 40).
2. A systematic screening for physiological abnormalities (e.g. glucose utilization, stress hormones, etc.) must be combined with every basic phenotyping work, even if the goal is primarily a behavioral one.
3. The screening of the general health, neurological status, basic physiological parameters, as well as behavioral assessment must be conducted repeatedly, since several changes may become overt in mice and rats at later stages of age (8, 38, 39, 42).
4. A behavioral phenotyping approach must be comprehensive and should not be hypothesis-driven, but should be composed of a complete test battery, which sufficiently detects even behavioral differences in domains that are out of the scope of the hypothesis.
5. At least two different tasks that test similar behavioral abilities but differ in their presuming non response-relevant requirements should be incorporated since these requirements may differ in regard to perception, motor abilities, motivation, or stress (34).

6. The tests chosen should be ethological-based, i.e. the right tasks for the right species must be designed (49, 50, 52, 53, 67).
7. Rearing conditions (numbers of pups, etc.) must be standardized and maternal behavior screened. Thus, every handling, especially postnatal manipulation, as well as environmental enrichment must be standardized/controlled. In the same line, (test) biographies of the animals must be considered as potentially interfering with the expression of a particular phenotype or its onset.
8. Social behaviors and social housing as well as its implications (social rank, aggression, isolation stress, etc.) must be considered wherever possible.
9. Apart from validation of a particular test, positive and negative controls as well as controls across different strains of rats should be included wherever possible in order to improve comparability with other laboratories and to validate the specific task applied.
10. Wherever possible, common test protocols (SOPs) should be used or developed in order to increase validity and reliability.

All in all, these requirements and limitations prompt us to conclude that "classical" phenotyping will be useful for validation purposes of potentially novel – yet to be developed – phenotyping approaches and may also serve as a secondary – in depth – screen after novel screening techniques have indicated a certain phenotype. However, it will not allow high throughput analysis.

*Ad (b) High throughput automated behavioral assessment as a primary screening technique*:

As to date, there is no perfect strategy of behavioral assessment fulfilling all requirements mentioned above, and the classical phenotyping approach using repeated application of test batteries across age does not allow high throughput. We therefore propose novel strategies and task developments using automated home-cage environments as primary screens. Apart from circumvention of the above-mentioned limitations of classical approaches, high-throughput behavioral screening is needed for endeavors to include effective phenotyping in large-scale mutagenesis and CNS-drug discovery programs (68) in which behavioral assessment represents a substantial bottleneck.

Improvements in the throughput of behavioral screening can be achieved by automated systems measuring multiple behaviors in parallel. Movements in that direction can be found at companies in the US (Hypnion, PsychoGenics) offering SCORE-2000 and SmartCube as automated screening systems, which cover many of the needs for mouse phenotyping and drug target validation (55, 57). The latter, however, is only used within PsychoGenics as part of the company's tool set and not available for the scientific community. In Europe, NewBehavior (IntelliCage) (69–73), Noldus (PhenoTyper), and TSE (LabMaster/PhenoMaster) provide products suitable for continuous monitoring of spontaneous, social, cognitive, emotional and physiological measures in home-cage-like environments for mice and are presently under development and validation for rats (www.ratstream.eu).

However, while it can be expected that these apparatus pick up novel behavioral dimension across time/age, and probably increase the speed of genetic/compound screening, with virtually no artifacts induced by handling and novelty stress, they need to be extensively validated. For maximal utility, a behavioral test apparatus should yield valid data for most of the commonly used inbred rodents with and without pharmacological treatment. Tests of simple, ubiquitous behaviors usually yield meaningful data for most mice and rats, especially when based on automated scoring or on simple physical measures that are likely to be replicable across laboratories. Extreme test scores resulting from nonperformance on a task can inflate the apparent reliability of a test, and devious adaptations to a task can undermine its validity. Thus, the optimal apparatus configuration/combination for certain genetic and/or pharmacological analyzes needs to be developed, but it appears to be clear that such screening systems should work in a home-cage-like environment.

Automated phenotyping cages screen the animals within home-cage-like environments using multiple-sensor and device-equipped dedicated animal housings, which continuously detect behavioral/physiological dimensions such as activity, feeding, drinking, activity, urine, fecal boli, anxiety, learning, metabolic performance, temperature, etc. It can be expected that these technologies will allow investigation of the rodents in a more home-cage-like environment such that their normal performances can optimally flourish, while their dysfunctions become more apparent. In contrast, it should be noted here that "classical" and "dedicated" environmental enrichment, in most cases, exerts neuroprotective-like effects and causes increased variation of various parameters within a cohort of animals. In addition, these fully automated units harvest large data sets online at a resource-effective manner and thus allow for functioning as high-throughput primary screens with a focus on different behavioral/physiological dimension, thereby complementing each other. Furthermore, by allowing data collection to take place around the clock in a familiar environment, more information is harvested per test animal under circumstances, which maximize the likelihood of natural behavior to take place, thereby representing a refinement of the use of laboratory animals for research.

*Ad (c) Data analysis and mining*:

Automated high throughput screens enable investigators to monitor animals over longer time scales, and experimental or genetic manipulations will be determined in the context of the integrated expression of multiple behavioral domains. Since this equipment per se collects hypothesis-independent data, analysis of these large data sets by complex statistic probably will reveal novel insights into the regulation of behavioral patterns. At least for construct validation, these behavioral patterns need to be established and validated in comparison with corresponding classical behavioral assessment. At later stage, environmental modifications may allow exploration of additional behavioral domains. However, establishment of stable behavioral patterns may need several weeks of monitoring. These large studies require analysis of terabytes of generated data, which in turn requires equipment, strategies, and know-how of specialists in bioinformatics.

### 2.1. Reasons for the Application of Automated High Throughput Screens in the Home Cage Environment

An anticipated increase in the number of transgenic rat models available in the near future causes predictable problems in "classical" behavioral phenotyping procedures using conventional behavioral test batteries due to limitation in resources, validity, reliability, throughput, and data mining. Phenotyping of a single mutant rat line requires 12–24 months for initial behavioral characterization and much longer for in-depth analysis (for example, *see*: (8, 38, 64, 74, 75)). Incomplete phenotyping due to a reduction of the numbers of tests applied, a limited observation period, limited complex statistics on large data sets, limited sensitivity of established testing procedures, and a lack of proper controls (litter mates, backcrosses) often cause false negative results as has been observed in past for several transgenic mice models.

The resources (space, personnel), equipment, and expertise (know-how, standardization, validation) necessary for a comprehensive phenotyping are not available at most centers, and routine procedures for such a phenotyping have not been established.

Finally, national, European, and international regulations for the use, transport, and hygiene of experimental animals may increase the administrative and technical burden for specialized research groups/centers to obtain and to test transgenic rodent models. Within the large array of possible behavioral testing procedures, those with a minimal stress such as "non-touch-within-home-cage"-tests are likely to meet the slightest resistance and will be favorable as "primary screens" to identify behaviorally functional mutations. Owing to ethical considerations, centers which generate large numbers of random mutations, may be asked to test their animals also for less obvious behavioral symptoms in hetero- as well as homozygous offspring.

## 3. Examples from Testing Transgenic Rats of Huntington's and Spinocerebellar Ataxia Type 17

### 3.1. A Proposed Test Battery for Repeated Testing and Screening of Rat Models

#### 3.1.1. SHIRPA[1] Equivalent in Rats

The SHIRPA protocol was first established in mice responding to a basic necessity for a comprehensive routine screening and testing protocol designed to identify and characterize phenotypic variations or disorders associated with the mouse genome (40). These have recently been further elaborated for a high throughput screening of mice (76, 77).

The SHIRPA procedure was set-up in three screening stages to individually test every single animal and to provide quantitative data in regard to general appearance, locomotor activity, sensory function, analgesia, histology or biochemistry, and other disease related behaviors at certain levels of test complexity (40). Because of the standardization of this test battery, results are directly comparable between animals or groups.

The first stage is the primary screen, in which the animals are observed with standard methods to define a behavioral and functional profile. The secondary screen involves a comprehensive behavioral test battery and pathological analysis. The last part of the SHIRPA screen, the tertiary screen, is more disease related and is either independently tailored but might also – depending on the facilities where the screen is running – refer to a behavioral phenotype detected in the primary and secondary screen or is rationally designed in its composition of tests according to a disease specific hypothesis.

We adapted and modified a SHIRPA-like procedure (*see* also, Fig. 24.1) in order to screen transgenic and mutant rat models for neurological diseases (rats transgenic for Huntington's disease (tgHD) and rats transgenic for Spinocerebellar ataxia type 17 (tgSCA17)) and other mutations (point mutation in the *dpp4* gene) (Table 24.1).

##### Primary Screen in Transgenic Rat Models

The primary screen proposed here is a condensed structured approach (64), which also consists of three parts: general health, neurological reflexes/sensory function, both being applied regularly (i.e. twice a month), and a pub/maternal behavior (78) screen.

Primary screening is initiated by a brief observation of the individual within the nest ecology. This is followed by transfer of the individual into a separate observation cage and by consecutive assessment of the appearance of fur and whiskers. Afterwards, body temperature is determined via rectal measurement. It should be noted here that the body temperature in rats (rodents) is highly variable depending on the stress the rat is experiencing (79); in

[1]SHIRPA = **S**mithKline Beecham Pharmaceuticals; **H**arwell, MRS Mouse Genome Centre and Mammalian Genetics Unit; **I**mperial College School of Medicine at St. Mary's; **R**oyal London Hospital, St. Bartholomew's and the Royal London School of Medicine; **P**henotype **A**ssessment.

**Table 24.1**
**Screening stage, behavioral/physiological domain, and the different corresponding experimental tasks**

| Screen | Behavioral/physiological domain | Experimental task |
|---|---|---|
| 1st | General health and neurological examination | Spontaneous behavior<br>Response to an unknown approaching object<br>Visual cliff<br>Several reflexes (e.g. eye blink and ear twitch) |
| 2nd | Motor function<br>Feeding behavior<br>Nociception<br>Anxiety<br>Learning and memory | Rotarod/Accelerod<br>Food and fluid consumption<br>Tail flick<br>Hot plate<br>Open field test Elevated plus maze<br>Social interaction test<br>(Passive avoidance)<br>Radial maze<br>Passive avoidance |
| 3rd | Tests related to symptoms of neuropsychiatric disorders | PPI (schizophrenic-like)<br>Porsolt swim test (depression-like)<br>Ethanol self-administration<br>Sedative effect of alcohol |

order to minimize this "stress-induced-hyperthermia," most researchers initiate their primary screening with the determination of rectal temperature.

Neurological reflexes and neuromuscular strength are evaluated by using the following tests: (a) *Balance*: The rat is placed in an empty cage, which is rapidly moved from side to side and then up and down. The normal postural reflex is to extend all four legs in order to maintain an upright, balanced position. (b) *Righting reflex*: The postural reflex is evaluated. The animal is turned on its back, and the time to right itself to an upright position is measured. Rats normally will right themselves up within a second. (c) *Eye blink reflex and ear twitch reflex*: These paradigms are measured by simply touching the eye with fibers tight from a cotton-tip swab and by slightly pinching the tip of the ear with a tweezers. (d) *Whisker-orienting reflex*: The whiskers of a freely moving animal are touched lightly. Normally, continuous movements of the whiskers will stop, and in many cases, the rat will turn its head to the side of the stimulus. The sensitivity of whiskers may also be evaluated using a touch test pen (from Frey filament), and positive stimulus-induced responsiveness is scored. (e) *Pupil constriction and dilation reflex*: This test determines the visual response of the pupil to light. In a dimly lit room, a penlight or a small flashlight beam is directed at the eye, and the constriction and following dilation (when the light is removed) of the pupils are observed (the test may be very difficult in albino rats). (f) For testing *neuromuscular strength* using the wire hang test, the rat is placed on a wire cage lid, and a stopwatch is started. The cage lid is gently turned upside down, approximately 20 cm above a mattress of a soft bedding material. The latency to fall onto the bedding is recorded, with a cut off time of 30 s (this test is highly age-dependent, as older rats fail to carry their body weight sufficiently). Still, it should be noted here that a differential motivation of the individual might cause substantial problems as well as differences in test performance. Usually, the experimenter has two possibilities: exclude nonmotivated individuals (if these are only rare cases) or increase motivation within the whole experimental cohort by conducting baited pretest runs.

Sensory functions are determined by screening visual and hearing abilities as well as olfaction. The visual cliff task and the ring paradigm (64) are conducted, and the acoustic startle may be used as advanced techniques for the evaluation of hearing abilities; the latter tests may also contribute to the secondary screen, as components of motor coordination are tested. For the visual cliff test, a piece of cardboard is placed under a clear Plexiglas plate; the cardboard is smaller than the Plexiglas, so there is a ledge, and the visual appearance of a cliff is generated. Normally, rats will stop at the 'edge' and explore the Plexiglas floor before walking on; handicapped rats may not stop at the edge and walk forward across the Plexiglas immediately. It should be noted here that the visual cliff reaction is confounded by the sensory input via the whiskers, and albino rats especially rely much more on sensory input via their whiskers, which may severely hamper the visual cliff test. The ring paradigm means that the rat is dangled by the tail into the middle of a plastic ring (diameter ca. 20 cm) with three convexities. The occurrence of the rat's reaching behavior with the forepaws in direction of one of the ring's convexities is observed and is theoretically fully dependent on optical information (64).

Hearing abilities may easily be checked using an acoustic startling stimulus induced e.g. by a clicking-metal-toy. The consecutive freezing reaction and eye blink represent indicators for functional hearing abilities. However, it should be noted that habituation by the experimental rat as well as other rats hearing the noise is very common. Therefore, every single rat is transported to a room next door and tested in a way that other rats are not habituated to the noise. Another test is the startle response (and prepulse inhibition (PPI) testing) using more sophisticated startle response systems.

Olfaction is tested by using the buried food olfactory test. This basic test measures the rat's ability to find a buried piece of food using olfactory cues and represents an option to screen for deficits in the olfactory system. To avoid problems associated with neophobia, rats were habituated to eat the new food by placing cookie pieces in the home-cage overnight. The rats were then food-deprived for 24 h and had to find the buried food in the center of a randomly selected rectangle.

A primary screening of rat pups may be accomplished as a very early screen useful to detect neurodevelopmental deficits. Testing PPI and in addition measurement of ultrasonic vocalization using bat-detector microphones and dedicated analysis software have been validated for this purpose.

#### Secondary Screen in Transgenic Rat Models

The secondary screen is aimed to test across different behavioral domains spanning motor function, cognition, emotion, and pain. The accelerod test has been validated to detect deficits in motor coordination and balance on an intra-individual repeated basis (80). Cognitive behavior can be observed using two-way-active avoidance tasks (associative learning/conditioning), novel object recognition test, radial maze (spatial learning), or the Skinner box (operant learning). For the emotional part of the screening battery, the social interaction test of anxiety (SI test) (81) or the elevated plus maze can be conducted, while only the SI test is suitable for intra-individual repeated testing. A hot plate analgesia meter and tail flick apparatus are used to test supraspinal and spinal pain perception in the animals (64).

The composition of the tertiary screen depends on the disease model and on the outcomes of the primary and secondary screens. For example, measuring biopotentials such as EEG and performing imaging studies such as PET or DTI can be accomplished to determine any neurological changes in the brain. Further behavioral tasks to evaluate disease-specific-like behaviors (e.g. depression by the Porsolt swim test) can be assessed additionally.

#### *3.1.2. Proposed Protocol for Accelerod Testing of Rats*

To determine fore and hind limb motor coordination and balance of rats, the accelerod testing has been established as a reliable and relatively sensitive procedure. Training and testing are conducted using commercially available Rotarod apparatus for rats (Fig. 24.2). This test is described in more detail here, as it can be repeatedly applied to the individual rat (64, 80). The Accelerod has been shown to be more sensitive than the Rotarod in detecting motor function deficits and in producing more consistent results.

A common apparatus for rats consists of a base platform and a rotating rod of 7.0–9.5 cm diameter with a nonskid surface. Three to four animals – depending on the size of animals and the distance of separations – can be tested simultaneously for their locomotor performance. The apparatus is generally applied for two different modes, i.e. the constant-speed "rotarod" modus and the accelerating-speed "accelerod" modus. In the constant mode, speed can be adjusted between 1 and 40/60 rpm. When operated in the accelerating modus, the rotor would accelerate from 1 to 40/60 rpm in a period of 5 min.

Procedure: Before experimental testing, rats have to be trained to run on the Accelerod in 4 training-trials per day on three consecutive days using a constant training-speed of 12/18 rpm.

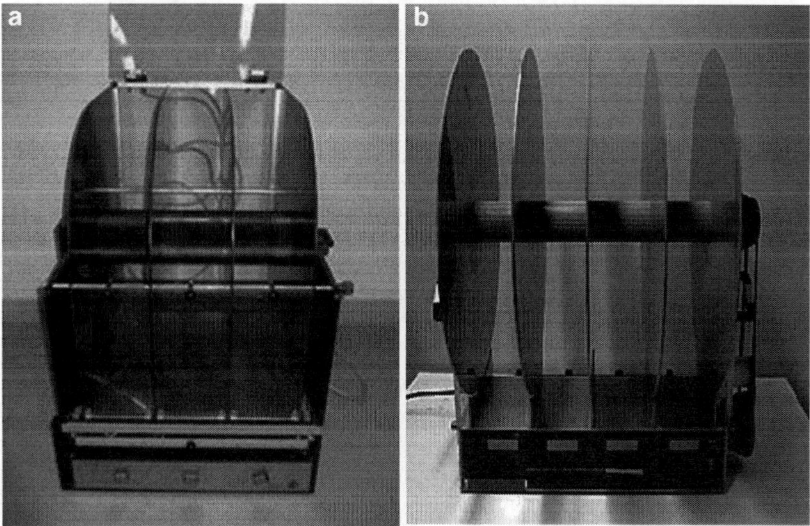


Fig. 24.2. To determine motor coordination and balance a Rotarod system for rats can be used. Several models are available on the market; depending on the size of the animals. (**a**) Three-place Rotarod for bigger rats. (**b**) Four-place system for smaller ones.

For this purpose, animals are placed on the rotating drum for a period of 2 min, and the time of their first falling off and the frequency of falling off the rod are recorded for each rat. After the training period, a steady baseline level of performance should be attained for all rats. On the day of testing, each rat starts with a speed of 1 rpm, which then accelerates gradually up to 40/60 rpm at 5 min. Two testing trials are given with one hour rest in between. The performance of the rats is measured as maximal time spent on the rod before falling off (s) and the maximum rotation per minute (rpm).

*3.1.3. Proposed Protocol for Startle Response (SR) and Prepulse Inhibition (PPI) Testing of Rats*

Basically, the combination of SR and PPI testing within a single set-up represents a multipurpose approach. The SR component of the combined testing protocol is attributable to the primary screening (sensory function, hearing ability), while the PPI testing provides measures of sensorimotoric gating (secondary screening, motor function). Testing PPI, however, also represents a tertiary screening method, since altered or loss of PPI represents a validated measure of schizophrenia-like symptoms in rodents (82–84). Generally, the prepulse inhibition (PPI) is a well-known phenomenon, which, when a startling noise is preceded 30 to 500 ms by a weaker prepulse, leads to an inhibition of the startle response (85). The stimulus is given acoustically, but light, air puff or electric stimuli can be used. The reduction of the amplitude of the startle reaction to the strong main stimulus reflects the ability of the nervous system to temporarily adapt to a strong sensory stimulus when a preceding weaker signal is given to activate central inhibitory processes of the individual. PPI can be applied to numerous species, e.g., mice, rats, and humans. Muscular contractions of the individual are measured indirectly via weight alteration using a piezoelectronical device within milliseconds (Fig. 24.3). Deficits in prepulse inhibition have been linked to abnormalities in sensorimotoric gaiting and this is seen in patients suffering from Schizophrenia and Huntington's disease, but also drugs and adverse experiences can induce such deficits, which are then commonly attributed to defects in cortical-striatal pathways.

Procedure: SR is measured using dedicated Startle Response Systems composed of several testing units. Animals are placed in testing chambers that are subsequently fixed to a piezo-accelerometer in a sound-attenuated chamber with loudspeakers inside (representative set-up is given in Fig. 24.3).

The test session is designed using several types of trials (Baseline, PPI-component, Prepulse-alone, SR-component) that are presented with a constant white background noise of 68 dB sound pressure level (SPL). Habituation and baseline determination of activity for a duration of 2 min with 68 dB were followed by 10 startle trials with a defined pulse (20 ms, 120 dB), which were presented at a randomly set intertrial interval of 6–12 s (Baseline determination component). For the assessment of PPI, animals receive three different trial types in a random order: (1) 10 pre-pulses of 72 dB, 76 dB, 80 dB, and 84 dB (duration 20 ms) followed after 100 ms by a startle pulse of 120 dB for 20 ms (PPI component), (2) pre-pulses (20 ms of 72/76/80/84 dB) without startle pulse, each three times randomly (PP alone component), (3) the 120 dB pulse (20 ms) alone 15 times randomly (SR component). All trials are programmed with an intertrial interval randomly set at 6 and 12 s with a white noise of 68 dB. The whole experiment consists of 77 trials and lasts about 14 min. SR response intensities are being averaged and expressed as arbitrary units. PPI is calculated as percent score: $(100-(\mathrm{PPI}\times 100)/\mathrm{SR})$.

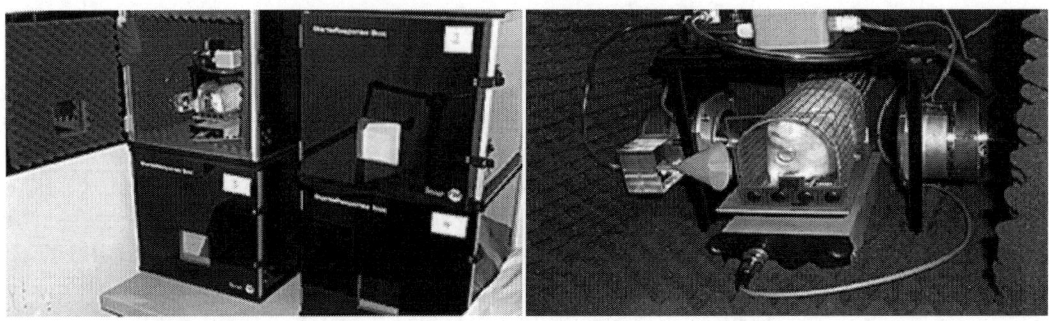

Fig. 24.3. Startle response and prepulse inhibition test. SR and PPI are measured with a Startle Response System. Animals are placed in wire-mesh cages that are fixed on a piezo-accelerometer in a sound-attenuated chamber with two loudspeakers inside. Different protocols can be conducted.

*3.1.4. Proposed Protocol for Social Interaction (SI) Test of Anxiety*

The SI test is carried out according to the method described by S. File (1998) with minor modifications. Two genotype-matched rats, with the interaction partners previously unknown to each other are paired, removed from different home cages, and exposed to a test arena (squared open field of 50 × 50 cm) (Fig. 24.4), which is placed inside a sound isolation box. The test starts at least one hour after onset of the dark-phase and the rats should be transported to the testing room 2 h before the test (79). Illumination of the open field is provided by a white photo bulb providing 170–195 lux, which can be adapted according to the specific need without losing construct validity (86). During test procedure, rats are monitored online by a video camera placed inside the sound isolation box. The duration of each test lasts 10 min during which the time spent in active social interaction (time in s) of the following socio-positive behaviors is scored: time spent sniffing, following, crawling under and over the other rat. Passive body contact such as resting and sleeping is not recorded. Additionally, we measure the duration of self-grooming of the rats (time in s).

## 4. Future Developments: Automated Screens

As it might have become obvious, the “classical” phenotyping approach, as described above, is useful for validation purposes and characterization of a limited number of rat models, however,

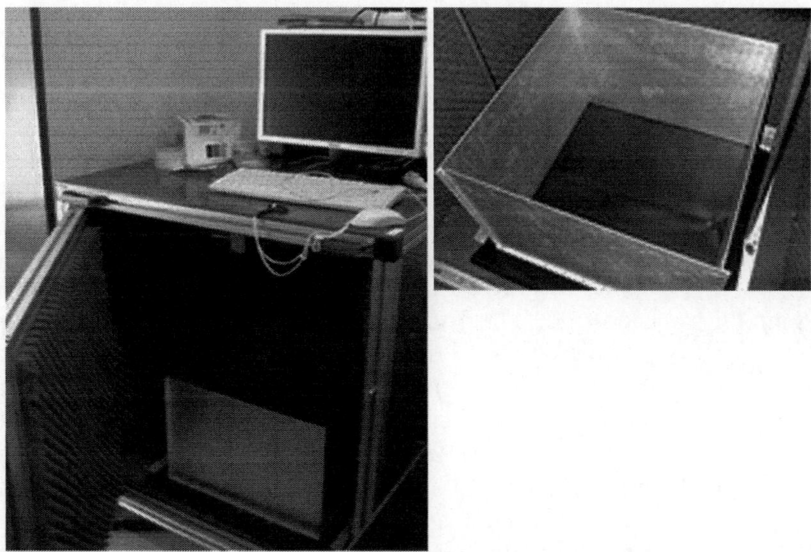

Fig. 24.4. Test box for the social interaction test of anxiety. The box is 50 × 50 cm, sound-isolated and illuminated. Two genotype-matched rats from different home-cages are tested for 10 min for their social behavior. The time they spend in social interaction and the time they spend for self-grooming is scored.

it will not allow high throughput analysis of, e.g., mutagenesis programs or larger numbers of transgenic rat models. Presently, therefore, automated behavioral assessment techniques potentially providing a higher throughput are developed and validated as general screening techniques. All these automated systems represent "intra-home-cage-like" devices, allowing continuous, online determination of multiple behaviors in parallel (www.ratstream.eu). Automated phenotyping will increase the number of parameters evaluated per single animal; thereby most likely allowing a reduction of the number of animals along with ethologically based "non-touch" experimental work.

### 4.1. Technical Developments

Three major developments for automated phenotyping in a home-cage-like environment are available at present in Europe: the PhenoMaster system for rats (TSE Systems), the IntelliCage for rats (NewBehavior), and the PhenoTyper (Noldus).

Presently, the PhenoMaster represents a multipurpose and modular system capable of screening singly kept rats in home-cage-like environments (Fig. 24.5a) for several parameters, such as food and fluid consumption, calorimetry, locomotor activity via a light beam-based activity frame, wheel running (Fig. 24.5b), and operant learning behavior (Fig. 24.5c) at a high temporal and spatial resolution. Each rat is screened individually and several devices run in parallel (Fig. 24.5a) (http://www.tse-systems.com/phenomaster/phenomaster.htm). The PhenoTyper has similar properties, but is based on video-tracking technology of single rats (http://www.noldus.com/animal-behavior-research/products/phenotyper). In contrast to the set-up of the other systems, IntelliCage for rats provides a social environment (Fig. 24.6a), where the animals live in groups and screening presently is especially focused on social as well as learning and memory behaviors (Fig. 24.6b). Tracking technology is based on transponders. The system for rats is designed out of four type IV cages connected through perspex tubes with each other, allowing free exploration of the habitat (http://www.newbehavior.com/products/icr).

### 4.2. Present Experience Using Automated Phenotyping Technology

So far the systems hold promise with specific advantages on each side. The IntelliCage provides a unique chance to detect completely novel qualities of behaviors as the animals are kept in social groups remaining totally undisturbed though being monitored for social and cognitive performance. However, space required for the set-up and throughput of the system are presently considered problematic. Monitoring social groups, in a way, represents a double-edged sword, as on the one hand ethological-based behaviors flourish in the group, most likely providing higher sensitivity in detecting pathological behaviors, but on the other hand the requirement of group housing limits

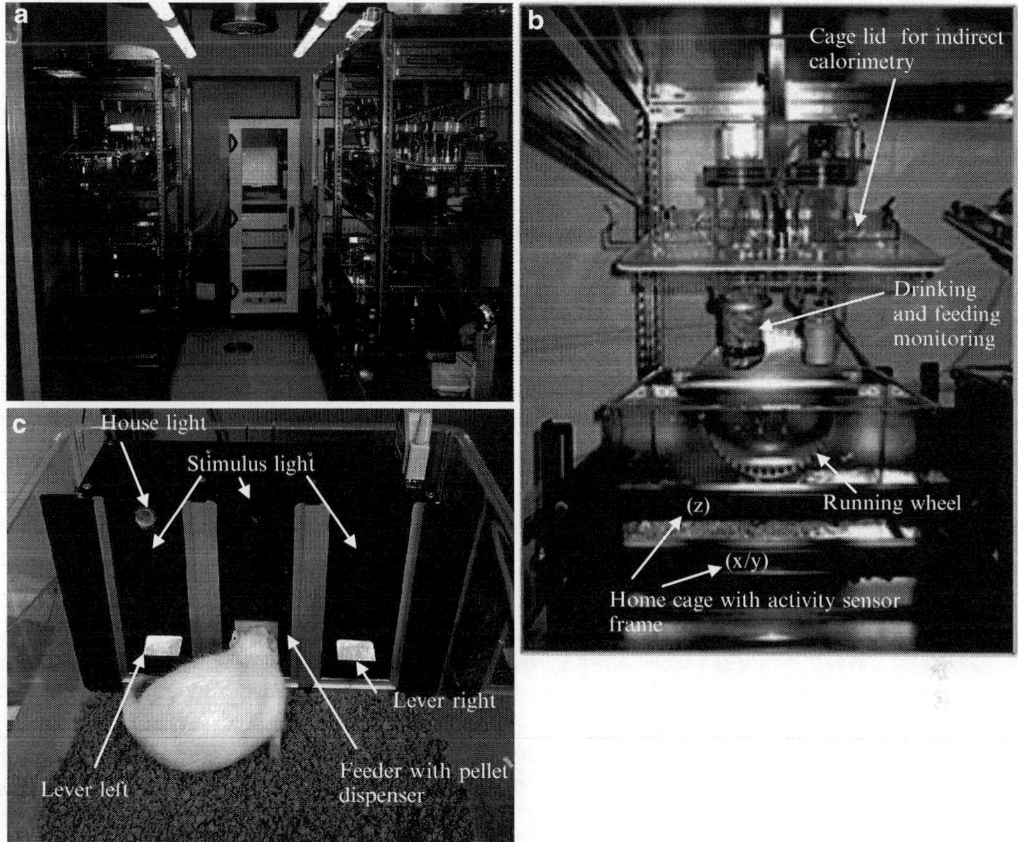


Fig. 24.5. Automated phenotyping. A: General set-up PhenoMaster, TSE Systems GmbH, Bad Homburg, Germany. (**a**) 12-cage set-up makes it possible to screen 12 animals in parallel for 72 h with high resolution. Circadian screening of food and fluid consumption, locomotor activity with rearing behavior, calorimetry, and optional wheel running or operant learning abilities. (**b**) Detailed view of the PhenoMaster home-cage. The cage lid containing food and fluid sensors and the equipment to measure indirect calorimetry can be mounted onto the rack roof, which makes it more comfortable to work with the system (i.e., clean the cages). The home cage is surrounded by a light beam-based activity frame, which detects the animal's activity in all three dimensions (xyz). In this modular concept, different components or modules can be integrated, such as wheels for wheel running or operant conditioning walls (Fig. 24.5c). (**c**) Operant conditioning wall for the PhenoMaster home-cage with two levers, a feeder with pellet dispenser for a food reward and several stimulus lights. A house light can indicate the starting of the trials.

throughput (groups have to be established and have to remain stable, furthermore drug screening appears to be more difficult). In contrast, the PhenoMaster approach allows higher throughput and manipulations of the individual animal, as they are kept singly. However, this implies the problem of social deprivation especially in rats (animal protection laws), and which must be avoided, thereby limiting the duration of experimental trials. Nonetheless, automated phenotyping of rats apparently provides a novel quality with several add-ons to the classical behavioral phenotyping approach.

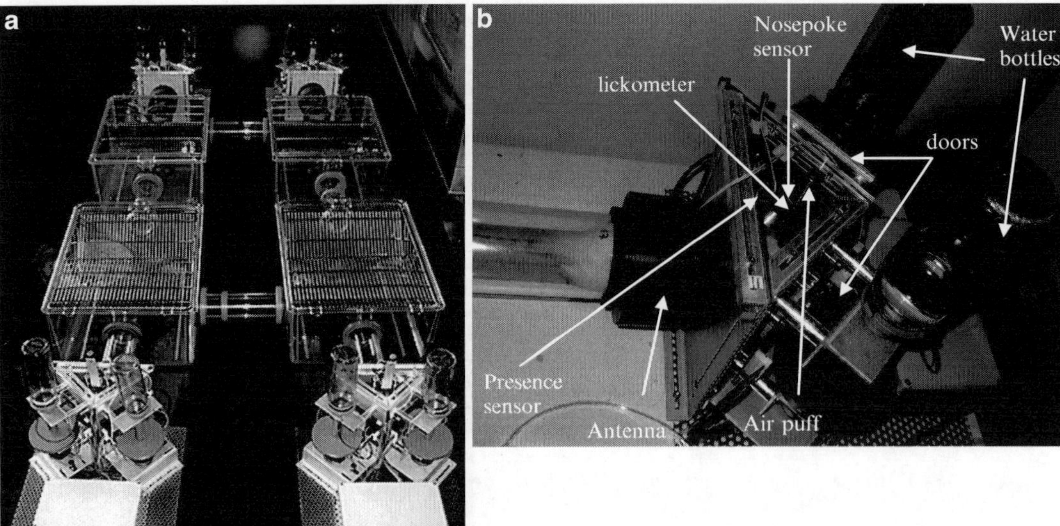


Fig. 24.6. Automated phenotyping. (**a**) General set-up IntelliCage New Behavior AG, Zürich, Switzerland. In this system, 10–12 animals can be screened in a social environment. Therefore, four type IV cages are connected with each other via acrylic plastic tubes. In each corner, an operant conditioning corner is connected containing water bottles as conditioning reward. (**b**) Detailed view of the operant corner with lickometer to measure the contacts with the water nipple, nosepoke holes as a manipulandum, where the rats have to make a response and doors will open and give access to fluid. A punishment can be conducted using the air puff. The antenna detects the transponder chip of the animal while a thermic sensor registers the actual presence of the animal.

## 5. Conclusion

It appears that "classical behavioral phenotyping" of rats, when following certain rules, represents today's standard, which however requires considerable resources and still does not provide highest reliability, validity, and cross-lab comparability. In the future, automated phenotyping may provide important add-ons, increased standardization, and throughput within more ethological-like conditions.

## Acknowledgments and Funding

European Community "RATSTREAM" #037846 Strep project

## References


1. Agmo A, Ellingsen E (2003) Relevance of non-human animal studies to the understanding of human sexuality. Scand J Psychol 44:293–301
2. Blanchard DC, Griebel G, Blanchard RJ (2001) Mouse defensive behaviors: pharmacological and behavioral assays for anxiety and panic. Neurosci Biobehav Rev 25:205–218
3. de Boer SF, van der Vegt BJ, Koolhaas JM (2003) Individual variation in aggression of feral rodent strains: a standard for the genetics of aggression and violence? Behav Genet 33:485–501
4. Abbott A (2004) Laboratory animals: the Renaissance rat. Nature 428:464–466
5. Kolb B (2005) Models and Tests, Neurological Models. In: Wihishaw IQ, Kolb B (eds) The behavior of the laboratory rat. Oxford University Press, New York, pp 449–461
6. Mullins JJ, Peters J, Ganten D (1990) Fulminant hypertension in transgenic rats harbouring the mouse Ren-2 gene. Nature 344:541–544
7. Hammer RE, Maika SD, Richardson JA, Tang JP, Taurog JD (1990) Spontaneous inflammatory disease in transgenic rats expressing HLA-B27 and human beta 2m: an animal model of HLA-B27-associated human disorders. Cell 63:1099–1112
8. von Horsten S, Schmitt I, Nguyen HP, Holzmann C, Schmidt T, Walther T et al (2003) Transgenic rat model of Huntington's disease. Hum Mol Genet 12:617–624
9. Liu L, Orozco IJ, Planel E, Wen Y, Bretteville A, Krishnamurthy P et al (2008) A transgenic rat that develops Alzheimer's disease-like amyloid pathology, deficits in synaptic plasticity and cognitive impairment. Neurobiol Dis 31:46–57
10. Vercammen L, Van der Perren A, Vaudano E, Gijsbers R, Debyser Z, Van den Haute C et al (2006) Parkin protects against neurotoxicity in the 6-hydroxydopamine rat model for Parkinson's disease. Mol Ther 14:716–723
11. Cenci MA, Lundblad M (2007) Ratings of L-DOPA-induced dyskinesia in the unilateral 6-OHDA lesion model of Parkinson's disease in rats and mice. Curr Protoc Neurosci Chapter 9:Unit 9 25
12. Dekundy A, Lundblad M, Danysz W, Cenci MA (2007) Modulation of L-DOPA-induced abnormal involuntary movements by clinically tested compounds: further validation of the rat dyskinesia model. Behav Brain Res 179:76–89
13. Kelly PH, Seviour PW, Iversen SD (1975) Amphetamine and apomorphine responses in the rat following 6-OHDA lesions of the nucleus accumbens septi and corpus striatum. Brain Res 94:507–522
14. Hudson JL, van Horne CG, Stromberg I, Brock S, Clayton J, Masserano J et al (1993) Correlation of apomorphine- and amphetamine-induced turning with nigrostriatal dopamine content in unilateral 6-hydroxydopamine lesioned rats. Brain Res 626:167–174
15. Jolicoeur FB, Rivest R, Drumheller A (1991) Hypokinesia, rigidity, and tremor induced by hypothalamic 6-OHDA lesions in the rat. Brain Res Bull 26:317–320
16. Jolicoeur FB, Rivest R, St-Pierre S, Drumheller A (1991) Antiparkinson-like effects of neurotensin in 6-hydroxydopamine lesioned rats. Brain Res 538:187–192
17. Thomas J, Wang J, Takubo H, Sheng J, de Jesus S, Bankiewicz KS (1994) A 6-hydroxydopamine-induced selective parkinsonian rat model: further biochemical and behavioral characterization. Exp Neurol 126:159–167
18. Ogata A, Tashiro K, Nukuzuma S, Nagashima K, Hall WW (1997) A rat model of Parkinson's disease induced by Japanese encephalitis virus. J Neurovirol 3:141–147
19. Betarbet R, Sherer TB, Greenamyre JT (2002) Animal models of Parkinson's disease. Bioessays 24:308–318
20. Aoki M, Ogasawara M, Matsubara Y, Narisawa K, Nakamura S, Itoyama Y et al (1993) Mild ALS in Japan associated with novel SOD mutation. Nat Genet 5:323–324
21. Rosen DR (1993) Mutations in Cu/Zn superoxide dismutase gene are associated with familial amyotrophic lateral sclerosis. Nature 364:362
22. Bates G, Harper P, Jones L (2002) Huntington's disease, vol 45, Oxford monographs on medical genetics. Oxford University Press, New York
23. Friedman MJ, Shah AG, Fang ZH, Ward EG, Warren ST, Li S et al (2007) Polyglutamine domain modulates the TBP-TFIIB interaction: implications for its normal function and neurodegeneration. Nat Neurosci 10:15 19–1528
24. Alves S, Regulier E, Nascimento-Ferreira I, Hassig R, Dufour N, Koeppen A et al (2008) Striatal and nigral pathology in a lentiviral rat model of Machado–Joseph disease. Hum Mol Genet 17:2071–2083
25. Reigel CE, Dailey JW, Jobe PC (1986) The genetically epilepsy-prone rat: an overview of seizure-prone characteristics and responsiveness to anticonvulsant drugs. Life Sci 39: 763–774

26. Owens D (2006) Spontaneous, surgically and chemically induced models of disease. In: Suckow MW, Weisbroth SH, Franklin CL (eds) The laboratory rat, 2nd edn. Elsevier Academic, Amsterdam, pp 711–726
27. Parent JM, Jessberger S, Gage FH, Gong C (2007) Is neurogenesis reparative after status epilepticus? Epilepsia 48(Suppl. 8):69–71
28. Miller DS, Bauer B, Hartz AM (2008) Modulation of P-glycoprotein at the blood-brain barrier: opportunities to improve central nervous system pharmacotherapy. Pharmacol Rev 60:196–209
29. Gerlai R (2002) Phenomics: fiction or the future? Trends Neurosci 25:506–509
30. Arndt SS, Surjo D (2001) Methods for the behavioural phenotyping of mouse mutants. How to keep the overview. Behav Brain Res 125:39–42
31. Blanchard DC, Griebel G, Blanchard RJ (2003) The mouse defense test battery: pharmacological and behavioral assays for anxiety and panic. Eur J Pharmacol 463:97–116
32. Blanchard RJ, Blanchard DC (2003) Bringing natural behaviors into the laboratory: a tribute to Paul MacLean. Physiol Behav 79:515–524
33. Crawley JN (2003) Behavioral phenotyping of rodents. Comp Med 53:140–146
34. Crawley JN, Belknap JK, Collins A, Crabbe JC, Frankel W, Henderson N et al (1997) Behavioral phenotypes of inbred mouse strains: implications and recommendations for molecular studies. Psychopharmacology (Berl) 132:107–124
35. Crawley JN, Paylor R (1997) A proposed test battery and constellations of specific behavioral paradigms to investigate the behavioral phenotypes of transgenic and knockout mice. Horm Behav 31:197–211
36. Hagan JJ, Harper AJ, Elliott H, Jones DN, Rogers DC (2000) Practical and theoretical issues in gene-targeting studies and their application to behaviour. Rev Neurosci 11:3–13
37. Hatcher JP, Jones DN, Rogers DC, Hatcher PD, Reavill C, Hagan JJ et al (2001) Development of SHIRPA to characterise the phenotype of gene-targeted mice. Behav Brain Res 125:43–47
38. Karl T, Hoffmann T, Pabst R, von Horsten S (2003) Behavioral effects of neuropeptide Y in F344 rat substrains with a reduced dipeptidyl-peptidase IV activity. Pharmacol Biochem Behav 75:869–879
39. Nguyen PV, Gerlai R (2002) Behavioural and physiological characterization of inbred mouse strains: prospects for elucidating the molecular mechanisms of mammalian learning and memory. Genes Brain Behav 1:72–81
40. Rogers DC, Fisher EM, Brown SD, Peters J, Hunter AJ, Martin JE (1997) Behavioral and functional analysis of mouse phenotype: SHIRPA, a proposed protocol for comprehensive phenotype assessment. Mamm Genome 8:711–713
41. Rogers DC, Jones DN, Nelson PR, Jones CM, Quilter CA, Robinson TL et al (1999) Use of SHIRPA and discriminant analysis to characterise marked differences in the behavioural phenotype of six inbred mouse strains. Behav Brain Res 105:207–217
42. Rogers DC, Peters J, Martin JE, Ball S, Nicholson SJ, Witherden AS et al (2001) SHIRPA, a protocol for behavioral assessment: validation for longitudinal study of neurological dysfunction in mice. Neurosci Lett 306:89–92
43. Surjo D, Arndt SS (2001) The mutant mouse behaviour network, a medium to present and discuss methods for the behavioural phenotyping. Physiol Behav 73:691–694
44. van der Staay FJ, Steckler T (2001) Behavioural phenotyping of mouse mutants. Behav Brain Res 125:3–12
45. van der Staay FJ, Steckler T (2002) The fallacy of behavioral phenotyping without standardisation. Genes Brain Behav 1:9–13
46. Wahlsten D, Rustay NR, Metten P, Crabbe JC (2003) In search of a better mouse test. Trends Neurosci 26:132–136
47. Churchill GA, Airey DC, Allayee H, Angel JM, Attie AD, Beatty J et al (2004) The collaborative cross, a community resource for the genetic analysis of complex traits. Nat Genet 36:1133–1137
48. Jansen RC (2003) Studying complex biological systems using multifactorial perturbation. Nat Rev Genet 4:145–151
49. Crabbe JC, Phillips TJ (2003) Mother nature meets mother nurture. Nat Neurosci 6:440–442
50. Crabbe JC, Wahlsten D (2003) Of mice and their environments. Science 299:1313–1314
51. Crabbe JC, Wahlsten D, Dudek BC (1999) Genetics of mouse behavior: interactions with laboratory environment. Science 284: 1670–1672
52. Gerlai R (2000) Targeting genes and proteins in the analysis of learning and memory: caveats and future directions. Rev Neurosci 11:15–26
53. Gerlai RT (2000) Genotype x environment interaction: an old problem in a new field. Trends Neurosci 23:450
54. Wahlsten D, Metten P, Phillips TJ, Boehm SL II, Burkhart-Kasch S, Dorow J et al (2003) Different data from different labs: lessons from studies of gene-environment interaction. J Neurobiol 54:283–311

55. Brunner D, Nestler E, Leahy E (2002) In need of high-throughput behavioral systems. Drug Discov Today 7:S107–112
56. Crabbe JC, Morris RG (2004) Festina lente: late-night thoughts on high-throughput screening of mouse behavior. Nat Neurosci 7:1175–1179
57. Tecott LH, Nestler EJ (2004) Neurobehavioral assessment in the information age. Nat Neurosci 7:462–466
58. Ganten D, Lindpaintner K, Ganten U, Peters J, Zimmermann F, Bader M et al (1991) Transgenic rats: new animal models in hypertension research. Invited lecture. Hypertension 17:843–855
59. Editorial (anonymous) (2008) Year of the rat. Nat Genet 40:513
60. Aitman TJ, Critser JK, Cuppen E, Dominiczak A, Fernandez-Suarez XM, Flint J et al (2008) Progress and prospects in rat genetics: a community view. Nat Genet 40:516–522
61. Saar K, Beck A, Bihoreau MT, Birney E, Brocklebank D, Chen Y et al (2008) SNP and haplotype mapping for genetic analysis in the rat. Nat Genet 40:560–566
62. Twigger SN, Pruitt KD, Fernandez-Suarez XM, Karolchik D, Worley KC, Maglott DR et al (2008) What everybody should know about the rat genome and its online resources. Nat Genet 40:523–527
63. Crawley JN (1999) Behavioral phenotyping of transgenic and knockout mice: experimental design and evaluation of general health, sensory functions, motor abilities, and specific behavioral tests. Brain Res 835:18–26
64. Karl T, Pabst R, von Horsten S (2003) Behavioral phenotyping of mice in pharmacological and toxicological research. Exp Toxicol Pathol 55:69–83
65. McIlwain KL, Merriweather MY, Yuva-Paylor LA, Paylor R (2001) The use of behavioral test batteries: effects of training history. Physiol Behav 73:705–717
66. Voikar V, Vasar E, Rauvala H (2004) Behavioral alterations induced by repeated testing in C57BL/6J and 129S2/Sv mice: implications for phenotyping screens. Genes Brain Behav 3:27–38
67. Gerlai R, Clayton NS (1999) Analysing hippocampal function in transgenic mice: an ethological perspective. Trends Neurosci 22:47–51
68. Hunter AJ, Nolan PM, Brown SD (2000) Towards new models of disease and physiology in the neurosciences: the role of induced and naturally occurring mutations. Hum Mol Genet 9:893–900
69. Dell'Omo G, Ricceri L, Wolfer DP, Poletaeva II, Lipp H (2000) Temporal and spatial adaptation to food restriction in mice under naturalistic conditions. Behav Brain Res 115:1–8
70. Dell'Omo G, Vannoni E, Vyssotski AL, Di Bari MA, Nonno R, Agrimi U et al (2002) Early behavioural changes in mice infected with BSE and scrapie: automated home cage monitoring reveals prion strain differences. Eur J NeuroSci 16:735–742
71. Galsworthy MJ, Amrein I, Kuptsov PA, Poletaeva P II, Zinn AR et al (2005) A comparison of wild-caught wood mice and bank voles in the intellicage: assessing exploration, daily activity patterns and place learning paradigms. Behav Brain Res 157:211–217
72. Lewejohann L, Skryabin BV, Sachser N, Prehn C, Heiduschka P, Thanos S et al (2004) Role of a neuronal small non-messenger RNA: behavioural alterations in BC1 RNA-deleted mice. Behav Brain Res 154:273–289
73. Vyssotski AL, Dell'Omo G, Poletaeva DL II, Vyssotsk LM, Klein R et al (2002) Long-term monitoring of hippocampus-dependent behavior in naturalistic settings: mutant mice lacking neurotrophin receptor TrkB in the forebrain show spatial learning but impaired behavioral flexibility. Hippocampus 12:27–38
74. Karl T, Chwalisz WT, Wedekind D, Hedrich HJ, Hoffmann T, Jacobs R et al (2003) Localization, transmission, spontaneous mutations, and variation of function of the Dpp 4 (Dipeptidyl-peptidase IV; CD26) gene in rats. Regul Pept 115:81–90
75. Karl T, Hoffmann T, Pabst R, von Horsten S (2003) Extreme reduction of dipeptidyl peptidase IV activity in F344 rat substrains is associated with various behavioral differences. Physiol Behav 80:123–134
76. Green EC, Gkoutos GV, Lad HV, Blake A, Weekes J, Hancock JM (2005) EMPReSS: European mouse phenotyping resource for standardized screens. Bioinformatics 21: 2930–2931
77. Mallon AM, Blake A, Hancock JM (2008) EuroPhenome and EMPReSS: online mouse phenotyping resource. Nucleic Acids Res 36:D715–718
78. Breivik T, Stephan M, Brabant GE, Straub RH, Pabst R, von Horsten S (2002) Postnatal lipopolysaccharide-induced illness predisposes to periodontal disease in adulthood. Brain Behav Immun 16:421–438
79. Kask A, Nguyen HP, Pabst R, von Horsten S (2001) Factors influencing behavior of group-housed male rats in the social interaction test: focus on cohort removal. Physiol Behav 74:277–282

80. Nguyen HP, Kobbe P, Rahne H, Worpel T, Jager B, Stephan M et al (2006) Behavioral abnormalities precede neuropathological markers in rats transgenic for Huntington's disease. Hum Mol Genet 15:3177–3194
81. File SE, Hyde JR (1978) Can social interaction be used to measure anxiety? Br J Pharmacol 62:19–24
82. Caine SB, Geyer MA, Swerdlow NR (1992) Hippocampal modulation of acoustic startle and prepulse inhibition in the rat. Pharmacol Biochem Behav 43:1201–1208
83. Koch M, Fendt M, Kretschmer BD (2000) Role of the substantia nigra pars reticulata in sensorimotor gating, measured by prepulse inhibition of startle in rats. Behav Brain Res 117:153–162
84. Swerdlow NR, Geyer MA, Braff DL (2001) Neural circuit regulation of prepulse inhibition of startle in the rat: current knowledge and future challenges. Psychopharmacology (Berl) 156:194–215
85. Swerdlow NR, Bakshi V, Geyer MA (1996) Seroquel restores sensorimotor gating in phencyclidine-treated rats. J Pharmacol Exp Ther 279:1290–1299
86. File SE, Seth P (2003) A review of 25 years of the social interaction test. Eur J Pharmacol 463:35–53

# Chapter 25

# Rat Genomics Applied to Psychiatric Research

**Marie-Pierre Moisan and André Ramos**

## Abstract

Psychiatric diseases are very debilitating and some of them highly prevalent (e.g., depression or anxiety). The rat remains one model of choice in this discipline to investigate the neural mechanisms underlying normal and pathological traits. Genomic tools are now applied to identify genes involved in psychiatric illnesses and also to provide new biomarkers for diagnostic and prognosis, new targets for treatment and more generally to better understand the functioning of the brain. In this report, we will review rat models, behavioral approaches used to model psychiatry-related traits and the major studies published in the field including genetic mapping of quantitative trait loci (QTL), transcriptomics, proteomics and transgenic models.

**Key words:** Behavior, Depression, Stress, Genetics, Alcohol, Addiction, Memory, Glucocorticoids, Anxiety, Gene profiling

## 1. Introduction

Neurosciences aim to understand the functioning of the central nervous system. Its medical applications fall mainly into psychiatric and neurological diseases that can be very frequent (e.g. depression) and/or very debilitating (neurodegenerative diseases). Neurological deficits are in general well-defined, not as influenced by environmental parameters as most psychiatric diseases and the genetic forms are caused by single or a small number of genes. Animal models have been developed to better understand the neurobiological mechanisms involved and for experimental therapeutic strategies. Conversely, psychiatric diseases are complex disorders with multigenic and environmental factors interacting along the life span of the individual. Consequently, the identification of genetic risk factors in human populations remains difficult in this domain. These facts have prompted scientists to develop animal models for which the environment as well as the genetic background can be manipulated

I. Anegon (ed.), *Rat Genomics: Methods and Protocols*, Methods in Molecular Biology, vol. 597
DOI 10.1007/978-1-60327-389-3_25, © Humana Press, a part of Springer Science+Business Media, LLC 2010

in a controllable manner. As the modeling of human psychiatric diseases in animals is difficult to achieve, scientists have focussed on traits related to essential behavioral components and endocrine or neuronal responses to challenging environments, that are conserved across species (1, 2). The heritability of traits related to psychiatric disorders has been recognized for more than 30 years with the selection of divergent animal lines, for example on emotionality traits (3). With the development of molecular genetic markers, it became possible to perform genetic mapping of quantitative trait loci (QTL) in contrasting animal lines using various genetic approaches. Regarding new genomic tools, transcriptomics, proteomics and metabonomics are now also applied to this field in rats as well as transgenesis or mutation induced by ENU (all these techniques are described in other chapters of this book). In this review, we will describe the rat models, the approaches specific to psychiatry-related traits and the major studies published in the field of psychogenomics.

## 2. Rat Models Used in Psychiatric Research

As for any phenotype, genetic mapping is performed in contrasting animal lines. Inbred lines are preferred in order to simplify the genetic studies as every animal of a given strain is genetically identical. Studies are performed either on rat strains divergently selected on a behavioral trait, for example the Wister-Kyoto Hyperactive and Wistar-Kyoto, selected for differential motor activity (WKHA/WKY) (4–6), the Roman High and Low Avoidance (RLA/RHA), selected for differential response to fear (3, 7), the Alcohol-preferring and nonpreferring (P/NP), selected for alcohol consumption (8, 9) and the High and Low Alcohol Sensitivity (HAS/LAS), selectively bred for differential sensitivity to ethanol administration (10). Bidirectional selection for a trait has the advantage to maximize the differences between strains but the selective breeding takes many years. An alternative to divergent selection is the screening of several strains for differential behavioral and/or physiologic responses to stress, prior to choosing a pair of contrasting rat lines to use in genomic studies. This is the case for the Spontaneously Hypertensive Rats and Lewis (SHR/LEW) pair of strains (11), Fisher and Wistar-Kyoto (F344/WKY) (12) and Wistar-Kyoto and High Ethanol Preferring (WKY/HEP) strains (13). Finally, a Recombinant Inbred (RI) panel has been used, the HxB/BxH derived from SHR and BN-*Lx* progenitors (14). The main advantage of RIs, in particular for a labile phenotype such as behavior, is that measures are done on groups of animals instead of on individuals and the genotyping is done only once. When a strong QTL has been detected, congenic strains are often derived from the parental strains to refine and to isolate the QTL in a new strain; the latter is also very useful in transcriptomic or proteomic studies to identify positional candidate genes. Alternatively to congenics, animal lines can be selected from the F2 population in which the QTL was detected using molecular markers flanking one or several QTLs.

For transcriptomics, proteomics or metabonomics, outbred models have been used with three to six animals per group, although inbred animals can help to reduce intragroup variability. Finally, few transgenic models of psychiatric diseases are available in the rat. More recently, large-scale mutagenesis studies, in which mutations are randomly induced by a chemical mutagen (N′-ethyl-N′-nitrosourea, ENU), have provided new rat genetic models (*see* Chapter 11).

## 3. Approaches Specific to Psychiatry-Related Traits

Stress, defined as a stimulus that threatens homeostasis, has long been implicated to play a role in the precipitation of emotional disorders such as depression and anxiety, in addiction and substance abuse as well as diseases associated with learning and memory. To survive challenging environments, the organism sets off a complex repertoire of behavioral and physiologic responses that reinstates homeostasis. Inadequate, excessive or prolonged reactions to stress may lead to disease (15). Thus, in order to identify genetic factors influencing psychiatric diseases, behavioral and physiologic responses to stress have been examined in animals.

A hallmark of the physiologic response to stress is the activation of the hypothalamo-pituitary-adrenal axis leading to the release of glucocorticoids (corticosterone in rat, cortisol in human) from the adrenal gland cortex and the autonomous nervous system, inducing epinephrine and norepinephrine discharge from the adrenal medulla as well as aldosterone through renin activation. As expected, these systems are often dysregulated in affective disorders. Thus, corticosteroid and catecholamine hormone levels and renin activity are measured in plasma or urine after an acute stress such as restraint for 20 min or after prolonged or chronic stress when rats have been submitted to a series of challenges. Thymus and adrenal gland weights at sacrifice indicate the long term sensitivity to glucocorticoid hormones.

In spite of the intrinsic difficulty of modelling human psychological traits and disorders, dozens of different behavioral tests have been conceived to investigate anxiety- and depression-related behaviors as well learning and memory and alcohol consumption or sensitivity in rats. Table 25.1 summarizes the main tests that have been used in genomic studies of psychiatry-related traits.

**Table 25.1**
**Main behavioural tests used in psychiatry-related traits**

| Test name | Related human condition | Based on | Main behavioural measures |
|---|---|---|---|
| Open field | Anxiety | Unconditioned exploration of a novel arena | – Central locomotion<br>– Defecation |
| Elevated plus maze | Anxiety | Unconditioned exploration of unprotected elevated alleys | – Time spent in the open arms<br>– % entries in the open arms |
| Acoustic startle | Anxiety | Startle response to acoustic stimulation | – Apparatus movement resulting from the startle response |
| Fear conditioning | Anxiety | Freezing behavior under exposure to a conditioned aversive stimulus | – Time spent freezing |
| Shuttle box | Anxiety | Active avoidance of a conditioned aversive stimulus | – Crossings between two box compartments during the conditioned stimulus |
| Defensive burying test | Anxiety | Active avoidance of an unconditioned aversive stimulus | – Latency to bury and duration of burying an electrified prod |
| Learned helplessness | Depression | Lack of response after exposure to repeated unescapable stress | – Failure to escape from an escapable footshock |
| Forced swim test | Depression | Lack of response after exposure to repeated unescapable stress | – Time of immobility in a water tank |
| Sucrose consumption | Depression | Anhedonia | – Preference for sucrose over water |
| Open field | Motor activity | Unconditioned exploration of a novel arena | – Total locomotion<br>– Peripheral locomotion<br>– Number of rearings |
| Activity cage | Motor activity | Unconditioned exploration of a novel cage | – Number of cage crossings<br>– Number of rearings |
| Prepulse inhibition | Sensory motor gating | Inhibition of a startle response (see above) by the prior presentation of a weak stimulus | – % reduction in the startle response |
| Morris water maze | Learning/ Memory | Ability to find a submerged platform located in a fixed position inside a water tank | – Distance traveled over trials in search of the platform<br>– Distance from platform once it has been removed |

(continued)

**Table 25.1 (continued)**

| Test name | Related human condition | Based on | Main behavioural measures |
|---|---|---|---|
| Social recognition task | Learning/ Memory | Ability to recognize a familiar conspecific | – Time spent in social-investigatory behavior after an inter-exposure interval |
| Ethanol self-administration | Alcoholism | Voluntary alcohol intake when two drinking bottles, one with water and one with alcohol, are available | – Alcohol consumption (g/kg/day)<br>– Ratio of alcohol to total fluids consumed (%) |
| Ethanol response | Alcoholism | Sensitivity to a standard dose of ethanol | – Loss of righting reflex<br>– Blood ethanol concentration |

The open field was developed for rats in 1934 as the first test to assess emotionality in animals (16). This test takes place in a large arena (much larger than the home cage) which is typically novel to the rat and brightly illuminated. During the following decades and mostly after the use of the first benzodiazepines to treat anxiety (1950s), several other tests have been created. Some are based on the unconditioned exploration of potentially dangerous areas, like the elevated plus maze, which contains two enclosed and two open elevated alleys (arms) interconnected by a platform and is currently the most popular test of anxiety (17). Others are based on conditioned responses to harmless stimuli (usually light or sound) that have been associated with painful or aversive stimuli (typically footshocks). It is not the aim of this chapter to provide extensive information about any of these tests, as interested readers can find excellent primary and review papers on this specific topic (18–23).

Because of its importance for public health policies, alcohol-related disorders are among the most studied topics in the field of behavior genetics. Most of the rodent experimental paradigms are designed to model two main features of alcohol-related disorders, namely self-administration and withdrawal. The former can be assessed using different protocols, but they very often involve offering the animals a free choice between two drinking bottles, one with water and one with alcohol. Another method uses operant conditioning, where the animal needs to perform a learned task in order to have access to ethanol. Finally, withdrawal symptoms, such as increased anxiety, tremor and convulsions, are evaluated after a varying period of exposure to alcohol. These general procedures are explained and discussed in details elsewhere (24–27) where the main methods for studying alcohol-related disorders from a genomic point of view, including the description of some of the rat strains currently available, are also reviewed.

## 4. Genetic Mapping Studies

Table 25.2 summarizes genetic mapping studies related to psychiatric diseases in rat models. The name of the QTL comes from the Rat Genome Database (RGD, http://www.rgd.mcw.edu) when filed; if not, we have suggested a name following the RGD nomenclature (names in gray color). To help the readers, the QTL names originally given in the associated publication are also provided. In our opinion some QTLs have been misnamed by RGD, so we listed them under a different phenotype. For example, traits such as open-field peripheral locomotion and rearings or entries in the closed arms of the elevated plus maze relate more to motor activity than anxiety, although none of these traits fall purely in one or the other category. Similarly, the intake of saccharin after a conditioned taste aversion relates to memory traits (*Sach6* and *Sach7*) and not to voluntary saccharin consumption such as the other *Sach* QTL. Finally, some QTL were given different names in RGD although they have been detected in the same cross; these QTLs were published in two reports because additional markers were added but we consider that they are identical (e.g. Alc9 = Alc5; Alc10 = Alc6, Alc11 = Alc7). In the same lines, when overlapping QTLs were detected in the same cross using similar measures, they are listed together (e.g. Alc14 = Alc16 for preference to 5% and 10% ethanol respectively) as well as measures from an independent cross but derived from the same parental strains (e.g. Alcrsp2 = Alcrsp17, Alcrsp3 = Alcrsp18, Alcrsp4 = Alcrsp19). Some of the listed QTLs were found to be sex or lineage-specific and some were transgressive (i.e. the increasing allele came from the low-phenotype parental strain), the details of which can be found in the associated publications. Of note, in spite of being all derived from a single F2 cross of 486 progeny, QTLs associated with behavioral and corticosterone stress responses appear disseminated in different sections of Table 25.2 and in five separated publications (28). Moreover, the column "overlapping QTL" refers to QTLs detected in the same region but it does not imply that the same genes are responsible for the overlapping QTLs, particularly in view of the very large confidence interval of some QTLs (usually the ones with lowest LOD scores). This information is still of

**Table 25.2**
**QTL mapping studies for psychiatry-related traits**

| Trait | Cross type ($n$) | Strains | QTL name and alias | Chr nb | Lod score max | $p$ value | References | Overlapping QTL |
|---|---|---|---|---|---|---|---|---|
| **Endocrine related stress response** | | | | | | | | |
| *HPA axis related* | | | | | | | | |
| Basal corticosterone | F2 (188) + congenics | F344 × LEW + DA. F344 | *Stresp4, Sr4* | 4 | | | (49) | Srcrt3, Sradr3, Foco1, Sach7 |
| Basal corticosterone | F2 (188) + congenics | F344 × LEW + DA. F344 | *Stresp5, Sr5* | 10 | 2.99 | 0.0025 | (49) | Stresp6, Stresp7, Anxrr26, Anxrr19 |
| Basal corticosterone | F2(327) | LH × LN | *Scort1* | 1 | 3.4 | | (50) | Despr9, Stresp9 |
| Basal corticosterone | F2(486) | F344 × WKY | *Stcrtb1* | 3 | 6.36 | 0.002 | (37) | Alc8, Alc19 |
| Basal corticosterone | F2(486) | F344 × WKY | *Stcrtb2* | 5 | 3.48 | 0.003 | (37) | Stresp17, Stresp18, Alcrsp3, Alcrsp8 |
| Corticosterone post-stress | F2(486) | F344 × WKY | *Srcrt1* | 2 | 3.66 | | (37) | Stresp13, Despr[illegible], Sach6, Alc15, Alcrsp2 |
| Corticosterone post-stress | F2(486) | F344 × WKY | *Srcrt2* | 3 | 2.78 | | (37) | Anxrr10, Despr11, Stresp16, Despr2 |
| Corticosterone post-stress | F2(486) | F344 × WKY | *Srcrt3* | 4 | 2.29 | | (37) | Sradr3, Stresp4, Sach7 |
| Corticosterone post-stress | F2(486) | F344 × WKY | *Srcrt4* | 6 | 6.39 | | (37) | Anxrr1, Anxrr5, Stresp14, Alcrsp4 |

(continued)

**Table 25.2 (continued)**

| Trait | Cross type (*n*) | Strains | QTL name and alias | Chr nb | Lod score max | *p* value | References | Overlapping QTL |
|---|---|---|---|---|---|---|---|---|
| Corticosterone post-stress | F2(486) | F344×WKY | *Srcrt5* | 15 | 3.38 | | (37) | |
| Aldosterone post-stress | F2 (122) | WKHA/BN | *Stresp15, Sr7* | 2 | 2 | 0.05 | (37) | Stresp14 |
| Aldosterone post-stress | F2 (122) | WKHA/BN | *Stresp17, Sr9* | 5 | 4.3 | 0.01 | (37) | Stcrb2, Stresp18, Alcrsp3, Alcrsp8 |
| Thymus weight | F2 (122) | WKHA/BN | *Stresp18, Sr10* | 5 | 2.9 | | (37) | Stresp17, Strcrb2, Alcrsp3, Alcrsp8 |
| Thymus weight | F2 (122) | WKHA/BN | *Stresp21, Sr13* | 10 | | | (37) | Stresp6, Anxrr25, Anxrr19, Alc9, Alc12 |
| Thymus weight | F2 (122) | WKHA/BN | *Stresp22, Sr14* | 16 | 3.3 | | (37) | Alc11, Sach2 |
| Adrenal gland weight | F2(486) | F344×WKY | *Sradr1* | 1 | 2.8 | | (37) | Foco2, Salc1, Anxrr13, Alcrsp1 |
| Adrenal gland weight | F2(486) | F344×WKY | *Sradr2* | 2 | 4.74 | | (37) | Stresp14, Anxrr7, Despr10, Sach6 |
| Adrenal gland weight | F2(486) | F344×WKY | *Sradr3* | 4 | 10.24 | | (37) | Stresp4, Srcrt3, Sach7 |
| Adrenal gland weight | F2(486) | F344×WKY | *Sradr4* | 4 | 6.19 | | (37) | |
| Adrenal gland weight | F2(486) | F344×WKY | *Sradr5* | 7 | 4.92 | | (37) | Activ5, Anxrr17, Anxrr8 |
| Adrenal gland weight | F2(486) | F344×WKY | *Sradr6* | 18 | 2.49 | | (37) | Anxrr11, Anxrr20, Ppulsi2, Sach3 |

| | | | | | | | | |
|---|---|---|---|---|---|---|---|---|
| Adrenal gland weight | F2 (227) | F344 × BN | *Gmadr1* | 19 | 5.83 | | (51) | |
| Renin activity | F2 (122) | WKHA/BN | *Stresp16, Sr8* | 3 | 4 | 0.001 | (36) | Despr2 |
| Renin activity | F2 (122) | WKHA/BN | *Stresp14, Sr6* | 2 | 5 | 0.001 | (36) | Anxrr7, Despr10, Sach6, Srad2, Stresp15 |
| Renin activity | F2 (122) | WKHA/BN | *Stresp19, Sr11* | 19 | 3.6 | 0.001 | (36) | |
| Renin activity | F2 (122) | WKHA/BN | *Stresp20, Sr12* | 8 | 3.7 | 0.05 | (36) | Activ1, Activ2, Anxrr15, Anxrr9, Stresp11 |
| Norepinephrine post-stress | F2 (189) | hHTg × BN | *Stresp6* | 10 | 4.17 | 0.022 | (52) | Stresp5, Stresp21, Anxrr26, Anxrr19 |
| Norepinephrine post-stress | F2 (189) | hHTg × BN | *Stresp7* | 10 | 3.52 | | (52) | Stresp5, Anxrr26 |
| Urinary excretion of norepinephrine post-stress | congenic | WKY.SHRSP | *Cstrr1* | 1 | | 0.0001 | (53) | Anxrr12, Memor3 |
| **Motor activity related** | | | | | | | | |
| Activity cages (crossing/1 h) | F2 (196) | WKHA/WKY | *Activ 1* | 8 | 15 | | (5, 6) | Stresp20, Activ2, Anxrr15, Anxrr9, Stresp11 |
| OFT: total locomotion | F2 (196) | WKHA/WKY | *Activ 1* | 8 | 9.3 | | (5, 6) | Stresp20, Activ2, Anxrr15, Anxrr9, Stresp11 |

(continued)

**Table 25.2 (continued)**

| Trait | Cross type (*n*) | Strains | QTL name and alias | Chr nb | Lod score max | *p* value | References | Overlapping QTL |
|---|---|---|---|---|---|---|---|---|
| Activity cages (rearings/1 h) | F2 (122) | WKHA/BN | *Activ 2* | 8 | 5 | 0.00004 | (6) | Activ1, Activ2, Anxrr15, Anxrr9, Stresp11 |
| Activity cages (rearings/1 h) | F2 (122) | WKHA/BN | *Activ 3* | 1 | 3.3 | 0.0013 | (6) | Scort1, Stresp9 |
| Activity cages (rearings/1 h) | F2 (122) | WKHA/BN | *Activ 4* | 3 | 3.3 | 0.00002 | (6) | Despr11, Sach1 |
| EPM: entries closed arms | RI (22) | H×B/B×H | *Anxrr10* | 3 | 3.9 | | (54) | Despr11 |
| EPM: entries closed arms | RI (22) | H×B/B×H | *Anxrr11* | 18 | 3.4 | | (54) | Sradr6, Anxrr20, Ppulsi2, Sach3 |
| EPM: entries closed arms | F2 (192) | SHR×LEW | *Activ 5* | 7 | 2.9 | 0.0052 | (11) | Anxrr17, Stresp8, Sradr5 |
| OFT: outer locomotion | F2 (192) | SHR×LEW | *Activ 6* | 5 | 3.7 | 0.0002 | (11) | Anxrr22 |
| OFT: total locomotion | F2 (196) | WKY×HEP | *Anxrr13* | 1 | 4.6 | | (13) | Alcrsp1 |
| OFT: rearings | F2 (908) | RLA×RHA | *Activ7* | 1 | 3.5 | | (7) | |
| OFT &DB: rearings | F2 (486) | F344×WKY | *Anxrr18, Rear1* | 2 | 2.98 | | (28) | |
| OFT &DB: rearings | F2 (486) | F344×WKY | *Anxrr19, Rear2* | 10 | 5.07 | | (28) | Stresp6, Stresp7, Anxrr24, Stresp5 |
| OFT : rearings | F2 (486) | F344×WKY | *Anxrr20, Rear3* | 18 | 3.04 | | (28) | Sradr6, Anxrr11, Ppulsi2, Sach3 |

| | | | | | | | | |
|---|---|---|---|---|---|---|---|---|
| OFT: rearings | F2 (486) | F344 × WKY | *Activ8* | 6 | 8.24 | | (28) | |
| **Anxiety related** | | | | | | | | |
| OFT: inner locomotion | F2 (192) | SHR × LEW | *Anxrr16, Ofil1* | 4 | 10.4 | 0.00001 | (11) | Alc18, Alc21, Alc14, Alc16, Sach4 |
| OFT: inner locomotion | F2 selected | SHR × LEW | *Anxrr16, Ofil1* | 4 | | 0.001 | (55) | |
| OFT: inner locomotion | F2 selected | WKY × HEP | *Anxrr16, Ofil1* | 4 | | | (56) | |
| OFT: inner locomotion | F2 selected | SHR × LEW | *Anxrr16, Ofil1* | 4 | | 0.05 | (57) | |
| OFT: inner locomotion | F2 (48) | SHR × LEW | *Anxrr17, Ofil2* | 7 | 3.66 | | (11) | Sradr5, Activ5, Anxrr6, Anxrr8, Stresp8, Alcrsp9 |
| OFT: inner locomotion | F2 selected | SHR × LEW | *Anxrr17, Ofil2* | 7 | | 0.001 | (55) | |
| EPM: %entries open arms | F2 (192) | SHR × LEW | *Anxrr1* | 6 | 2.8 | 0.0017 | (11) | Srcrt4, Anxrr5, Stresp14, Alcrsp4 |
| OFT: peripheral locomotion & defecation | F2 (908) | RLA × RHA | *Anxrr21* | 3 | 5.04 & 3.18 | | (7) | |
| Shuttle box: avoidance, latency, intertrial corssing | F2 (908) | RLA × RHA | *Anxrr22* | 5 | 9.47; 6.12 ; 6.46 | | (7) | Activ6 |
| Fear conditioning: cue & context | F2 (908) | RLA × RHA | *Anxrr22* | 5 | 3.49 & 4.46 | | (7) | Activ6 |
| EPM: entries open arms | F2 (908) | RLA × RHA | *Anxrr22* | 5 | 4.08 | | (7) | Activ6 |
| OFT: peripheral locomotion & grooming & rearings | F2 (908) | RLA × RHA | *Anxrr22* | 5 | 3.14; 3.05; 4.63 | | (7) | Activ6 |

(continued)

**Table 25.2 (continued)**

| Trait | Cross type (*n*) | Strains | QTL name and alias | Chr nb | Lod score max | *p* value | References | Overlapping QTL |
|---|---|---|---|---|---|---|---|---|
| Activity cages (crossing/30 min) & defecation | F2 (908) | RLA × RHA | *Anxrr23* | 6 | 7.45; 3.36 | | (7) | |
| Shuttle box: avoidance, latency, intertrial corssing | F2 (908) | RLA × RHA | *Anxrr24* | 10 | 4.1; 3.1; 4.00 | | (7) | Stresp5, Stresp6, Stresp7, Anxrr19 |
| Fear conditioning: context | F2 (908) | RLA × RHA | *Anxrr24* | 10 | 5.95 | | (7) | |
| Acoustic startle response | F2 (908) | RLA × RHA | *Anxrr24* | 10 | 3.53 | | (7) | |
| Activity cages (crossing/30 min) & defecation | F2 (908) | RLA × RHA | *Anxrr24* | 10 | 3.26 | | (7) | |
| EPM: closed arms entries | F2 (908) | RLA × RHA | *Anxrr25* | 15 | 3.43 | | (7) | |
| OFT: inner locomotion & grooming | F2 (908) | RLA × RHA | *Anxrr25* | 15 | 3.45 &3.06 | | (7) | |
| Acoustic startle response | F2 (908) | RLA × RHA | *Anxrr25* | 15 | 4.83 | | (7) | |
| Grooming & defecation | F2 (908) | RLA × RHA | *Anxrr26* | 19 | 3.95 &3.3 | | (7) | |
| Activity cages (crossing/30 min) & defecation | F2 (908) | RLA × RHA | *Anxrr27* | X | 3.9 & 6.18 | | (7) | |
| Anxiety related | | | | | | | | |
| EPM: time in center | F2 | WKY × HEP | *Anxrr12* | 1 | 4.9 | 0.00001 | (13) | Cstrr1, Memor3 |
| EPM: time open arms | F2 | WKY × HEP | *Anxrr14* | 1 | 4 | 0.00035 | (13) | |

| | | | | | | | | |
|---|---|---|---|---|---|---|---|---|
| EPM: time in center | F2 | WKY × HEP | *Anxrr15* | 8 | 3.6 | 0.005 | (13) | Activ1, Activ2, Stresp20, Anxrr9, Stresp11 |
| EPM: %entries open arms | RI (22) | H × B/B × H | *Anxrr3* | 2 | 3.8 | | (54) | |
| EPM: %entries open arms | RI (22) | H × B/B × H | *Anxrr4* | 5 | 5.1 | | (54) | |
| EPM: %entries open arms | RI (22) | H × B/B × H | *Anxrr5* | 6 | 5.9 | | (54) | Anxrr1, Srcrt4, Stresp14, Alcrsp4 |
| EPM: %entries open arms | RI (22) | H × B/B × H | *Anxrr6* | 7 | 4.1 | | (54) | |
| EPM: %entries open & closed arms | RI (22) | H × B/B × H | *Anxrr7* | 2 | 4.4 | | (54) | Sradr2, Stresp14, Despr10, Sach6 |
| EPM: %entries open & closed arms | RI (22) | H × B/B × H | *Anxrr8* | 7 | 4.1 | | (54) | Sradr5, Anxrr17, Anxrr8, Alcrsp9 |
| EPM: %entries open & closed arms | RI (22) | H × B/B × H | *Anxrr9* | 8 | 6.1 | | (54) | Activ1, Activ2, Anxrr15, Stresp20, Stresp11 |
| DB: latency & duration of burying prod | F2 (486) | F344 × WKY | *Stresp1* | X | 4.96 | 0.00001 | (12) | Stresp2, Memor1, Memor2, Memor14, Memor15 |
| DB: approach to prod | F2 (486) | F344 × WKY | *Stresp2* | X | 3.4 | 0.0004 | (12) | Stresp1, Stresp3, Memor1, Memor2, Memor13, Memor14, Memor15 |

(continued)

**Table 25.2 (continued)**

| Trait | Cross type (*n*) | Strains | QTL name and alias | Chr nb | Lod score max | *p* value | References | Overlapping QTL |
|---|---|---|---|---|---|---|---|---|
| DB: approach to prod | F2 (486) | F344 × WKY | *Stresp3* | X | 4.61 | 0.0066 | (12) | Stresp2, Memor1, Memor13, Memor15 |
| DB: nb of shocks | F2 (486) | F344 × WKY | *Stresp8* | 7 | 4.37 | 0.0008 | (58) | |
| DB: approach to prod | F2 (486) | F344 × WKY | *Stresp9* | 1 | 6.13 | 0.0035 | (58) | Scort1 |
| DB: approach to prod | F2 (486) | F344 × WKY | *Stresp10* | 6 | 6.83 | 0.0066 | (58) | Despr12, Alc17, Sach5 |
| DB: approach to prod | F2 (486) | F344 × WKY | *Stresp11* | 8 | 6.83 | 0.0019 | (58) | Activ1, Activ2, Anxrr15, Anxrr9, Stresp20 |
| DB: approach to prod | F2 (486) | F344 × WKY | *Stresp12* | 13 | 3.35 | | (58) | |
| DB: latency to burying prod | F2 (486) | F344 × WKY | *Stresp13* | 2 | 4.56 | 0.0011 | (58) | Despr1 |
| DB: latency to burying prod | F2 (486) | F344 × WKY | *Stresp10* | 6 | 3.55 | | (58) | Anxrr1, Anxrr5, Srcrt4, Alcrsp4 |
| **Depression related** | | | | | | | | |
| FST: immobility level | F2 (486) | F344 × WKY | *Despr1, Imm1* | 2 | | 0.0035 | (59) | Stresp13 |
| FST: immobility level | F2 (486) | F344 × WKY | *Despr2, Imm2* | 3 | | 0.0028 | (59) | Stresp16 |
| FST: immobility level | F2 (486) | F344 × WKY | *Despr3, Imm3* | 5 | | 0.0002 | (59) | |
| FST: immobility level | F2 (486) | F344 × WKY | *Despr4, Imm4* | 8 | | 0.0036 | (59) | |
| FST: immobility level | F2 (486) | F344 × WKY | *Despr5, Imm5* | 9 | | 0.0017 | (59) | Alcrsp13 |

| | | | | | | | | |
|---|---|---|---|---|---|---|---|---|
| FST: immobility level | F2 (486) | F344 × WKY | *Despr6, Imm6* | 16 | | 0.0067 | (59) | Despr7, Alc11, Alc13 |
| FST: immobility & climbing | F2 (486) | F344 × WKY | *Despr7* | 16 | | 0.016 | (59) | Despr6, Alc11, Alc13 |
| FST: climbing level | F2 (486) | F344 × WKY | *Despr8, Climb1* | 1 | | 3.41E-05 | (59) | |
| FST: climbing level | F2 (486) | F344 × WKY | *Despr9, Climb2* | 1 | | 0.00019 | (59) | Scort1, Activ3 |
| FST: climbing level | F2 (486) | F344 × WKY | *Despr10, Climb3* | 2 | | 2.5E-06 | (59) | Sradr2, Sresp14, Anxrr7, Sach6 |
| FST: climbing level | F2 (486) | F344 × WKY | *Despr11, Climb4* | 3 | | 0.00003 | (59) | |
| FST: climbing level | F2 (486) | F344 × WKY | *Despr12, Climb5* | 6 | | 0.0012 | (59) | Stresp10, Alc17, Sach5 |
| FST: climbing level | F2 (486) | F344 × WKY | *Despr13, Climb6* | 15 | | 0.0022 | (59) | Alcrsp14 |
| FST: swimming level | F2 (486) | F344 × WKY | *Despr14, Climb7* | 8 | | 0.0056 | (59) | |
| FST: swimming level | F2 (486) | F344 × WKY | *Despr15, Climb8* | 20 | | 0.0027 | (59) | |
| **Sensory motor gating related** | | | | | | | | |
| Prepulse inhibition | N2 (160) | BN × WKY | *Ppulsi1* | 2 | 3.63 | | (60) | |
| Prepulse inhibition | N2 (160) | BN × WKY | *Ppulsi2* | 18 | 2.71 | | (60) | Sradr6, Anxrr20, Anxrr11, Sach3 |
| Prepulse inhibition | F2 selected | SHR × LEW | *Anxrr17, Ofil2* | 7 | | | (61) | Srcrt4, Anxrr5, Stresp14, Alcrsp4 |

(continued)

**Table 25.2 (continued)**

| Trait | Cross type (*n*) | Strains | QTL name and alias | Chr nb | Lod score max | *p* value | References | Overlapping QTL |
|---|---|---|---|---|---|---|---|---|
| **Memory related** | | | | | | | | |
| MWM: acquisition performance. | F2 (200) | Dahl R×Dahl S | *Memor1, Nav-1* | X | 1.9 | | (62) | Stresp1, Stresp3, Memor1, Memor2, Memor13, Memor14, Memor15 |
| MWM: spatial accuracy | F2 (200) | Dahl R×Dahl S | *Memor2, Nav-2* | X | 2.5 | | (62) | Stresp1, Stresp2, Memor1, Memor14, Memor15 |
| MWM: acquisition performance. | F2 (200) | Dahl R×Dahl S | *Memor3, Nav-3* | 1 | 4.5 | | (62) | Cstrr1, Anxrr12 |
| MWM: acquisition performance. | F2 (200) | Dahl R×Dahl S | *Memor4, Nav-4* | 8 | 2.5 | | (62) | |
| MWM: acquisition performance. | F2 (200) | Dahl R×Dahl S | *Memor5, Nav-5* | 17 | 5.3 | | (62) | |
| MWM: acquisition performance. | F2 (200) | Dahl R×Dahl S | *Memor6, Nav-6* | 20 | 3.1 | | (62) | Memor12 |
| MWM: acquisition performance. | F2 (200) | Dahl R×Dahl S | *Memor7, Nav-7* | 11 | interacting | | (62) | |
| MWM: spatial accuracy | F2 (200) | Dahl R×Dahl S | *Memor8, Nav-8* | 2 | 3 | | (62) | Sach6 |
| MWM: spatial accuracy | F2 (200) | Dahl R×Dahl S | *Memor9, Nav-9* | 2 | 2.7 | | (62) | Sach6, Alc15, Alcrsp2 |

| | | | | | | | |
|---|---|---|---|---|---|---|---|
| MWM: spatial accuracy | F2 (200) | Dahl R×Dahl S | *Memor10, Nav-10* | 3 | 2 | (62) | |
| MWM: spatial accuracy | F2 (200) | Dahl R×Dahl S | *Memor11, Nav-11* | 9 | 2.5 | (62) | Memor16, Memor17 |
| MWM: spatial accuracy | F2 (200) | Dahl R×Dahl S | *Memor12, Nav-12* | 20 | 2.4 | (62) | Memor6 |
| Social recognation task | F2 (178) | Dahl R×Dahl S | *Memor13, SR-1* | X | 3.2 | (62) | Stresp2, Stresp3, Memor1, Memor14, Memor15 |
| Social recognation task | F2 (178) | Dahl R×Dahl S | *Memor14, SR-2* | X | 4.5 | (62) | Stresp1, Stresp2, Memor1, Memor2, Memor13, Memor15 |
| Social recognation task | F2 (178) | Dahl R×Dahl S | *Memor15, SR-3* | X | 4.4 | (62) | Stresp1, Stresp2, Stresp3 Memor1, Memor2, Memor13, Memor14 |
| Social recognation task | F2 (178) | Dahl R×Dahl S | *Memor16, SR-14* | 9 | 1.9 | (62) | Memor11 |
| Social recognation task | F2 (178) | Dahl R×Dahl S | *Memor17, SR-15* | 9 | 2.2 | (62) | Memor11 |
| Conditioned taste aversion: saccharin + LiCl injection | RI (24) + congenic SHR-2 | H×B/B×H | *Sach6* | 2 | 4.9 | (14) | Memor8, Despr10, Anxrr7, Stresp14, Sradr2 |

(continued)

**Table 25.2 (continued)**

| Trait | Cross type (*n*) | Strains | QTL name and alias | Chr nb | Lod score max | *p* value | References | Overlapping QTL |
|---|---|---|---|---|---|---|---|---|
| Conditioned taste aversion: saccharin + LiCl injection | RI (24) | H × B/B × H | *Sach7* | 4 | 5.1 | | (14) | |
| **Alcohol & saccharin consumption related** | | | | | | | | |
| Alcohol preference (10%) | F2(384) | P × NP | *Alc18 = Alc21, Eoh1* | 4 | 9.2 | | (8, 9) | Anxrr16, Alc21, Alc14, Alc16, Sach4 |
| Alcohol preference (10%) | congenics | P.NPchr4, NP.PChr4 | *Alc21* | 4 | | 0.0005 | (63) | Alc18, Anxrr16, Alc14, Alc16, Sach4 |
| Alcohol preference (5% & 10%) | F2 (196) | HEP × WKY | *Alc14 = Alc16, CoEt5* | 4 | 7.6 & 3.6 | | (13) | Alc18, Alc21, Anxrr16, Alc16, Sach4 |
| Saccharin preference 7,5 mM | F2 (196) | HEP × WKY | *Sach4* | 4 | 4.9 | | (13) | Alc18, Alc21, Alc14, Alc16, Anxrr16 |
| Alcohol preference (10%) | F2(384) | P × NP | *Alc19, Eoh2* | 3 | 2.5 | | (9) | |
| Alcohol preference (10%) | F2(384) | P × NP | *Alc20, Eoh3* | 8 | 2 | | (9) | |
| Alcohol consumption (10%) | F2(459) | HAD1 × LAD1 | *Alc4* | 5 | 3.5 | | (64) | |
| Alcohol consumption (10%) | F2(459) | HAD1 × LAD1 | *Alc9 = Alc5* | 10 | 2.2 | 0.01 | (65, 66) | Alc12 |
| Alcohol consumption (10%) | F2(428) | HAD2 × LAD2 | *Alc12* | 10 | 4.5 | 0.0001 | (66) | Alc9 |
| Alcohol consumption (10%) | F2(459) | HAD1 × LAD1 | *Alc10 = Alc6* | 12 | 6.2 | 0.0001 | (65, 66) | Alcrsp5, Alcrsp15 |

| Alcohol consumption (10%) | F2(459) | HAD1×LAD1 | *Alc11*=*Alc7* | 16 | 3.2 | 0.001 | (65, 66) | Alc13, Despr6, Despr7, Sach2, Stresp22 |
|---|---|---|---|---|---|---|---|---|
| Alcohol consumption (10%) | F2(428) | HAD2×LAD2 | *Alc13* | 16 | 3.1 | 0.001 | (66) | Alc11, Despr6, Despr7 |
| Alcohol consumption (10%) | F2(384) | P×NP | *Alc8* | 3 | 3.5 | 0.03 | (65) | Alc19, Sach1 |
| Saccharin preference 0,1%w/v | F2(381) | P×NP | *Sach1* | 3 | 2.7 | 0.02 | (65) | Alc8, Alc19 |
| Saccharin preference 0,1%w/v | F2(381) | P×NP | *Sach2* | 16 | 2.5 | 0.02 | (65) | Alc11, Stresp22 |
| Saccharin preference 0,1%w/v | F2(381) | P×NP | *Sach3* | 18 | 3.9 | 0.01 | (65) | Sradr6, Anxrr20, Ppulsi2, Anxrr11 |
| Alcohol consumption (10%) | F2 (196) | HEP×WKY | *Alc15* | 2 | 4.1 | | (13) | |
| Alcohol preference (5%) | F2 (196) | HEP×WKY | *Alc17* | 6 | 3.1 | | (13) | Stresp10, Despr12, Sach5 |
| Saccharin preference 7,5 mM | F2 (196) | HEP×WKY | *Sach5* | 6 | 3.9 | | (13) | Stresp10, Despr12, Alc17 |
| Alcohol response: LORR | F2 (363) | HAS1×LAS1 | *Alcrsp1* | 1 | 2.2 | | (10) | |
| Alcohol response: dBEC | F3 selected | HAS1×LAS1 | *Alcrsp1* | 1 | | | (67) | |
| Alcohol response: LORR | F2 (363) | HAS1×LAS1 | *Alcrsp2* | 2 | 3.5 | | (10) | Alc15, Sach6, Memor9, Srcrt1 |
| Alcohol response: LORR | F2 (450) +con-genic | HAS1×LAS1 | *Alcrsp2* = *Alcrsp17* | 2 | 2.9 | | (68) | Alc15 |
| Alcohol response: LORR | F2 (363) | HAS1×LAS1 | *Alcrsp3* | 5 | 2.8 | | (10) | Alcrsp8 |

(continued)

**Table 25.2 (continued)**

| Trait | Cross type (*n*) | Strains | QTL name and alias | Chr nb | Lod score max | *p* value | References | Overlapping QTL |
|---|---|---|---|---|---|---|---|---|
| Alcohol response: LORR | F2 (450) +congenic | HAS1 × LAS1 | *Alcrsp3* = *Alcrsp18* | 5 | 1.7 | | (68) | Stresp17, Stresp18, Alcrsp3, Stcrb2 |
| Alcohol response: LORR | F2 (363) | HAS1 × LAS1 | *Alcrsp4* | 6 | 2.2 | | (10) | Anxrr1, Anxrr5, Stresp14, Srcrt4 |
| Alcohol response: dBEC | F4 selected | HAS1 × LAS1 | *Alcrsp4* = *Alcrsp19* | 6 | | | (67) | |
| Alcohol response: LORR | F2 (363) | HAS1 × LAS1 | *Alcrsp5* | 12 | 2.4 | | (10) | Alc10, Alcrsp15 |
| Alcohol response: LORR | F2 (450) | HAS1 × LAS1 | *Alcrsp5?* | 12 | 4.9 | | (68) | |
| Alcohol response: dBEC | F3 selected | HAS1 × LAS1 | *Alcrsp5?* | 12 | | | (67) | |
| Alcohol response: LORR | F2 (363) | HAS1 × LAS1 | *Alcrsp6* | 13 | 2.9 | | (10) | |
| Alcohol response: LORR | F2 (450) | HAS1 × LAS1 | *Alcrsp6* | 13 | 3.7 | | (68) | |
| Alcohol response: dBEC | F4 selected | HAS1 × LAS1 | *Alcrsp6* | 13 | | | (67) | |
| Alcohol response: LORR | F2 (363) | HAS1 × LAS1 | *Alcrsp7* | 20 | 2.2 | | (10) | |
| Alcohol response: BEC | F2 (363) | HAS1 × LAS1 | *Alcrsp8* | 5 | 2.5 | | (10) | Stresp17, Stresp18, Alcrsp3, Stcrb2 |
| Alcohol response: BEC | F2 (363) | HAS1 × LAS1 | *Alcrsp9* | 7 | 3 | | (10) | Sradr5, Anxrr17, Anxrr8 |
| Alcohol response: BEC | F2 (363) | HAS1 × LAS1 | *Alcrsp10* | 11 | 2.8 | | (10) | |
| Alcohol response: BEC | F2 (363) | HAS1 × LAS1 | *Alcrsp11* | 14 | 2.5 | | (10) | |

| | | | | | | | |
|---|---|---|---|---|---|---|---|
| Alcohol response: NTR1-Nac | F2 (363) | HAS1 × LAS1 | *Alcrsp12* | 9 | 2.4 | (10) | |
| Alcohol response: NTR1-CP | F2 (363) | HAS1 × LAS1 | *Alcrsp13* | 9 | 3.2 | (10) | Despr5 |
| Alcohol response: NTR1-CP | F2 (363) | HAS1 × LAS1 | *Alcrsp14* | 15 | 3.6 | (10) | Despr13 |
| Alcohol response: BEC | F2 (450) | HAS1 × LAS1 | *Alcrsp15* | 12 | 2 | (68) | Alc10, Alcrsp5 |
| Alcohol response: BEC | F2 (450) | HAS1 × LAS1 | *Alcrsp16* | 13 | 3.8 | (68) | |

*OFT* open field test, *EPM* elevated plus maze, *DB* defense burying test, *FST* forced swimming test, *MWM* Morris water maze, *LORR* loss of righting reflex, *BEC* blood ethanol concentration, *NTR1* neurotensin receptor 1, *Nac* nucleus accumbens, *CP* caudate putamen, Strains: *F344* Fischer-344, *LEW* Lewis, *DA* Dahl, *LH* Lyon hypertensive, *LN* Lyon nonhypertensive, *WKY* Wistar-kyoto, *WKHA* Wistar-kyoto hyperactive, *BN* Brown Norway, *hHTg* hereditary hypertriglyceridemic, *SHRSP* spontaneously hypertensive stroke prone, HxB or BxH RI from BN and SHR: *SHR* spontaneously hypertensive, *HEP* high ethanol preferring, *RLA* Roman low avoidance, *RHA* Roman High avoidance, *Dahl R* Dahl resistant, *Dahl S* Dahl sensitive, *P* alcohol preferring, *NP* alcohol nonpreferring, *HAD* high alcohol drinking, *LAD* low alcohol drinking, *HAS* high alcohol sensitive, *LAS* low alcohol sensitive interest as some QTL "hot spots" sometimes appear as affecting related traits.

The most robust QTLs are of course the ones with highly significant LOD score (>4.5), replicated in an independent cross, confirmed in a congenic or F2 selected strain and overlapping with QTLs associated to related traits originating from independent tests. From Table 25.2, one can find some QTLs that respond to two or three of these criteria e.g. for endocrine measures after stress, motor activity, anxiety or alcohol consumption. Few positional candidate genes for a given QTL have been tested, except the alpha-synuclein gene in *Alc18*, which was associated with alcohol-drinking in P and NP rats (29) and was also found to cosegregate with higher inner locomotion in the open field test in an overlapping QTL associated to anxiety (*Anxrr16*) (30). The potential relationship between polymorphisms in the alpha-synuclein gene and alcohol drinking was corroborated in a clinical study, where eight SNPs mapped across this gene were associated with alcohol craving in humans (31), which is a good example of transposition of results from rat studies to human research. A second line of investigation focused on the putative role of neuropeptide Y (NPY), which is also located near the QTL *Alc18*, and on the differences observed between P and NP rats. Interestingly, the gene coding for NPY has been suggested to have a common effect on both anxiety- and alcohol-related disorders through the action of cAMP responsive element-binding protein (CREB), which shows like NPY a lower expression in the amygdala of P rats (32). Moreover, ethanol administration reduced anxiety and increased CREB function in the amygdala of P, but not of NP rats (32), a result confirmed in outbred Long-Evans rats (33). Differences in the expression of NPY-related genes were further confirmed by a recent microarray study looking at different brain areas of P and NP rats (34). Finally, in humans, a SNP in the NPY gene (at -602) has been recently found to be associated with alcohol dependence in a Nordic population (35).

Other serious candidate genes include the gene encoding the angiotensin 1B receptor for a QTL associated to renin activity after stress (Stresp14) (36), and the gene encoding transcortin for poststress corticosterone levels (Srcrt4) (37) that was previously found for the same trait in a pig genetic study (38).

## 5. Transcriptomic and Proteomic Studies

DNA microarrays technology involves large-scale monitoring of relative differences in RNA concentrations between samples, which are assumed to represent changes in function. The expression of thousands of genes can be analyzed simultaneously from a given tissue (reviewed by N. Mei in this book). This technology has opened new perspectives in the field of psychiatric research with the aim to identify genes that confer risk for a particular disease or trait and for the identification of genes that contribute to the regulation of the neuronal pathways implicated in normal and pathological state of the same traits. Once more, the rat has remained one model of choice, by circumventing the limitations inherent to human studies and by providing a large number of behavioral paradigms and pharmacological manipulations established in this species. Transcriptomics holds great potential to identify QTL candidate genes in particular by comparing a strain to its QTL congenic as in the study by (39). Proteomic analysis in neuroscience and psychiatry (sometimes called psychoproteomics) is still in its infancy but is rapidly advancing (40). Proteomics are complementary to transcriptomics analysis as some regulation appeared to occur only at the protein levels e.g. ΔFosB in drug addiction (41) or BDNF in resilience to stress (42) and also because of the highest complexity of the proteome compared to the transcriptome in terms of number of proteins versus transcripts, protein modifications or range of expression from fento to millimolar (40). In classical protocols, protein profiles are visualized by 2D gel electrophoresis, based on the isoelectric point and molecular mass of the proteins. The spots corresponding to differentially expressed proteins are then isolated from the gels and identified by mass spectrometry and protein databases. These approaches have been used by several groups to investigate the molecular basis of depression and antidepressant responses, alcoholism, drug addiction and toxicity or learning and memory. Table 25.3 summarises a number of these transcriptomics and proteomics studies with a focus on the approaches rather than the results, which would be too long to elaborate. However, recent reviews are available such as (40), (43). As general comments, it is apparent that in all these studies, changes in gene or protein levels are subtle, which fits with the low percentage of variability explained by a single QTL that is usually reported. When several tissues have been examined in a given study, the gene or protein changes were found to be region-specific.

Metabonomic approaches were first used in characterizing metabolic responses to toxicity and disease in mammals. Recently, this approach has been used in combination with hematological analysis to investigate the biochemical effects of acute and chronic psychological stress in rats (44). Plasma samples of rats collected at different time points after stress sessions were submitted to $^{1}H$ nuclear magnetic resonance (NMR) spectroscopy that provides, without a priori hypothesis, metabolic profiles of low molecular weight metabolites in a given fluid. The information obtained through these high-resolution spectra is then analyzed by chemo-

**Table 25.3**
**Examples of transcriptomic, proteomic and metabonomic studies on psychiatry-related traits**

| Disease/Phenotype | Rat strain(s) | Experimental groups | Tissues | No. of transcripts screened (chips name) | Publications |
|---|---|---|---|---|---|
| Depression: antidepressant response | Sprague-Dawley (adult male) | – Learned helplessness + saline<br>Learned helplessness + imipramine<br>– Learned helplessness + fluoxetine<br>– Unstressed controls + saline | Prefrontal cortex<br>Hippocampus | 8,000 transcripts (Affymetrix U34A) | (69) |
| Depression: sensitivity vs resilience | Sprague-Dawley (adult male) | – Learned helplessness sensitives<br>– Learned helplessness resilients<br>– Stressed (restrain)<br>– Naïve controls | Hippocampus | 8,000 transcripts (Affymetrix U34A) | (70) |
| Depression: social dominance-submission | Long-Evans (adult male) | – Resident/intruder dominants–<br>Resident/intruder subordinates<br>– Controls | Neocortex | 1,178 CNS mRNA-Home-made | (71) |
| Depression: inherited genetic differences | WKY & Sprague-Dawley | – WKY (depression-sensitive)<br>– Sprague-Dawley (controls) | Locus coeruleus-Dorsal raphe | 1,323 transcripts (Affymetrix U34A) | (72) |
| Depression/Anxiety: chronic mild stress | Wistar (adult male) | – Chronic mild stress treated<br>– Untreated controls | Frontal cortex | 30,200 transcripts (Affymetrix 230 2.0) | (73) |
| Depression: antidepressants response | Wistar | – Type 5 metabotropic glutamate receptor antagonists treated<br>– Untreated controls | Cortex | 30,200 transcripts (Affymetrix 230 2.0) | (74) |
| Depression: antidepressant response | Sprague-Dawley (adult male) | – Venlafaxine treated<br>– Fluoxetine treated<br>– Untreated controls | Hippocampus | Whole tissue proteins | (75) |
| Depression: antidepressant response | Sprague-Dawley (adult male) | – Fluoxetine treated Neurokinin 1 receptor antagonist treated<br>– Corticotropin-releasing factor 1 antagonist treated<br>– Controls | Frontal cortex<br>Hippocampus | Whole tissue proteins | (76) |

| | | | | | |
|---|---|---|---|---|---|
| Depression/Anxiety: chronic unpredictable stress | Sprague-Dawley (adult male) | – Chronic stress treated<br>– Untreated controls | Hippocampus | Whole tissue proteins | (77) |
| Depression/Anxiety: restraint stress | | – Restraint stress treated<br>– Unstressed controls | Hippocampus | Whole tissue proteins | (78) |
| Depression/Anxiety: acute & chronic stress | Sprague-Dawley (adult male) | – Basal condition<br>– Acute stress<br>– Chronic stress | – | Plasma metabolites | (44) |
| Alcoholism: inherited genetic differences | Alko, alcohol (AA) Alko,non alcohol (ANA) | – Alko, alcohol (AA)<br>– Alko, nonalcohol (ANA) | Cortex | 1,176 transcripts | (79) |
| Alcoholism: inherited genetic differences | P/NP | – P (alcohol preferring)<br>– NP (alcohol nonpreferring) | Nucleus accumbens<br>Caudate putamen<br>Frontal cortex<br>Hippocampus<br>Amygdala | 8,000 transcripts (Affymetrix U34A) | (34, 80) |
| Alcoholism: congenic with QTL associated to alcohol preference | P/NP | – NP (nonpreferring)<br>– NP.P (congenic with QTL on chr4 for alcohol preference) | Nucleus accumbens<br>Caudate putamen<br>Frontal cortex<br>Hippocampus<br>Amygdala | 30,200 transcripts (Affymetrix 230 2.0) | (39) |
| Alcoholism: vulnerability of hippocampus during adolescence | Wistar (young & adult males) | – Adolescent rats (27-day-old) with access to beer (4.44%ethanol v/v)<br>– Adult rats (55-day-old) with access to beer (4.44%ethanol v/v)<br>– Age-matched controls | Huippocampus | Whole tissue proteins | (81) |

(continued)

**Table 25.2 (continued)**

| Disease/Phenotype | Rat strain(s) | Experimental groups | Tissues | No. of transcripts screened (chips name) | Publications |
|---|---|---|---|---|---|
| Drug abuse: individual differences | Sprague-Dawley (adult male) | – Low responders basal conditions<br>– High responders basal conditions<br>– Low responders socially defeated<br>– High responders socially defeated | Hippocampus | 1,200 brain transcripts (Affymetrix rat neurobiology U34A) | (82) |
| Drug abuse: escalation of cocaine use | Wistar (adult male) | – Escalated cocaine intake<br>– Stable cocaine intake<br>– Cocaine naive | Prefrontal cortex, nucleus accumbens, septum, lateral hypothalamus, amygdala, ventral tegmental area | 1,200 brain transcripts (Affymetrix rat neurobiology U34A) | (83) |
| Drug abuse: chronic morphine & morphine withdrawal | Sprague-Dawley (adult male) | – Chronic morphine + saline treated<br>– Chronic morphine + antagonist<br>– Sham treated + saline<br>– Sham-treated + antagonist | Locus coeruleus-Ventral tegmental area | 8,000 transcripts (Affymetrix U34A) | (84) |
| Drug abuse: methamphetamine toxicity | Sprague-Dawley (adult male) | – Methamphetamine treated<br>– Saline controls | Cortex | Whole tissue proteins | (85) |
| Drug abuse: methamphetamine toxicity | Wistar (adult male) | – Methamphetamine treated<br>– Saline controls | Frontal cortex<br>Hippocampus<br>Striatum | Whole tissue proteins | (86) |
| Learning/memory: age-related spatial navigation | F344 (young & adult males) | – 3-month-old rat trained on Morris maze<br>– 24-month-old rats unimpaired learners<br>– 24-month-old impaired learners<br>– Age-matched controls | Hippocampus: CA1 and dentate gyrus | 30,200 transcripts (Affymetrix 230 2.0) | (87) |

metric models (mathematical and statistical methods applied to chemistry), to identify the metabolites associated with the response to treatment, i.e. stress in this study. Such metabonomic studies are promising as they are noninvasive and may provide predictive biomarkers of diseases.

## 6. Transgenics and ENU Induced Mutations

A rat transgenic model that expresses a human mutant of the amyloid precursor protein (APPswe), driven by the platelet-derived growth factor promoter, was developed on a Fischer-344 background (45). These TgAPPswe rats exhibit a small increase in brain APP mRNA and peptide levels. Unexpectedly, the cognitive performance of TgAPPswe rats was better than the age-matched Fischer-344 controls in social and Morris water maze tasks. The authors concluded that APP expression has to reach a threshold to become pathologic. Later, the role of adenosine receptors in the central nervous system was evaluated by the creation of a transgenic rat overexpressing human adenosine 2A receptors mainly in the cerebral cortex, the hippocampus and the cerebellum by using the neuron-specific enolase promoter on a Sprague-Dawley background. These TGR(NSEhA2A) rats were submitted to a battery of behavioral tests to assess learning and memory, spontaneous motor activity and anxiety. Deficits were detected in working memory estimated by a poorer performance in the Morris water maze, in a 6 arm radial maze and in recognition of a novel object. No difference in activity and anxious-like behaviors were found compared to control animals (46).

Also noteworthy is the recent study involving a rat line carrying a null mutation in the serotonin transporter gene. These knockout animals were generated by ENU-induced mutagenesis followed by genotypic selection of the mutant target gene (47). When submitted to a battery of emotionality tests, the homozygous knockout rats were more anxious-like than the wild types in the open field, elevated plus maze and novelty suppressed feeding (males only) and more depressive-like in the forced swim and sucrose consumption tests. Moreover, mutant animals had higher levels of extracellular serotonin, as expected (48).

## 7. Conclusion

Genomics methods and tools have largely been used in the field of psychiatric research. As technologies are continuously improving, new studies are expected that would explore for example the role of DNA methylation, chromatin structure or mRNA processing in gene expression changes in brain after a given stimulus or in response to a treatment. Moreover, the activation of protease systems called "degradome" in response to drug intake for various structural proteins will assist in the development of disease-specific biomarkers. The rat will probably remain a model of choice in this field besides the mouse because some behavioral paradigms have been elaborated only in this species and its bigger size renders a number of manipulations much easier than in the mouse.

## References


1. Mormede P, Courvoisier H, Ramos A, Marissal-Arvy N, Ousova O, Desautes C et al (2002) Molecular genetic approaches to investigate individual variations in behavioral and neuroendocrine stress responses. Psychoneuroendocrinology 27:563–583
2. Kas MJ, Fernandes C, Schalkwyk LC, Collier DA (2007) Genetics of behavioural domains across the neuropsychiatric spectrum; of mice and men. Mol Psychiatry 12:324–330
3. Broadhurst PL (1975) The Maudsley reactive and nonreactive strains of rats: a survey. Behav Genet 5:299–319
4. Hendley ED, Ohlsson WG (1991) Two new inbred rat strains derived from SHR: WKHA, hyperactive, and WKHT, hypertensive, rats. Amer J Physiol 261:H583–H589
5. Moisan MP, Courvoisier H, Bihoreau MT, Gauguier D, Hendley ED, Lathrop M et al (1996) A major quantitative trait locus influences hyperactivity in the WKHA rat. Nat Genet 14:471–473
6. Moisan MP, Llamas B, Cook MN, Mormede P (2003) Further dissection of a genomic locus associated with behavioral activity in the Wistar-Kyoto hyperactive rat, an animal model of hyperkinesis. Mol Psychiatry 8:348–352
7. Fernandez-Teruel A, Escorihuela RM, Gray JA, Aguilar R, Gil L, Gimenez-Llort L et al (2002) A quantitative trait locus influencing anxiety in the laboratory rat. Genome Res 12:618–626
8. Carr LG, Foroud T, Bice P, Gobbett T, Ivashina J, Edenberg H et al (1998) A quantitative trait locus for alcohol consumption in selectively bred rat lines. Alcohol Clin Exp Res 22:884–887
9. Bice P, Foroud T, Bo R, Castelluccio P, Lumeng L, Li TK et al (1998) Genomic screen for QTLs underlying alcohol consumption in the P and NP rat lines. Mamm Genome 9:949–955
10. Radcliffe RA, Erwin VG, Draski L, Hoffmann S, Edwards J, Deng XS et al (2004) Quantitative trait loci mapping for ethanol sensitivity and neurotensin receptor density in an F2 intercross derived from inbred high and low alcohol sensitivity selectively bred rat lines. Alcohol Clin Exp Res 28:1796–1804
11. Ramos A, Moisan MP, Chaouloff F, de C, de P (1999) Identification of female-specific QTLs affecting an emotionality-related behavior in rats. Mol Psychiatry 4:453–462
12. Ahmadiyeh N, Churchill GA, Shimomura K, Solberg LC, Takahashi JS, Redei EE (2003) X-linked and lineage-dependent inheritance of coping responses to stress. Mamm Genome 14:748–757
13. Terenina-Rigaldie E, Moisan MP, Colas A, Beauge F, Shah KV, Jones BC et al (2003) Genetics of behaviour: phenotypic and molecular study of rats derived from high- and low-alcohol consuming lines. Pharmacogenetics 13:543–554
14. Bielavska E, Kren V, Musilova A, Zidek V, Pravenec M (2002) Genome scanning of the HXB/BXH sets of recombinant inbred strains of the rat for quantitative trait loci associated with conditioned taste aversion. Behav Genet 32:51–56
15. McEwen BS (2003) Mood disorders and allostatic load. Biol Psychiatry 54:200–207
16. Hall CS (1934) Emotional behavior in the rat I. Defecation and urination as measures of individual differences in emotionality. J Comp Psychol 18:385–395
17. Carobrez AP, Bertoglio LJ (2005) Ethological and temporal analyses of anxiety-like behavior: the elevated plus-maze model 20 years on. Neurosci Biobehav Rev 29:1193–1205

18. Lister RG (1990) Ethologically-based animal models of anxiety disorders. Pharmacol Ther 46:321–340
19. Fernandez-Teruel A, Gimenez-Llort L, Escorihuela RM, Gil L, Aguilar R, Steimer T et al (2002) Early-life handling stimulation and environmental enrichment: are some of their effects mediated by similar neural mechanisms? Pharmacol Biochem Behav 73: 233–245
20. Ramos A, Mormède P (1998) Stress and emotionality: a multidimensional and genetic approach. Neurosci Biobehav Rev 22:33–57
21. Prut L, Belzung C (2003) The open field as a paradigm to measure the effects of drugs on anxiety-like behaviors: a review. Eur J Pharmacol 463:3–33
22. El Yacoubi M, Vaugeois JM (2007) Genetic rodent models of depression. Curr Opin Pharmacol 7:3–7
23. Sousa N, Almeida OF, Wotjak CT (2006) A hitchhiker's guide to behavioral analysis in laboratory rodents. Genes Brain Behav 5(Suppl 2): 5–24
24. Crabbe JC, Phillips TJ (2004) Pharmacogenetic studies of alcohol self-administration and withdrawal. Psychopharmacology (Berl) 174: 539–560
25. Rodd ZA, Bertsch BA, Strother WN, Le Niculescu H, Balaraman Y, Hayden E et al (2007) Candidate genes, pathways and mechanisms for alcoholism: an expanded convergent functional genomics approach. Pharmacogenomics J 7:222–256
26. Overstreet DH, Rezvani AH, Djouma E, Parsian A, Lawrence AJ (2007) Depressive-like behavior and high alcohol drinking co-occur in the FH/WJD rat but appear to be under independent genetic control. Neurosci Biobehav Rev 31:103–114
27. Colombo G, Lobina C, Carai MA, Gessa GL (2006) Phenotypic characterization of genetically selected Sardinian alcohol-preferring (sP) and -non-preferring (sNP) rats. Addict Biol 11:324–338
28. Baum AE, Solberg LC, Churchill GA, Ahmadiyeh N, Takahashi JS, Redei EE (2006) Test- and behavior-specific genetic factors affect WKY hypoactivity in tests of emotionality. Behav Brain Res 169:220–230
29. Liang T, Spence J, Liu L, Strother WN, Chang HW, Ellison JA et al (2003) alpha-Synuclein maps to a quantitative trait locus for alcohol preference and is differentially expressed in alcohol-preferring and -nonpreferring rats. Proc Natl Acad Sci USA 100:4690–4695
30. Chiavegatto S, Izidio GS, Mendes-Lana A, Aneas I, Freitas TA, Torrao AS et al (2009) Expression of alpha-synuclein is increased in the hippocampus of rats with high levels of innate anxiety. Mol Psychiatry 14: 894–905
31. Foroud T, Wetherill LF, Liang T, Dick DM, Hesselbrock V, Kramer J et al (2007) Association of alcohol craving with alpha-synuclein (SNCA). Alcohol Clin Exp Res 31:537–545
32. Pandey SC, Zhang H, Roy A, Xu T (2005) Deficits in amygdaloid cAMP-responsive element-binding protein signaling play a role in genetic predisposition to anxiety and alcoholism. J Clin Invest 115:2762–2773
33. Primeaux SD, Wilson SP, Bray GA, York DA, Wilson MA (2006) Overexpression of neuropeptide Y in the central nucleus of the amygdala decreases ethanol self-administration in "anxious" rats. Alcohol Clin Exp Res 30:791–801
34. Kimpel MW, Strother WN, McClintick JN, Carr LG, Liang T, Edenberg HJ et al (2007) Functional gene expression differences between inbred alcohol-preferring and -non-preferring rats in five brain regions. Alcohol 41:95–132
35. Mottagui-Tabar S, Prince JA, Wahlestedt C, Zhu G, Goldman D, Heilig M (2005) A novel single nucleotide polymorphism of the neuropeptide Y (NPY) gene associated with alcohol dependence. Alcohol Clin Exp Res 29:702–707
36. Llamas B, Contesse V, Guyonnet-Duperat V, Vaudry H, Mormede P, Moisan MP (2005) QTL mapping for traits associated with stress neuroendocrine reactivity in rats. Mamm Genome 16:505–515
37. Solberg LC, Baum AE, Ahmadiyeh N, Shimomura K, Li R, Turek FW et al (2006) Genetic analysis of the stress-responsive adrenocortical axis. Physiol Genomics 27:362–369
38. Ousova O, Guyonnet-Duperat V, Iannuccelli N, Bidanel JP, Milan D, Genet C et al (2004) Corticosteroid binding globulin: a new target for cortisol-driven obesity. Mol Endocrinol 18:1687–1696
39. Carr LG, Kimpel MW, Liang T, McClintick JN, McCall K, Morse M et al (2007) Identification of candidate genes for alcohol preference by expression profiling of congenic rat strains. Alcohol Clin Exp Res 31:1089–1098
40. Kobeissy FH, Sadasivan S, Liu J, Gold MS, Wang KK (2008) Psychiatric research: psychoproteomics, degradomics and systems biology. Expert Rev Proteomics 5:293–314

41. Nestler EJ (2001) Psychogenomics: opportunities for understanding addiction. J Neurosci 21:8324–8327
42. Krishnan V, Han MH, Graham DL, Berton O, Renthal W, Russo SJ et al (2007) Molecular adaptations underlying susceptibility and resistance to social defeat in brain reward regions. Cell 131:391–404
43. McClung CA, Nestler EJ (2008) Neuroplasticity mediated by altered gene expression. Neuropsychopharmacology 33:3–17
44. Teague CR, Dhabhar FS, Barton RH, Beckwith-Hall B, Powell J, Cobain M et al (2007) Metabonomic studies on the physiological effects of acute and chronic psychological stress in Sprague-Dawley rats. J Proteome Res 6:2080–2093
45. Ruiz-Opazo N, Kosik KS, Lopez LV, Bagamasbad P, Ponce LR, Herrera VL (2004) Attenuated hippocampus-dependent learning and memory decline in transgenic TgAPPswe Fischer-344 rats. Mol Med 10:36–44
46. Gimenez-Llort L, Schiffmann SN, Shmidt T, Canela L, Camon L, Wassholm M et al (2007) Working memory deficits in transgenic rats overexpressing human adenosine A2A receptors in the brain. Neurobiol Learn Mem 87:42–56
47. Homberg JR, Olivier JD, Smits BM, Mul JD, Mudde J, Verheul M et al (2007) Characterization of the serotonin transporter knockout rat: a selective change in the functioning of the serotonergic system. Neuroscience 146:1662–1676
48. Olivier JD, Van Der Hart MG, Van Swelm RP, Dederen PJ, Homberg JR, Cremers T et al (2008) A study in male and female 5-HT transporter knockout rats: an animal model for anxiety and depression disorders. Neuroscience 152:573–584
49. Potenza MN, Brodkin ES, Joe B, Luo X, Remmers EF, Wilder RL et al (2004) Genomic regions controlling corticosterone levels in rats. Biol Psychiatry 55:634–641
50. Bilusic M, Bataillard A, Tschannen MR, Gao L, Barreto NE, Vincent M et al (2004) Mapping the genetic determinants of hypertension, metabolic diseases, and related phenotypes in the lyon hypertensive rat. Hypertension 44:695–701
51. Marissal-Arvy N, Lombes M, Petterson J, Moisan MP, Mormede P (2004) Gain of function mutation in the mineralocorticoid receptor of the Brown Norway rat. J Biol Chem 279:39232–39239
52. Klimes I, Weston K, Gasperikova D, Kovacs P, Kvetnansky R, Jezova D et al (2005) Mapping of genetic determinants of the sympathoneural response to stress. Physiol Genomics 20:183–187
53. Cui ZH, Ikeda K, Kawakami K, Gonda T, Nabika T, Masuda J (2003) Exaggerated response to restraint stress in rats congenic for the chromosome 1 blood pressure quantitative trait locus. Clin Exp Pharmacol Physiol 30:464–469
54. Conti LH, Jirout M, Breen L, Vanella JJ, Schork NJ, Printz MP (2004) Identification of quantitative trait Loci for anxiety and locomotion phenotypes in rat recombinant inbred strains. Behav Genet 34:93–103
55. Mormede P, Moneva E, Bruneval C, Chaouloff F, Moisan MP (2002) Marker-assisted selection of a neuro-behavioural trait related to behavioural inhibition in the SHR strain. an animal model of ADHD. Genes Brain Behav 1:111–116
56. Terenina-Rigaldie E, Jones BC, Mormede P (2003) Pleiotropic effect of a locus on chromosome 4 influencing alcohol drinking and emotional reactivity in rats. Genes Brain Behav 2:125–131
57. Vendruscolo LF, Terenina-Rigaldie E, Raba F, Ramos A, Takahashi RN, Mormede P (2006) Evidence for a female-specific effect of a chromosome 4 locus on anxiety-related behaviors and ethanol drinking in rats. Genes Brain Behav 5:441–450
58. Ahmadiyeh N, Churchill GA, Solberg LC, Baum AE, Shimomura K, Takahashi JS et al (2005) Lineage is an epigenetic modifier of QTL influencing behavioral coping with stress. Behav Genet 35:189–198
59. Solberg LC, Baum AE, Ahmadiyeh N, Shimomura K, Li R, Turek FW et al (2004) Sex- and lineage-specific inheritance of depression-like behavior in the rat. Mamm Genome 15:648–662
60. Palmer AA, Breen LL, Flodman P, Conti LH, Spence MA, Printz MP (2003) Identification of quantitative trait loci for prepulse inhibition in rats. Psychopharmacology (Berl) 165:270–279
61. Vendruscolo LF, Terenina-Rigaldie E, Raba F, Ramos A, Takahashi RN, Mormede P (2006) A QTL on rat chromosome 7 modulates prepulse inhibition, a neuro-behavioral trait of ADHD, in a Lewis × SHR intercross. Behav Brain Funct 2:21
62. Ruiz-Opazo N, Tonkiss J (2006) Genome-wide scan for quantitative trait loci influencing spatial navigation and social recognition memory in Dahl rats. Physiol Genomics 26:145–151
63. Carr LG, Habegger K, Spence JP, Liu L, Lumeng L, Foroud T (2006) Development of

congenic rat strains for alcohol consumption derived from the alcohol-preferring and non-preferring rats. Behav Genet 36:285–290

64. Foroud T, Bice P, Castelluccio P, Bo R, Miller L, Ritchotte A et al (2000) Identification of quantitative trait loci influencing alcohol consumption in the high alcohol drinking and low alcohol drinking rat lines. Behav Genet 30:131–140
65. Foroud T, Bice P, Castelluccio P, Bo R, Ritchotte A, Stewart R et al (2002) Mapping of QTL influencing saccharin consumption in the selectively bred alcohol-preferring and -nonpreferring rat lines. Behav Genet 32:57–67
66. Foroud T, Ritchotte A, Spence J, Liu L, Lumeng L, Li TK et al (2003) Confirmation of alcohol preference quantitative trait loci in the replicate high alcohol drinking and low alcohol drinking rat lines. Psychiatr Genet 13:155–161
67. Radcliffe RA, Bludeau P, Deng XS, Erwin VG, Deitrich RA (2007) Short-term selection for acute ethanol tolerance and sensitization from an F2 population derived from the high and low alcohol-sensitive selectively bred rat lines. Alcohol 41:557–566
68. Radcliffe RA, Bludeau P, Asperi W, Fay T, Deng XS, Erwin VG et al (2006) Confirmation of quantitative trait loci for ethanol sensitivity and neurotensin receptor density in crosses derived from the inbred high and low alcohol sensitive selectively bred rat lines. Psychopharmacology (Berl) 188:343–354
69. Nakatani N, Aburatani H, Nishimura K, Semba J, Yoshikawa T (2004) Comprehensive expression analysis of a rat depression model. Pharmacogenomics J 4:114–126
70. Kohen R, Kirov S, Navaja GP, Happe HK, Hamblin MW, Snoddy JR et al (2005) Gene expression profiling in the hippocampus of learned helpless and nonhelpless rats. Pharmacogenomics J 5:278–291
71. Kroes RA, Panksepp J, Burgdorf J, Otto NJ, Moskal JR (2006) Modeling depression: social dominance-submission gene expression patterns in rat neocortex. Neuroscience 137:37–49
72. Pearson KA, Stephen A, Beck SG, Valentino RJ (2006) Identifying genes in monoamine nuclei that may determine stress vulnerability and depressive behavior in Wistar-Kyoto rats. Neuropsychopharmacology 31:2449–2461
73. Orsetti M, Di Brisco F, Canonico PL, Genazzani AA, Ghi P (2008) Gene regulation in the frontal cortex of rats exposed to the chronic mild stress paradigm, an animal model of human depression. Eur J NeuroSci 27:2156–2164
74. Gass JT, Olive MF (2008) Transcriptional profiling of the rat frontal cortex following administration of the mGlu5 receptor antagonists MPEP and MTEP. Eur J Pharmacol 584:253–262
75. Khawaja X, Xu J, Liang JJ, Barrett JE (2004) Proteomic analysis of protein changes developing in rat hippocampus after chronic antidepressant treatment: Implications for depressive disorders and future therapies. J Neurosci Res 75:451–460
76. Carboni L, Vighini M, Piubelli C, Castelletti L, Milli A, Domenici E (2006) Proteomic analysis of rat hippocampus and frontal cortex after chronic treatment with fluoxetine or putative novel antidepressants: CRF1 and NK1 receptor antagonists. Eur Neuropsychopharmacol 16:521–537
77. Mu J, Xie P, Yang ZS, Yang DL, Lv FJ, Luo TY et al (2007) Neurogenesis and major depression: implications from proteomic analyses of hippocampal proteins in a rat depression model. Neurosci Lett 416:252–256
78. Kim HG, Kim KL (2007) Decreased hippocampal cholinergic neurostimulating peptide precursor protein associated with stress exposure in rat brain by proteomic analysis. J Neurosci Res 85:2898–2908
79. Worst TJ, Tan JC, Robertson DJ, Freeman WM, Hyytia P, Kiianmaa K et al (2005) Transcriptome analysis of frontal cortex in alcohol-preferring and nonpreferring rats. J Neurosci Res 80:529–538
80. Edenberg HJ, Strother WN, McClintick JN, Tian H, Stephens M, Jerome RE et al (2005) Gene expression in the hippocampus of inbred alcohol-preferring and -nonpreferring rats. Genes Brain Behav 4:20–30
81. Hargreaves GA, Quinn H, Kashem MA, Matsumoto I, McGregor IS (2008) Proteomic analysis demonstrates adolescent vulnerability to lasting hippocampal changes following chronic alcohol consumption. Alcohol Clin Exp Res 33:86–94
82. Kabbaj M, Evans S, Watson SJ, Akil H (2004) The search for the neurobiological basis of vulnerability to drug abuse: using microarrays to investigate the role of stress and individual differences. Neuropharmacology 47(Suppl. 1):111–122
83. Ahmed SH, Lutjens R, van der Stap LD, Lekic D, Romano-Spica V, Morales M et al (2005) Gene expression evidence for remodeling of lateral hypothalamic circuitry in cocaine addiction. Proc Natl Acad Sci USA 102:11533–11538

84. McClung CA, Nestler EJ, Zachariou V (2005) Regulation of gene expression by chronic morphine and morphine withdrawal in the locus ceruleus and ventral tegmental area. J Neurosci 25:6005–6015
85. Kobeissy FH, Warren MW, Ottens AK, Sadasivan S, Zhang Z, Gold MS et al (2008) Psychoproteomic analysis of rat cortex following acute methamphetamine exposure. J Proteome Res 7:1971–1983
86. Li X, Wang H, Qiu P, Luo H (2008) Proteomic profiling of proteins associated with methamphetamine-induced neurotoxicity in different regions of rat brain. Neurochem Int 52:256–264
87. Burger C, Lopez MC, Baker HV, Mandel RJ, Muzyczka N (2008) Genome-wide analysis of aging and learning-related genes in the hippocampal dentate gyrus. Neurobiol Learn Mem 89:379–396

# Chapter 26

# Genomics Studies of Immune-Mediated Diseases Using the BN–LEW Rat Model

**Isabelle Bernard, Gilbert J. Fournié, and Abdelhadi Saoudi**

## Abstract

LEW and BN rats, that behave in opposite ways for their susceptibility to various immune-mediated diseases, provide a powerful model to investigate the molecular and genetic bases of immune system physiology and dysregulation. Using this model, we addressed the question of the genetic control of central nervous system autoimmunity, of xenobiotic-induced allergic diseases, and of T cell subsets that differ by their cytokine profiles. By linkage analysis and genetic dissection, using a panel of congenic rats, we identified a 120 Kb region on chromosome 9 that controls all these phenotypes, indicating that this region contains a gene or set of genes that plays an important role in the immune system homeostasis and susceptibility to immune mediated diseases. In this review, we will describe these rat genomics studies and will discuss the cellular and genetic factors that may be involved in the differences between these rat strains.

**Key words:** Rats, Autoimmunity, Allergy, T cells, Cytokines, Th1/Th2 cells

## 1. Introduction

The etiology of immune-mediated diseases, such as autoimmune and allergic diseases, is unknown but epidemiological studies have shown that both genetic and environmental factors influence disease susceptibility. Although extensive efforts have been put into linkage and association studies in humans, only a limited number of genes modulating these diseases have unambiguously been identified so far. Several factors hamper these studies, particularly genetic heterogeneity of human population, varialility of penetrance, multiplicity of involved genes, epistatic interactions between genes and complexity of interactions between genes, and environmental. Studies on animal models representative of human pathologies can overcome some of these difficulties, as they can be

I. Anegon (ed.), *Rat Genomics: Methods and Protocols*, Methods in Molecular Biology, vol. 597
DOI 10.1007/978-1-60327-389-3_26, © Humana Press, a part of Springer Science+Business Media, LLC 2010

conducted in inbred strains in a controlled environment. In this respect, Brown–Norway (BN) and Lewis (LEW) rat strains represent models of choice to study immune-mediated diseases from the cellular and genetic points of view. Our data as well as data from other groups have clearly shown that BN and LEW rats have opposite susceptibility to experimental immune-mediated diseases (Table 26.1 and Fig. 26.1). LEW rats are mainly susceptible to Th1-mediated autoimmune diseases, while BN rats are highly susceptible to Th2-mediated immune diseases (1). Interestingly, these two rat strains differ also with respect to organ transplantation (2) and susceptibility to some infectious agents, such as Toxoplasma gondi (3,4) or Chlamydia pneumoniae (5). Therefore, these two rat strains behave in an opposite way concerning their immune responses to self and foreign antigens and have been used to investigate the pathophysiology and the genetic control of immune-mediated diseases. In this review, we will first introduce the different T cell subsets and their functions. We will, thereafter, present our results on the differences between LEW and BN rats with respect to their susceptibility to auto-antigen induced-central nervous system autoimmunity and xenobiotic induced allergy. Finally, we will discuss the cellular and genetic factors that may be involved in these differences.

**Table 26.1**
**LEW and BN rats differs by their immune system and their susceptibility to immune mediated immunopathological manifestations**

| | Lewis | Brown–Norway |
|---|---|---|
| Allergic-disorders after injections of $HgCl_2$, gold salts | Resistant | Susceptible |
| Encephalomyelitis[a] | Susceptible | Resistant |
| Uveoretinitis[a] | Susceptible | Resistant |
| Graft-versus-host disease | Acute (Th1) | Chronic (Th2) |
| Immune response after stimulation in vitro[b] | IFNγ > IL-4 | IL-4 > IFNγ |
| CD45RC T cell subsets[c] | $RC^{high} > RC^{low}$ | $RC^{low} > RC^{high}$ |

[a]Experimental autoimmune encephalomyelitis and uveoretinitis are considered as a Th1-dependent autoimmune diseases

[b]T cells were stimulated in vitro with T cell mitogen, ConA, in mixte lymphocyte reaction or with anti-TCR and anti-CD28 mAbs

[c]BN rats, that are highly susceptible to type-2 cytokine-mediated immunological disorders, exhibit a high proportion of CD4R5RC$^{low}$ CD4 and CD8 T cells, producing IL-4, IL-13, IL-10, and IL-17 as compared with LEW rats. Conversely, LEW rats, that are highly susceptible to develop type-1 cytokine-mediated autoimmune diseases, have a preponderance of the CD45RC$^{high}$ CD4 and CD8 T cells producing mainly IFNγ

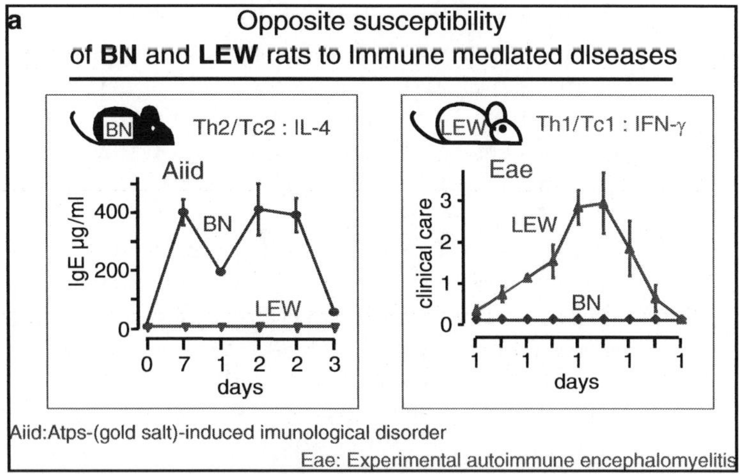


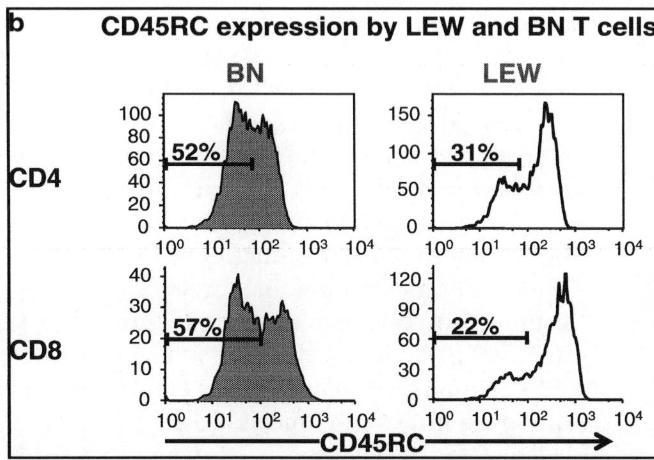


Fig. 26.1. BN and LEW rats behave in opposite way for their susceptibility to immune-mediated diseases and for the repartition of CD4 and CD8 T cell subsets. (**a**, *left panel*): BN rats are prone to develop type-2 IL-4-mediated immune responses (Th:T helper; Tc: T cytotoxic). To induce Aiid, LEW, and BN rats were s.c. injected with Aurothiopropanolsulfanate (ATPS) at a dose of 2 mg/100 g.b.w. three times a week for 5 weeks. Serum IgE concentration was determined by ELISA. Only BN rats, but not LEW rats develop Aiid as evidenced by IgE response. (**a**, *right panel*): LEW rats are prone to develop type 1 IFN-γ-mediated immune responses. To induce active EAE, rats were injected in the hind footpads with 10 µg of guinea pig MBP emulsified in CFA. Animals were scored daily for clinical signs of disease on a scale from 0 to 5 depending on severity. LEW rats are susceptible to EAE while BN rats are resistant. (**b**) Inverse proportion of CD4 and CD8 T cell subsets, as defined by the level of CD45RC expression, in BN and LEW rats. Peripheral blood leukocytes from naive BN (*left panels*) and LEW (*right panels*) rats were incubated with anti-TCRαβ, anti-CD45RC and anti-CD4 or anti-CD8 mAbs and analyzed by flow cytometry. The results represent histograms of the CD45RC expression by CD4 T cells (*top panels*) and CD8 T cells (*bottom panels*). The value on the top of each histogram represents the % of CD45RC CD4 or CD8 T cell subpopulations.

## 2. T Cell Subsets and Their Functions

CD4 T cells play a central role in orchestrating immune responses through their capacity to provide help and regulate other cells of the adaptive or innate immune systems. The T helper (Th)-cell population comprises functionally distinct subsets that are characterized by the patterns of lymphokines, and they produce following activation. Although these subsets were first identified by in vitro analysis of murine T cell clones, similar subsets exist in vivo in mice (6), rats (7, 8), and humans (9). Activated effector CD4 T cells have been subclassified on the basis of their phenotypic pattern of cytokine production into at least three subsets: Th1, Th2, and Th17. Th1 cells were defined on the basis of their production of IFNγ, a cytokine necessary for the clearance of certain intracellular pathogens. Th2 cells produce IL-4, IL-5, and IL-13, which promote IgG1 and IgE class switch and eosinophil recruitment, and are involved in enhancing the clearance of parasites. Th17 are characterized by the production of IL-17 and IL-6 (10), and they are involved in the control of certain categories of pathogens not covered by Th1 and Th2, such as extracellular bacterial pathogens (11, 12). The development of these functionally distinct T cells is influenced by several factors, such as cytokines present in the T cell microenvironment during antigen presentation and initiation of T cell responses (12, 13). IL-12, produced by activated macrophages and dendritic cells, is the principal Th1-inducing cytokine, while IL-4 is mainly responsible for the differentiation of Th2 cells (13–15). Finally, TGF-β and IL-6 favors the differentiation of Th17 cells (16). Numerous studies have shown that the effector functions of CD8 T cells overlap those of CD4 T cells much more than previously anticipated. There is now accumulating evidence showing that different CD8 T cell subsets exist, Tc1, Tc2, and Tc17 with similar cytokine profiles than their CD4 counterparts. The benefits of adaptive CD4 and CD8 T cell responses, however, come at a cost. Inappropriate or poorly controlled effector T cells can be deleterious when directed against self or ubiquitous environmental antigen. In this setting, persistent effector T cell responses drives chronic inflammatory disorders such as autoimmunity, allergy, or atopy (17, 18).

## 3. LEW and BN Rats Differ by Their Susceptibility to Immune-mediated Diseases

### *3.1. Susceptibility of LEW and BN Rats to Th1-mediated Autoimmune Diseases: LEW Rats are Highly Susceptible While BN Rats are Resistant*

Experimental autoimmune encephalomyelitis (EAE), a useful animal model for multiple sclerosis, is a T cell-mediated autoimmune disease of the central nervous system (19–21). EAE arises as a consequence of the breakdown of self-tolerance induced by immunization of susceptible animals with myelin-derived antigens emulsified in CFA. Potentially auto-aggressive cells are present in the healthy immune repertoire. Upon activation in the periphery, they circulate in the bloodstream to reach their target organ where they mount an attack against the local milieu, the starting point of a pathogenic inflammatory reaction. The clinical course of EAE varies with the animal model and the used protocol. In most models, acute EAE is followed either by a permanent remission and concomitant resistance to further disease induction, or by a transient remission followed by subsequent relapses and remissions (19, 22, 23). Although recent studies have suggested the involvement of Th17 in the development of EAE (16, 24, 25), a large body of evidence shows a pivotal role of Th1-type immune response in the induction of EAE (26, 27). It has been recently demonstrated that Th1 cells have a preferential ability to access the noninflamed CNS and that the pathology they cause promotes the subsequent infiltration of Th17 cells (27). In contrast, Th2 cells specific for encephalitogenic peptides and producing IL-4 and IL-10 are unable to induce EAE in immunocompetent animals (28, 29). EAE is genetically controlled. Up to date, more than 40 rodent quantitative trait loci (QTLs) regulating different EAE phenotypes have been identified (30–33).

LEW rats immunized with myelin basic protein (MBP) exhibit an acute and transient disease and affected animals recover completely within one week after disease onset (Fig. 26.1a). After the recovery phase, attempts to induce further episodes of the disease are unsuccessful (22). BN rats are resistant to develop EAE (Fig. 26.1a), and this resistance involves MHC and non-MHC genes. Indeed, BN.1L rats, a BN strain congenic for the MHC region of the LEW strain, remains resistant to EAE (34). The analysis of the cytokine repertoire revealed that the natural resistance of BN.1L rats to develop clinical EAE is associated with a decreased production of the typical type-1 cytokine, IFN-γ and increased productions of typical Th2 cytokine IL-4 and regulatory cytokine TGF-β. However, IL-17 production has not been analyzed in any of these studies. The involvement of IL-4 and TGF-β in the resistance of BN.1L rats to the induction of EAE has been tested by treating the rats with neutralizing mAbs. Neutralization of endogenous TGF-β, but not IL-4, renders BN.1L rats susceptible to clinical EAE. The neutralization of endogenous TGF-β affect neither the proliferation of MBP-specific T cells nor the cytokine repertoire upon *in vitro* restimulation of lymph node cells with MBP (35). Therefore, the mechanism whereby TGF-β mediates resistance to EAE in BN.1L rats remains to be determined.

#### 3.2. Susceptibility of LEW and BN Rats to Th2-mediated Autoimmune Diseases: BN Rats are Highly Susceptible Whereas LEW Rats are Resistant

Chronic injections of nontoxic doses of mercuric chloride ($HgCl_2$) induces a Th2 cell-dependent polyclonal B cell activation in susceptible BN rats, evidenced by an increase in serum IgE and IgG1 concentration (36–40). These animals produce various autoantibodies, including anti-glomerular basement membrane antibodies associated with the development of a glomerulopathy (41, 42). They may also exhibit arthritis and gut vasculitis (43, 44). The disease is transient and disappears in surviving animals within 1 month, even if injections of $HgCl_2$ are pursued. Rats are resistant to rechallenge. In contrast to BN rats, LEW rats are not only resistant to $HgCl_2$-induced autoimmunity, but also develop a generalized immune suppression (45) that protects against autoimmune diseases such as EAE (46). Under regular injections of gold salts, BN rats develop the same immunological disorders as $HgCl_2$-treated BN rats (Fig. 26.1a). Lewis rats are resistant but do not develop immunosuppression after gold salt administration (47).

The differences between BN and LEW rats in their susceptibility to develop immunological disorders after gold salt or $HgCl_2$ administration could result from the opposite polarization of their immune responses. Indeed, in vitro as well as in vivo studies have shown that these conditions of stimulation favors IL-4 responses in BN rat T cells. It was also found that IFN-γ production contributes, at least in part, to the resistance of LEW rats to gold salts-induced immunological disorders (48). Therefore, a given xenobiotic agent can triggers different levels of cytokine expression according to the genomic origin of the cells, which is associated to marked differences in immune responses in vivo. These observations prompted us to investigate the genetic control of in vivo responsiveness to gold salts.

### 3.3. LEW and BN Rats Exhibit Several Differences Concerning Their Immune System

The comparison of the T cell compartments between LEW and BN rats revealed several important differences. (1) BN rats have less T cells than LEW rats, and this is due to a MHC-dependent defect in the CD8 T cell compartment (49). (2) In vitro as well as in vivo studies have shown a higher IL-4 response and a lower IFN-γ response in T cells from BN rats as compared with those from LEW rats, and these differences are MHC-independent. These differences concerns both CD4 and CD8 T cell compartment with a major contribution of CD8 T cells (50). The defective IFN-γ production and increased IL-4 expression by the BN CD8 T cell compartment may account for the susceptibility of this rat strain to Th2-mediated immune diseases. In agreement with this hypothesis, we have shown that BN.1L rats, which have three times more CD8 T cells than BN rats, develop a higher IgE response (Th2 response) after gold salt treatment. In addition, the depletion of CD8 T cells in BN-1L rats suppresses this IgE response (unpublished data).

In the rat, the level of expression of the CD45RC isoform divides both CD4 and CD8 T lymphocytes into two subpopulations (Fig. 26.1b). The $CD45RC^{high}$ T cell subset produces preferentially type-1 cytokines while the production of type-2 cytokines and IL-17 are restricted to the $CD45RC^{low}$ subset (7, 51, 52). There is functional inter-regulation between the $CD45RC^{high}$ and $CD45RC^{low}$ CD4 and CD8 T cell subsets. In nude rats, the adoptive transfer of $CD45RC^{high}$ CD4 T cells from congenic euthymic donors induces a fatal wasting disease which can be prevented by cotransfering the $CD45RC^{low}$ CD4 T cell subset (53). In type-2 cytokine dependent mercuric chloride-induced disease, a protective role has been ascribed to the CD45RC$^{high}$ CD4 T cells (54). In a CD4-dependent Graft-versus-Host disease model, the cotransfer of CD45RC$^{high}$ CD8 T cell subset exacerbates this pathology while the cotransfer of CD45RC$^{low}$ CD8 T cell subset is preventive (55).

The relative proportion of these CD45RC cell subsets differs between LEW and BN rats. BN rats exhibit a higher proportion of CD4 and CD8 CD4R5RC$^{low}$ T cells, producing IL-4, IL-13, IL-10, and IL-17 compared with LEW rats (Fig. 26.1b). Conversely, LEW rats have a preponderance of the CD4 and CD8 CD45RC$^{high}$ T cells producing mainly IFNγ. These data suggest that the CD45RC$^{high}$/CD45RC$^{low}$ ratio, both within the CD4 and CD8 T cell compartments, may be a marker for the predisposition to develop type-1 or type-2 immune responses and subsequently for the susceptibility to associated immunopathological manifestations. Indeed, the non-MHC encoded resistance to develop EAE in BN.1L rats is associated with a higher percentage of CD45RC$^{low}$ T cells than in susceptible LEW rats (35). Similar results were obtained in radiation bone marrow chimeras between LEW, BN.1L and (LEW × BN) F1 rats, and these results showed that the resistance to develop EAE is associated with a high frequency of the CD4 and CD8 T CD45RC$^{low}$ cell subsets (unpublished data). Furthermore, in (LEW × BN) F2 rats, there is a correlation between the CD45RC$^{high}$/CD45RC$^{low}$ ratio and IgE response after gold salt treatment. The highest IgE response was observed in the rats that have more CD45RC$^{low}$ T cells (51, 52).

## 4. Linkage Analyses Using LEW × BN Intercross

The genetic control of Th2-mediated disorders triggered by gold salts, and particularly of the IgE response, has been studied in F2 hybrids obtained by intercrossing (LEW × BN) F1 rats (56–58). Three susceptibility loci named *Aiid3*, *Aiid2* and *Aiid1*, respectively (*Aiid*: Atps-induced immunological disorder locus; Atps: aurothiopropanol sulfonate Atps), have been identified on chromosomes 9, 10, and 20, respectively (Fig. 26.2). *Aiid1*, that contains the MHC region, controls the kinetics of the IgE response and the anti-DNA antibody response. *Aiid2* and *Aiid3* control the IgE response (13 and 31 % of the genetic variance, respectively) and the antilaminin response. We also studied the genetic control of CD45RC subsets within CD4 and CD8 T cell compartments using (LEW × BN) F2 rats. This linkage analysis led to the identification of two loci, *Cec1*, and *Cec2* (*Cec* for CD45RC expression on T cells) on chromosomes 9 and 20 (51, 52), which overlap respectively with *Aiid3* and *Aiid1* (Fig. 26.2). An independent genome-wide search

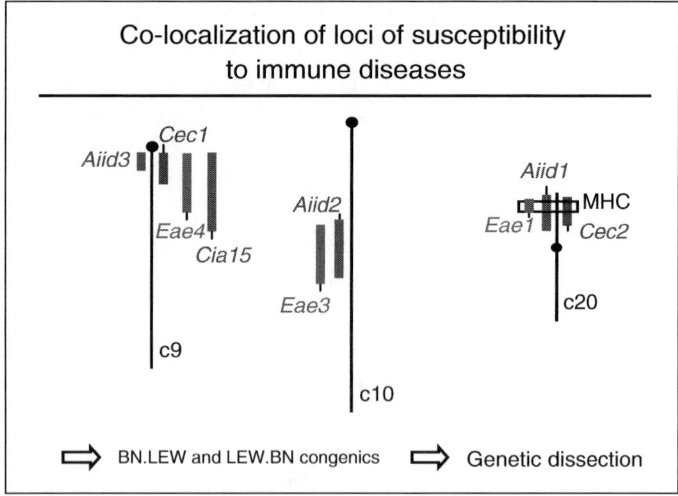


Fig. 26.2. Colocation of several susceptibility loci on rat chromosomes 9, 10 and 20. *Vertical bars on the left* of chromosomes 9, 10, and 20 (c9, c10, c20) show the colocalization of various susceptibility loci that have been found by linkage analyses of F2 hybrids rat populations in these chromosomes. *Aiid1*, *Aiid2*, and *Aiid3* are loci for susceptibility to develop IgE response in rats injected with aurothiopropanol sulfonate (Atps). *Eae4, Eae3, and Eae-1* are susceptibility loci to experimental autoimmune encephalomyelitis (EAE). *Cec1 and Cec-2* represent the loci that have been found to control the proportion of CD45RC$^{high}$ and CD45RC$^{low}$ T cells, two subsets of CD4+ and CD8+ T cells with different cytokine profiles and functions. *Cia15* controls joint/bone inflammation in a collagen-induced model of arthritis. MHC: Major Histocompatibility Complex. These data suggest that these chromosomal regions contain genes or set of genes that could play major roles in T cell functions and immune system homeostasis. The extent to which all these phenotypes are under the same genetic control is currently under investigation by genetic dissection using new congenic lines.

for loci controlling the susceptibility to EAE in a (LEW × BN) F2 population has identified a locus, *Eae3*, that overlaps with *Aiid2* (33). Moreover, *Aiid3/Cec1* intervals are fully included within *Cia15* (collagen-induced arthritis QTL 15, (59)) and *Eae4* (Experimental allergic encephalomyelitis QTL 4, (31)), two loci that control autoimmune diseases. *Cia15* controls joint/bone inflammation in a collagen-induced model of arthritis and was described in a cross between the BBDR and BN rat strains. *Eae4* controls brain/spinal cord inflammation in experimental autoimmune encephalomyelitis. Initially described in a DA x BN rat strain cross, we have recently shown that it also controls the susceptibility of LEW rats to EAE. Collectively, these data indicate that the identified loci contains a gene, or a set of closely linked genes, that play a major role in T cell functions and immune system homeostasis. The extent to which all these phenotypes are under the same genetic control remains to be determined.

## 5. Genetic Dissection Using LEW/BN Congenic Lines

To confirm and narrow down the loci intervals for the further positional cloning of the genes, we created a set of reciprocal BN/LEW lines congenic for *Aiid2* and *Aiid3*. Using these congenic lines, we found that *Aiid3* exerts a major effect on Aiid as compared with *Aiid2* (60). Using new recombinant congenic lines and a genetic map with high density in microsatellite markers, we have been recently able to narrow down the *Aiid3/Cec1* region to a 120 Kb interval (Fig. 26.3). Since this region was not correctly annotated in the rat genome databases, we then constructed a physical map using linkage analyses with rat/hamster radiation hybrids and by using comparative genomics. This 120 Kb region contains four candidates genes that are involved in

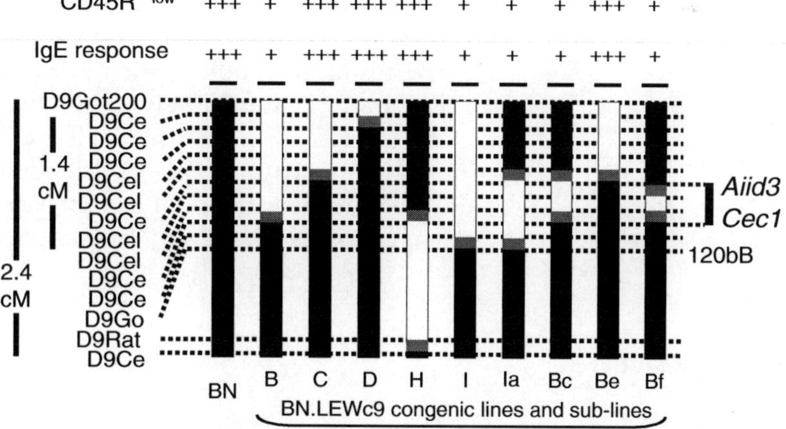


Fig. 26.3 Refinement Aiid3 and Cec1 to 120 Kb region using interval-specific BN.LEWc9 congenic lines and sublines. CD45RC expression and IgE reponse according to the genetic maps of the BN parental strain and of 9 BN.LEWc9 congenic strains and substrains. For CD45RC$^{low}$; +++ indicates that the CD4 and CD8 CD45RC$^{low}$ populations are predominant; + indicates the contrary, i.e., that the CD45RC$^{high}$ populations are predominant. For IgE response, +++ indicates a high response as observed in the BN parental strain; + indicate a 80–90% decrease in the IgE response . The location of D9Got200, D9Got5, and D9Cel4 is at scale; location of other markers is not at scale. D9Cel are microsatellite markers developed for this study and not yet described in databases. The *white* and *black* colors indicate, respectively, the LEW and BN genomes (*grey*: undetermined genomic origin; intervals in which recombination has occurred). Genetic maps of the lines in which decreased proportion of CD45RC$^{low}$ subset and down-regulation of gold salt induced IgE response are observed localized precisely region that controls these two phenotypes within an interval of 120 kilobases indicated by the *vertical bar on the right*.

**Table 26.2**
**Candidate genes in the *Aiid3/Cec1* 120 kilobases interval**

| Genes[a] | Synonyms | Function, biological process |
|---|---|---|
| *Tnfsf14* | HVEM-L, LIGHT | Cytokine activity, TNF receptor binding |
| *C3* | Complement factor | Complement activation, immune/inflammatory responses |
| *Trip10* | Cip4 | Signal transduction, actin filament organization |
| | Cdc42-interacting protein 4 | |
| *Vav1* | | T cell activation and differentiation, |
| | | Signal transduction in haematopoietic cells |

[a]*See* detailed information with references at:
http://www.informatics.jax.org/searches/accession_report.cgi?id=MGI%3A. 1355317 (Tnfsf14), 88227 (C3), 2146901 (Trip10), 98923 (Vav1)

immune pathways, particularly in T cell activation and function (Table 26.2). Among these, *Vav1* appeared to be the stronger candidate, since Vav1 is constitutively and specifically expressed in cells of haematopoietic origin. We found a nonsynonymous polymorphism in the first exon that could affect the function of this signal transducer. Studies aimed at assessing the precise role of *Vav1* in immune homeostasis and immune-mediated diseases are in progress. Recent results have indicated that the 120 kb interval controls the regulatory Foxp3 T cell population, thus establishing a possible mechanistic link between Vav1 or other genes of the interval and the susceptibility to immune mediated diseases.

## 6. Conclusions

The rat is the second most frequently used animal model (after the mouse) in the field of immunological research. However, it is the major model for physiological investigation, and it provides the most relevant models for the genetic study of common multifactorial diseases. In this respect, BN and LEW rats represent models of choice to study immune-mediated diseases from the cellular and genetic points of view. By using this model, we identified a 120 Kb interval that controls the proportion of CD45RC T cells subsets, the susceptibility to Th2-mediated immunological disorders induced by xenobiotics and central nervous system inflammation. These data suggest that this interval contains a gene, or set of genes, of general importance. We have evidence that supports a major role of the *Vav1*, in particular, the identification of a polymorphism that may affect the function of this molecule.

This guanine nucleotide exchange factor is a key signal transducer downstream of the TCR and thus plays a central role in the regulation of lymphocytic responses, either under physiological or pathological conditions. However, several questions remain to be answered. In particular, the conclusive proof for the role of the identified *Vav1* polymorphism on T cell function will require further functional studies. Similarly, the role of this polymorphism in the various models of immune-mediated diseases studied so far remains to be confirmed. Finally, it should be taken into account that the implication of other genes within the *Aiid3/Cec1* 120 kilobases interval has not been completely excluded.

## Acknowledgments

The authors thanks the members of Autoimmunity and Immunoregulation team as well as the member of Gilbert Fournié team for their helpful comments and suggestions. We would like also to thank all the dedicated people who participated in this work, in particular, Philippe Druet, Dominique Lagrange, Lucette Pelletier, Magali Mas, Jean-François Subra, Pierre Cavallès, Bastien Cautain, Emmanuel Xystrakis, Céline Colacios, Anne Dejean, Laurence Ordonez, Christine Duthoit, Lucille Lamouroux, Olivier Papapietro, and Audrey Casemayou . We present our apologies to colleagues whose work could not be adequately cited or discussed, due to space limitations. This work was supported by "Institut National de la Santé et de la Recherche Médicale (INSERM)," the Conseil Général de Région Midi-Pyrénées, by grants from " Agence Nationale de Recherche," "Association pour la Recherche sur la Sclérose en Plaques," "Fondation pour la Recherche Médicale," "Association Française contre les myopathies," and Arthrithis foundation Courtin.

### References

1. Fournie GJ, Cautain B, Xystrakis E, Damoiseaux J, Mas M, Lagrange D, Bernard I, Subra JF, Pelletier L, Druet P, Saoudi A (2001) Cellular and genetic factors involved in the difference between Brown Norway and Lewis rats to develop respectively type-2 and type-1 immune-mediated diseases. Immunol Rev 184:145–160
2. Murase N, Demetris AJ, Woo J, Tanabe M, Furuya T, Todo S, Starzl TE (1993) Graft-versus-host disease after brown Norway-to-Lewis and Lewis-to-Brown Norway rat intestinal transplantation under FK506. Transplantation 55:1–7
3. Sergent V, Cautain B, Khalife J, Deslee D, Bastien P, Dao A, Dubremetz JF, Fournie GJ, Saoudi A, Cesbron-Delauw MF (2005) Innate refractoriness of the Lewis rat to toxoplasmosis is a dominant trait that is intrinsic to bone marrow-derived cells. Infect Immun 73:6990–6997
4. Cavailles P, Sergent V, Bisanz C, Papapietro O, Colacios C, Mas M, Subra JF, Lagrange D, Calise M, Appolinaire S, Faraut T, Druet P,

Saoudi A, Bessieres MH, Pipy B, Cesbron-Delauw MF, Fournie GJ (2006) The rat Toxo1 locus directs toxoplasmosis outcome and controls parasite proliferation and spreading by macrophage-dependent mechanisms. Proc Natl Acad Sci U S A 103:744–749

5. Inman RD, Chiu B (2006) Early cytokine profiles in the joint define pathogen clearance and severity of arthritis in Chlamydia-induced arthritis in rats. Arthritis Rheum 54:499–507
6. Mosmann TR, Coffman RL (1989) TH1 and TH2 cells: different patterns of lymphokine secretion lead to different functional properties. Annu Rev Immunol 7:145
7. McKnight AJ, Barclay AN, Mason DW (1991) Molecular cloning of rat interleukine 4 cDNA and analysis of the cytokine repertoire of subsets of CD4+ T cells. Eur J Immunol 21: 1187–1194
8. Fowell D, McKnight AJ, Powrie F, Dyke R, Mason D (1991) Subsets of CD4+ T cells and their role in the induction and prevention of autoimmunity. Immunol Rev 123:37–64
9. Del Prete G, De Carli M, Ricci M, Romagnani S (1991) Helper activity for immunoglobulin synthesis of T helper type 1 (Th1) and Th2 human T cell clones: the help of Th1 clones is limited by their cytolytic capacity. J Exp Med 174:809–813
10. Bettelli E, Oukka M, Kuchroo VK (2007) T(H)-17 cells in the circle of immunity and autoimmunity. Nat Immunol 8:345–350
11. Weaver CT, Harrington LE, Mangan PR, Gavrieli M, Murphy KM (2006) Th17: an effector CD4 T cell lineage with regulatory T cell ties. Immunity 24:677–688
12. Weaver CT, Hatton RD, Mangan PR, Harrington LE (2007) IL-17 family cytokines and the expanding diversity of effector T cell lineages. Annu Rev Immunol 25:821–852
13. O'Garra A (1998) Cytokines induce the development of functionally heterogeneous T helper cell subsets. Immunity 8:275–283
14. Manetti R, Parronchi P, Giudizi MG, Piccinni M-P, Maggi E, Trinchieri G, Romagnani S (1993) Natural killer cell stimulatory factor (interleukin 12, IL-12) induces T helper type 1 (Th1)-specific immune responses and inhibits the development of IL-4-producing Th cells. J Exp Med 177:1199–1204
15. Swain SL, Weinberg AD, English M, Huston G (1990) IL-4 directs the development of Th2-like helper effectors. J Immunol 145:3796
16. Bettelli E, Carrier Y, Gao W, Korn T, Strom TB, Oukka M, Weiner HL, Kuchroo VK (2006) Reciprocal developmental pathways for the generation of pathogenic effector TH17 and regulatory T cells. Nature 441:235–238
17. Druet P, Ramanathan S, Pelletier L (1996) TH1 and TH2 lymphocytes in autoimmunity. Adv Nephrol Necker Hosp 25:217–241
18. Akdis CA (2006) Allergy and hypersensitivity: mechanisms of allergic disease. Curr Opin Immunol 18:718–726
19. Raine CS (1984) Analysis of autoimmune demyelination: its impact upon multiple sclerosis. Lab Invest 50:608–635
20. Steinman L (1996) Multiple sclerosis: a coordinated immunological attack against myelin in the central nervous system. Cell 95:299–302
21. Storch MK, Stefferl A, Brehm U, Weissert R, Wallstrom E, Kerschensteiner M, Olsson T, Linington C, Lassmann H (1998) Autoimmunity to myelin oligodendrocyte glycoprotein in rats mimics the spectrum of multiple sclerosis pathology. Brain Pathol 8:681–694
22. Mason D (1991) Genetic variation in the stress response: susceptibilty to experimental allergic encephalomyelitis and implications for human inflammatory disease. Immunol Today 12:57–60
23. Swanborg RH (2001) Experimental autoimmune encephalomyelitis in the rat: lessons in T-cell immunology and autoreactivity. Immunol Rev 184:129–135
24. Cua DJ, Sherlock J, Chen Y, Murphy CA, Joyce B, Seymour B, Lucian L, To W, Kwan S, Churakova T, Zurawski S, Wiekowski M, Lira SA, Gorman D, Kastelein RA, Sedgwick JD (2003) Interleukin-23 rather than interleukin-12 is the critical cytokine for autoimmune inflammation of the brain. Nature 421:744–748
25. Langrish CL, Chen Y, Blumenschein WM, Mattson J, Basham B, Sedgwick JD, McClanahan T, Kastelein RA, Cua DJ (2005) IL-23 drives a pathogenic T cell population that induces autoimmune inflammation. J Exp Med 201:233–240
26. Ando DG, Clayton J, Kono D, Urban JL, Sercarz EE (1989) Encephalithogenic T cells in the B10.PL model of experimental allergic encephlomyelitis (EAE) are of the Th-1 lymphokine subtype. Cell Immunol 124:132–143
27. O'Connor RA, Prendergast CT, Sabatos CA, Lau CW, Leech MD, Wraith DC, Anderton SM (2008) Cutting edge: Th1 cells facilitate the entry of Th17 cells to the central nervous system during experimental autoimmune encephalomyelitis. J Immunol 181:3750–3754

28. Cua DJ, Hinton DR, Stohlman SA (1995) Self-antigen-induced Th2 responses in experimental allergic encephalomyelitis (EAE)-resistant mice. J Immunol 155:4052–4059
29. Lafaille JJ, Van de Keere F, Hsu AL, Baron JL, Haas W, Raine CS, Tonegawa S (1997) Myelin basic protein-specific T helper (Th2) cells cause experimental autoimmune encephalomyelitis in immunodeficient hosts rather than protect them from the disease. J Exp Med 186:307–312
30. Dahlman I, Lorentzen JC, de Graaf KL, Stefferl A, Linington C, Luthman H, Olsson T (1998) Quantitative trait loci disposing for both experimental arthritis and encephalomyelitis in the DA rat; impact on severity of myelin oligodendrocyte glycoprotein-induced experimental autoimmune encephalomyelitis and antibody isotype pattern. Eur J Immunol 28:2188–2196
31. Dahlman I, Jacobsson L, Glaser A, Lorentzen JC, Andersson M, Luthman H, Olsson T (1999) Genome-wide linkage analysis of chronic relapsing experimental autoimmune encephalomyelitis in the rat identifies a major susceptibility locus on chromosome 9. J Immunol 162:2581–2588
32. Dahlman I, Wallstrom E, Weissert R, Storch M, Kornek B, Jacobsson L, Linington C, Luthman H, Lassmann H, Olsson T (1999) Linkage analysis of myelin oligodendrocyte glycoprotein-induced experimental autoimmune encephalomyelitis in the rat identifies a locus controlling demyelination on chromosome 18. Hum Mol Genet 8:2183–2190
33. Roth MP, Viratelle C, Dolbois L, Delverdier M, Borot N, Pelletier L, Druet P, Clanet M, Coppin H (1999) A genome-wide search identifies two susceptibility loci for experimental autoimmune encephalomyelitis on rat chromosomes 4 and 10. J Immunol 162:1917–1922
34. Happ MP, Wettstein P, Dietzschold B, Heber-Katz E (1988) Genetic control of the development of experimental allergic encephalomyelitis in rats: separation of MHC and non-MHC effects. J Immunol 141:1489–1494
35. Cautain B, Damoiseaux J, Bernard I, van Straaten H, van Breda Vriesman P, Boneu B, Druet P, Saoudi A (2001) Essential role of TGF-beta in the natural resistance to experimental allergic encephalomyelitis in rats. Eur J Immunol 31:1132–1140
36. Gillespie KM, Qasim FJ, Tibbats LM, Thiru S, Oliveira DBG, Mathieson PW (1995) Interleukin-4 gene expression in mercury-induced autoimmunity. Scand J Immunol 41:268–272
37. Gillespie KM, Saoudi A, Kuhn J, Whittle CJ, Druet P, Bellon B, Mathieson PW (1996) Th1/Th2 cytokine gene expression after mercuric chloride in susceptible and resistant rat strains. Eur J Immunol 10:2388–2392
38. Dubey C, Bellon B, Hirsch F, Kuhn J, Vial M-C, Goldman M, Druet P (1991) Increased expression of class II major histocompatibility complex molecules on B cells in rats susceptible or resistant to HgCl2-induced autoimmunity. Clin Exp Immunol 86:118–125
39. Prigent P, Saoudi A, Pannetier C, Graber P, Bonnefoy Y, Druet P, Hirsch F (1995) Mercuric chloride, a chemical responsible for Th2-mediated autoimmunity in Brown–Norway rats, directly triggers T cells to produce IL-4. J Clin Invest 96:1484–1489
40. Saoudi A, Kuhn J, Huygen K, de Kozak Y, Velu T, Goldman M, Druet P, Bellon B (1993) TH2 activated cells prevent experimental autoimmune uveoretinitis, a TH1-dependent autoimmune disease. Eur J Immunol 23: 3096–3103
41. Goldman M, Druet P, Gleichmann E (1991) $T_H2$ cells in systemic autoimmunity: insights from allogeneic diseases and chemically-induced autoimmunity. Immunol Today 12:223–227
42. Badou A, Pelletier L, Druet P (1997) Th1/Th2 balance in the control of drug-induced autoimmunity. In: Adorini L, Landes RG (eds) vol 8. Company, Austin, Texas, USA, pp 53–69
43. Kiely PDW, Thiru S, Oliveira DBG (1995) Inflammatory polyarthritis induced by mercuric chloride in the Brown Norway rat. Lab Invest 73:284–293
44. Kiely PD, Pecht I, Oliveira DB (1997) Mercuric chloride-induced vasculitis in the Brown Norway rat: alpha beta T cell-dependent and -independent phases: role of the mast cell. J Immunol 159:5100–5106
45. Pelletier L, Pasquier R, Rossert J, Druet P (1987) $HgCl_2$ induces non specific immunosuppression in LEW rats. Eur J Immnol 17:49–54
46. Pelletier L, Rossert J, Pasquier R, Villarroya H, Belair M-F, Vial M-C, Oriol R, Druet P (1988) Effect of HgCl2 on experimental allergic encephalomyelitis in Lewis rats. HgCl2-induced downmodulation of the disease. Eur J Immunol 18:243–247
47. Tournade H, Pelletier L, Guéry JC, Pasquier R, Nochy D, Hinglais N, Guilbert B, Druet P (1991) Experimental gold-induced autoimmunity. Nephrol Dial Transplant 6:621–630
48. Savignac M, Badou A, Delmas C, Subra JF, De Cramer S, Paulet P, Cassar G, Druet P, Saoudi A, Pelletier L (2001) Gold is a T cell

polyclonal activator in BN and LEW rats but favors IL-4 expression only in autoimmune prone BN rats. Eur J Immunol 31: 2266–2276

49. Damoiseaux JG, Cautain B, Bernard I, Mas M, van Breda Vriesman PJ, Druet P, Fournie G, Saoudi A (1999) A dominant role for the thymus and MHC genes in determining the peripheral CD4/CD8 T cell ratio in the rat. J Immunol 163:2983–2989

50. Cautain B, Damoiseaux J, Bernard I, Xystrakis E, Fournie E, van Breda Vriesman P, Druet P, Saoudi A (2002) The CD8 T cell compartment plays a dominant role in the deficiency of Brown-Norway rats to mount a proper type 1 immune response. J Immunol 168:162–170

51. Subra JF, Cautain B, Xystrakis E, Mas M, Lagrange D, van der Heijden H, van de Gaar MJ, Druet P, Fournie GJ, Saoudi A, Damoiseaux J (2001) The balance between CD45RChigh and CD45RClow CD4 T cells in rats is intrinsic to bone marrow-derived cells and is genetically controlled. J Immunol 166:2944–2952

52. Xystrakis E, Cavailles P, Dejean AS, Cautain B, Colacios C, Lagrange D, van de Gaar MJ, Bernard I, Gonzalez-Dunia D, Damoiseaux J, Fournie GJ, Saoudi A (2004) Functional and genetic analysis of two CD8 T cell subsets defined by the level of CD45RC expression in the rat. J Immunol 173:3140–3147

53. Powrie F, Mason D (1990) OX-22$^{high}$ CD4+ T cells induce wasting disease with multiple organ pathology: prevention by the OX-22$^{low}$ subset. J Exp Med 172:1701–1708

54. Mathieson PW, Thiru S, Oliveira DBG (1993) Regulatory role of OX22$^{high}$ T cells in mercury-induced autoimmunity in the Brown Norway rat. J Exp Med 177:1309–1316

55. Xystrakis E, Dejean AS, Bernard I, Druet P, Liblau R, Gonzalez-Dunia D, Saoudi A (2004) Identification of a novel natural regulatory CD8 T-cell subset and analysis of its mechanism of regulation. Blood 104:3294–3301

56. Mas M, Subra JF, Lagrange D, Pilipenko-Appolinaire S, Kermarrec N, Gauguier D, Druet P, Fournie GJ (2000) Rat chromosome 9 bears a major susceptibility locus for IgE response. Eur J Immunol 30:1698–1705

57. Kermarrec N, Blanpied C, Pelletier L, Feingold N, Mandet C, Druet P, Hirsch F (1995) Genetic study of gold-salt-induced immune disorders in the rat. Nephrol Dial Transplant 10:2187–2191

58. Kermarrec N, Dubay C, De Gouyon B, Blanpied C, Gauguier D, Gillespie K, Druet P, Lathrop M, Hirsch F (1996) Serum IgE concentration and other immune manifestations of treatment with gold salts are linked to MHC and IL-4 regions in the rat. Genomics 31:111–114

59. Furuya T, Salstrom JL, McCall-Vining S, Cannon GW, Joe B, Remmers EF, Griffiths MM, Wilder RL (2000) Genetic dissection of a rat model for rheumatoid arthritis: significant gender influences on autosomal modifier loci. Hum Mol Genet 9:2241–2250

60. Mas M, Cavailles P, Colacios C, Subra JF, Lagrange D, Calise M, Christen MO, Druet P, Pelletier L, Gauguier D, Fournie GJ (2004) Studies of congenic lines in the Brown Norway rat model of Th2-mediated immunopathological disorders show that the aurothiopropanol sulfonate-induced immunological disorder (Aiid3) locus on chromosome 9 plays a major role compared to Aiid2 on chromosome 10. J Immunol 172:6354–6361

# Chapter 27

# Rat Models of Cardiovascular Diseases

**Michael Bader**


## Abstract

In cardiovascular research, the rat has been the main model of choice for decades. Experimental procedures were developed to generate cardiovascular disease states in this species, such as systemic and pulmonary hypertension, cardiac hypertrophy and failure, myocardial infarction, and stroke. Furthermore, rats have been bred, which spontaneously develop such diseases. They became extremely valuable models to understand the genetics of these diseases, since powerful genomic tools are now available for the rat. One of these tools is transgenic technology, which has allowed the creation of even more disease models in the rat. This review summarizes the experimental, genetic, and transgenic rat models for cardiovascular diseases.


**Key words:** Systemic hypertension, Pulmonary hypertension, Cardiac hypertrophy, Heart failure, Myocardial infarction, Stroke, Transgenic rat

## 1. Introduction

The rat has been the animal model of choice for more than a century in cardiovascular research (1). Numerous experimental and genetic rat models for cardiovascular diseases have been developed and extensively analyzed. Furthermore, transgenic technology has allowed the generation of additional rat models (2–7). In the last decade, these genetic and transgenic rat models were instrumental to identify genes involved in the pathogenesis of cardiovascular diseases (1). However, in the same time, the significance of the rat has declined compared to the mouse due to the lack of germline-competent embryonic stem cells and therefore of the technology for the targeted alterations of the genome. Novel developments in stem cell and transgenic technology will, however, allow to specifically downregulate or even target genes in this species (8–11).

I. Anegon (ed.), *Rat Genomics: Methods and Protocols*, Methods in Molecular Biology, vol. 597
DOI 10.1007/978-1-60327-389-3_27, © Humana Press, a part of Springer Science+Business Media, LLC 2010

This review will summarize the most frequently used experimental, genetic, and transgenic rat models of cardiovascular diseases. For more comprehensive information, the reader is referred to earlier reviews about these subjects (1, 7, 12–14).

## 2. Experimental Rat Models

### *2.1. Experimental Models for Systemic Hypertension*

#### *2.1.1. DOCA-Salt Hypertension*

The desoxycorticosterone acetate (DOCA)/salt model of hypertension is based on the volume retaining effect of DOCA in conjunction with high salt. Usually, the animals are submitted to uninephrectomy followed by the subcutaneous implantation of a 50 mg DOCA tablet and by salt loading with 1% NaCl in the drinking water. After about 1 week, systolic blood pressure starts to increase, and reaches about 180 mmHg after 6 weeks of treatment (15). The model is accompanied by a drastically reduced plasma renin-angiotensin system, and is therefore considered to be angiotensin II independent. However, there are discussions about the involvement of the renin-angiotensin system in the brain (16).

#### *2.1.2. NO Blockade*

Chronic blockade of nitric oxide (NO) generation by inhibiting NO synthases is a common way to induce hypertension in animals. The most frequently used substances for this purpose are L-NAME (NG-nitro-l-arginine methyl ester) given at 20–40 mg/kg body weight, intraperitoneally or orally, and Nitro-l-arginine (17–19). Blood pressure increases after 2 days of treatment and reaches about 180 mmHg systolic at 4 weeks. There is evidence that not only NO-depletion in the vascular wall leading to endothelial dysfunction, but also central mechanisms regulating the sympathetic tone as well as the renin-angiotensin system may be involved in the pathogenesis of the model (20).

#### *2.1.3. Renovascular Hypertension*

Induction of renal ischemia by renal artery stenosis induces hypertension in the rat (21, 22). This model has first been described by Harry Goldblatt's group in the 1930s (23), and is therefore often called "Goldblatt" hypertension. Two basic models are performed. In the two-kidney one-clip model (2K1C), both kidneys are preserved, but the renal artery of one of them is constricted. The ischemic kidney secretes renin, which leads to elevated blood pressure. The increasing blood pressure causes an augmented (pressure) natriuresis in the intact contralateral kidney precluding sodium retention. In the one-kidney one-clip (1K1C) model, the contralateral kidney is removed and sodium retention is observed followed by an inhibition of renin secretion and low plasma renin levels. In both models, blood pressure increases slowly and reaches highest levels of around 180 mmHg systolic about 3 weeks after surgery.

#### 2.1.4. Angiotensin II Infusion

Chronic infusion of low doses of the vasoconstrictor peptide angiotensin II (50–200 ng/kg min) by osmotic minipump is another classical model for hypertension in the rat (24). Such a ("slow-pressor") dose leads to a slowly developing hypertension over the course of 8–13 days. This effect is not so dependent on the vascular actions of the peptide, but mediated by the activation of the brain renin-angiotensin system (25).

### 2.2. Experimental Models for Pulmonary Hypertension

#### 2.2.1 Monocrotaline

Monocrotaline is a pyrrolizidine alkaloid extracted from the seeds of Crotalaria spectabilis that induces pulmonary hypertension in rats (26). After a single intraperitoneal injection of 60 mg/kg, the substance is metabolized and thereby activated in the liver. The resulting pyrrols have a very short half-life of a few seconds and thus mainly affect the pulmonary circulation. After about 2 days, there is first evidence of endothelial cell damage and after 4 weeks the disease has fully developed with increased pulmonary artery and right ventricular pressure followed by right ventricular hypertrophy (27).

#### 2.2.2. Hypoxia

Chronic hypoxia increases pulmonary artery pressure and induces a remodeling of the pulmonary vessels, finally resulting in pulmonary hypertension (28). In rats, hypoxia is produced either by putting the animals in a hypobaric chamber (pressure around 400 mmHg) for about 3 weeks or in a chamber, in which the $O_2$ concentration is maintained at 10% for several hours per day for 1–2 weeks.

### 2.3. Experimental Models for Myocardial Infarction

#### 2.3.1. Coronary Artery Ligation

The temporary or permanent ligation of a coronary artery, mostly the left anterior descending artery (LAD), creates a classical model for myocardial infarction followed by heart failure (29, 30). The pathogenesis of the rat model is comparable to the disease in humans in particular when reperfusion is allowed (31).

### 2.4. Experimental Models for Cardiac Hypertrophy

#### 2.4.1. Aortic Banding

The coarctation of the aorta (aortic banding) is a surgical method frequently used to induce pressure-overload hypertrophy of the heart (32). Either the thoracic (trans-aortic constriction or TAC-model) or the abdominal aorta may be constricted to about 30% of its original diameter (33). Depending on this placement of the coarctation, the model is less or more severe. Constriction of the ascending aorta leads to a pure pressure overload model of very progressive cardiac hypertrophy. In contrast, abdominal aortic banding between the two renal arteries creates a 2K1C-like situation (s. above) with hypoperfusion of the distal kidney, increase in renin and an angiotensin II induced hypertension and cardiac hypertrophy.

#### 2.4.2. Aortocaval Shunt

A surgically installed fistula between the aorta and the vena cava causes a drastic volume overload (about 50% of total cardiac output) on the heart, which results in marked left and right ventricular hypertrophy 2 months after surgery (34, 35). This aortocaval shunt method has become the most frequently used technique to induce volume overload hypertrophy followed by heart failure in rats (6).

#### 2.4.3. Isoproterenol

Adrenergic stimulation of the heart leads to cardiomyocyte hypertrophy and cardiac fibrosis independent of hemodynamic changes (36). This mechanism is exploited to induce cardiac hypertrophy in rats, mostly treating the animals with the β-adrenoceptor agonist isoproterenol by minipumps or by daily injections for 7–10 days (37). Cardiac weight increases by more than 50%, but this increase is reversible after cessation of treatment.

### 2.5. Experimental Models for Stroke

#### 2.5.1. Middle Cerebral Artery Occlusion

Strokes of the cortex can be induced in rats by occlusion of the middle cerebral artery (MCAO) either by surgery or by an intraluminal thread (38). Large parts of the neocortex get underperfused and necrotic. These occlusions can be made permanent or reversible. However, due to collaterals, the occlusion of a single artery often leads to small and variable infarcts. The additional ligation of both common carotid arteries, the three vessel occlusion model (3VO), yields more consistent results (39, 40).

## 3. Genetic Rat Models

### 3.1. Genetic Models for Systemic Hypertension

#### 3.1.1. Spontaneously Hypertensive Rats

Spontaneously hypertensive rats were developed in the early 1960s in Kyoto by Aoki and Okamoto (41). They are the most studied genetically hypertensive model so far. Hypertension develops with age and is sex-specifically different. Commonly, Wistar Kyoto (WKY) rats are used at normotensive controls, despite that they have not been bred in parallel to the SHR but about 10 years later. Therefore, there are numerous genetic differences, not only related to hypertension. Furthermore, in the meantime SHR have been split into several distinct substrains bred at different locations with different genotypes and phenotypes (42). Nevertheless, SHR rats have been successfully employed to discover genes involved in hypertension and cardiac hypertrophy, such as the ones for osteoglycin and CD36 (43, 44).

#### 3.1.2. Milan Hypertensive Rats

In the 1980s, a genetically hypertensive rat strain has been bred by Guiseppe Bianchi in Milan, the Milan hypertensive strain (MHS) (45). Blood pressure starts to rise at 8 weeks of age and reaches maximum systolic levels of about 175 mmHg already at 10 weeks of age (46). Thus, the Milan hypertensive rats are a model of mild hypertension. One gene, which may be involved in the pathogenesis of hypertension in this model, is α-adducin regulating ion channel functions (47).

*3.1.3. Lyon Hypertensive Rats*

In Lyon, a genetically hypertensive rat (Lyon hypertensive (LH) rat) was generated by selective breeding, together with two control strains, one normotensive (LN) and one with low blood pressure (LL) (48). Blood pressure starts to increase at about 6 weeks of age, reaching levels of 175 mmHg systolic at adult age. Thus, LH rats are also a model of mild hypertension. Despite that these animals show low plasma renin activity, hypertension and end-organ damage are highly sensitive to blockade of the renin-angiotensin system, arguing in favor of a pathophysiological relevance for angiotensin II in this model (49).

*3.1.4. New Zealand Hypertensive Rats*

In the 1950s, a genetically hypertensive (GH) rat strain was generated at the University of Otago (50). Already directly after birth, these animals show an increased blood pressure, which then rises quickly reaching maximum values of above 200 mmHg at 10 weeks of age (51, 52).

*3.1.5. Prague Hypertensive Rats*

In Prague, a hypertriglyceridemic model was bred from a single parental pair of Wistar rats in the late 1980s, which later turned out to be also hypertensive, the Prague hypertensive rat (PHR) and its normotensive line, PNR (53, 54). They develop relatively mild hypertension with systolic blood pressures of around 170 mmHg.

*3.1.6. Dahl Salt-Sensitive Rat*

Lewis Dahl bred two rat strains in the 1960s, one of which developed hypertension on a 8% NaCl diet (Dahl-S), the other one was resistant to this treatment (Dahl-R) (55). When high salt is given at weaning, the Dahl-S rats develop fulminant hypertension and die at 16 weeks of age; when it is started later, the animals survive and reach systolic blood pressure levels of about 185 mmHg (56). The animals also get hypertensive under normal and even low salt diets, but only at several months of age. In contrast, Dahl-R rats stay normotensive under high salt diet. These animals are ideal models to study the interaction of salt intake and genetics in blood pressure regulation (57).

*3.1.7. San Juan Hypertensive Rats*

About 20 years ago, a novel rat strain for salt-sensitive hypertension was attempted to be bred from Munich Wistar rats in Puerto Rico analogously to the Dahl strain (58). However, the resulting San Juan hypertensive (SJH) animals turned out to get equally hypertensive even without salt loading. The animals show an increased number of superficial glomeruli in the kidney and develop systolic blood pressures of about 200 mmHg at 10 weeks of age.

#### 3.1.8. Sabra Rats

Another strain of rats with salt-sensitive hypertension was bred in Israel by Ben-Ishay et al. (59). The original Sabra salt-sensitive and resistant strains were further inbred by Yagil et al. yielding the SBH/y and SBN/y strains respectively (60). The SBH/y strain only becomes hypertensive under a high-salt diet mostly achieved by 1% NaCl in the drinking water and the subcutaneous implantation of a DOCA pellet. Under these conditions, SBH/y rats reach systolic blood pressure above 200 mmHg. Also this strain is now an important model to study the genetics of salt-sensitivity (61).

#### 3.1.9. Inherited Stress-Induced Arterial Hypertensive Rats

In Novosibirsk, Markel developed a strain of rats by inbreeding, which presents inherited stress-induced arterial hypertension (ISIAH) (62). Mild emotional stress, e.g., handling or swimming, applied daily in an unpredictable manner induces a rise in blood pressure in this model, starting at 6 weeks of age and reaching 180 mmHg systolic at about 6 months of age (63). This model is not widely used but may be suitable to study the link between stress and hypertension.

### 3.2. Genetic Models for Pulmonary Hypertension

#### 3.2.1. Fawn-Hooded Rats

The fawn-hooded rat discovered in 1947 (64) has a platelet storage disease comparable to the Hermansky–Pudlak syndrome in humans (65). The genetic defect resides on chromosome 1 and involves the gene for rab38 (66). The animals exhibit increased plasma endothelin and serotonin levels. As a consequence, they spontaneously develop pulmonary hypertension, which is markedly aggravated by mild hypoxia. There are several possibilities for the mechanisms causing the disease. Serotonin is known to be an important pathophysiological factor for pulmonary hypertension (67). It may be covalently linked by transglutaminases to small G-proteins in pulmonary vascular cells, a process called "serotonylation" (68), and thereby activate these proteins, which have been shown to be involved in pulmonary hypertension in fawn-hooded rats (69). On the other hand, a mitochondrial defect was described to be involved in the pathogenesis of the disease in these animals (70). However, since serotonin is degraded in mitochondria by monoamine oxidases both mechanisms may be linked.

In addition, these animals also get hypertensive and develop a fatal renal pathophysiology (71).

### 3.3. Genetic Models for Heart Failure

#### 3.3.1. Spontaneously Hypertensive Heart Failure Rat

A cross of the Koletsky and the SHR rat has been inbred to yield the spontaneously hypertensive heart failure rat (SHHF) in late 1980s (72). SHHF animals develop hypertension and cardiac hypertrophy at 3–5 months of age and by 16–20 months of age congestive heart failure featuring the typical hallmarks of the human disease. SHHF rats were used for recent genetic studies, which led to the discovery of epoxide hydrolase as a susceptibility factor for heart failure (73).

### 3.4. Genetic Models for Stroke

#### 3.4.1. Stroke-Prone Spontaneously Hypertensive Rat

The stroke-prone spontaneously hypertensive rat (SHRSP) is a substrain generated from the SHR in the 1970s by selecting the second generation offspring of animals, which died from stroke at later age, for further breeding (74). Stroke occurs frequently but stochastically in these animals until 9 months of age, when they are fed a high-salt, low-protein, low potassium diet (75, 76). Alterations in the cerebrovasculature are more important for this phenotype than the increase in blood pressure. Nevertheless, hypertension is as pronounced as in the SHR (around 200 mmHg systolic) and accompanied by the typical signs of the human disease, such as endothelial dysfunction and end-organ damage. This model has also been already successfully applied to discover genes involved in hypertension by genetics, such as GSTM1 (77), and by transgenic complementation, such as ACE2 (78).

## 4. Transgenic Models for Hypertension

### 4.1. TGR(mREN2)27

The TGR(mREN2)27 rat carries the mouse renin Ren-2 as transgene (2). These rats produce high amounts of renin in tissues, which results in an increase of angiotensin II and aldosterone concentrations. The heterozygous TGR(mREN2)27 rats become markedly hypertensive and show all signs of hypertensive end-organ damage, such as cardiac hypertrophy and fibrosis as well as renal damage (79). They have become a frequently used hypertension model with systolic values of around 240 mmHg. The homozygous TGR(mREN2)27 rats with the double gene dose are even more hypertensive and mostly die at around 12 weeks of age (80). Nevertheless, they have been used as the heart failure model to study genes predisposing to this disease (81).

### 4.2. Transgenic Rats With Inducible Renin Expression

The same mouse Ren-2 gene has been used to generate a transgenic rat with inducible hypertension (82). To this purpose, the promoter of the cytochrome P450 enzyme Cyp1A1 was employed which can be induced mainly in the liver by treatment of the animals with a xenobiotic drug, indole-3 carbinol. After application of the drug, prorenin and renin levels get dose-dependently and reversibly upregulated, and lead to an increase in blood pressure and vascular injury, which seem to be salt-sensitive (83–85). This model has been very useful to study the mechanisms of renin and angiotensin II induced end-organ damage.

### 4.3. Human Renin/Human Angiotensinogen Double Transgenic Rats

Two transgenic rat lines, TGR(hREN) and TGR(hAOGEN), have been generated, expressing the genes for human renin and angiotensinogen, respectively (4). Due to the species-specificity of the renin-angiotensin system, angiotensin II synthesis and cardiovascular physiology in these animals is unaffected by the human transgenes. By breeding of female TGR(hREN) with male TGR(hAOGEN), double-transgenic rats (dTGR) expressing human renin and angiotensinogen can be generated, which produce high amounts of angiotensin II and develop fulminant hypertension followed by overt organ damage of heart and kidney comparable to hypertensive patients (86). They have been used in several studies to analyze the pathogenetic mechanisms of end-organ damage and are ideal models to study human renin inhibitors, since this novel class of antihypertensive drugs acts species-specifically (87).

Interestingly, the opposite cross, male TGR(hREN) with female TGR(hAOGEN), is a model for pregnancy-induced hypertension and preeclampsia, since the dams become hypertensive and proteinuric at day 13 of pregnancy (88).

### 4.4. Other Transgenic Rat Models

There are numerous other transgenic rat lines with alterations in genes involved in cardiovascular regulation ((7), http://www.ifr26.nantes.inserm.fr/ITERT/transgenese-rat/liste_rats.php). However, none of these lines is used routinely in more than one laboratory. Therefore, they are not further described in this review.

## References

1. Aitman TJ, Critser JK, Cuppen E, Dominiczak AF, Fernandez XM, Flint J, Gauguier D, Geurts AM, Gould M, Harris PC, Holmdahl R, Huebner N, Iszvak Z, Jacob H, Kuramoto T, Kwitek AE, Marrone A, Mashimo T, Moreno-Quinn C, Mullins J, Mullins LJ, Olsson T, Riley L, Saar K, Serikawa T, Shul JD, Szpirer C, Twigger SN, Voigt B, Worley K (2008) Progress and prospects in rat genetics: a community view. Nat Genet 40:516–522
2. Mullins JJ, Peters J, Ganten D (1990) Fulminant hypertension in transgenic rats harbouring the mouse Ren-2 gene. Nature 344:541–544
3. Hammer RE, Maika SD, Richardson JA, Tang J, Taurog JD (1990) Spontaneous inflammatory disease in transgenic rats expressing HLA-B27 and human β2m: an animal model of HLA-B27-associated human disorders. Cell 63:1099–1112
4. Ganten D, Wagner J, Zeh K, Bader M, Michel J-B, Paul M, Zimmermann F, Ruf P, Hilgenfeldt U, Ganten U, Kaling M, Bachmann S, Fukamizu A, Mullins JJ, Murakami K (1992) Species specificity of renin kinetics in transgenic rats harboring the human renin and angiotensinogen genes. Proc Natl Acad Sci USA 89:7806–7810
5. Silva JA Jr, Araujo RC, Baltatu O, Oliveira SM, Tschöpe C, Fink E, Hoffmann S, Plehm R, Chai KX, Chao L, Chao J, Ganten D, Pesquero JB, Bader M (2000) Reduced cardiac hypertrophy and altered blood pressure control in transgenic rats with the human tissue kallikrein gene. FASEB J 14:1858–1860
6. Langenickel T, Buttgereit J, Pagel-Langenickel I, Lindner M, Beuerlein K, Al-Saadi N, Plehm R, Popova E, Tank J, Dietz R, Willenbrock R, Bader M (2006) Cardiac hypertrophy in transgenic rats expressing a dominant negative mutant of the natriuretic peptide receptor B. Proc Natl Acad Sci USA 103:4735–4740
7. Bader M, Bohnemeier H, Zollmann FS, Lockley-Jones OE, Ganten D (2000) Transgenic animals in cardiovascular disease research. Exp Physiol 85:713–731
8. Buehr M, Meek S, Blair K, Yang J, Ure J, Silva J, McLay R, Hall J, Ying QL, Smith A (2008) Capture of authentic embryonic stem cells from rat blastocysts. Cell 135:1287–1298
9. Li P, Tong C, Mehrian-Shai R, Jia L, Wu N, Yan Y, Maxson RE, Schulze EN, Song H, Hsieh CL, Pera MF, Ying QL (2008) Germline competent embryonic stem cells derived from rat blastocysts. Cell 135:1299–1310

10. Herold MJ, van den BJ, Seibler J, Reichardt HM (2008) Inducible and reversible gene silencing by stable integration of an shRNA-encoding lentivirus in transgenic rats. Proc Natl Acad Sci U S A 105:18507–18512
11. Kotnik K, Popova E, Todiras M, Mori MA, Alenina N, Seibler J, Bader M (2009) Inducible transgenic rat model for diabetes mellitus based on shRNA-mediated gene knockdown. PLoS One 4:e5124
12. Doggrell SA, Brown L (1998) Rat models of hypertension, cardiac hypertrophy and failure. Cardiovasc Res 39:89–105
13. Pinto YM, Paul M, Ganten D (1998) Lessons from rat models of hypertension: from Goldblatt to genetic engineering. Cardiovasc Res 39:77–88
14. Yagil Y, Yagil C (2001) Genetic models of hypertension in experimental animals. Exp Nephrol 9:1–9
15. De Champlain J, Krakoff LR, Axelrod J (1967) Catecholamine metabolism in experimental hypertension in the rat. Circ Res 20:136–145
16. Schenk J, McNeill JH (1992) The pathogenesis of DOCA-salt hypertension. J Pharmacol Toxicol Methods 27:161–170
17. Rees DD, Palmer RM, Schulz R, Hodson HF, Moncada S (1990) Characterization of three inhibitors of endothelial nitric oxide synthase in vitro and in vivo. Br J Pharmacol 101:746–752
18. Johnson RA, Freeman RH (1992) Sustained hypertension in the rat induced by chronic blockade of nitric oxide production. Am J Hypertens 5:919–922
19. Ribeiro MO, Antunes E, de Nucci G, Lovisolo SM, Zatz R (1992) Chronic inhibition of nitric oxide synthesis. A new model of arterial hypertension. Hypertension 20:298–303
20. Campbell DJ (2006) L-NAME hypertension: trying to fit the pieces together. J Hypertens 24:33–36
21. Martinez-MaldonadoM(1991)Pathophysiology of renovascular hypertension. Hypertension 17:707–719
22. Pickering TG (1989) Renovascular hypertension: etiology and pathophysiology. Semin Nucl Med 19:79–88
23. Goldblatt H, Lynch J, Hanzal RF, Summerville WW (1934) The production of persistent elevation of systolic blood pressure by means of renal ischemia. J Exp Med 59:347–379
24. Dickinson CJ, Lawrence JR (1963) A slowly developing pressor response to small concentrations of angiotensin. Its bearing on the pathogenesis of chronic renal hypertension. Lancet 1:1354–1356
25. Baltatu O, Silva JA Jr, Ganten D, Bader M (2000) The brain renin-angiotensin system modulates angiotensin II-induced hypertension and cardiac hypertrophy. Hypertension 35:409–412
26. Masugi Y, Oami H, Aihara K, Hashimoto K, Hakozaki T (1965) Renal and pulmonary vascular changes induced by Crotalaria spectabilis in rats. Acta Pathol Jpn 15:407–415
27. Meyrick B, Gamble W, Reid L (1980) Development of Crotalaria pulmonary hypertension: hemodynamic and structural study. Am J Physiol 239:H692–H702
28. Stenmark KR, Fagan KA, Frid MG (2006) Hypoxia-induced pulmonary vascular remodeling: cellular and molecular mechanisms. Circ Res 99:675–691
29. Johns TN, Olson BJ (1954) Experimental myocardial infarction. I. A method of coronary occlusion in small animals. Ann Surg 140:675–682
30. Goldman S, Raya TE (1995) Rat infarct model of myocardial infarction and heart failure. J Card Fail 1:169–177
31. Wayman NS, McDonald MC, Chatterjee PK, Thiemermann C (2003) Models of coronary artery occlusion and reperfusion for the discovery of novel antiischemic and antiinflammatory drugs for the heart. Methods Mol Biol 225:199–208
32. Nair KG, Cutilletta AF, Zak R, Koide T, Rabinowitz M (1968) Biochemical correlates of cardiac hypertrophy. I. Experimental model; changes in heart weight, RNA content, and nuclear RNA polymerase activity. Circ Res 23:451–462
33. Barbosa ME, Alenina N, Bader M (2005) Induction and analysis of cardiac hypertrophy in transgenic animal models. Methods Mol Med 112:339–352
34. Flaim SF, Minteer WJ, Nellis SH, Clark DP (1979) Chronic arteriovenous shunt: evaluation of a model for heart failure in rat. Am J Physiol 236:H698–H704
35. Garcia R, Diebold S (1990) Simple, rapid, and effective method of producing aortocaval shunts in the rat. Cardiovasc Res 24:430–432
36. Zierhut W, Zimmer HG (1989) Significance of myocardial alpha- and beta-adrenoceptors in catecholamine-induced cardiac hypertrophy. Circ Res 65:1417–1425
37. Stanton HC, Brenner G, Mayfield ED Jr (1969) Studies on isoproterenol-induced cardiomegaly in rats. Am Heart J 77:72–80
38. Longa EZ, Weinstein PR, Carlson S, Cummins R (1989) Reversible middle cerebral artery occlusion without craniectomy in rats. Stroke 20:84–91

39. Hiramatsu K, Kassell NF, Goto Y, Soleau S, Lee KS (1993) A reproducible model of reversible, focal, neocortical ischemia in Sprague-Dawley rat. Acta Neurochir (Wien) 120:66–71
40. Yanamoto H, Nagata I, Niitsu Y, Xue JH, Zhang Z, Kikuchi H (2003) Evaluation of MCAO stroke models in normotensive rats: standardized neocortical infarction by the 3VO technique. Exp Neurol 182:261–274
41. Okamoto K, Aoki K (1963) Development of a strain of spontaneously hypertensive rats. Jpn Circ J 27:282–293
42. Louis WJ, Howes LG (1990) Genealogy of the spontaneously hypertensive rat and Wistar-Kyoto rat strains: implications for studies of inherited hypertension. J Cardiovasc Pharmacol 16(Suppl 7):S1–S5
43. Petretto E, Sarwar R, Grieve I, Lu H, Kumaran MK, Muckett PJ, Mangion J, Schroen B, Benson M, Punjabi PP, Prasad SK, Pennell DJ, Kiesewetter C, Tasheva ES, Corpuz LM, Webb MD, Conrad GW, Kurtz TW, Kren V, Fischer J, Hubner N, Pinto YM, Pravenec M, Aitman TJ, Cook SA (2008) Integrated genomic approaches implicate osteoglycin (Ogn) in the regulation of left ventricular mass. Nat Genet 40:546–552
44. Pravenec M, Churchill PC, Churchill MC, Viklicky O, Kazdova L, Aitman TJ, Petretto E, Hubner N, Wallace CA, Zimdahl H, Zidek V, Landa V, Dunbar J, Bidani A, Griffin K, Qi N, Maxova M, Kren V, Mlejnek P, Wang J, Kurtz TW (2008) Identification of renal Cd36 as a determinant of blood pressure and risk for hypertension. Nat Genet 40:952–954
45. Bianchi G, Ferrari P, Barber BR (1984) The Milan hypertensive strain. In: de Jong W (ed) Experimental and genetic models of hypertension. Elsevier Science, Oxford, pp 328–349
46. Menini S, Ricci C, Iacobini C, Bianchi G, Pugliese G, Pesce C (2004) Glomerular number and size in Milan hypertensive and normotensive rats: their relationship to susceptibility and resistance to hypertension and renal disease. J Hypertens 22:2185–2192
47. Bianchi G, Tripodi G (2003) Genetics of hypertension: the adducin paradigm. Ann N Y Acad Sci 986:660–668
48. Vincent M, Dupont J, Sassard J (1979) Simultaneous selection of spontaneously hypertensive, normotensive and lowtensive rats. Jpn Heart J 20(S1):135–137
49. Sassard J, Lo M, Liu KL (2003) Lyon genetically hypertensive rats: an animal model of "low renin hypertension". Acta Pharmacol Sin 24:1–6
50. Smirk FH, Hall WH (1958) Inherited hypertension in rats. Nature 182:727–728
51. Jones DR, Dowd DA (1970) Development of elevated blood pressure in young genetically hypertensive rats. Life Sci 9:247–250
52. Ledingham JM, Laverty R (1998) Renal afferent arteriolar structure in the genetically hypertensive (GH) rat and the ability of losartan and enalapril to cause structural remodelling. J Hypertens 16:1945–1952
53. Vrana A, Kazdova L (1990) The hereditary hypertriglyceridemic nonobese rat: an experimental model of human hypertriglyceridemia. Transplant Proc 22:2579
54. Heller J, Hellerova S, Dobesova Z, Kunes J, Zicha J (1993) The Prague Hypertensive Rat: a new model of genetic hypertension. Clin Exp Hypertens 15:807–818
55. Dahl LK, Heine M, Tassinari L (1962) Role of genetic factors in susceptibility to experimental hypertension due to chronic excess salt ingestion. Nature 194:480–482
56. Rapp JP (1982) Dahl salt-susceptible and salt-resistant rats. A review. Hypertension 4: 753–763
57. Bashyam H (2007) Lewis Dahl and the genetics of salt-induced hypertension. J Exp Med 204:1507
58. Rodriguez-Sargent C, Cangiano JL, Fernandez-Repollet E, Estape-Wainwright E, Torres-Negron I, Martinez-Maldonado M (1988) A new model of genetic hypertension in rats with superficial glomeruli. J Hypertens Suppl 6:S29–S32
59. Ben Ishay D, Saliternik R, Welner A (1972) Separation of two strains of rats with inbred dissimilar sensitivity to Doca-salt hypertension. Experientia 28:1321–1322
60. Yagil C, Katni G, Rubattu S, Stolpe C, Kreutz R, Lindpaintner K, Ganten D, Ben Ishay D, Yagil Y (1996) Development, genotype and phenotype of a new colony of the Sabra hypertension prone (SBH/y) and resistant (SBN/y) rat model of salt sensitivity and resistance. J Hypertens 14:1175–1182
61. Yagil Y, Yagil C (1998) Genetic basis of salt-susceptibility in the Sabra rat model of hypertension. Kidney Int 53:1493–1500
62. Markel AL (1985) Experimental model of inherited arterial hypertension, conditioned by stress. Izvestia Akad Nauk SSSR Seria Biol 3:466–469
63. Maslova LN, Bulygina VV, Markel AL (2002) Chronic stress during prepubertal development: immediate and long-lasting effects on arterial blood pressure and anxiety-related behavior. Psychoneuroendocrinology 27:549–561

64. Castle WE, King HD (1947) Linkage studies of the rat. VIII. Fawn, a new colour dilution gene. J Hered 38:341–344
65. Prieur DJ, Meyers KM (1984) Genetics of the fawn-hooded rat strain. The coat color dilution and platelet storage pool deficiency are pleiotropic effects of the autosomal recessive red-eyed dilution gene. J Hered 75:349–352
66. Oiso N, Riddle SR, Serikawa T, Kuramoto T, Spritz RA (2004) The rat Ruby ( R) locus is Rab38: identical mutations in Fawn-hooded and Tester-Moriyama rats derived from an ancestral Long Evans rat sub-strain. Mamm Genome 15:307–314
67. Morecroft I, Dempsie Y, Bader M, Walther DJ, Kotnik K, Loughlin L, Nilsen M, MacLean MR (2007) Effect of tryptophan hydroxylase 1 deficiency on the development of hypoxia-induced pulmonary hypertension. Hypertension 49:232–236
68. Walther DJ, Peter JU, Winter S, Höltje M, Paulmann N, Grohmann M, Vowinckel J, Alamo-Bethencourt V, Wilhelm CS, Ahnert-Hilger G, Bader M (2003) Serotonylation of small GTPases is a signal transduction pathway that triggers platelet α-granule release. Cell 115:851–862
69. Nagaoka T, Gebb SA, Karoor V, Homma N, Morris KG, McMurtry IF, Oka M (2006) Involvement of RhoA/Rho kinase signaling in pulmonary hypertension of the fawn-hooded rat. J Appl Physiol 100:996–1002
70. Bonnet S, Michelakis ED, Porter CJ, Andrade-Navarro MA, Thebaud B, Bonnet S, Haromy A, Harry G, Moudgil R, McMurtry MS, Weir EK, Archer SL (2006) An abnormal mitochondrial-hypoxia inducible factor-1alpha-Kv channel pathway disrupts oxygen sensing and triggers pulmonary arterial hypertension in fawn hooded rats: similarities to human pulmonary arterial hypertension. Circulation 113:2630–2641
71. Brown DM, Provoost AP, Daly MJ, Lander ES, Jacob HJ (1996) Renal disease susceptibility and hypertension are under independent genetic control in the fawn-hooded rat. Nat Genet 12:44–51
72. Mccune S, Baker PB, Stills FH (1990) SHHF/Mcc-cp rat: model of obesity, non-insulin dependent diabetes, and congestive heart failure. ILAR News 32:23–27
73. Monti J, Fischer J, Paskas S, Heinig M, Schulz H, Gosele C, Heuser A, Fischer R, Schmidt C, Schirdewan A, Gross V, Hummel O, Maatz H, Patone G, Saar K, Vingron M, Weldon SM, Lindpaintner K, Hammock BD, Rohde K, Dietz R, Cook SA, Schunck WH, Luft FC, Hubner N (2008) Soluble epoxide hydrolase is a susceptibility factor for heart failure in a rat model of human disease. Nat Genet 40:529–537
74. Okamoto H, Yamori Y, Nagaoka A (1974) The establishment of the stroke prone hypertensive rat. Circ Res 34(Suppl I):I-143–I-153
75. Yamori Y, Horie R, Tanase H, Fujiwara K, Nara Y, Lovenberg W (1984) Possible role of nutritional factors in the incidence of cerebral lesions in stroke-prone spontaneously hypertensive rats. Hypertension 6:49–53
76. Smeda JS (1989) Hemorrhagic stroke development in spontaneously hypertensive rats fed a North American, Japanese-style diet. Stroke 20:1212–1218
77. McBride MW, Brosnan MJ, Mathers J, McLellan LI, Miller WH, Graham D, Hanlon N, Hamilton CA, Polke JM, Lee WK, Dominiczak AF (2005) Reduction of Gstm1 expression in the stroke-prone spontaneously hypertension rat contributes to increased oxidative stress. Hypertension 45:786–792
78. Rentzsch B, Todiras M, Iliescu R, Popova E, Campos LA, Oliveira ML, Baltatu OC, Santos RA, Bader M (2008) Transgenic ACE2 overexpression in vessels of SHRSP rats reduces blood pressure and improves endothelial function. Hypertension 52:967–973
79. Böhm M, Lippoldt A, Wienen W, Ganten D, Bader M (1996) Reduction of cardiac hypertrophy in TGR(mREN2)27 by angiotensin II receptor blockade. Mol Cell Biochem 163–164:217–221
80. Lee MA, Böhm M, Paul M, Bader M, Ganten U, Ganten D (1996) Physiological characterization of the hypertensive transgenic rat TGR(mREN2)27. Am J Physiol 270:E919–E929
81. Sharma UC, Pokharel S, van Brakel TJ, van Berlo JH, Cleutjens JP, Schroen B, Andre S, Crijns HJ, Gabius HJ, Maessen J, Pinto YM (2004) Galectin-3 marks activated macrophages in failure-prone hypertrophied hearts and contributes to cardiac dysfunction. Circulation 110:3121–3128
82. Kantachuvesiri S, Fleming S, Peters J, Peters B, Brooker G, Lammie AG, McGrath I, Kotelevtsev Y, Mullins JJ (2001) Controlled hypertension, a transgenic toggle switch reveals differential mechanisms underlying vascular disease. J Biol Chem 276:36727–36733
83. Peters B, Grisk O, Becher B, Wanka H, Kuttler B, Ludemann J, Lorenz G, Rettig R, Mullins JJ, Peters J (2008) Dose-dependent titration of prorenin and blood pressure in Cyp1a1ren-2 transgenic rats: absence of prorenin-induced glomerulosclerosis. J Hypertens 26:102–109

84. Mitchell KD, Bagatell SJ, Miller CS, Mouton CR, Seth DM, Mullins JJ (2006) Genetic clamping of renin gene expression induces hypertension and elevation of intrarenal Ang II levels of graded severity in Cyp1a1-Ren2 transgenic rats. J Renin Angiotensin Aldosterone Syst 7:74–86

85. Howard LL, Patterson ME, Mullins JJ, Mitchell KD (2005) Salt-sensitive hypertension develops after transient induction of ANG II-dependent hypertension in Cyp1a1-Ren2 transgenic rats. Am J Physiol Renal Physiol 288:F810–F815

86. Luft FC, Mervaala E, Müller DN, Gross V, Schmidt F, Park JK, Schmitz C, Lippoldt A, Breu V, Dechend R, Dragun D, Schneider W, Ganten D, Haller H (1999) Hypertension-induced end-organ damage: a new transgenic approach to an old problem. Hypertension 33: 212–218

87. Pilz B, Shagdarsuren E, Wellner M, Fiebeler A, Dechend R, Gratze P, Meiners S, Feldman DL, Webb RL, Garrelds IM, Jan Danser AH, Luft FC, Muller DN (2005) Aliskiren, a human renin inhibitor, ameliorates cardiac and renal damage in double-transgenic rats. Hypertension 46:569–576

88. Bohlender J, Ganten D, Luft FC (2000) Rats transgenic for human renin and human angiotensinogen as a model for gestational hypertension. J Am Soc Nephrol 11: 2056–2061

# Chapter 28

# Use of Rat Genomics for Investigating the Metabolic Syndrome


## Michal Pravenec



## Abstract

The spontaneously hypertensive rat (SHR) is the most widely used animal model of essential hypertension and accompanying metabolic disturbances. In this model, the use of whole genome sequencing and gene expression profiling techniques, linkage and correlation analyses in recombinant inbred strains, and in vitro and in vivo functional studies in congenic and transgenic lines has recently enabled molecular identification of quantitative trait loci (QTLs) relevant to the metabolic syndrome: (1) a deletion variant in *Cd36* (fatty acid translocase) responsible for QTLs on chromosome 4 associated with dyslipidemia, insulin resistance and hypertension, (2) mutated *Srebf1* (sterol regulatory element binding factor 1) as a QTL on chromosome 10 influencing dietary-induced changes in hepatic cholesterol levels, and (3) *Ogn* (osteoglycin) as a QTL on chromosome 17 associated with left ventricular hypertrophy. In addition, selective replacement of the mitochondrial genome of the SHR with the mitochondrial genome of the Brown Norway rat influenced several major metabolic risk factors for type 2 diabetes and provided evidence that spontaneous variation in the mitochondrial genome per se can promote systemic metabolic disturbances relevant to the pathogenesis of metabolic syndrome. Owing to recent progress in the development of rat genomic resources, the pace of QTL identification and discovery of new disease mechanisms can be expected to accelerate in the near future.



**Key words:** Spontaneously hypertensive rat, Recombinant inbred strains, Gene expression profiles, Genetical genomics, Conplastic strains


## 1. Introduction

The metabolic syndrome affects upwards of 15–25% of the adult population in developed countries and is characterized by the clustering of multiple risk factors for diabetes and cardiovascular disease (1–3). The metabolic syndrome epidemic appears to be a result of profoundly maladaptive diets and lifestyles together with genetic susceptibility factors operating within predisposing environments. The identification of such genetic susceptibility factors


I. Anegon (ed.), *Rat Genomics: Methods and Protocols*, Methods in Molecular Biology, vol. 597
DOI 10.1007/978-1-60327-389-3_28, © Humana Press, a part of Springer Science+Business Media, LLC 2010

is expected to advance understanding of the etiology of the metabolic syndrome and ultimately help guide development of improved approaches to prevention and therapy of this highly common disorder.

Recently, genome wide association studies (GWAS) have been applied to analyses of common diseases, including obesity, essential hypertension, and type 2 diabetes. These studies have yielded exciting new results and have also posed questions regarding data analyses, interpretation, and clinical significance (4–6). Significant associations of single nucleotide polymorphisms (SNPs) with specific genes reported in GWAS have also raised questions about the need and usefulness of animal models for genetic studies of common diseases. For instance, are animal models still necessary when responsible genes can be now identified in GWAS? The answer is yes, because it is likely that in most cases, GWAS that screen common variants will not contribute to understanding the pathophysiology of common diseases but only uncover the effects of closely linked functional variants (6). For instance, some significantly associated SNPs are located in unannotated genes or even in noncoding sequences. Thus, even when an association to disease has apparently been localized to a single gene, in the absence of additional experimental data, it is usually impossible to definitively declare that it is the actual causal gene. The in vivo functional roles of genes implicated by GWAS studies can only be directly investigated in animal models.

## 2. Recent Advances in Rat Genetics and Genomics

Comparative genomics is an essential tool for genetic dissection of multifactorial diseases and much of our current knowledge concerning the genes that regulate clinically important pathophysiological traits has been derived from rodent studies (7). The laboratory rat is the most widely used animal model for human common diseases (8). In the last decade, there has been an extraordinary increase in rat genome resources (9). Rat genome resources include 7.5× sequence of the Brown Norway (BN/Mcwi) rat genome. Sequencing of other strains, including the spontaneously hypertensive rat (SHR/Ola) is presently underway using Illumina/Solexa paired-end sequencing techniques (Aitman, Hübner, Pravenec, Kurtz, et al., unpublished results). Millions of SNPs between different rat strains are now available for linkage analyses using new methods of efficient genotyping based on genechip microarrays (10). In addition, over 800,000 ESTs and 5,000 annotated rat gene sequences are available for functional analyses of candidate genes. Development of new methodologies for high throughput phenotyping, such as expression profiling, are becoming routinely used. Most of these genetic and phenotypic traits are available to the scientific community in databases, such as Ensembl (http://www.ensembl.org), the Rat Genome Database (http://www.rgd.mcw.edu), eQTL Explorer (http://www. web.bioinformatics.ic.ac.uk/eqtlexplorer) or GeneNetwork (http://www.genenetwork.org). Additional online rat genetic resources have been recently reviewed by Twigger et al. (11).

## 3. Genetic Analyses of Complex Metabolic Traits in Rat Models

In multifactorially determined polygenic traits such as metabolic syndrome and its individual components, there is no direct relationship between pathophysiological phenotypes and genotypes. An individual may carry a predisposing allele without being affected (low penetrance) or may be affected (a phenocopy) while carrying a protective allele of a given gene. The reason for such uncertainty is related to complex interactions among many genes (epistasis) and not fully understood interactions of genetic background with environmental factors, including epigenetic effects (methylome) (12). In animal models, genetic determinants of complex traits are identified by linkage analyses as quantitative trait loci (QTLs), which are usually relatively large chromosome segments containing often hundreds of candidate genes. Identification of QTLs at the molecular level as mutations within specific genes (referred to as quantitative trait genes or QTGs) represents a very difficult task (7).

Traditional approaches to identify QTL at the molecular level have involved the use of linkage analysis followed by time-consuming positional mapping studies in congenic strains and sublines. For example, this strategy has enabled isolation of blood pressure regulatory QTLs within approximately 1 Mbp differential chromosome segments of "minimal" congenic strains (13–15). To speed up this process, investigators can derive congenic sublines from consomic strains (*see* specialized strains described in Box 1), use marker-assisted selection to develop "speed" congenic strains, (16) or superovulation techniques to derive "supersonic" congenic strains (17).

A useful strategy for QTL identification at the molecular level is represented by the analysis of "intermediate" phenotypes, which are related to complex pathophysiological traits, but have a relatively simple genetic determination. For instance, transcript abundance might represent a useful intermediate phenotype between variability at the DNA level and complex physiological phenotypes such as blood pressure or insulin resistance. Gene expression profiling of tissues relevant to the pathogenesis of metabolic syndrome can be combined with linkage and correlation analyses. Integrating transcriptional profiling and linkage analyses (also called "genetical genomics") offers the advantage to identify genetic determinants that are responsible for variability in gene expression, and to distinguish whether they regulate transcript level of the gene itself in *cis* (*cis*-regulated expression QTLs or *cis*-eQTLs) or of another gene in *trans* (*trans*-eQTLs) (18). *Cis*-eQTLs, which are located in the same chromosomal regions as QTL associated with physiological traits, are then prioritized as candidate genes for follow-up sequencing and functional studies. Additional supporting evidence for the most promising candidate genes can be obtained by correlation analyses of hemodynamic or metabolic phenotypes with transcript abundance to identify quantitative trait transcripts (QTT) (19). These approaches are especially effective in recombinant inbred (RI) strains (*see* Box 1) because of the cumulative nature of all results that can be used for combined gene expression profiling, linkage, and correlation analyses (20).

The current review is focused on genetic analysis of the SHR, a model of essential hypertension and associated metabolic disturbances (21). Recent studies in this model, including whole genome sequencing, gene expression profiling in multiple tissues relevant to the pathogenesis of the metabolic syndrome as well as derivation of specialized strains, including congenic, conplastic, recombinant inbred strains, and transgenic lines, illustrate the use of rat genome tools and resources for analyses of the metabolic syndrome. Additional rat models of type 2 diabetes, such as the Goto-Kakizaki strain, Cohen rat, Otsuka Long-Evans Fatty Tokushima (OLEFT) rats, Zucker diabetic fatty rats, hereditary hypertriglyceridemic (HHTG), or Polydactylous (PD) rats, were described in several recent reviews (22–26).

**Box 1 Specialized Strains**

*Congenic strains*

An inbred strain that contains chromosome segment from another strain, but which is otherwise identical to the original inbred strain. Congenic strains are derived by backcrossing to a parental inbred strain for at least ten generations while selecting for heterozygosity at a particular locus. After ten generations of backcrossing, heterozygotes are intercrossed and homozygous animals of N10F1 are selected to maintain the new congenic strain.

*Consomic strains*

An inbred strain that contains an entire single chromosome from another strain.

(continued)

**Box 1** (continued)

*Conplastic strains*

Conplastic strains are two inbred strains that have identical nuclear genomes but different mitochondrial genomes. These strains are derived by backcross breeding techniques in which the mitochondrial genome of a donor strain is transferred onto the nuclear genetic background of recipient strain.

*Recombinant inbred (RI) strains*

RI strains are derived from an outcross between two highly inbred progenitor strains. F1 hybrids are intercrossed to produce a large set of F2 animals. The F2 animals are chosen at random to serve as the founders for new inbred strains. The offspring from each F2 founder pair are chosen randomly for brother–sister mating to produce the next generation. The same process is repeated at each subsequent generation until at least 20 sequential rounds of strict brother–sister matings have been completed, and a new set of RI strains is established.

## 4. The Spontaneously Hypertensive Rat and Derived Strains: A Model System for Genetic Analysis of the Metabolic Syndrome

SHR not only exhibits high blood pressure, but is also susceptible to dietary-induced features of insulin resistance and dyslipidemia (21). Multiple QTLs associated with blood pressure and metabolic traits have been reported in the SHR (Rat Genome Database at http://www.rgd.mcw.edu), however, identification of these putative QTLs at the molecular level is a very difficult task because of the complexity of the hemodynamic and metabolic phenotypes. It is therefore hoped that such polygenic traits might be subdivided into intermediate phenotypes characterized by oligogenic or even Mendelian inheritance. Examples of intermediate phenotypes with relatively simple genetic determination are gene expression levels that are regulated in *cis* by variation in or near the genes themselves (18, 20). Additional levels of complexity represent physiological phenotypes on cellular, tissue, and organ levels. Systemic phenotypes such as blood pressure exhibit the highest level of complexity. Combining linkage and correlation analyses at several levels of complexity requires extensive phenotyping that is often impossible in conventional genetically segregating populations, but is feasible in RI strains where phenotypic results can be accumulated over time (27). Using this strategy, the first QTLs associated with complex traits in the SHR have been recently identified at the molecular level: variant *Cd36* (fatty acid translocase) contributing to insulin resistance, dyslipidemia and hypertension (28–31), variant *Srebf1* (sterol regulatory element binding factor 1) gene influencing hepatic cholesterol levels (32), and variant *Ogn* (osteoglycin) influencing left ventricle hypertrophy (33).

### 4.1. Recombinant Inbred Strains for the Analysis of Genes and Functional Networks Predisposing to Metabolic Syndrome

Recombinant inbred (RI) strains (*see* Box 1) combine the advantages of inbreeding and gene segregation. An important feature of RI strains is that, because they are inbred and genetically defined populations, repeated assays can be made so that phenotypes of each strain are systematically characterized with a rigor that is often impossible in conventional segregating populations. Moreover, the acquired data are cumulative across assays, studies, and research groups and therefore large datasets can be analyzed and trait relationships discovered that might not have been expected otherwise. This cumulative nature of studies in RI strains is an enormous advantage and an essential feature for the analysis of complex traits. For genetic and correlation analyses of spontaneous hypertension and metabolic defects in the SHR, the BXH/HXB sets of RI strains were developed by reciprocal crossings of the SHR/Ola and the Brown Norway (BN-*Lx*/Cub) strains (34). At present, 20 HXB and 10 BXH RI strains are available. The current map of RI strains contains more than 3,000 markers, mostly microsatellites and SNPs (10, 20) and over 200 physiological phenotypes, mostly hemodynamic (radiotelemetry blood pressures and heart rates), metabolic and biochemical traits (GeneNetwork at http://www.genenetwork.org; Pravenec et al., unpublished) are available for linkage and correlation analyses. In 2005, the BXH/HXB panel of RI strains was used to map the major genetic determinants of gene expression in SHR for 15,923 genes in two of the key tissues relevant for the pathophysiology of the metabolic syndrome: kidney and fat. After assessment of genome-wide significance and accounting for multiple testing and false discovery rates (FDR), over 1,000 expression QTLs (eQTLs) were found in kidney and fat. These eQTLs represent a large source of attractive candidate genes for the scores of physiological QTLs (pQTLs) that have been mapped in SHR (20, 21); http://www.genenetwork.org). Recently, gene expression profiles, using Affymetrix GeneChip Arrays, were determined in additional tissues relevant to the pathogenesis of metabolic syndrome, including adrenals (Aitman, Hübner, O`Connor, Pravenec et al., unpublished results), left ventricles (33), whole brains (PhenoGen Informatics at phenogen.uchsc.edu), liver, and soleus muscle (Aitman, Hübner, Pravenec et al., unpublished).

The GeneNetwork (http://www.genenetwork.org) is a public resource that combines genomic and phenotypic information, including gene expression profiles, about RI strains with fast software for identification of pQTLs and eQTLs, using genetical genomics and QTT approaches (35). The current version of the BXH/HXB database at the GeneNetwork contains a genetic map based on more than 1,000 microsatellites and about 90 physiological traits and gene expression profiles determined in retroperitoneal fat and kidney.

The limitations of RI strains for mapping of QTLs associated with complex traits are widely recognized, especially in regard to progenitor strains and to small numbers of RI strains in individual sets (36). On the other hand, RI strains can be used with advantage to map QTLs with relatively large effects and for mapping less complex, essentially monogenic *cis*-acting eQTLs (20, 37). In addition, the unique genetic constitution of RI strains can be used in a powerful way for network (correlation) analyses, because the primary goal in such studies is to measure the tendency of different traits to cosegregate rather than to map genes. Thus, because of their unique patterns of genetic randomization, RI strains provide useful statistical power to measure tendencies of traits to cosegregate (38).

### 4.2. Identification of *Cd36* as a Gene Causing Defective Fatty Acid and Glucose Metabolism, and Increased Blood Pressure in the SHR

Linkage analysis of the BXH/HXB RI strains revealed significant QTLs near the centromeric end of rat chromosome 4 that were associated with in vitro phenotypes of insulin resistance and dyslipidemia, specifically with insulin-stimulated glucose uptake and isoproterenol-induced lipolysis in isolated adipocytes (39). Additional suggestive QTLs associated with serum HDL2 phospholipids (40) and blood pressure (41) were mapped to the same chromosome region. In follow-up studies, a combination of cDNA microarrays and congenic mapping was used to identify a defective SHR gene, *Cd36* (also known as *Fat*, as it encodes fatty acid translocase), at the peak of these QTL linkages (28, 29). Sequence analysis revealed that the SHR *Cd36* cDNA contains multiple variants, caused by unequal genomic recombination of a duplicated ancestral gene (42). The SHR defect in *Cd36* has been conclusively shown to result in disturbances in lipid and glucose metabolism based on transgenic rescue studies using several transgenic strains of SHR derived to express wild type *Cd36* (30).

To search for QTLs involved in the renal pathogenesis of essential hypertension, we combined genome-wide eQTL and QTT analyses of the kidney transcriptome in the RI strains. This strategy identified inherited variation in the renal expression of *Cd36* as a possible determinant of inherited variation in the risk for hypertension: *cis*-regulated *Cd36* transcripts in kidneys of RI strains exhibited the most significant correlations with blood pressure (31). Expression of wild type *Cd36* in several transgenic lines of SHR showed inconsistent blood pressure effects with only a single transgenic line showing a clear reduction in blood pressure.

This transgenic line proved to be the only transgenic line that showed significant expression of wild type *Cd36* in the kidney. This raised the possibility that in the SHR progenitor strain, deficient renal expression of wild type *Cd36* might be contributing to increased blood pressure. To investigate whether selective lack of wild type *Cd36* in the kidney might be sufficient to promote increased blood pressure, we performed renal transplantation experiments using donor kidneys from either the mutant SHR progenitor strain that lacks wild type *Cd36* or from the transgenic strain of SHR with robust renal expression of wild type *Cd36*. Because the wild type *Cd36* transgene was expressed on the genetic background of the highly inbred SHR progenitor strain lacking wild type *Cd36*, all donor kidneys were genetically identical except for the presence or absence of wild type *Cd36*. For transplant recipients, we used bilaterally nephrectomized SHR congenic rats expressing wild type *Cd36* that are otherwise genetically identical to the SHR progenitor strain except for loci linked to *Cd36* on rat chromosome 4. Thus, this design enabled us to compare blood pressures in two groups of immunologically tolerant transplant recipients that differed only in the renal expression of wild type *Cd36* (Fig. 28.1). The blood pressure of the recipients that received a donor kidney lacking wild type *Cd36* was significantly greater than the blood pressure of the recipients that received a donor kidney expressing wild type *Cd36* (31).

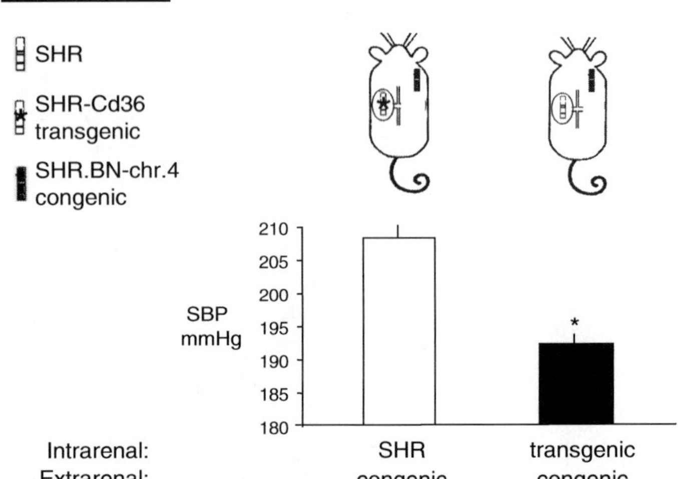


Fig. 28.1. Systolic blood pressures (SBP) in two groups of immunologically tolerant transplant recipients that differ only in the renal expression of wild type *Cd36*. Bilaterally nephrectomized SHR congenic rats expressing wild type *Cd36* and transplanted with kidneys from the SHR progenitor strain harboring mutant *Cd36* showed significantly increased SBP when compared to recipients transplanted with kidneys from a transgenic strain of SHR that renally expressed wild type transgenic *Cd36*.

The CD36 fatty acid transporter is expressed in several regions of the kidney, including on capillary endothelium in the renal medulla. CD36 is also known to co-localize with endothelial nitric oxide synthase (eNOS) in caveolae of endothelial cells and it has been shown that CD36 is a determinant of eNOS activation by fatty acids. Given that reduced nitric oxide activity in the renal medulla has been implicated in the pathogenesis of hypertension, it is reasonable to speculate that mutations in *Cd36* might influence the regulation of blood pressure through nitric oxide-related pathways in the kidney (31). These results not only demonstrate that *Cd36* mutations within the kidney can increase blood pressure, they collectively indicate that naturally occurring variation in a single gene can promote clustering of all the main hemodynamic and biochemical disturbances that characterize the metabolic syndrome (Fig. 28.2).

## 5. Conplastic Rats: The Role of Mitochondrial DNA in the Pathogenesis of Metabolic Syndrome

Recently, the relationship of mitochondrial DNA (mtDNA) variants to metabolic risk factors for diabetes and other common diseases has begun to attract increasing attention. Substitution of different mitochondrial genomes on the same nuclear genetic background in conplastic strains provides a way to unambiguously isolate effects of the mitochondrial genome on complex traits. Recently, we derived the SHR/OlaIpcv-mt$^{BN/Crl}$ conplastic strain by transferring the mitochondrial genome from the BN/Crl

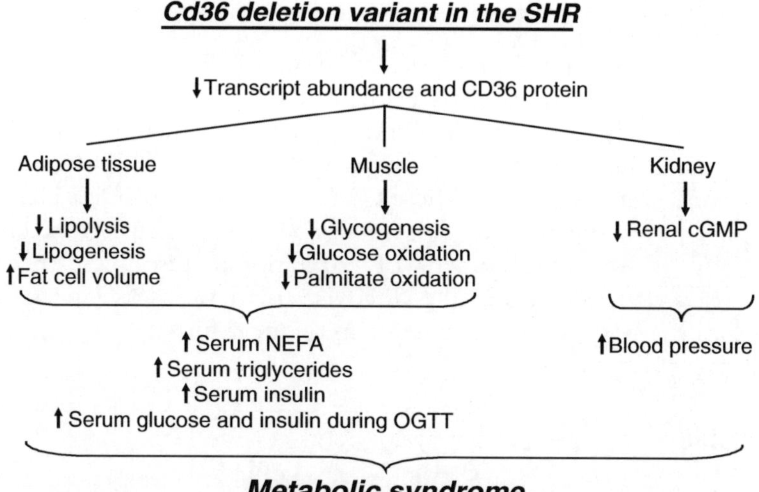


Fig. 28.2. The *Cd36* deletion variant is associated with metabolic syndrome in the SHR. The deletion variant of *Cd36* in adipose and muscle tissue is responsible for dyslipidemia, impaired glucose tolerance and insulin resistance. In kidneys, the deletion variant of *Cd36* is responsible for decreased levels of the major nitric oxide second messenger cyclic GMP, suggesting that renal defects in *Cd36* may be affecting blood pressure through nitric oxide-related pathways.

strain onto genetic background of the SHR using a "supersonic" breeding method (17, 43, 44). These conplastic strains of rats, with identical nuclear genomes but divergent mitochondrial genomes that encode amino acid differences in proteins of oxidative phosphorylation, exhibited differences in major metabolic risk factors for type 2 diabetes, including significant strain differences in glucose and insulin levels during oral glucose tolerance testing and strain differences in skeletal muscle glycogen and ATP content (44). These results demonstrate that selective replacement of the mitochondrial genome of the SHR with the mitochondrial genome of the BN rat influences several major metabolic risk factors for type 2 diabetes and provides evidence that spontaneous variation in the mitochondrial genome per se can promote systemic metabolic disturbances relevant to the pathogenesis of metabolic syndrome. The next step will be to determine the effects on cellular metabolism of the specific mtDNA variants encoding amino acid substitutions in proteins of oxidative phosphorylation that were linked to biochemical features of the metabolic syndrome in the conplastic strains. One approach to this task could involve rescue studies in which allotopic gene expression techniques are used to create cell lines with identical nuclear genomes that differ in the expression of only a single protein encoded by the mitochondrial genome (45). Any phenotypic differences between these cell lines could then be related to differences in activity of the candidate mitochondrial protein of interest.

## 6. Summary and Conclusions

The rat is an important animal model for genetic analysis of common diseases as evidenced by the mapping of multiple QTLs associated with blood pressure and metabolic traits in the SHR and other strains during the past 20 years. However, pinpointing the identity of these QTLs at the molecular level has proven to be a difficult, but not impossible, task. Thanks to recent progress in the development of rat genomic resources, the pace of this research and the discovery of new disease mechanisms can be expected to accelerate in the near future.

## Acknowledgments

This work was supported by grants 1P05ME791 and 1M6837805002 from the Ministry of Education of the Czech Republic, grants 301/06/0028 and 301/08/0166 from the Grant Agency of the Czech Republic, grant IAA500110604 from the Grant Agency of the Academy of Sciences of the Czech Republic, and by the European Commission within the Sixth Framework Programme through the Integrated Project EURATools (contract no. LSHG-CT-2005-019015). M.P. is an international research scholar of the Howard Hughes Medical Institute.

## References

1. Lakka HM, Laaksonen DE, Lakka TA, Niskanen LK, Kumpusalo E, Tuomilehto J et al (2002) The metabolic syndrome and total and cardiovascular disease mortality in middle-aged men. JAMA 288:2709–2716
2. Park YW, Zhu S, Palaniappan L, Heshka S, Carnethon MR, Heymsfield SB (2003) The metabolic syndrome: prevalence and associated risk factor findings in the US population from the Third National Health and Nutrition Examination Survey, 1988–1994. Arch Intern Med 163:427–436
3. Laaksonen DE, Lakka HM, Niskanen LK, Kaplan GA, Salonen JT, Lakka TA (2002) Metabolic syndrome and development of diabetes mellitus: application and validation of recently suggested definitions of the metabolic syndrome in a prospective cohort study. Am J Epidemiol 156:1070–1077
4. Groop L, Lyssenko V (2008) Genes and type 2 diabetes mellitus. Curr Diab Rep 8:192–197
5. Zeggini E, Scott LJ, Saxena R, Voight BF, Marchini JL, Hu T et al (2008) Meta-analysis of genome-wide association data and large-scale replication identifies additional susceptibility loci for type 2 diabetes. Nat Genet 40:638–645
6. Bodmer W, Bonilla C (2008) Common and rare variants in multifactorial susceptibility to common diseases. Nat Genet 40:695–701
7. Glazier AM, Nadeau JH, Aitman TJ (2002) Finding genes that underlie complex traits. Science 298:2345–2349
8. Jacob HJ (1999) Functional genomics and rat models. Genome Res 9:1013–1016
9. Aitman TJ, Critser JK, Cuppen E, Dominiczak A, Fernandez-Suarez XM, Flint J et al (2008) Progress and prospects in rat genetics: a community view. Nat Genet 40:516–522
10. Consortium STAR, Saar K, Beck A, Bihoreau MT, Birney E, Brocklebank D, Chen Y et al (2008) SNP and haplotype mapping for genetic analysis in the rat. Nat Genet 40:560–566
11. Twigger SN, Pruitt KD, Fernández-Suárez XM, Karolchik D, Worley KC, Maglott DR et al (2008) What everybody should know about the rat genome and its online resources. Nat Genet 40:523–527
12. Butcher LM, Beck S (2008) Future impact of integrated high-throughput methylome analyses on human health and disease. J Genet Genomics 35:391–401
13. Saad Y, Garrett MR, Manickavasagam E, Yerga-Woolwine S, Farms P, Radecki T et al (2007) Fine-mapping and comprehensive transcript analysis reveals nonsynonymous variants within a novel 1.17 Mb blood pressure QTL region on rat chromosome 10. Genomics 89:343–353
14. Lee SJ, Liu J, Westcott AM, Vieth JA, DeRaedt SJ, Yang S et al (2006) Substitution mapping in dahl rats identifies two distinct blood pressure quantitative trait loci within 1.12- and 1.25-mb intervals on chromosome 3. Genetics 174:2203–2213
15. Seda O, Liska F, Sedová L, Kazdová L, Krenová D, Kren V (2005) A 14-gene region of rat chromosome 8 in SHR-derived polydactylous congenic substrain affects muscle-specific insulin resistance, dyslipidaemia and visceral adiposity. Folia Biol (Praha) 51:53–61
16. Fournie GJ (2009) Generation of consomic and congenic rat strains (including speed congenics). Method Mol Biol current issue
17. Landa V, Zidek V, Pravenec M (2009) Generation of rat conplastic strains using superovulation. Method Mol Biol current issue
18. Jansen RC, Nap JP (2001) Genetical genomics: the added value from segregation. Trends Genet 17:388–391
19. Passador-Gurgel G, Hsieh WP, Hunt P, Deighton N, Gibson G (2007) Quantitative trait transcripts for nicotine resistance in Drosophila melanogaster. Nat Genet 39: 264–268
20. Hübner N, Wallace CA, Zimdahl H, Petretto E, Schulz H, Maciver F et al (2005) Integrated transcriptional profiling and linkage analysis for identification of genes underlying disease. Nat Genet 37:243–253
21. Pravenec M, Zídek V, Landa V, Simáková M, Mlejnek P, Kazdová L et al (2004) Genetic analysis of "metabolic syndrome" in the

spontaneously hypertensive rat. Physiol Res 53(Suppl 1):S15–S22

22. Srinivasan K, Ramarao P (2007) Animal models in type 2 diabetes research: an overview. Indian J Med Res 125:451–472
23. Chen D, Wang MW (2005) Development and application of rodent models for type 2 diabetes. Diabetes Obes Metab 7:307–317
24. Weksler-Zangen S, Yagil C, Zangen DH, Ornoy A, Jacob HJ, Yagil Y (2001) The newly inbred cohen diabetic rat: a nonobese normolipidemic genetic model of diet-induced type 2 diabetes expressing sex differences. Diabetes 50:2521–2529
25. Sedová L, Kazdová L, Seda O, Krenová D, Kren V (2000) Rat inbred PD/cub strain as a model of dyslipidemia and insulin resistance. Folia Biol (Praha) 46:99–106
26. Vrána A, Kazdová L (1990) The hereditary hypertriglyceridemic nonobese rat: an experimental model of human hypertriglyceridemia. Transplant Proc 22:2579
27. Williams RW, Gu J, Qi S, Lu L (2001) The genetic structure of recombinant inbred mice: high-resolution consensus maps for complex trait analysis. Genome Biol 2:RESEARCH004
28. Aitman TJ, Glazier AM, Wallace CA, Cooper LD, Norsworthy PJ, Wahid FN et al (1999) Identification of Cd36 (Fat) as an insulin-resistance gene causing defective fatty acid and glucose metabolism in hypertensive rats. Nat Genet 21:76–83
29. Pravenec M, Zidek V, Simakova M, Kren V, Krenova D, Horky K et al (1999) Genetics of Cd36 and the clustering of multiple cardiovascular risk factors in spontaneous hypertension. J Clin Invest 103:1651–1657
30. Pravenec M, Landa V, Zidek V, Musilova A, Kren V, Kazdova L et al (2001) Transgenic rescue of defective Cd36 ameliorates insulin resistance in spontaneously hypertensive rats. Nat Genet 27:156–158
31. Pravenec M, Churchill PC, Churchill MC, Viklicky O, Kazdova L, Aitman TJ et al (2008) Identification of renal Cd36 as a determinant of blood pressure and risk for hypertension. Nat Genet 40:952–954
32. Pravenec M, Kazdova L, Landa V, Zidek V, Mlejnek P, Simakova M et al (2008) Identification of mutated Srebf1 as a QTL influencing risk for hepatic steatosis in the spontaneously hypertensive rat. Hypertension 51:148–153
33. Petretto E, Sarwar R, Grieve I, Lu H, Kumaran MK, Muckett PJ et al (2008) Integrated genomic approaches implicate osteoglycin (Ogn) in the regulation of left ventricular mass. Nat Genet 40:546–552
34. Pravenec M, Klír P, Kren V, Zicha J, Kunes J (1989) An analysis of spontaneous hypertension in spontaneously hypertensive rats by means of new recombinant inbred strains. J Hypertens 7:217–221
35. Wang J, Williams RW, Manly KF (2003) WebQTL: web-based complex trait analysis. Neuroinformatics 1:299–308
36. Darvasi A (1998) Experimental strategies for the genetic dissection of complex traits in animal models. Nat Genet 18:19–24
37. Chesler EJ, Lu L, Shou S, Qu Y, Gu J, Wang J et al (2005) Complex trait analysis of gene expression uncovers polygenic and pleiotropic networks that modulate nervous system function. Nat Genet 37:233–242
38. Nadeau JH, Burrage LC, Restivo J, Pao YH, Churchill G, Hoit BD (2003) Pleiotropy, homeostasis, and functional networks based on assays of cardiovascular traits in genetically randomized populations. Genome Res 13:2082–2091
39. Aitman TJ, Gotoda T, Evans AL, Imrie H, Heath KE, Trembling PM et al (1997) Quantitative trait loci for cellular defects in glucose and fatty acid metabolism in hypertensive rats. Nat Genet 16:197–201
40. Bottger A, van Lith HA, Kren V, Krenová D, Bílá V, Vorlícek J et al (1996) Quantitative trait loci influencing cholesterol and phospholipid phenotypes map to chromosomes that contain genes regulating blood pressure in the spontaneously hypertensive rat. J Clin Invest 98:856–862
41. Pravenec M, Gauguier D, Schott JJ, Buard J, Kren V, Bila V et al (1995) Mapping of quantitative trait loci for blood pressure and cardiac mass in the rat by genome scanning of recombinant inbred strains. J Clin Invest 96:1973–1978
42. Glazier AM, Scott J, Aitman TJ (2002) Molecular basis of the Cd36 chromosomal deletion underlying SHR defects in insulin action and fatty acid metabolism. Mamm Genome 13:108–113
43. Behringer R (1998) Supersonic congenics? Nat Genet 18:108
44. Pravenec M, Hyakukoku M, Houstek J, Zidek V, Landa V, Mlejnek P et al (2007) Direct linkage of mitochondrial genome variation to risk factors for type 2 diabetes in conplastic strains. Genome Res 17:1319–1326
45. Manfredi G, Fu J, Ojaimi J, Sadlock JE, Kwong JQ, Guy J et al (2002) Rescue of a deficiency in ATP synthesis by transfer of MTATP6, a mitochondrial DNA-encoded gene, to the nucleus. Nat Genet 30:394–399

# Chapter 29

# Genomic Research in Rat Models of Kidney Disease

**Yoram Yagil and Chana Yagil**

## Abstract

Current understanding of the mechanisms underlying renal disease in humans is incomplete. Consequently, our ability to prevent the occurrence of renal disease or treat kidney disease once it develops is limited. There are objective difficulties in investigating kidney disease directly in humans, leading investigators to resort to experimental animal models that simulate renal disease in humans. Animal models have thus been a tool of major importance in the study of normal renal physiology and have been crucial in shedding light on the complex mechanisms involved in normal kidney function and in our current understanding of and ability to treat renal disease. Among the animal models, rat has been the preferred and most commonly used species for the investigation of renal disease. This chapter reviews what has been achieved over the years, using rat as a tool for the investigation of renal disease in humans, focusing on the contribution of rat genetics and genomics to the elucidation of the mechanisms underlying the pathophysiology of the major types of renal disease, including primary and secondary renal diseases.

**Key words:** Kidneys, Disease, Primary kidney disease, Secondary kidney disease, Function, Experimental models, Rodents, Rat, Genetic basis, Pathophysiology, Mechanisms, Genomics, Genes, Transcriptomics, Expression, Proteomics

## 1. Introduction

The investigation of renal disease, whether primary or secondary, has been ongoing for several decades and a vast amount of knowledge has accrued over the years, some of which has been successfully translated into clinical practice. Current understanding of the mechanisms underlying renal disease remains, nonetheless, incomplete. Consequently, our ability to prevent the occurrence of renal disease or treat renal disease once it develops is limited, as it is evident from the continuously increasing number of patients who develop chronic kidney disease that often leads to end stage renal failure (http://www.usrds.org/2008/view/esrd_02.asp).

I. Anegon (ed.), *Rat Genomics: Methods and Protocols*, Methods in Molecular Biology, vol. 597
DOI 10.1007/978-1-60327-389-3_29, 

There is, therefore, dire need to continue in depth investigation of the pathophysiology underlying diseases in the kidney.

There are objective difficulties in the continuing investigation of kidney disease in humans. One is inherent in the sheer definition of "kidney disease", a general term that encompasses a large variety of clinical and pathological entities that affect the kidneys, all of which ultimately culminate in renal failure. There are many types and subtypes of renal disease, and although there may be common pathophysiological pathways, each needs to be investigated in separate, a huge task by itself. A second difficulty is the fact that the kidneys are not readily accessible for studying, and much of the data reflecting renal disease are derived from surrogates of renal disease, such as protein excretion or indices of glomerular filtration rate, all of which are not truly satisfactory. Histological assessment of renal tissue is possible through a kidney biopsy, but that is an invasive procedure that in most cases cannot be ethically justified if performed solely for research purposes. The difficulties in investigating renal disease directly in humans have led investigators to resort to experimental models of renal disease, in the hope that such models faithfully simulate and represent renal disease in humans.

Animal models have been a tool of major importance in the study of normal renal physiology and have been crucial in shedding light on the complex mechanisms involved in the normal daily function of the kidney, whether at the whole organ, cellular, or molecular level. Animal models simulating renal disease have also contributed immensely to our understanding of renal disease and to our current ability to treat it. Among the animal models that have been used to study renal disease, there stand out rodents, dogs, and pigs in the order of importance. Among rodents, traditional physiological and pathophysiological studies have been successfully carried out in the rat, and to a lesser extent in the mouse. The ease of use of the rat for phenotyping of the normal and the diseased kidney have rendered it as the preferred species for investigation of the kidney. In recent years, technological advances that have improved the ability to phenotype the mouse kidney and the availability of knock-out technology in the mouse but not until recently in the rat have rendered the mouse as an attractive alternative to the rat. Nonetheless, the rat continues to be the preferred and most commonly used experimental model for the investigation of renal disease.

Research on kidney disease in experimental models has been based in the past on traditional physiology and histopathology. The results of this research have had a major impact on our current knowledge of how the kidney functions in both health and disease, a level of understanding which remains nonetheless incomplete and unsatisfactory. With the advent of genomics and the sequencing of several mammalian genomes, there has been much hope that advanced technological tools that came along with the genomic revolution would allow a major leap forward in our understanding of disease in all its forms, including kidney disease. There has been, over the years, extensive research into the pathophysiology of renal disease using the traditional deductive candidate gene approach that is based on what is known about the pathophysiology of renal disease. There has also been abundant inductive research using a genomic and transcriptomic approach, with linkage analysis in crosses between informative rat strains, targeted construction of consomic and congenic strains, differential gene expression profiling and more recently proteomics. Gain of function studies by overexpressing select candidate genes in the form of transgenic rats (1) has been successfully achieved in select experimental models, but the loss of function studies have not been feasible until most recently, as knockout technology is only currently beginning to evolve in the rat (http://knockoutrat.org/index.php?iid=8).

In this chapter, we aim to review what has been achieved using the rat as a tool for the investigation of renal disease in humans, focusing on the contribution of rat genetics and genomics to the elucidation of the mechanisms underlying the pathophysiology of the major types of renal disease, including primary and secondary renal diseases. We will not address in this review the extensive genetic work that has been carried out deductively through the candidate gene approach and will focus primarily on the inductive type of research that incorporates genomics and transcriptomics. Furthermore, it is clear that a large number of genomic studies have been carried out in additional models of kidney disease, and it would be presumptuous to claim to have covered them all. We do hope, however, to provide sufficient coverage to provide the reader with an objective overview of the range of research that has been carried out in this field, and of the achievements attained so far.

## 2. Pitfalls and Limitations in Investigating Renal Disease

Prior to evaluating the contribution of genetic/genomic research to our understanding of renal disease, it is important to comprehend the limitations of such research in rodent models and its applicability to human disease.

The study of renal disease in experimental animal models has been based on the premise that it faithfully reproduces the human disease. To prove reproducibility of the human disease in the animal model, one has to show similarity to the human disease in the inciting etiological factors, in the pathophysiology of the disease, in the resulting histopathology of the lesion within the affected organ, in the level of renal dysfunction and in the course of the disease. One prerequisite to show such reproducibility is the ability to accurately define the phenotype, and translated into clinical terms, there needs to be an accurate and reproducible way to phenotype renal disease. A detailed discussion of these features is of essence.

In terms of inciting etiological factors, it is difficult to generalize as to the similarity between animal models and kidney disease in humans, as very often the etiology of the disease is yet unknown in either humans or animal models. It is also unclear as to whether the controlled environment and diet in experimental models of kidney disease, or the insults that generate the renal injury in animal models, truly reproduce the etiological factors in human disease. Thus, in terms of etiology, these experimental models are of value possibly, but findings in the animal need to be validated back in humans.

In terms of pathophysiology of renal disease, even though the anatomy of the rat kidney is somewhat different from that of humans, the similarity in the structure and function of the nephron in humans and rats is remarkable. Therefore, any pathophysiology of renal disease detected in the rat may be highly relevant to the human disease, although validation in humans remains a necessary step.

As a correlate to the similarity of the human and rat nephron, the histology of the rat kidney, in terms of differentiation between the medulla and the cortex, and the appearance of the glomerulus and tubuli, is also remarkably similar to that of humans. The histopathology of renal lesions that typifies human disease can thus be easily sought in the rat models, and glomerular, tubular, and interstitial disease can be sought in the rat kidney which allegedly simulated the kidney disease in humans. A significant difficulty arises, however, when one realizes that the histology of the renal lesion in the rat does not necessarily reproduce the exact or full lesion in humans. This holds true for diabetic nephropathy, in which some of the models reproduce part but not all the features of the diabetic histopathological lesions. Although some experimental models of diabetes exhibit thickening of the basement membrane and mild diffuse focal glomerulosclerosis which represent some of the changes observed in diabetic nephropathy in humans, they do not consistently develop glomerular hypertrophy, mesangial expansion, or progress to severe global glomerulosclerosis with nodule formation leading to end-stage renal failure. A similar argument also applies to experimental models of focal and segmental glomerulosclerosis in which the histopathological lesions are not uniform in all the models, exhibiting at least three different types of glomerular lesions (2). Thus, experimental models of glomerulosclerosis can reproduce parts of but often not the full spectrum of the lesions observed in the human diseased kidney.

Beyond the issue of whether histopathological changes in the rat models accurately and fully reproduce those changes seen in human disease, genomic studies often require a quantitative assessment of such changes. A quantitative assessment of the extent of renal damage can be attained in animal models by computerized morphometry which provides quantitative measures of normal and disease components of the kidney histology. The difficulty in applying this methodology to genomic studies lies in its labor-intensiveness, high cost, and low availability.

In terms of accurately defining the phenotype of the disease under investigation, assessing the level of renal function or dysfunction in the rat presents a certain problem. In humans, renal function is usually assessed by the glomerular filtration rate which can either be measured and derived from plasma creatinine and creatinine excretion in 24 h urine collections or can be estimated by using formulas such as the Cockcroft-Gault or Modification of Diet in Renal Disease (MDRD) formula, with all their limitations (3). Acute studies of renal function in humans, which are rarely carried out nowadays, can be carried out by using inulin for glomerular filtration rate measurements and PAH for renal plasma flow. Cystatin is a relatively new method of assessing renal function by measuring plasma levels, without the need for urine measurements. In the rat, the assessment of glomerular filtration rate is similar to that in humans and can be derived from 24 h urine collections and plasma creatinine levels. The inherent difficulties and inaccuracies of such measurements in rodents are, however, well recognized by all investigators. Unfortunately, in the rat, no formulas are available for estimating glomerular filtration rate based on plasma creatinine. Cystatin levels may equally serve as a useful tool in the rat as in humans, but the utility of this mode of measurement in experimental models remains to be validated. Assessment of renal dysfunction, a surrogate and consequence of renal disease, can be obtained in the rat by following sequentially serum creatinine levels, or by documenting a decline in creatinine clearance. As it is not possible to derive estimated glomerular filtration rate in the rat, creatinine clearance relies heavily on the accuracy of urine collection, which is often inaccurate in the rat. The use of blood urea nitrogen (BUN) as an alternative measure reflecting glomerular filtration is highly inaccurate, as urea levels are dependent on a number of variables that do not necessarily reflect renal function.

Another important surrogate of renal disease is proteinuria, reflecting either the leak of protein from the glomerular microcirculation into the tubular space, lack of adequate reabsorption of protein from the tubular lumen into the tubular epithelial cells, or both. Just as proteinuria is an important surrogate of renal disease in humans, it is used similarly in the rat. One problem in the study of proteinuria in the rat is rather low absolute level of protein excretion even in pathological states. Another problem is the length of time required for proteinuria to develop in experimental models simulating human disease. Although in humans, renal disease is usually progressive over many months to many years while in rat models, proteinuria usually develops within weeks or months, the cost of maintaining laboratory animals for this period of time may be prohibitive. Investigators have resorted to uninephrectomy as a means to resolve these problems, as uninephrectomy has been shown to both augment the level of proteinuria and to accelerate its development. A difficulty may arise in interpreting the results obtained in the uninephrectomy model of renal disease, as one has to differentiate between the contribution of uninephrectomy per se to the renal disease from that of the underlying process initiating the renal disease, a task that is often quite complicated.

Despite all the reservations discussed above as to the use of rat models to investigate human kidney disease, the rat continues to serve as an important tool for genetic and genomic studies of the kidney. Among the methods that are most commonly used in defining the extent of the renal lesion, proteinuria is used most often as the surrogate of renal injury, followed by glomerular filtration rate, and to an even lesser extent the histopathological changes within the kidney which are scarcely used.

## 3. Genomic Investigation of Kidney Disease in the Rat

### 3.1. Primary Renal Disease

There are a number of primary renal diseases in humans that have been reproduced to a greater or lesser extent in the rat, most notable among them are glomerulosclerosis, IgA nephropathy, and membranous nephropathy (Heyman nephritis). Genomic tools have been extensively applied in several animal models in the investigation of the pathophysiology underlying glomerulosclerosis, and to a far lesser extent in one model of IgA nephropathy.

#### 3.1.1. Glomerulosclerosis

Glomerulosclerosis is a clinical entity with histopathological features that are common to many nephropathies which eventually lead to end stage renal failure. Glomerulosclerosis may be truly a primary renal disease or develop secondarily to or in conjunction with hypertension, depending on the model. In some rat models of glomerulosclerosis, the renal lesions develop spontaneously without the development of hypertension, whereas in others, glomerulosclerosis develops along with the appearance of hypertension, and it is often difficult to differentiate between the two entities. Irrespective of whether glomerulosclerosis is related or not to hypertension, the contribution of genomic studies in the rat to our understanding of glomerulosclerosis has been very substantial, the most commonly used experimental models of glomerulosclerosis being the Fawn-hooded Hypertensive (FHH) rat, the Munich Wistar Fromter (MWF) rat, the Sabra Salt-sensitive Hypertension (SBH/y) prone rat, the Dahl Salt-sensitive (DS) rat, and the Buffalo Mna (Buffalo/Mna) rat.

#### Fawn-Hooded Rat

The fawn hooded rat expresses a renal lesion that is manifested histologically by focal and segmental glomerulosclerosis (4). The animals become hypertensive, develop proteinuria, and eventually renal failure. Genetic research in this model has been directed at detecting the mechanisms underlying proteinuria, renal failure and focal and segmental glomerulosclerosis. In a backcross involving the fawn-hooded hypertensive rats (FHH) and normotensive August-Copenhagen-Irish (ACI) rats, Brown et al. (5) reported genetic linkage between a quantitative trait locus (QTL) on chromosome 1 (RNO1) and markers of renal disease, including proteinuria, plasma creatinine, and a macroscopic renal index suggestive of renal damage. They termed this locus as *Rf-1*, Rf standing for renal failure (5). Provoost et al. (6) subsequently confirmed this locus by constructing a congenic strain in which part of the QTL interval from FHH was introgressed onto the ACI genetic background and reporting that a subsection of their QTL (Rf1B) was indeed associated with proteinuria and glomerulosclerosis. Brown et al. (5) reported in the same cross on a second locus (*Rf-2*) on RNO1 which was found to be weakly linked with plasma creatinine and the macroscopic renal index. *Rf-2* on RNO1 is concordant to human chromosome 19q13, which incorporates a yet unidentified gene underlying a monogenic form of focal segmental glomerulosclerosis (7). The homologous region in mice is linked to glomerulonephritis susceptibility in a model for systemic lupus erythematosus (8). Shiozawa et al. (9) repeated the cross and established that *Rf-1* and *Rf-2* were strongly linked to proteinuria and glomerulosclerosis. They detected this time three additional QTLs: *Rf-3* on RNO3, *Rf-4* on RNO1, and *Rf-5* on RNO17. *Rf-3* and *Rf-4* showed suggestive linkage to proteinuria but only hints of linkage to glomerulosclerosis. The rat QTL on RNO3 is concordant with kidney disease QTL identified in Pima Indians (10). Shiozawa et al. (9) went further and detected a significant and important interaction between these three loci, at least with respect to the level of proteinuria. Rangel-Filho et al. (11) identified within the *Rf-2* region the Rab-38 gene as a highly likely candidate contributing to proteinuria in the FHH rat. They also found in FHH a spontaneous protein null mutation within Rab-38 which prevents translation of the protein. Such mutation does not alter glomerular proteinuria but would be expected to diminish tubular protein reabsorption.

This was the first identification of a gene responsible for a QTL linked to proteinuria in an animal model of renal disease.

Munich Wistar Fromter (MWF)

This model, which exhibits spontaneous glomerulosclerosis, proteinuria, and moderate hypertension, all possibly related to a reduced number of nephrons (12–14), has been extensively utilized in the genetic dissection of renal disease. Schulz et al. (14) investigated the genetic basis of proteinuria in this model by crossing MWF with Lewis (Lew) rat. They detected linkage for proteinuria at 4 loci: QTL1 on RNO1 which colocalizes with Rf-2, QTL2 on RNO6, QTL3 on RNO12 and QTL4 on RNO17 which colocalizes with Rf-5. Notable in this cross was the relatively low level of proteinuria, which nonetheless allowed the detection of these 4 QTL. In this cross, no putative linkage was detected for glomerulosclerosis. In a separate study, Schulz et al. (15) used a different cross between MWF and SHR to dissect the genetic basis of proteinuria and renal interstitial fibrosis, the latter being a phenotypic feature consistent with evolving renal insufficiency. Eight QTLs were detected this time with suggestive or significant linkage on RNO 2, 4, 6, 7, 9, 15, and X. Once again, no linkage was detected to the glomerular sclerosis index, but suggestive linkage was detected for interstitial fibrosis on RNO6. Schulz et al. (16) pursued these findings by developing a congenic strain in which RNO6 from SHR was introgressed onto the genetic background of MWF, resulting in marked attenuation of proteinuria and RIF and demonstrating the importance of this genetic locus with respect to these two phenotypes. Schulz et al. (17) constructed an additional consomic strain in which RNO8 from SHR was introgressed into the genetic background of MWF, resulting in a marked reduction in proteinuria. Kreutz et al. (18) used a microarray platform to compare glomerular gene expression between MWF and Wistar, and found that diminished expression of C1q and CD24 in the glomerular epithelium was associated with glomerular proteinuria, rendering these two candidate genes for glomerular dysfunction with proteinuria. How and whether these two genes induce proteinuria, or are they merely markers of proteinuria, remains to be clarified.

Sabra Hypertension Prone (SBH/y)

The SBH/y rat is a model of spontaneous glomerulosclerosis and proteinuria that is unrelated to salt loading or hypertension (19). Yagil et al. (19) studied the level of proteinuria in consomic strains in which RNO1 and RNO17 had been introgressed from SBN/y onto the genetic background of SBH/y for another unrelated project, and found a marked reduction in the level of proteinuria without affecting the level of glomerulosclerosis in the RNO1 consomic. These findings suggest that RNO1 and RNO17 harbor genes that affect proteinuria, and that proteinuria is dissociated from glomerulosclerosis in this model. Yagil et al. (20) later dissected the genetic basis of proteinuria in the Sabra model using an F2 cross between SBH/y and the Sabra salt and hypertension resistant SBN/y strain, detecting three QTLs that were associated with the development of proteinuria: on RNO2 (*SUP2*), RNO17 (*SUP17*), and RNO20 (*SUP20*). Interestingly, no proteinuria-related QTL was detected on RNO1. An additional QTL was detected on RNO3 (*SUP3*) which conferred protection from proteinuria. Consomic strains in which RNO2, RNO17, and RNO20 had been introgressed from SBN/y onto the genetic background of SBH/y had significantly less proteinuria than SBH/y, thus confirming the presence of proteinuria related genes on RNO2 and RNO20 as well (unpublished data).

#### Dahl Salt Sensitive (SS)

The Dahl rat, which is salt sensitive similar to the Sabra rat, develops proteinuria at a young age and prior to the development of hypertension, without any impairment in renal function (21). The development of hypertension aggravates proteinuria and causes an impairment in renal function (21). The glomerular damage in SS is characterized by mesangial expansion and fibrinoid arteriolar and glomerular necrosis and is thus not a true model of glomerulosclerosis. Poyan Mehr et al. (22) studied the genetic basis of proteinuria in the SS rat using an F2 cross between SS and the spontaneously hypertension rat (SHR) which is resistant to proteinuria. They identified seven suggestive or significant proteinuria-related QTLs on RNO2, RNO6, RNO8, RNO9, RNO10, RNO11, and RNO19, demonstrating the polygenetic basis of proteinuria in the SS rat as in all other models of glomerulosclerosis. The QTL for albuminuria on RNO19 is concordant with the mouse region containing the *Os* mutation, which causes glomerular hypertrophy, severe glomerulosclerosis, and a 50% reduction in nephron number in mice (23). Garrett et al. (24) investigated in parallel the genetic basis of proteinuria in SS but using a backcross with SHR and found nine proteinuria-related QTLs on RNO1, RNO2, RNO6, RNO8, RNO9, RNO10, RNO11, RNO13, and RNO19. Most of the proteinuria-related QTLs colocalized with QTL for kidney lesions. Cowley et al. (25) confirmed the relevance of the QTL on RNO13 by constructing a consomic strain in which RNO13 was introgressed from the Brown Norway (BN) rat into the genomic SS background, resulting in a marked reduction in the level of proteinuria. Interestingly, the consomic strain was not protected from the glomerular injury as evidenced by histology. The albuminuria-related QTL on RNO13 is concordant with a systemic lupus erythematosus glomerulonephritis susceptibility locus (*Sle1*) in mice (8) and a monogenic form of familial focal segmental glomerulosclerosis in humans (26). As the previous studies had been carried out in male rats only, Moreno et al. (27) focused on the gender effect and identified in F2 females resulting from a cross between SS and Brown Norway

(BN) rats proteinuria and glomerular injury-related QTLs on RNO2 and RNO11, which overlaps with some of the QTLs detected in males. It is noteworthy that despite abundant work using the Dahl model of glomerulosclerosis, none of these studies in this model have yet yielded any definitive candidate genes for proteinuria or glomerulosclerosis in that model.

Buffalo/ Mna Rat

The Buffalo/Mna rat is a model that develops spontaneously focal and segmental glomerulosclerosis, along with proteinuria (2, 28). In a cross between BUF/Mna and WKY rats, a proteinuria-related QTL was detected by linkage analysis on RNO13 (29). Akiyama et al. (30) subsequently fine-mapped the QTL with SNPs and identified the actin-related protein 3 (Arp3) as a candidate gene within the Pur1 QTL, suggesting that a disturbance in actin assembly may lead to glomerulosclerosis.

Nephrotoxic Nephritis

The nephrotoxic nephritis model consists of Wistar Kyoto (WKY) rats that are injected nephrotoxic serum and that develop an autoimmune crescentic type of glomerulonephritis with proteinuria, with similarity to human focal and segmental glomerulosclerosis (31). Lewis, Brown Norway, and Wistar rats injected with the same serum do not develop crescents nor proteinuria (31). Aitman et al. (31) used this model in an F2 cross between WKY and Lewis rats. The investigators detected by linkage analysis two QTLs which were associated with crescent formation and proteinuria, one on RNO13 (Crgn1) and the other on RNO16 (Crgn2). The investigators identified within Crgn1 QTL the activatory Gc receptor for IgG (*Fcgr3*) as a major candidate gene affecting the development of glomerulosclerosis.

*3.1.2. IgA Nephropathy*

IgA nephropathy is one of the most common types of primary glomerulopathies. Its pathophysiology is incompletely understood. An experimental model of IgA nephropathy is the Lewis rat which is injected a single dose of anti-Thy1 antibody (32). The animal develops the so-called *anti-Thy-1 glomerulonephritis*, which is a model of complement-mediated mesangial damage that is followed by mesangial hypercellularity and extracellular matrix deposition, simulating IgA nephropathy in humans (32, 33). In the Lewis/ Mollegard (Lew/Moll) substrain, the lesion heals spontaneously within 4 weeks, whereas in the Lewis/ Mastricht (Lew/Maa) substrain, proteinuria develops and the animals go on to progressive glomerulosclerosis (32). IJpelaar et al. (33) most recently investigated the genetic basis of susceptibility to this type of renal injury in a cross between the two substrains and identified a QTL on RNO1 (*GS1*) that was linked to progressive glomerulosclerosis after acute glomerulonephritis. No candidate genes have been positively identified yet within this QTL.

## 3.2. Secondary Renal Disease

### 3.2.1. Diabetic Nephropathy

Diabetic nephropathy is the secondary renal disease that has been studied the most in the rat. Diabetic nephropathy develops in genetic models of spontaneous diabetes, in models of obesity-related diabetes, in models of diet-induced diabetes, and in models of chemically-induced diabetes. The histological lesions within the glomerulus in those animal models are quite similar to the early stages of diabetic nephropathy in humans, with thickening of the basement membrane and mild glomerulosclerosis; the characteristic changes of later stage of diabetic nephropathy, including glomerular hypertrophy, mesangial expansion, and progressive glomerulosclerosis, are often present only in part (34), and do not exhibit the complete features of diabetic nephropathy as it is expressed in the diseased human kidney. Nonetheless, rodent models of diabetic nephropathy appear to be very useful and informative in the ongoing attempts to elucidate the pathophysiology of diabetic nephropathy.

#### 3.2.1.1. Spontaneous Diabetes

Diabetes can develop spontaneously in the rat, or it can be induced by dietary manipulations or injection of pancreo-toxic agents. Both types of diabetes develop end-organ damage in the kidney, expressing various features of diabetic nephropathy as is observed in humans. Among the genetic models that develop spontaneous diabetes with nephropathy and in which genomic studies have been carried out stand out the Goto–Kakizaki (GK), the Zucker rats, the T2DN, and the BioBreeding rats.

*Goto–Kakizaki rat*

The GK rat is a nonobese normotensive model of type 2 diabetes that develops glomerular hypertrophy and thickening of basement membrane but does not develop proteinuria, glomerulosclerosis, interstitial fibrosis or impairment of renal function (35). With superimposed hypertension, however, these animals develop progressive nephropathy (36). The nonhypertensive diabetic GK model, thus, represents a type of "subclinical" nephropathy due to prolonged hyperglycemia, without overt renal disease (35). Page et al. (37) studied the nonhypertensive GK model. They used differential display, prior to the advent of high throughput DNA microarrays, to isolate genes that show transcriptional changes in the kidney during the development of diabetes. They successfully identified 8 candidate cDNA fragments, *CDK 1–8* that are allegedly involved in the development of the diabetic changes within the kidney. Page et al. (38) pursued these findings and established the identity of *CDK4* as the rat beta-defensin-1 gene (*rBD-1*). They further studied gene expression of biglycan and TGF-β1 in the kidneys of GK rats and found they were upregulated, as they have been found in other experimental models of diabetes and in humans (38). Malik et al. (39) identified the *CDK7* as a thiol-related gene with a putative role in oxidative stress. The role of these two candidate genes in the pathophysiology of diabetes requires further validation and investigation.

*Zucker (fatty) Rat*

The Zucker rat is a genetic model which expresses obesity, hyperphagia and hyperinsulinemia (40) which have been attributed to a spontaneous mutation in the leptin receptor gene (*fa*) (41). The Zucker Diabetic Fatty (ZDF) rat is a model of type 2 diabetes that was derived by selective inbreeding of the hyperglycemic Zucker Fatty rat and which develops spontaneous overt diabetes (42). Schafer et al. (43) investigated the renal lesion in the ZDF rat and found histological lesions resembling the typical lesions in human diabetic nephropathy. Erderly et al. (44) also studied the renal injury in ZDF rats and found a progressive decline in glomerular filtration rate (creatinine clearance), increasing proteinuria, and abnormal histology (44) which, however, simulated those seen in focal and segmental glomerulosclerosis and not those seen in diabetic nephropathy in humans (44). The investigators concluded that the nephropathy in ZDF can, therefore, not be considered as a model simulating the typical human diabetic nephropathy. Interestingly, Baylis et al. (45) found that the insulin sensitizing agent rosiglitazone decreased proteinuria and attenuated the structural changes within the kidneys of diabetic ZDF rats (45), suggesting that the nephropathy in ZDF is related to diabetes, but perhaps not with the typical histopathological changes seen in humans. Using the ZDF rat as a model of kidney disease that simulates at least in part the renal lesions seen in diabetes, Niehof and Borlak (46) identified, using expression assays and si-RNA, *HNF4α* and the calcium channel *TRPC1* as novel candidate genes for diabetic nephropathy. *HNF4α* is an orphan nuclear receptor which is an essential transcription factor and master regulatory protein in the control of gene expression of a wide range of enzymes involved in metabolism (46). This gene is known to control glucose metabolism and participates in glucose dependent insulin secretory pathways (47). The second gene *TRCP1* encodes for a nonselective cation channel (46). How both of these genes are involved in the pathophysiology of diabetic nephropathy remains to be resolved. Zhang et al. (48) pursued the findings with respect to *TRPC1* and attempted, unsuccessfully so far, to confirm in humans the association between polymorphisms within the *TRPC1* gene and diabetic nephropathy.

*T2DN*

The T2DN strain is genetic model of diabetic nephropathy that was derived by Nobrega et al. (49) by crossing between FHH and GK. This strain develops spontaneous type 2 diabetes, along with progressive proteinuria and a decline in renal function. The histopathological changes within the kidney are reminiscent of diabetic nephropathy. Genomic studies are ongoing in this strain to elucidate the pathophysiological mechanisms involved.

*BioBreeding (BB) Rats*

This model of spontaneous diabetes develops only mild histopathology resembling human diabetic nephropathy, including increased mesangial volume and glomerular basement membrane thickness, which can be detected in older animals only (50). The findings of hyperfiltration and mild glomerular morphological changes in diabetic BioBreeding rats are thus similar to the abnormalities seen in the early stages of human diabetic nephropathy. Hsieh et al. (51) used DNA microarrays to study gene expression in the proximal tubules of diabetic BB rats expressing this mild type of diabetic nephropathy. They found that Osteopontin (OPN) was significantly upregulated in the diabetic rats and that this increased expression was mediated via reactive oxygen species (ROS), activation of the renin-angiotensin system (RAS), protein kinase C β1 signaling, and TGF-β1 expression. Their findings suggest that OPN is directly related to tubulointerstitial injury in diabetic nephropathy.

3.2.1.2. Induced Diabetes

There are other models of diabetes in which the disease is induced experimentally and in which diabetic nephropathy develops. Among these models, there stands out the Cohen Diabetic rat in which diabetes is induced by a custom-prepared diet and the Streptozotocin rat in which diabetes is induced by Streptozotocin injection resulting in injury to the endocrine pancreas.

*Cohen Diabetic Rat*

This is a genetic model of diet-induced diabetes that is composed of two strains: the Cohen Diabetic sensitive (CDs) strain which when provided regular rat chow does not develop diabetes, but which invariably develops diabetes when provided a custom-prepared "diabetogenic" diet, and the Cohen Diabetic resistant (CDr) strain (52). The CDs strain fed diabetogenic diet develops over time diabetic nephropathy that is expressed histologically and functionally, but with no proteinuria. Yagil et al. (53) described in this model glomerular histopathological changes that are reminiscent of human diabetic nephropathy, including thickening of the glomerular basement membrane, mesangial expansion and glomerulosclerosis with deterioration of renal function over time. The outstanding feature in this model is that proteinuria does not develop (53). This model thus reproduces the non-proteinuric variety of diabetic nephropathy described in humans (54). Genomic studies on the genetic basis of the diabetic nephropathy in this model are currently ongoing.

*Streptozotocin (STZ) Rat*

Injection of Streptozotocin to the rat induces insulin-deficient type 1 diabetes. The diabetic animals develop a nephropathy which shows resemblance to diabetic nephropathy in humans, although the typical lesions seen in humans are not uniformly present, and their severity is usually milder in the rat (55). STZ has been injected in Sprague Dawley (SD), Wistar Kyoto (WKY) and the spontaneously hypertensive rat (SHR), inducing in all diabetes and diabetic nephropathy which can usually begin to be studied, with all its limitations, already 3 weeks after the injection. Morrison et al. (56) used the STZ-SD rat to validate the identification of dysregulated genes detected by DNA microarrays in mesangial cells in culture that had been exposed in vitro to glucose. They identified upregulation of the thiol antioxidative pathway, which they interpreted as an adaptational response of mesangial cells to glucose which may fulfill a critical role in the development of diabetic nephropathy. Hsieh et al. (51) injected STZ to Wistar rats, inducing diabetic nephropathy; they studied gene expression in the proximal tubules using DNA microarrays and found upregulation of OPN, as in the BB rat. Han et al. (57) used the STZ-SD rat to validate gene expression patterns in glucose stimulated podocytes; they demonstrated upregulation of heme oxygensase-1 (HO-1), vascular endothelial growth factor A (VEGF-A) and thorombospondin (TSP-1) and down regulation of angiotensin converting enzyme 2 (ACE2) and peroxisomal proliferator activator receptor γ (PPARγ). It is yet unclear how these candidate genes are involved in the pathophysiology of diabetic nephropathy.

### *3.3. Other Kidney Diseases*

There are other kidney diseases that do not fall exactly into the category of primary or secondary renal disease in which investigators have made use of genomic tools to investigate the underlying pathophysiology. An example is renal stone disease. Nephrolithiasis is a major health problem in humans that is often due to excessive calcium excretion (idiopathic hypercalciuria) and that carries a family history and, therefore, a genetic component in a substantial number of patients. To investigate the pathophysiology of hypercalciuria, Hoopes et al. (58) used an F2 cross between the genetic hypercalciuric stone-forming (GSH) rats (59), and the normocalciuric WKY rat. They detected significant linkage for hypercalciuria on RNO1 (*HC1*), and suggestive linkage on RNO4, 7, 10, and 14 (58). Hoopes et al. (60) subsequently confirmed the validity of the hypercalciuria-related *HC1* QTL on RNO1 by constructing a congenic strain in which the QTL was introgressed from GSH onto the WKY background. Candidate genes remain to be identified within this QTL.

## 4. Conclusions and Perspectives

The promise of animal research has been that genetic studies in the rat (and mouse) would render identification of candidate genes easier and more successful because of a seemingly lesser genetic complexity and the availability of improved genetic tools and a controllable environment (61, 62). Has this promise materialized? Genetic and genomic studies in the rat have identified multiple QTLs for a number of phenotypes that are related to several kidney diseases. In a number of studies, QTLs have led to the identification of high priority candidate genes and proteins. Several of these kidney disease-related QTLs in the rat are concordant with kidney disease loci in mice and humans, suggesting that conserved disease genes underlie these QTL and implying that kidney disease QTL in animal models can predict the locations of disease genes in humans (61). This concordance further validates the use of animal models in searching for disease related genes in humans. And yet, have genomic studies in animal models truly furthered our understanding of disease afflicting the kidneys, beyond the level of understanding achieved by traditional physiology or studies based on the traditional candidate gene approach? A lucid assessment of the gains achieved so far from genomic studies in the rat and mouse reveals that much knowledge has accumulated on the genetics of chronic kidney disease, and yet that we are still very far from unraveling the complexity of kidney function in health and disease. It is now clear that the pathophysiology underlying renal disease is not less complex than the huge complexity of the mammalian genome. Novel approaches and application of advanced technologies are required, including, for example, the investigation of the role of the recently uncovered microRNAs (63) and the advent of knockout technology in the rat, before at least some of the expectations from genomic research in animal models that investigators have had so far difficulty to deliver can be fulfilled.

## References

1. Tesson L, Cozzi J, Menoret S et al (2005) Transgenic modifications of the rat genome. Transgenic Res 14:531–46
2. Howie AJ, Kizaki T, Beaman M et al (1989) Different types of segmental sclerosing glomerular lesions in six experimental models of proteinuria. J Pathol 157:141–51
3. Steven L, Perrone R (2008) Assessment of kidney function: Serum creatinine; BUN and GFR. In: Basow D (ed) UpToDate, MA: Waltham.
4. Simons JL, Provoost AP, Anderson S et al (1993) Pathogenesis of glomerular injury in the fawn-hooded rat: early glomerular capillary hypertension predicts glomerular sclerosis. J Am Soc Nephrol 3:1775–82
5. Brown DM, Provoost AP, Daly MJ, Lander ES, Jacob HJ (1996) Renal disease susceptibility and hypertension are under independent genetic control in the fawn-hooded rat. Nat Genet 12:44–51
6. Provoost AP, Shiozawa M, Van Dokkum RPE, Jacob HJ (2002) Transfer of the Rf-1 region from FHH onto the ACI background increases susceptibility to renal impairment. Physiol Genomics 8:123–9
7. Mathis BJ, Kim SH, Calabrese K et al (1998) A locus for inherited focal segmental glomerulosclerosis maps to chromosome 19q13. Kidney Int 53:282–6
8. Morel L, Rudofsky UH, Longmate JA, Schiffenbauer J, Wakeland EK (1994) Polygenic control of susceptibility to murine systemic lupus erythematosus. Immunity 1:219–29
9. Shiozawa MASA, Provoost AP, Dokkum RPEV, Majewski RR, Jacob HJ (2000) Evidence of gene–gene interactions in the

genetic susceptibility to renal impairment after unilateral nephrectomy. J Am Soc Nephrol 11:2068–78

10. Imperatore G, Hanson RL, Pettitt DJ, Kobes S, Bennett PH, Knowler WC (1998) Sib-pair linkage analysis for susceptibility genes for microvascular complications among Pima Indians with type 2 diabetes. Pima Diabetes Genes Group. Diabetes 47:821–30
11. Rangel-Filho A, Sharma M, Datta YH et al (2005) RF-2 gene modulates proteinuria and albuminuria independently of changes in glomerular permeability in the fawn-hooded hypertensive rat. J Am Soc Nephrol 16: 852–6
12. Remuzzi A, Puntorieri S, Mazzoleni A, Remuzzi G (1988) Sex related differences in glomerular ultrafiltration and proteinuria in Munich-Wistar rats. Kidney Int 34:481–6
13. Fassi A, Sangalli F, Maffi R et al (1998) Progressive glomerular injury in the MWF rat is predicted by inborn nephron deficit. J Am Soc Nephrol 9:1399–406
14. Schulz A, Litfin A, Kossmehl P, Kreutz R (2002) Genetic dissection of increased urinary albumin excretion in the munich wistar fromter rat. J Am Soc Nephrol 13:2706–14
15. Schulz A, Standke D, Kovacevic L et al (2003) A major gene locus links early onset albuminuria with renal interstitial fibrosis in the MWF rat with polygenetic albuminuria. J Am Soc Nephrol 14:3081–9
16. Schulz A, Weiss J, Schlesener M et al (2007) Development of overt proteinuria in the Munich Wistar Fromter rat Is suppressed by replacement of chromosome 6 in a consomic rat strain. J Am Soc Nephrol 18:113–21
17. Schulz A, Hansch J, Kuhn K et al (2008) Nephron deficit is not required for progressive proteinuria development in the Munich Wistar Fromter rat. Physiol Genomics 35: 30–5
18. Kreutz R, Schulz A, Sietmann A et al (2007) Induction of C1q expression in glomerular endothelium in a rat model with arterial hypertension and albuminuria. J Hypertens 25:2308–16
19. Yagil C, Sapojnikov M, Katni G et al (2002) Proteinuria and glomerulosclerosis in the Sabra genetic rat model of salt susceptibility. Physiol Genomics 9:167–78
20. Yagil C, Sapojnikov M, Wechsler A, Korol A, Yagil Y (2006) Genetic dissection of proteinuria in the Sabra rat. Physiol Genomics 25:121–33
21. Sterzel RB, Luft FC, Gao Y et al (1988) Renal disease and the development of hypertension in salt-sensitive Dahl rats. Kidney Int 33: 1119–29
22. Poyan Mehr A, Siegel AK, Kossmehl P et al (2003) Early onset albuminuria in Dahl rats is a polygenetic trait that is independent from salt loading. Physiol Genomics 14:209–16
23. He C, Esposito C, Phillips C et al (1996) Dissociation of glomerular hypertrophy, cell proliferation, and glomerulosclerosis in mouse strains heterozygous for a mutation (Os) which induces a 50% reduction in nephron number. J Clin Invest 97:1242–9
24. Garrett MR, Dene H, Rapp JP (2003) Time-course genetic analysis of albuminuria in Dahl salt-sensitive rats on low-salt diet. J Am Soc Nephrol 14:1175–87
25. Cowley AW Jr, Roman RJ, Kaldunski ML et al (2001) Brown Norway chromosome 13 confers protection from high salt to consomic Dahl S rat. Hypertension 37:456–61
26. Winn MP, Conlon PJ, Lynn KL et al (1999) Linkage of a gene causing familial focal segmental glomerulosclerosis to chromosome 11 and further evidence of genetic heterogeneity. Genomics 58:113–20
27. Moreno C, Dumas P, Kaldunski ML et al (2003) Genomic map of cardiovascular phenotypes of hypertension in female Dahl S rats. Physiol Genomics 15:243–57
28. Nakamura T, Oite T, Shimizu F et al (1986) Sclerotic lesions in the glomeruli of Buffalo/Mna rats. Nephron 43:50–5
29. Murayama S, Yagyu S, Higo K et al (1998) A genetic locus susceptible to the overt proteinuria in BUF/Mna rat. Mamm Genome 9:886–8
30. Akiyama K, Morita H, Suetsugu S et al (2008) Actin -related protein 3 (Arp3) is mutated in proteinuric BUF/Mna rats. Mamm Genome 19:41–50
31. Aitman TJ, Dong R, Vyse TJ et al (2006) Copy number polymorphism in Fcgr3 predisposes to glomerulonephritis in rats and humans. Nature 439:851–5
32. Ketteler M, Westenfeld R, Gawlik A et al (2002) Nitric oxide synthase isoform expression in acute versus chronic anti-Thy 1 nephritis. Kidney Int 61:826–33
33. IJpelaar DHT, Schulz A, Aben J et al (2008) Genetic predisposition for glomerulonephritis-induced glomerulosclerosis in rats is linked to chromosome 1. Physiol Genomics 35:173–81
34. Susztak K, Sharma K, Schiffer M, McCue P, Ciccone E, Bottinger EP (2003) Genomic strategies for diabetic nephropathy. J Am Soc Nephrol 14:S271–S278

35. Phillips AO, Baboolal K, Riley S et al (2001) Association of prolonged hyperglycemia with glomerular hypertrophy and renal basement membrane thickening in the Goto Kakizaki model of non-insulin-dependent diabetes mellitus. Am J Kidney Dis 37:400–10
36. Janssen U, Riley SG, Vassiliadou A, Floege J, Phillips AO (2003) Hypertension superimposed on type II diabetes in Goto Kakizaki rats induces progressive nephropathy. Kidney Int 63:2162–70
37. Page R, Morris C, Williams J, von Ruhland C, Malik AN (1997) Isolation of diabetes-associated kidney genes using differential display. Biochem Biophys Res Commun 232:49–53
38. Page RA, Malik AN (2003) Elevated levels of beta defensin-1 mRNA in diabetic kidneys of GK rats. Biochem Biophys Res Commun 310:513–21
39. Malik AN, Rossios C, Al Kafaji G, Shah A, Page RA (2007) Glucose regulation of CDK7, a putative thiol related gene, in experimental diabetic nephropathy. Biochem Biophys Res Commun 357:237–44
40. Bray GA (1977) The Zucker-fatty rat: a review. Fed Proc 36:148–53
41. Phillips MS, Liu Q, Hammond HA et al (1996) Leptin receptor missense mutation in the fatty Zucker rat. Nat Genet 13:18–9
42. Etgen GJ, Oldham BA (2000) Profiling of Zucker diabetic fatty rats in their progression to the overt diabetic state. Metabolism 49:684–8
43. Schafer S, Linz W, Bube A et al (2003) Vasopeptidase inhibition prevents nephropathy in Zucker diabetic fatty rats. Cardiovasc Res 60:447–54
44. Erdely A, Freshour G, Maddox DA, Olson JL, Samsell L, Baylis C (2004) Renal disease in rats with type 2 diabetes is associated with decreased renal nitric oxide production. Diabetologia 47:1672–6
45. Baylis C, Atzpodien EA, Freshour G, Engels K (2003) Peroxisome proliferator-activated receptor )gamma) agonist provides superior renal protection versus angiotensin-converting enzyme inhibition in a rat model of type 2 diabetes with obesity. J Pharmacol Exp Ther 307:854–60
46. Niehof M, Borlak J (2008) HNF4{alpha} and the Ca-channel TRPC1 are novel disease candidate genes in diabetic nephropathy. Diabetes 57:1069–77
47. Wang H, Maechler P, Antinozzi PA, Hagenfeldt KA, Wollheim CB (2000) Hepatocyte nuclear factor 4alpha regulates the expression of pancreatic beta -cell genes implicated in glucose metabolism and nutrient-induced insulin secretion. J Biol Chem 275:35953–9
48. Zhang D, Freedman BI, Flekac M et al (2008) Evaluation of genetic association and expression reduction of TRPC1 in the development of diabetic nephropathy. Am J Nephrol 29:244–51
49. Nobrega MA, Fleming S, Roman RJ et al (2004) Initial characterization of a rat model of diabetic nephropathy. Diabetes 53:735–42
50. Feld LG, Cachero S, Ellis EN et al (1995) Resistance to glomerular injury in the diabetic biobreeding rat. Exp Physiol 80:991–1000
51. Hsieh TJ, Chen R, Zhang SL et al (2006) Upregulation of osteopontin gene expression in diabetic rat proximal tubular cells revealed by microarray profiling. Kidney Int 69:1005–15
52. Weksler-Zangen S, Yagil C, Zangen DH, Ornoy A, Jacob HJ, Yagil Y (2001) The newly inbred Cohen diabetic rat: a nonobese normolipidemic genetic model of diet-induced type 2 diabetes expressing sex differences. Diabetes 50:2521–9
53. Yagil C, Barak A, Ben Dor D et al (2005) Nonproteinuric diabetes-associated nephropathy in the Cohen rat model of type 2 diabetes. Diabetes 54:1487–96
54. MacIsaac RJ, Tsalamandris C, Panagiotopoulos S, Smith TJ, McNeil KJ, Jerums G (2004) Nonalbuminuric renal insufficiency in type 2 diabetes. Diabetes Care 27:195–200
55. Tesch GH, Allen TJ (2007) Rodent models of streptozotocin-induced diabetic nephropathy. Nephrology (Carlton) 12:261–6
56. Morrison J, Knoll K, Hessner MJ, Liang M (2004) Effect of high glucose on gene expression in mesangial cells: upregulation of the thiol pathway is an adaptational response. Physiol Genomics 17:271–82
57. Han SH, Yang S, Jung DS et al (2008) Gene expression patterns in glucose-stimulated podocytes. Biochem Biophys Res Commun 370:514–8
58. Hoopes RR Jr, Reid R, Sen S et al (2003) Quantitative trait loci for hypercalciuria in a rat model of kidney stone disease. J Am Soc Nephrol 14:1844–50
59. Bushinsky DA, Favus MJ (1988) Mechanism of hypercalciuria in genetic hypercalciuric rats. Inherited defect in intestinal calcium transport. J Clin Invest 82:1585–91
60. Hoopes RR Jr, Middleton FA, Sen S et al (2006) Isolation and confirmation of a calcium excretion quantitative trait locus on chromosome 1 in genetic hypercalciuric stone-forming congenic rats. J Am Soc Nephrol 17:1292–304

61. Korstanje R, DiPetrillo K (2004) Unraveling the genetics of chronic kidney disease using animal models. Am J Physiol Renal Physiol 287:F347–F352
62. Liang M, Cowley AW Jr, Hessner MJ, Lazar J, Basile DP, Pietrusz JL (2005) Transcriptome analysis and kidney research: toward systems biology. Kidney Int 67: 2114–22
63. Tian Z, Greene AS, Pietrusz JL, Matus IR, Liang M (2008) MicroRNA-target pairs in the rat kidney identified by microRNA microarray, proteomic, and bioinformatic analysis. Genome Res 18:404–11

# Chapter 30

# Cancer Research in Rat Models

## Claude Szpirer

## Abstract

Rat has been the major model species used in several biomedical fields, notably in drug development and toxicology, including carcinogenicity testing. Rat is also a useful model in basic cancer research. Several rat models of monogenic (Mendelian) human hereditary cancers are available. Some were obtained spontaneously, while others were generated either by mutagenesis of tumor suppressor genes or by transgenesis of activated oncogenes (transgenesis can be performed efficiently in the rat). In addition, among the hundreds of inbred rat strains that have been isolated, some are highly susceptible or resistant to certain types of cancer, and these divergent phenotypes were shown to be polygenic. Numerous quantitative trait loci (QTLs) controlling cancer susceptibility/resistance have been defined in linkage analyses, and several of these QTLs were physically demonstrated in congenic strains. These studies led, in particular, to rapid translation to the human, with the identification of loci controlling susceptibility to a form of multiple endocrine neoplasia (monogenic trait) and to breast cancer (polygenic disease). The biology of cancer resistance has also been analyzed, and in some (but not all) cases, it was linked to regression of preneoplasic lesions. Rat tumors have been the subject of various types of analyses, and these studies led to important conclusions, including that tumors can be classified on the basis of the identity of the inducing agent, thereby suggesting that analyses of human tumors may be valuable in determining retrospectively the role of specific carcinogens in the formation of human cancers, and of human breast cancer in particular.

**Key words:** Rat, Cancer, Carcinogenesis, Tumor, QTL

## 1. Introduction

Rat has been the major model species used in several biomedical fields, notably in drug development and toxicology, including carcinogenicity testing ((1–4) and other chapters in this book). In cancer research, as in other fields, no animal model recapitulates all features of the human situation, and rat is, with mouse, a mammalian model that may be relevant or faithful. Mouse has been the

I. Anegon (ed.), *Rat Genomics: Methods and Protocols*, Methods in Molecular Biology, vol. 597
DOI 10.1007/978-1-60327-389-3_30, © Humana Press, a part of Springer Science+Business Media, LLC 2010

model of choice in experimental genetics, including in cancer genetics (5, 6), owing to the early development of mouse genetics and to the fact that mouse turned out to be the best mammalian model to generate targeted mutants (7, 8). However, in the context of considerable progress in rat genetics and genomic resources over the past years, genetic studies in the rat model are fully justified ((4, 9, 10) and other chapters in this book). In some instances, rat can provide one with an apparently more satisfactory model. For instance, mouse carrying a defective *Apc* gene develop intestinal tumors, while inactivating mutations in the homologue gene lead to colon tumors in the rat, as in the human ((11) and references therein). With respect to mammary cancer, rat and human carcinomas show similar developments and histopathologic features (12, 13). Rat mammary tumors are strongly hormone-dependent for both induction and growth (12, 14, 15), thus resembling human breast tumors. In addition, no virus appears to be involved in rat and human mammary carcinogenesis (unlike mouse mammary carcinogenesis). Hepatocarcinogenesis has been actively studied in rat, which is susceptible to many carcinogenic regimens, while mouse is less susceptible to some hepatocarcinogens such as aflatoxin (16, 17).

In this review, I will first give a brief overview on the use of the rat in nongenetic cancer research and then focus on genetic aspects of cancer research in this model, summarizing the genetic analyses of cancer susceptibilities and then reviewing the studies on somatic anomalies of rat tumors (Subheading 30.4).

## 2. Toxicology, Carcinogenesis

Hundreds of inbred rat strains have been isolated, and not unexpectedly, some of these strains are highly susceptible or resistant to certain cancers. This phenotype is organ-specific. For instance, the COP strain is resistant to mammary and liver cancer, but is susceptible to leukemia and thymus tumors (18, 19). While toxicology testing and carcinogenesis screening have been based on the use of outbred stocks (a choice which has been questioned) (3, 20), basic cancer research has rather taken advantage of these inbred strains, along with some outbred stocks such as SPRD and new transgenic models (21, 22) not only to study the mechanisms of carcinogenesis (23–34), but also to identify the target cells of carcinogens (35), and to analyze the effects of protective compounds (36–46). More recently, inbred strains diverging in specific cancer susceptibility have been used to dissect the genetic bases of this phenotype (*see* Subheadings 30.3.1 and 30.3.2).

## 3. Hereditary Aspects

Several types of human cancer are inherited in a dominant, monogenic (Mendelian) manner, as a result of rare, germinal mutations inactivating tumor suppressor genes (10, 47–49). A larger fraction of cancers are influenced by both multiple low penetrance genes and nongenetic factors and thus behave as quantitative traits. Inbred rodent strains are helpful as models of both situations.

### 3.1. Monogenic Cancers

Several rat models of monogenic hereditary cancers are available. Some were obtained spontaneously, while others were generated either by mutagenesis of tumor suppressor genes (and subsequent screening) or by transgenesis of activated oncogenes (transgenesis can be performed efficiently in rat ((9, 21) and other chapters of this book)).

#### 3.1.1. Colon Cancer (Apc Gene)

A mutagen-induced nonsense allele of the *Apc* gene on an inbred F344 genetic background has been isolated by Amos-Landgraf and colleagues (11). The mutant rats (named *Pirc*) develop multiple neoplasms in the colon and the small intestine, a distribution better simulating the human situation than the murine *Apc* mutants. This feature, as well as other aspects of the *Pirc* rat, suggests that this new mutant will be a useful model of human colon cancer.

#### 3.1.2. Mammary Cancer (Brca1 and Brca2 Genes)

Gould and coworkers developed a method to produce knock-out (KO) rats and isolated mutants in the two tumor suppressor genes *Brca1* and *Brca2*, the homologues of the major breast cancer predisposing human genes (50–52). Interestingly, while KO *Brca2* mutation in homozygous mice lead to limited survival of approximately 3 months, the *Brca2–/–* rats are 100% viable, and vast majority of them live to over 1 year of age (51). These new rat mutants provide a means to study the role of the BRCA proteins in increasing cancer susceptibility.

On the other hand, Tsuda and coworkers (53, 54) generated transgenic rats carrying the human *HRAS* proto-oncogene and showed that these rats are highly susceptible to induction of mammary carcinogenesis, thereby providing a new model of human breast cancer. Interestingly, carcinogen-induced mutations were detected in the transgene of a large fraction of carcinomas, while no mutations were found in the endogeneous *Hras* gene (54).

#### 3.1.3. Multiple Endocrine Neoplasia (Cdkn1b Gene)

MENX is a recessive multiple endocrine neoplasia (MEN)-like syndrome in the rat, similar to the human MEN types 1 and 2 (due to mutations in the *MEN1* and *RET* genes, respectively) (55).

MENX is caused by a mutation in the *Cdkn1b* gene, encoding the cyclin-dependent kinase inhibitor p27 (Kip1), and interestingly, a form of human MEN has subsequently been shown to be due to a nonsense mutation in the human homologue gene (56).

*3.1.4. Renal Cancer (Tsc2 and Bdh Genes)*

Hereditary renal carcinoma in the Ekker rat is a model of dominantly inherited Mendelian predisposition to a specific cancer and was found to be determined by germ-line mutation of the tuberous sclerosis 2 (*Tsc2*) gene, a typical tumor suppressor gene (57, 58). This model has been used successfully to identify the coding sequences responsible for *Tsc2*-mediated tumor suppression (59) as well as the modifier genes affecting tumor incidence and tumor size (60, 61).

Another model of hereditary renal cell carcinoma is the Nihon rat. This rat is mutated in the *Bhd* gene, the rat homologue of the human *BHD* gene (62, 63). Mutations in the *BHD* gene cause the Bird–Hogg–Dubé syndrome, a rare autosomal dominant disease characterized by several features including renal neoplasm, and the Nihon rat is thus an interesting model for this disease.

*3.1.5. Teratoma*

A model for teratoma was obtained through a spontaneous mutation (*ter*) WKY/Ttm strain. In the WKY/Ztm-ter rat, both sexes are affected by the mutation (64). This is in contrast to the mouse in which a mutation in the *Dnd1* gene, controlling teratoma susceptibility, is expressed in males alone (65).

***3.2. Polygenic Cancers: Quantitative Trait Loci Analyses***

Several mouse and rat strains that show a wide range of tissue-specific cancer susceptibility are available, and crosses involving such strains demonstrated that cancer susceptibility is a polygenic trait. These strains thus constitute an excellent resource to identify cancer susceptibility genes and to analyze the mechanisms underlying tumor development (6, 66–68). The studies carried out in the rat have been recently reviewed (4, 10, 69) and will thus be simply summarized here in the form of a table (Table 30.1) and briefly commented below.

These studies uncovered a bunch of QTLs controlling several types of cancers, most of which were chemically-induced. The role of several of these QTLs was physically demonstrated in congenic strains (*see* references in Table 30.1). Remarkably, when several inbred strains diverging in susceptibility to a given type of cancer were intercrossed in different combinations, some QTLs were found to be shared between strains (thus appearing in different crosses), while others ones were specific to one cross or limited to a few crosses (10, 70). The latter QTLs reflect genetic heterogeneity, i.e. the fact that the same trait is influenced by distinct loci in different strains and suggest that distinct mechanisms may operate in different strains to influence one and the same phenotype. On the other hand, shared QTLs open the door to

**Table 30.1**
**QTL analyses of polygenic cancers in rat models**

| Cancer type | Rat strains, susceptible/resistant | Number of QTLs | Comments | Ref. |
|---|---|---|---|---|
| Colorectal cancer | F344, WF/ACI | 10 | (1) | (82) |
| Endometrial cancer | BDII/BN, SPRD-Cu3 | 5 | Spontaneous tumors; (1) | (70) |
| Liver nodules and hepatocarcinoma | F344/BN, COP, DRH | >20 | The *Drh2* QTL controls tumor formation; the other ones control nodule formation and/or remodeling; (1;2;3) | (80) |
| Mammary cancer, chemically-induced | WF, SPRD-Cu3/COP, WKY | ~15 | The role of one QTL was translated to the human; (1;2;3); resistance associated with differentiation (WKY) | (10, 75) |
| Mammary cancer, estrogen-induced | ACI/COP, BN | 7 | (1;2) | (69) |
| Neural tumors | BDIX/BDIV | 7 | (3) | (119) |
| Pituitary tumors | F344, ACI/BN, COP | | (1;2) | (120, 121) |
| Prostate cancer | ACI/F344 | 4 | Spontaneous tumors | (122) |
| Stomach cancer | ACI/BUF | 4 | | (81) |
| Testicular tumors | ACI, F344/– | 1 | QTL identified in a cross between 2 susceptible strains | (122) |
| T-lymphomas | F344/ME/Stm | 3 | | (123) |
| Tongue cancer | DA/WF | 8 | Tumors show loss of heterozygosity at some QTLs; (2) | (124, 125) |

(1) Distinct QTLs identified in different crosses. (2) The role of some QTLs was physically demonstrated in congenic strains. (3) Resistance to tumor formation is, in some instances, associated with formation and regression of preneoplasic lesions

the multiple cross mapping strategy and haplotype mapping, which are proven to be efficient approaches to finely map QTLs and to identify the underlying genes (71).

In addition, (1) most of the cancer susceptibility QTLs cover large chromosome intervals and some of these loci turned out to be clusters of "sub-QTLs", each of which exhibits a (very) low penetrance, (2) some QTLs include cryptic (transgressive) alleles, and (3) QTLs interact with one another through various epistatic effects (72–77). Cryptic alleles could be defined as protective (resistant) alleles present in the genome of otherwise susceptible individuals or, vice versa, as susceptibility alleles present in the genome of otherwise resistant organisms (77). This phenomenon has been named "transgressive segregation" (78). Cryptic/transgressive alleles are a source of some nomenclature difficulties. Some authors designated such transgressive QTLs as cancer susceptibility QTLs (because the cryptic allele present in the resistant strain is a susceptibility allele), while designating standard QTLs as cancer resistance QTLs (79–82). This dichotomic nomenclature does not seem to be justified. Indeed, it is the variant alleles, rather than the loci that have positive or negative effects on blood pressure, or confer cancer susceptibility or resistance. With respect to cancer QTL names, it seems arbitrary to choose between the terms susceptibility and resistance since these terms cover comparative effects, but it would probably be preferable to choose the term susceptibility, which has been more widely used (for rules of nomenclature, *see* http://www.informatics.jax.org/mgihome/nomen/).

It should be stressed that so far, a single rat cancer susceptibility QTL has been mapped to a very small interval (*Mcs5a*, impacting on mammary cancer susceptibility), and remarkably, the homologue human chromosome region could rapidly be shown to also behave as a susceptibility locus (75). This important result demonstrates that translation to the human can be achieved with success and contributes to validade rat genetics in cancer research.

Another value of the rat model (and of other animal models) is that mechanisms can be studied in a way that is, naturally, out of reach in the human. In this respect, a remarkable finding of rat cancer research is that cancer resistance could be associated in several instances with the elimination of premalignant cells (18). This holds true for resistance to mammary cancer of the COP rats (83, 84), to carcinogen-induced hepatic lesions in rats resistant to liver tumors (85) and to schwannomas (86). However, this is not a universal rule. Indeed, in the WKY rats, which are resistant to mammary cancer, no preneoplasic lesions are induced by chemical carcinogen treatment, and resistance seems to be associated with precocious differentiation (87).

## 4. Cancer Somatic Genetics and Analysis of Rat Tumors

All classical genetic and epigenetic tumor-associated anomalies have been reported in rat tumors, namely:

- Nonrandom chromosome aberrations, identified by conventional cytogenetics as far back as the early 1970 (88, 89);
- Specific oncogene mutations (one of the first activated oncogenes discovered in DNA transfection experiments, i.e., *Neu/Erbb2*, was isolated from BDIX rat neuronal and glial tumors (90);

- Allelic imbalance or loss of heterozygosity, indicative of the involvement of tumor suppressor genes (91, 92);
- Specific gene amplification (93–95); in addition to conventional cytogenetics, comparative genomic hybridization (CGH) has also been used more recently to detect specific gene gains, or losses, in rat tumors (96–99);
- Alterations in DNA methylation and histone modifications, also detected at an early stage of carcinogenesis (100–103);
- Specific changes in the gene expression pattern (34, 104).

These studies led to important observations and conclusions. First, similarities were found in the genetic anomalies of rat and human tumors, such as myelomas (105), lung tumors (97), neuroblastomas (94), and endometrial or bladder tumors (34, 96), suggesting common carcinogenesis mechanisms for tumor formation in both species, and thereby validating the use of the rat in the elucidation of the basic mechanisms of tumorigenesis. Naturally, this holds true for mouse tumors (94, 104, 106, 107). Comparative oncogenomic studies of mouse and human tumors led to the identification of new oncogenes driving liver cancer and melanoma development (108, 109). It is remarkable that similarities across the three species are found in tumors that have distinct etiologies.

Another major conclusion that emerged quite early from these studies is the notion that tumor features can also be specifically linked to etiology. Mitelman and coworkers (110) reported that fibrosarcomas induced in the same (rat or Chinese hamster) strain either by Rous sarcoma virus or by DMBA show distinct nonrandom chromosome variation. Along similar lines, Christian and coworkers (99) reported that 2-amino-1-methyl-6-phenylimidazo(4,5-b)pyridine (PhIP)-induced mammary tumors show specific chromosome losses, while no consistent pattern of chromosome change could be found in three DMBA-induced tumors. In addition, N-methyl-nitrosurea (NMU)-induced mammary tumors in several strains (BUF, SPRD, F344) were found to frequently carry mutations in the *Hras* gene (111, 112), while DMBA-induced (SPRD) or spontaneous (F344) tumors show unfrequent or no mutation in this gene, nor in the *Kras* or *Nras* genes ((113) and references therein). Different carcinogens generate distinct DNA lesions. In combination with differential repair and cell-specific DNA damage response (114), they may thus promote the growth and the selection of cells, the genome of which has been differentially damaged. Gene expression profiling analyses also showed that histologically similar spontaneous and DMBA-induced mammary cancers of SPRD rats can be separated (115). Similarly, gene expression profiling of rat mammary carcinomas induced by two distinct carcinogens could be classified in two groups that coincide with the identity of the carcinogen (116, 117).

In the mouse, mammary tumors initiated by transgenic expression of distinct oncogenes have also been shown to exhibit oncogene-specific patterns of gene expression (118). These results thus indicate that in one and the same organ, tumors can be classified on the basis of the identity of the inducing agent. They strongly suggest that gene expression analyses (in conjunction with genomic analyses) may be valuable in determining retrospectively the role of specific carcinogens in the formation of human cancers, and of human breast cancer in particular.

## Acknowledgments

Recent work done in the author laboratory was supported by the Fund for Scientific Medical Research (FRSM, 3.4517.05), the Fund for Collective Fundamental research (FRFC, 2.4565.04), the National Fund for Scientific Research (FNRS, Télévie, 7.4620.07 and 7.4530.06), and the FP6 program EURATools. The author is a Research Director of the FNRS (Belgium).

### References

1. Jacob HJ (1999) Functional genomics and rat models. Genome Res 9:1013–1016
2. Hamm TE, King-Herbert A, Vasbinder MA (2006) Toxicology. In: Suckow MA, Weisbroth SH, Franklin CL (eds) The laboratory rat, 2nd edn. Elsevier, Amsterdam, pp 803–816
3. Festing MF (1997) Fat rats and carcinogenesis screening. Nature 388:321–322
4. Szpirer C, Levan G (2009) Rat gene mapping and genomics. In: Denny P, Kole C (eds) Genome mapping and genomics in laboratory animals. Springer (in press)
5. Balmain A, Nagase H (1998) Cancer resistance genes in mice: models for the study of tumour modifiers. Trends Genet 14:139–144
6. Demant P (2003) Cancer susceptibility in the mouse: genetics, biology and implications for human cancer. Nat Rev Genet 4:721–734
7. Abbott A (2007) Biologists claim Nobel prize with a knock-out. Nature 449:642
8. van der Weyden L, Adams DJ, Bradley A (2002) Tools for targeted manipulation of the mouse genome. Physiol Genomics 11:133–164
9. Aitman TJ, Critser JK, Cuppen E, Dominiczak A, Fernandez-Suarez XM, Flint J, Gauguier D, Geurts AM, Gould M, Harris PC, Holmdahl R, Hubner N, Izsvak Z, Jacob HJ, Kuramoto T, Kwitek AE, Marrone A, Mashimo T, Moreno C, Mullins J, Mullins L, Olsson T, Pravenec M, Riley L, Saar K, Serikawa T, Shull JD, Szpirer C, Twigger SN, Voigt B, Worley K (2008) Progress and prospects in rat genetics: a community view. Nat Genet 40:516–522
10. Szpirer C, Szpirer J (2007) Mammary cancer susceptibility: human genes and rodent models. Mamm Genome 18:817–831
11. Amos-Landgraf JM, Kwong LN, Kendziorski CM, Reichelderfer M, Torrealba J, Weichert J, Haag JD, Chen KS, Waller JL, Gould MN, Dove WF (2007) A target-selected Apc-mutant rat kindred enhances the modeling of familial human colon cancer. Proc Natl Acad Sci U S A 104:4036–4041
12. Russo J, Gusterson BA, Rogers AE, Russo IH, Wellings SR, van Zwieten MJ (1990) Comparative study of human and rat mammary tumorigenesis. Lab Invest 62:244–278
13. Thompson HJ, Singh M (2000) Rat models of premalignant breast disease. J Mammary Gland Biol Neoplasia 5:409–420
14. Welsch CW (1985) Host factors affecting the growth of carcinogen-induced rat mammary carcinomas: a review and tribute to Charles Brenton Huggins. Cancer Res 45:3415–3443
15. Blakely CM, Stoddard AJ, Belka GK, Dugan KD, Notarfrancesco KL, Moody SE, D'Cruz

CM, Chodosh LA (2006) Hormone-induced protection against mammary tumorigenesis is conserved in multiple rat strains and identifies a core gene expression signature induced by pregnancy. Cancer Res 66:6421–6431

16. Sell S (1993) The role of determined stem-cells in the cellular lineage of hepatocellular carcinoma. Int J Dev Biol 37:189–201
17. Sell S (2003) Mouse models to study the interaction of risk factors for human liver cancer. Cancer Res 63:7553–7562
18. Wood GA, Korkola JE, Archer MC (2002) Tissue-specific resistance to cancer development in the rat: phenotypes of tumor-modifier genes. Carcinogenesis 23:1–9
19. Greenhouse DG, Festing MF, Hasan S, Cohen AL (1990) Catalogue of inbred strains of rats and mutants. In: Hedrich HJ (ed) Genetic monitoring of inbred strains of rats. Gustav Fischer Verlag, Stuttgard, New York, pp 410–480
20. King-Herbert A, Thayer K (2006) NTP workshop: animal models for the NTP rodent cancer bioassay: stocks and strains – should we switch? Toxicol Pathol 34:802–805
21. Tesson L, Cozzi J, Menoret S, Remy S, Usal C, Fraichard A, Anegon I (2005) Transgenic modifications of the rat genome. Transgenic Res 14:531–546
22. Dann CT (2007) New technology for an old favorite: lentiviral transgenesis and RNAi in rats. Transgenic Res 16:571–580
23. Aidoo A, Morris SM, Casciano DA (1997) Development and utilization of the rat lymphocyte hprt mutation assay. Mutat Res 387:69–88
24. Blanco D, Vicent S, Fraga MF, Fernandez-Garcia I, Freire J, Lujambio A, Esteller M, Ortiz-de-Solorzano C, Pio R, Lecanda F, Montuenga LM (2007) Molecular analysis of a multistep lung cancer model induced by chronic inflammation reveals epigenetic regulation of p16 and activation of the DNA damage response pathway. Neoplasia 9:840–852
25. Dragan YP, Hully JR, Nakamura J, Mass MJ, Swenberg JA, Pitot HC (1994) Biochemical events during initiation of rat hepatocarcinogenesis. Carcinogenesis 15:1451–1458
26. Dybdahl M, Risom L, Moller P, Autrup H, Wallin H, Vogel U, Bornholdt J, Daneshvar B, Dragsted LO, Weimann A, Poulsen HE, Loft S (2003) DNA adduct formation and oxidative stress in colon and liver of Big Blue rats after dietary exposure to diesel particles. Carcinogenesis 24:1759–1766
27. Grippo PJ, Sandgren EP (2005) Modeling pancreatic cancer in animals to address specific hypotheses. Methods Mol Med 103:217–243
28. Nagao M, Ushijima T, Toyota M, Inoue R, Sugimura T (1997) Genetic changes induced by heterocyclic amines. Mutat Res 376:161–167
29. Jenkins S, Rowell C, Wang J, Lamartiniere CA (2007) Prenatal TCDD exposure predisposes for mammary cancer in rats. Reprod Toxicol 23:391–396
30. Fielden MR, Brennan R, Gollub J (2007) A gene expression biomarker provides early prediction and mechanistic assessment of hepatic tumor induction by nongenotoxic chemicals. Toxicol Sci 99:90–100
31. Nioi P, Pardo ID, Sherratt PJ, Snyder RD (2008) Prediction of non-genotoxic carcinogenesis in rats using changes in gene expression following acute dosing. Chem Biol Interact 172:206–215
32. Ohnishi T, Fukamachi K, Ohshima Y, Jiegou X, Ueda S, Iigo M, Takasuka N, Naito A, Fujita K, Matsuoka Y, Izumi K, Tsuda H (2007) Possible application of human c-Ha-ras proto-oncogene transgenic rats in a medium-term bioassay model for carcinogens. Toxicol Pathol 35:436–443
33. Ryu JY, Lee BM, Kacew S, Kim HS (2007) Identification of differentially expressed genes in the testis of Sprague-Dawley rats treated with di(n-butyl) phthalate. Toxicology 234:103–112
34. Williams PD, Lee JK, Theodorescu D (2008) Molecular credentialing of rodent bladder carcinogenesis models. Neoplasia 10: 838–846
35. Sell S (2007) Stem cells in hepatocarcinogenesis – the liver is the exception that proves the rule. Cellscience Reviews 3:302–341
36. Pascale RM, Simile MM, De Miglio MR, Feo F (2002) Chemoprevention of hepatocarcinogenesis: S-adenosyl-L-methionine. Alcohol 27:193–198
37. Manjanatha MG, Shelton S, Bishop ME, Lyn-Cook LE, Aidoo A (2006) Dietary effects of soy isoflavones daidzein and genistein on 7, 12-dimethylbenz[a]anthracene-induced mammary mutagenesis and carcinogenesis in ovariectomized Big Blue transgenic rats. Carcinogenesis 27: 2555–2564
38. Reddy BS, Hirose Y, Cohen LA, Simi B, Cooma I, Rao CV (2000) Preventive potential of wheat bran fractions against experimental colon carcinogenesis: implications for human colon cancer prevention. Cancer Res 60:4792–4797

39. Stone WL, Krishnan K, Campbell SE, Qui M, Whaley SG, Yang H (2004) Tocopherols and the treatment of colon cancer. Ann N Y Acad Sci 1031:223–233
40. Banerjee S, Bueso-Ramos C, Aggarwal BB (2002) Suppression of 7, 12-dimethylbenz(a) anthracene-induced mammary carcinogenesis in rats by resveratrol: role of nuclear factor-kappaB, cyclooxygenase 2, and matrix metalloprotease 9. Cancer Res 62:4945–4954
41. Burns FJ, Tang MS, Frenkel K, Nadas A, Wu F, Uddin A, Zhang R (2007) Induction and prevention of carcinogenesis in rat skin exposed to space radiation. Radiat Environ Biophys 46:195–199
42. Eason RR, Till SR, Frank JA, Badger TM, Korourian S, Simmen FA, Simmen RC (2006) Tumor-protective and tumor-promoting actions of dietary whey proteins in an N-methyl-N-nitrosourea model of rat mammary carcinogenesis. Nutr Cancer 55:171–177
43. Su Y, Simmen FA, Xiao R, Simmen RC (2007) Expression profiling of rat mammary epithelial cells reveals candidate signaling pathways in dietary protection from mammary tumors. Physiol Genomics 30:8–16
44. Whitsett TG Jr, Lamartiniere CA (2006) Genistein and resveratrol: mammary cancer chemoprevention and mechanisms of action in the rat. Expert Rev Anticancer Ther 6:1699–1706
45. Woditschka S, Haag JD, Waller JL, Monson DM, Hitt AA, Brose HL, Hu R, Zheng Y, Watson PA, Kim K, Lindstrom MJ, Mau B, Steele VE, Lubet RA, Gould MN (2006) Neu-induced retroviral rat mammary carcinogenesis: a novel chemoprevention model for both hormonally responsive and nonresponsive mammary carcinomas. Cancer Res 66:6884–6891
46. Toden S, Bird AR, Topping DL, Conlon MA (2007) High red meat diets induce greater numbers of colonic DNA double-strand breaks than white meat in rats: attenuation by high-amylose maize starch. Carcinogenesis 28:2355–2362
47. Turnbull C, Hodgson S (2005) Genetic predisposition to cancer. Clin Med 5:491–498
48. Balmain A, Gray J, Ponder B (2003) The genetics and genomics of cancer. Nat Genet 33(Suppl):238–244
49. Hecht F (2007) Familial cancer syndromes: catalog with comments. Cytogenet Genome Res 118:222–228
50. Zan Y, Haag JD, Chen KS, Shepel LA, Wigington D, Wang YR, Hu R, Lopez-Guajardo CC, Brose HL, Porter KI, Leonard RA, Hitt AA, Schommer SL, Elegbede AF, Gould MN (2003) Production of knockout rats using ENU mutagenesis and a yeast-based screening assay. Nat Biotechnol 21:645–651
51. Cotroneo MS, Haag JD, Zan Y, Lopez CC, Thuwajit P, Petukhova GV, Camerini-Otero RD, Gendron-Fitzpatrick A, Griep AE, Murphy CJ, Dubielzig RR, Gould MN (2007) Characterizing a rat Brca2 knockout model. Oncogene 26:1626–1635
52. Smits BM, Cotroneo MS, Haag JD, Gould MN (2007) Genetically engineered rat models for breast cancer. Breast Dis 28:53–61
53. Asamoto M, Ochiya T, Toriyama-Baba H, Ota T, Sekiya T, Terada M, Tsuda H (2000) Transgenic rats carrying human c-Ha-ras proto-oncogenes are highly susceptible to N-methyl-N-nitrosourea mammary carcinogenesis. Carcinogenesis 21:243–249
54. Tsuda H, Fukamachi K, Ohshima Y, Ueda S, Matsuoka Y, Hamaguchi T, Ohnishi T, Takasuka N, Naito A (2005) High susceptibility of human c-Ha-ras proto-oncogene transgenic rats to carcinogenesis: a cancer-prone animal model. Cancer Sci 96: 309–316
55. Falchetti A, Marini F, Tonelli F, Brandi ML (2005) Lessons from genes mutated in multiple endocrine neoplasia (MEN) syndromes. Ann Endocrinol (Paris) 66:195–205
56. Pellegata NS, Quintanilla-Martinez L, Siggelkow H, Samson E, Bink K, Hofler H, Fend F, Graw J, Atkinson MJ (2006) Germline mutations in p27Kip1 cause a multiple endocrine neoplasia syndrome in rats and humans. Proc Natl Acad Sci U S A 103:15558–15563
57. Yeung RS, Xiao GH, Jin F, Lee WC, Testa JR, Knudson AG (1994) Predisposition to renal carcinoma in the Eker rat is determined by germ-line mutation of the tuberous sclerosis 2 (TSC2) gene. Proc Natl Acad Sci U S A 91:11413–11416
58. Kobayashi T, Hirayama Y, Kobayashi E, Kubo Y, Hino O (1995) A germline insertion in the tuberous sclerosis (Tsc2) gene gives rise to the Eker rat model of dominantly inherited cancer. Nat Genet 9:70–74
59. Momose S, Kobayashi T, Mitani H, Hirabayashi M, Ito K, Ueda M, Nabeshima Y, Hino O (2002) Identification of the coding sequences responsible for Tsc2-mediated tumor suppression using a transgenic rat system. Hum Mol Genet 11:2997–3006
60. Yeung RS, Gu H, Lee M, Dundon TA (2001) Genetic identification of a locus, Mot1, that affects renal tumor size in the rat. Genomics 78:108–112

61. Kikuchi Y, Sudo A, Mitani H, Hino O (2004) Presence of a modifier gene(s) affecting early renal carcinogenesis in the Tsc2 mutant (Eker) rat model. Int J Oncol 24:75–80
62. Kouchi M, Okimoto K, Matsumoto I, Tanaka K, Yasuba M, Hino O (2006) Natural history of the Nihon (Bhd gene mutant) rat, a novel model for human Birt-Hogg-Dube syndrome. Virchows Arch 448:463–471
63. Togashi Y, Kobayashi T, Momose S, Ueda M, Okimoto K, Hino O (2006) Transgenic rescue from embryonic lethality and renal carcinogenesis in the Nihon rat model by introduction of a wild-type Bhd gene. Oncogene 25:2885–2889
64. Langer B, Dorsch M, Gartner K, Wedekind D, Kamino K, Hedrich HJ (2004) WKY/Ztm-ter: a new rat inbred strain on the WKY/Ztm genetic background with congenital teratomas. Lab Anim 38:425–431
65. Youngren KK, Coveney D, Peng X, Bhattacharya C, Schmidt LS, Nickerson ML, Lamb BT, Deng JM, Behringer RR, Capel B, Rubin EM, Nadeau JH, Matin A (2005) The Ter mutation in the dead end gene causes germ cell loss and testicular germ cell tumours. Nature 435:360–364
66. Mao JH, Balmain A (2003) Genomic approaches to identification of tumour-susceptibility genes using mouse models. Curr Opin Genet Dev 13:14–19
67. Frese KK, Tuveson DA (2007) Maximizing mouse cancer models. Nat Rev Cancer 7:645–658
68. Peters LL, Robledo RF, Bult CJ, Churchill GA, Paigen BJ, Svenson KL (2007) The mouse as a model for human biology: a resource guide for complex trait analysis. Nat Rev Genet 8:58–69
69. Shull JD (2007) The rat oncogenome: comparative genetics and genomics of rat models of mammary carcinogenesis. Breast Dis 28:69–86
70. Roshani L, Mallon P, Sjostrand E, Wedekind D, Szpirer J, Szpirer C, Hedrich HJ, Klinga-Levan K (2005) Genetic analysis of susceptibility to endometrial adenocarcinoma in the BDII rat model. Cancer Genet Cytogenet 158:137–1341
71. Park YG, Zhao X, Lesueur F, Lowy DR, Lancaster M, Pharoah P, Qian X, Hunter KW (2005) Sipa1 is a candidate for underlying the metastasis efficiency modifier locus Mtes1. Nat Genet 37:1055–10562
72. Schaffer BS, Lachel CM, Pennington KL, Murrin CR, Strecker TE, Tochacek M, Gould KA, Meza JL, McComb RD, Shull JD (2006) Genetic bases of estrogen-induced tumorigenesis in the rat: mapping of loci controlling susceptibility to mammary cancer in a Brown Norway × ACI intercross. Cancer Res 66:7793–7800
73. Haag JD, Shepel LA, Kolman BD, Monson DM, Benton ME, Watts KT, Waller JL, Lopez-Guajardo CC, Samuelson DJ, Gould MN (2003) Congenic rats reveal three independent Copenhagen alleles within the Mcs1 quantitative trait locus that confer resistance to mammary cancer. Cancer Res 63:5808–5812
74. Samuelson DJ, Aperavich BA, Haag JD, Gould MN (2005) Fine mapping reveals multiple loci and a possible epistatic interaction within the mammary carcinoma susceptibility quantitative trait locus, Mcs5. Cancer Res 65:9637–9642
75. Samuelson DJ, Hesselson SE, Aperavich BA, Zan Y, Haag JD, Trentham-Dietz A, Hampton JM, Mau B, Chen KS, Baynes C, Khaw KT, Luben R, Perkins B, Shah M, Pharoah PD, Dunning AM, Easton DF, Ponder BA, Gould MN (2007) Rat Mcs5a is a compound quantitative trait locus with orthologous human loci that associate with breast cancer risk. Proc Natl Acad Sci U S A 104:6299–6304
76. Cotroneo MS, Merry GM, Haag JD, Lan H, Shepel LA, Gould MN (2006) The Mcs7 quantitative trait locus is associated with an increased susceptibility to mammary cancer in congenic rats and an allele-specific imbalance. Oncogene 25:5011–5017
77. Quan X, Laes JF, Stieber D, Riviere M, Russo J, Wedekind D, Coppieters W, Farnir F, Georges M, Szpirer J, Szpirer C (2006) Genetic identification of distinct loci controlling mammary tumor multiplicity, latency, and aggressiveness in the rat. Mamm Genome 17:310–321
78. Solberg LC, Baum AE, Ahmadiyeh N, Shimomura K, Li R, Turek FW, Churchill GA, Takahashi JS, Redei EE (2004) Sex- and lineage-specific inheritance of depression-like behavior in the rat. Mamm Genome 15:648–662
79. De Miglio MR, Canzian F, Pascale RM, Simile MM, Muroni MR, Calvisi D, Romeo G, Feo F (1999) Identification of genetic loci controlling hepatocarcinogenesis on rat chromosomes 7 and 10. Cancer Res 59: 4651–4655
80. Feo F, De Miglio MR, Simile MM, Muroni MR, Calvisi DF, Frau M, Pascale RM (2006) Hepatocellular carcinoma as a complex polygenic disease. Interpretive analysis of recent developments on genetic predisposition. Biochim Biophys Acta 1765:126–147

81. Ushijima T, Yamamoto M, Suzui M, Kuramoto T, Yoshida Y, Nomoto T, Tatematsu M, Sugimura T, Nagao M (2000) Chromosomal mapping of genes controlling development, histological grade, depth of invasion, and size of rat stomach carcinomas. Cancer Res 60:1092–1096
82. De Miglio MR, Virdis P, Calvisi DF, Mele D, Muroni MR, Frau M, Pinna F, Tomasi ML, Simile MM, Pascale RM, Feo F (2007) Identification and chromosome mapping of loci predisposing to colorectal cancer that control Wnt/beta-catenin pathway and progression of early lesions in the rat. Carcinogenesis 28:2367–2374
83. Korkola JE, Archer MC (1999) Resistance to mammary tumorigenesis in Copenhagen rats is associated with the loss of preneoplastic lesions. Carcinogenesis 20:221–227
84. Harris SR, Mehta RS, Hartle DK, Broderson JR, Bunce OR (1994) Failure of high fat diets to promote mammary cancers in spontaneously hypertensive rats. Cancer Lett 87:9–15
85. De Miglio MR, Simile MM, Muroni MR, Calvisi DF, Virdis P, Asara G, Frau M, Bosinco GM, Seddaiu MA, Daino L, Feo F, Pascale RM (2003) Phenotypic reversion of rat neoplastic liver nodules is under genetic control. Int J Cancer 105:70–75
86. Kindler-Röhrborn A, Kind AB, Koelsch BU, Fischer C, Rajewsky MF (2000) Suppression of ethylnitrosourea-induced schwannoma development involves elimination of neu/erbB-2 mutant premalignant cells in the resistant BDIV rat strain. Cancer Res 60:4756–4760
87. Lella V, Stieber D, Riviere M, Szpirer J, Szpirer C (2007) Mammary cancer resistance and precocious mammary differentiation in the WKY rat: identification of 2 quantitative trait loci. Int J Cancer 121:1738–1743
88. Mitelman F (1971) The chromosomes of fifty primary Rous rat sarcomas. Hereditas 69:155–186
89. Levan G, Ahlstrom U, Mitelman F (1974) The specificity of chromosome A2 involvement in DMBA-induced rat sarcomas. Hereditas 77:263–280
90. Shih C, Padhy LC, Murray M, Weinberg RA (1981) Transforming genes of carcinomas and neuroblastomas introduced into mouse fibroblasts. Nature 290:261–264
91. Nordlander C, Karlsson S, Karlsson A, Sjoling A, Winnes M, Klinga-Levan K, Behboudi A (2007) Analysis of chromosome 10 aberrations in rat endometrial cancer-evidence for a tumor suppressor locus distal to Tp53. Int J Cancer 120:1472–1481
92. Koelsch BU, Kindler-Rohrborn A, Held S, Zabel S, Rajewsky MF (2002) Loss of heterozygosity in malignant rat schwannomas chemically induced in hybrids of inbred rat strains with differential tumor susceptibility. Carcinogenesis 23:1033–1037
93. Adamovic T, Trosso F, Roshani L, Andersson L, Petersen G, Rajaei S, Helou K, Levan G (2005) Oncogene amplification in the proximal part of chromosome 6 in rat endometrial adenocarcinoma as revealed by combined BAC/PAC FISH, chromosome painting, zoo-FISH, and allelotyping. Genes Chromosomes Cancer 44:139–153
94. Lastowska M, Chung YJ, Cheng Ching N, Haber M, Norris MD, Kees UR, Pearson AD, Jackson MS (2004) Regions syntenic to human 17q are gained in mouse and rat neuroblastoma. Genes Chromosomes Cancer 40:158–163
95. Karlsson A, Helou K, Walentinsson A, Hedrich HJ, Szpirer C, Levan G (2001) Amplification of Mycn, Ddx1, Rrm2, and Odc1 in rat uterine endometrial carcinomas. Genes Chromosomes Cancer 31:345–356
96. Samuelson E, Nordlander C, Levan G, Behboudi A (2008) Amplification studies of MET and Cdk6 in a rat endometrial tumor model and their correlation to human type I endometrial carcinoma tumors. Adv Exp Med Biol 617:511–517
97. Dano L, Guilly MN, Muleris M, Morlier JP, Altmeyer S, Vielh P, El-Naggar AK, Monchaux G, Dutrillaux B, Chevillard S (2000) CGH analysis of radon-induced rat lung tumors indicates similarities with human lung cancers. Genes Chromosomes Cancer 29:1–8
98. Walentinsson A, Sjoling A, Helou K, Klinga-Levan K, Levan G (2000) Genomewide assessment of genetic alterations in DMBA-induced rat sarcomas: cytogenetic, CGH, and allelotype analyses reveal recurrent DNA copy number changes in rat chromosomes 1, 2, 4, and 7. Genes Chromosomes Cancer 28:184–195
99. Christian AT, Snyderwine EG, Tucker JD (2002) Comparative genomic hybridization analysis of PhIP-induced mammary carcinomas in rats reveals a cytogenetic signature. Mutat Res 506–507:113–119
100. Bagnyukova TV, Tryndyak VP, Montgomery B, Churchwell MI, Karpf AR, James SR, Muskhelishvili L, Beland FA, Pogribny IP (2008) Genetic and epigenetic changes in rat preneoplastic liver tissue induced by 2-acetylaminofluorene. Carcinogenesis 29:638–646

101. Kovalchuk O, Tryndyak VP, Montgomery B, Boyko A, Kutanzi K, Zemp F, Warbritton AR, Latendresse JR, Kovalchuk I, Beland FA, Pogribny IP (2007) Estrogen-induced rat breast carcinogenesis is characterized by alterations in DNA methylation, histone modifications and aberrant microRNA expression. Cell Cycle 6:2010–2018
102. Loree J, Koturbash I, Kutanzi K, Baker M, Pogribny I, Kovalchuk O (2006) Radiation-induced molecular changes in rat mammary tissue: possible implications for radiation-induced carcinogenesis. Int J Radiat Biol 82:805–815
103. Shimizu K, Onishi M, Sugata E, Sokuza Y, Mori C, Nishikawa T, Honoki K, Tsujiuchi T (2007) Disturbance of DNA methylation patterns in the early phase of hepatocarcinogenesis induced by a choline-deficient L-amino acid-defined diet in rats. Cancer Sci 98:1318–1322
104. Wang W, Wyckoff JB, Goswami S, Wang Y, Sidani M, Segall JE, Condeelis JS (2007) Coordinated regulation of pathways for enhanced cell motility and chemotaxis is conserved in rat and mouse mammary tumors. Cancer Res 67:3505–3511
105. Sumegi J, Spira J, Bazin H, Szpirer J, Levan G, Klein G (1983) Rat c-myc oncogene is located on chromosome 7 and rearranges in immunocytomas with t(6:7) chromosomal translocation. Nature 306:497–498
106. Klein G (1979) Lymphoma development in mice and humans: diversity of initiation is followed by convergent cytogenetic evolution. Proc Natl Acad Sci U S A 76:2442–2446
107. Maser RS, Choudhury B, Campbell PJ, Feng B, Wong KK, Protopopov A, O'Neil J, Gutierrez A, Ivanova E, Perna I, Lin E, Mani V, Jiang S, McNamara K, Zaghlul S, Edkins S, Stevens C, Brennan C, Martin ES, Wiedemeyer R, Kabbarah O, Nogueira C, Histen G, Aster J, Mansour M, Duke V, Foroni L, Fielding AK, Goldstone AH, Rowe JM, Wang YA, Look AT, Stratton MR, Chin L, Futreal PA, DePinho RA (2007) Chromosomally unstable mouse tumours have genomic alterations similar to diverse human cancers. Nature 447:966–971
108. Kim M, Gans JD, Nogueira C, Wang A, Paik JH, Feng B, Brennan C, Hahn WC, Cordon-Cardo C, Wagner SN, Flotte TJ, Duncan LM, Granter SR, Chin L (2006) Comparative oncogenomics identifies NEDD9 as a melanoma metastasis gene. Cell 125:1269–1281
109. Zender L, Spector MS, Xue W, Flemming P, Cordon-Cardo C, Silke J, Fan ST, Luk JM, Wigler M, Hannon GJ, Mu D, Lucito R, Powers S, Lowe SW (2006) Identification and validation of oncogenes in liver cancer using an integrative oncogenomic approach. Cell 125:1253–1267
110. Mitelman F, Mark J, Levan G, Levan A (1972) Tumor etiology and chromosome pattern. Science 176:1340–1341
111. Sakai H, Ogawa K (1991) Mutational activation of c-Ha-ras genes in intraductal proliferation induced by N-nitroso-N-methylurea in rat mammary glands. Int J Cancer 49:140–144
112. Zarbl H, Sukumar S, Arthur AV, Martin-Zanca D, Barbacid M (1985) Direct mutagenesis of Ha-ras-1 oncogenes by N-nitroso-N-methylurea during initiation of mammary carcinogenesis in rats. Nature 315:382–385
113. Samuelson E, Nilsson J, Walentinsson A, Szpirer C, Behboudi A (2008) Absence of Ras mutations in rat DMBA-induced mammary tumors. Mol Carcinog 48:150–155
114. Engelbergs J, Thomale J, Rajewsky MF (2000) Role of DNA repair in carcinogen-induced ras mutation. Mutat Res 450:139–153
115. Marxfeld H, Grenet O, Bringel J, Staedtler F, Harleman JH (2006) Differentiation of spontaneous and induced mammary adenocarcinomas of the rat by gene expression profiling. Exp Toxicol Pathol 58:151–161
116. Kuramoto T, Morimura K, Yamashita S, Okochi E, Watanabe N, Ohta T, Ohki M, Fukushima S, Sugimura T, Ushijima T (2002) Etiology-specific gene expression profiles in rat mammary carcinomas. Cancer Res 62:3592–3597
117. Shan L, Yu M, Snyderwine EG (2005) Gene expression profiling of chemically induced rat mammary gland cancer. Carcinogenesis 26:503–509
118. Desai KV, Xiao N, Wang W, Gangi L, Greene J, Powell JI, Dickson R, Furth P, Hunter K, Kucherlapati R, Simon R, Liu ET, Green JE (2002) Initiating oncogenic event determines gene-expression patterns of human breast cancer models. Proc Natl Acad Sci U S A 99:6967–6972
119. Koelsch BU, Fischer C, Neibecker M, Schmitt N, Schmidt O, Rajewsky MF, Kindler-Rohrborn A (2006) Gender-specific polygenic control of ethylnitrosourea-induced oncogenesis in the rat peripheral nervous system. Int J Cancer 118:108–114
120. Pandey J, Wendell DL (2006) Angiogenesis and capillary maturation phenotypes associated with the Edpm3 locus on rat chromosome 3. Mamm Genome 17:49–57

121. Shull JD, Lachel CM, Murrin CR, Pennington KL, Schaffer BS, Strecker TE, Gould KA (2007) Genetic control of estrogen action in the rat: mapping of QTLs that impact pituitary lactotroph hyperplasia in a BN×ACI intercross. Mamm Genome 18:657–669

122. Yamashita S, Suzuki S, Nomoto T, Kondo Y, Wakazono K, Tsujino Y, Sugimura T, Shirai T, Homma Y, Ushijima T (2005) Linkage and microarray analyses of susceptibility genes in ACI/Seg rats: a model for prostate cancers in the aged. Cancer Res 65:2610–2616

123. Lu LM, Shisa H, Tanuma J, Hiai H (1999) Propylnitrosourea-induced T-lylphomas in LEXF RI strains of rats: genetic analysis. Br J Cancer 80:855–861

124. Hirano M, Tanuma J, Hirayama Y, Ohyama M, Semba I, Wakusawa S, Shisa H, Hiai H, Kitano M (2006) A speed congenic rat strain bearing the tongue cancer susceptibility locus Tscc1 from Dark-Agouti rats. Cancer Lett 231:185–191

125. Ogawa K, Tanuma J, Hirano M, Hirayama Y, Semba I, Shisa H, Kitano M (2006) Selective loss of resistant alleles at p15INK4B and p16INK4A genes in chemically-induced rat tongue cancers. Oral Oncol 42:710–717

# Index